Quizzes: These mid-chapter quizzes allow you to periodically check your understanding of the material covered before moving on to new concepts.

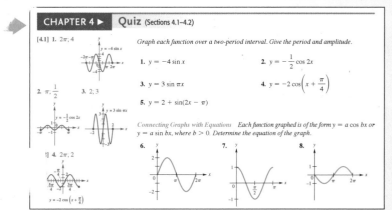

CHAPTER 4 ▶ Quiz (Sections 4.1–4.2)

[4.1] **1.** 2π; 4

Graph each function over a two-period interval. Give the period and amplitude.

1. $y = -4 \sin x$

2. $y = -\dfrac{1}{2} \cos 2x$

3. $y = 3 \sin \pi x$

4. $y = -2 \cos\left(x + \dfrac{\pi}{4}\right)$

2. π, $\dfrac{1}{2}$ **3.** 2; 3

5. $y = 2 + \sin(2x - \pi)$

] 4. 2π; 2

Connecting Graphs with Equations   Each function graphed is of the form $y = a \cos bx$ or $y = a \sin bx$, where $b > 0$. Determine the equation of the graph.

6.   **7.**   **8.**

Examples: The step-by-step solutions incorporate additional side comments and section references to previously covered material. Pointers in the examples provide on-the-spot reminders and warnings about common pitfalls.

▶ EXAMPLE 1   GRAPHING $y = \sin(x - d)$

Graph $y = \sin\left(x - \dfrac{\pi}{3}\right)$ over one period.

Solution   *Method 1*   For the argument $x - \dfrac{\pi}{3}$ to result in all possible values throughout one period, it must take on all values between 0 and 2π, inclusive. Therefore, to find an interval of one period, we solve the three-part inequality

$$0 \le x - \frac{\pi}{3} \le 2\pi \quad \text{(Appendix A)}$$

$$\frac{\pi}{3} \le x \le \frac{7\pi}{3}. \qquad \text{Add } \tfrac{\pi}{3} \text{ to each part.}$$

Chapter 4 Summary

KEY TERMS

4.1 periodic function period sine wave (sinusoid) amplitude	**4.2** phase shift argument **4.3** vertical asymptote **4.4** addition of ordinates	**4.5** simple harmonic motion frequency	damped oscillatory motion

QUICK REVIEW

CONCEPTS	EXAMPLES
4.1 Graphs of the Sine and Cosine Functions	
4.2 Translations of the Graphs of the Sine and Cosine Functions	

Sine and Cosine Functions

$y = \sin x$ $y = \cos x$

Domain: $(-\infty, \infty)$
Range: $[-1, 1]$
Amplitude: 1
Period: 2π

Domain: $(-\infty, \infty)$
Range: $[-1, 1]$
Amplitude: 1
Period: 2π

The graph of

$y = c + a \sin b(x - d)$ or $y = c + a \cos b(x - d)$,

$b > 0$, has

1. amplitude $|a|$,   2. period $\dfrac{2\pi}{b}$,
3. vertical translation c units up if $c > 0$ or $|c|$ units down if $c < 0$, and
4. phase shift d units to the right if $d > 0$ or $|d|$ units to the left if $d < 0$.

See pages 150 and 161 for a summary of graphing techniques.

Graph $y = \sin 3x$.

$y = \sin 3x$

period: $\dfrac{2\pi}{3}$ amplitude: 1
domain: $(-\infty, \infty)$ range: $[-1, 1]$

Graph $y = -2 \cos x$.

$y = -2 \cos x$

period: 2π amplitude: 2
domain: $(-\infty, \infty)$ range: $[-2, 2]$

Chapter Summary: Each chapter ends with an extensive Summary, featuring a section-by-section list of Key Terms, New Symbols, and a Quick Review of important Concepts, presented alongside corresponding Examples. A comprehensive set of Review Exercises and a Chapter Test are also provided to make test preparation easy.

Trigonometry

Ninth Edition

DATE DUE

OCT 0 4 2016	

Trigonometry

Ninth Edition

Margaret L. Lial
American River College

John Hornsby
University of New Orleans

David I. Schneider
University of Maryland

With the assistance of
Callie J. Daniels
St. Charles Community College

Boston San Francisco New York
London Toronto Sydney Tokyo Singapore Madrid
Mexico City Munich Paris Cape Town Hong Kong Montreal

Executive Editor: Anne Kelly

Project Manager: Joanne Dill

Editorial Assistant: Leah Goldberg

Senior Managing Editor: Karen Wernholm

Senior Production Supervisor: Sheila Spinney

Cover Designer: Joyce Cosentino Wells

Photo Researcher: Beth Anderson

Digital Assets Manager: Marianne Groth

Media Producer: Carl Cottrell

Software Development: Monica Salud, MathXL; Mary Durnwald, TestGen

Executive Marketing Manager: Becky Anderson

Marketing Coordinator: Bonnie Gill

Senior Author Support/Technology Specialist: Joe Vetere

Senior Prepress Supervisor: Caroline Fell

Rights and Permissions Advisor: Dana Weightman

Manufacturing Manager: Evelyn Beaton

Text Design, Production Coordination, Composition, and Illustrations: Nesbitt Graphics, Inc.

Cover Photo: Trellis in the garden. © Getty Images/Yukari Ochiai

Library of Congress Cataloging-in-Publication Data

Lial, Margaret L.

   Margaret L. Lial, John Hornsby, David Schneider.

      p. cm.

   Includes index.

   High school binding ISBN 0-13-135480-9; ISBN 978-0-13-135480-7

   1. Algebra—Textbooks.   I. Hornsby, E. John.   II. Schneider, David I.   III. Title.

   College Edition binding ISBN 0-321-52885-9; ISBN 978-0-321-52885-8

QA154.3.L538   2009

512.9—dc22

                                      2007060064

1 2 3 4 5 6 7 8 9 10—DOW—10 09 08

Contents

Preface

It is with great pleasure that we offer the ninth edition of *Trigonometry*. Over the years, the text has been shaped and adapted to meet the changing needs of both students and educators, and this edition faithfully continues that process. As always, we have taken special care to respond to the specific suggestions of users and reviewers through enhanced discussions, new and updated examples and exercises, helpful features, and an extensive package of supplements and study aids. We believe the result is an easy-to-use, comprehensive text that is the best edition yet.

Students planning to continue their study of mathematics in calculus, statistics, or other disciplines, will benefit from the text's student-oriented approach. In particular, we have added NEW! pointers to examples, included NEW! Quizzes in each chapter, revised 25% of the exercises, and added NEW! Connecting Graphs with Equations problems to make the book even more accessible for students. Additionally, we believe instructors will welcome the expanded *Annotated Instructor's Edition* that now includes an extensive set of NEW! Classroom Examples, in addition to helpful Teaching Tips and on-page answers for almost all text exercises.

This text is part of a series that also includes the following books:

- *College Algebra, Tenth Edition,* by Lial, Hornsby, Schneider

- *College Algebra and Trigonometry, Fourth Edition,* by Lial, Hornsby, Schneider

- *Precalculus, Fourth Edition,* by Lial, Hornsby, Schneider.

Features

We are pleased to offer the following features, each of which is designed to increase ease-of-use by students and actively engage them in learning mathematics.

Chapter Openers These provide a motivating application topic that is tied to the chapter content, plus a list of sections and any Quizzes or Summary Exercises in the chapter. The openers for Chapters 1, 6, and 7 are entirely new to this edition.

Examples As in the previous editions, we have continued to add new examples and polish existing ones. The step-by-step solutions now incorporate additional side comments and more section references to previously covered material. NEW! *Pointers* in the examples provide students with on-the-spot reminders and warnings about common pitfalls. Selected examples continue to provide graphing calculator solutions alongside traditional algebraic solutions. *The graphing calculator solutions can be easily omitted if desired.*

Now Try Exercises To actively engage students in the learning process, each example concludes with a reference to one or more parallel odd-numbered exercises from the corresponding exercise set. In this way, students are able to immediately apply and reinforce the concepts and skills presented in the examples.

Real-Life Applications   We have incorporated many applied examples and exercises from fields such as business, pop culture, sports, the life sciences, and environmental studies that show the relevance of trigonometry to daily life. Applications that feature mathematical modeling are labeled with a *Modeling* head. All applications are titled, and a comprehensive Index of Applications is included at the back of the text.

Function Boxes   Special function boxes offer a comprehensive, visual introduction to each circular and inverse circular function and also serve as an excellent resource for student reference and review throughout the course. Each function box includes a table of values alongside traditional and calculator graphs, as well as the domain, range, and other specific information about the function.

Figures and Photos   Today's students are more visually oriented than ever. As a result, we have made a concerted effort to include new mathematical figures, diagrams, tables, and graphs whenever possible. In this edition, we have added more than 20 NEW! figures to the exposition and improved or revised an additional 25 figures. Drawings of famous mathematicians accompany historical exposition. More photos than ever now accompany selected applications in examples and exercises.

Use of Technology   As in the previous edition, we have integrated the use of graphing calculators where appropriate, although graphing technology is not a central feature of this text. We continue to stress that graphing calculators are an aid to understanding and that students must master the underlying mathematical concepts. We have included graphing calculator solutions for selected examples and continue to mark all graphing calculator notes and exercises that use graphing calculators with an icon ⊟ for easy identification and added flexibility. ***This graphing calculator material is optional and can be omitted without loss of continuity.***

Cautions and Notes   We often give students warnings of common errors and emphasize important ideas in **Caution** and **Note** comments that appear throughout the exposition.

Looking Ahead to Calculus   These margin notes offer glimpses of how the trigonometric topics currently being studied are used in calculus.

Connections   Connections boxes continue to provide connections to the real world or to other mathematical concepts, as well as historical background and thought-provoking questions for writing, class discussion, or group work. As requested by users, this edition features several NEW! Connections. The authors wish to thank Professor Raja Khoury of Colin County Community College and Professor M. R. Khadivi of Jackson State University for their suggestion to include the material found in the new Connections in Section 3.3.

Exercise Sets   *The exercise sets have been a primary focus of this revision.* Approximately 25% of the exercises are NEW!. As a result, the text includes more problems than ever to provide students with ample opportunities to practice, apply, connect, and extend concepts and skills. We have included writing exercises 🗒 and optional graphing calculator problems ⊟, as well as multiple-choice, matching, true/false, and completion problems. Exercises marked

Concept Check focus on mathematical thinking and conceptual understanding. By special request, NEW! *Connecting Graphs with Equations* problems provide students with opportunities to write equations for given graphs.

Relating Concepts Exercises   Appearing in selected exercise sets, these sets of problems help students tie together topics and develop problem-solving skills as they compare and contrast ideas, identify and describe patterns, and extend concepts to new situations. These exercises make great collaborative activities for pairs or small groups of students. All answers to these problems appear in the answer section at the back of the student book.

Solutions to Selected Exercises   Exercise numbers enclosed in a blue circle, such as 11. , indicate that a complete solution for the problem is included at the back of the text. These solutions are given for selected exercises that extend the skills and concepts presented in the section examples—actually providing students with a pool of examples for different and/or more challenging problems.

NEW! **Quizzes**   To allow students to periodically check their understanding of material covered, a short Quiz now appears in each chapter. All answers, with corresponding section references, appear in the answer section at the back of the student text.

Summary Exercises   These sets of in-chapter exercises give students the all-important *mixed* review problems they need to synthesize concepts and select appropriate solution methods. All answers to these problems appear in the answer section at the back of the student book.

Chapter Reviews   One of the most popular features of the text, each chapter ends with an extensive Summary, featuring a section-by-section list of Key Terms, New Symbols, and a Quick Review of important Concepts, presented alongside corresponding Examples. A comprehensive set of Review Exercises and a Chapter Test are also provided. Based on feedback from our users, ***all Chapter Tests have been revised and expanded.***

Quantitative Reasoning   These end-of-chapter problems enable students to apply trigonometric concepts to life situations, such as examining patterns for utility bill fluctuations or improving athletic performance. A photo highlights each problem.

Glossary   As an additional student study aid, a comprehensive glossary of key terms from throughout the text is provided at the back of the book.

Content Changes

You will find many places in the text where we have polished individual presentations and added examples, exercises, and applications based on reviewer feedback. Some of the changes you may notice include the following:

- There are over 600 new and updated exercises, many of which are devoted to skill development, as well as new Concept Check, Connecting Graphs with Equations, and Quiz problems.

- New Connections have been incorporated in Sections 1.2, 3.3, 5.3, and 7.3.

- Every section in Chapter 4 on Graphs of the Circular Functions now features new *Connecting Graphs with Equations* exercises, including a new example in Section 4.4.

- Former Section 4.3 has been expanded and divided into two sections. Section 4.3 covers graphs of the tangent and cotangent functions, while new Section 4.4 now presents the graphs of the secant and cosecant functions.

- Chapter 7 includes a new set of *Summary Exercises on Applications of Trigonometry and Vectors.*

- Additionally, the following topics have been expanded:

  Coverage of complementary, supplementary, and quadrantal angles (Section 1.1)
  Examples of signs and ranges of trigonometric function values (Section 1.4)
  Discussion of significant digits (Section 2.4)
  Presentation of radian measure and the unit circle (Sections 3.1 and 3.3)
  Coverage of reduction formulas (Section 5.3)
  Graphs of inverse circular functions (Section 6.1)
  Proof of Heron's formula (Section 7.3)

Supplements

For a comprehensive list of the supplements and study aids that accompany *Trigonometry,* Ninth Edition, see pages xvi–xviii.

Acknowledgments

Previous editions of this text were published after thousands of hours of work, not only by the authors, but also by reviewers, instructors, students, answer checkers, and editors. To these individuals and all those who have worked in some way on this text over the years, we are most grateful for your contributions. We could not have done it without you. We especially wish to thank the following individuals who provided valuable input into this edition of the text.

Leigh Ellen Cooley, *Bridgewater Raynham Regional High School*
Patrick E. Hasenstab, *O'Fallon Township High School*
James Patrick Herrington, *O'Fallon Township High School*
Thomas E. Hottinger, *Archbishop Hoban High School*
Susan M. Kinney, *Bridgewater Raynham Regional High School*
Catherine D. Weber, *Weir High School*

Over the years, we have come to rely on an extensive team of experienced professionals. Our sincere thanks go to these dedicated individuals at Pearson Addison-Wesley, who worked long and hard to make this revision a success: Greg Tobin, Anne Kelly, Becky Anderson, Sheila Spinney, Karen Wernholm, Joyce Wells, Bonnie Gill, Joanne Dill, Leah Goldberg, and Carl Cottrell.

We are truly blessed to have Callie Daniels as a contributor to this edition, and we hope that this collaboration will last for many years to come. Terry McGinnis continues to provide invaluable behind-the-scenes guidance—we have come to rely on her expertise during all phases of the revision process. Janette Krauss, Joanne Boehme, and Nesbitt Graphics, Inc. provided excellent production work. Special thanks to Abby Tanenbaum for preparing the new Classroom Examples for the *Annotated Instructor's Edition*. Douglas Ewert, Perian Herring, and Lauri Semarne did an outstanding job accuracy checking. Lucie Haskins prepared the accurate, useful index, and Becky Troutman compiled the comprehensive Index of Applications.

As an author team, we are committed to providing the best possible text to help instructors teach and students succeed. As we continue to work toward this goal, we would welcome any comments or suggestions you might have via e-mail to *math@aw.com*.

Margaret L. Lial
John Hornsby
David I. Schneider

Student Supplements	Instructor Supplements

The following student supplements are available for purchase:

Student's Solutions Manual

- By Beverly Fusfield
- Provides detailed solutions to all odd-numbered text exercises
  ISBN: 0-321-53040-3 & 978-0-321-53040-0

Graphing Calculator Manual

- By Darryl Nester, *Bluffton University*
- Provides instructions and keystroke operations for the TI-83/84 Plus, TI-85, TI-86, and TI-89
  ISBN: 0-321-53042-X & 978-0-321-53042-4

Video Lectures on CD with Optional Captioning

- Feature Quick Reviews and Example Solutions. Quick Reviews cover key definitions and procedures from each section. Example Solutions walk students through the detailed solution process for every example in the textbook.
- Ideal for distance learning or supplemental instruction at home or on campus
- Include optional text captioning
  ISBN: 0-321-53053-5 & 978-0-321-53053-0

Additional Skill and Drill Manual

- By Cathy Ferrer, *Valencia Community College*
- Provides additional practice and test preparation for students
  ISBN: 0-321-53052-7 & 978-0-321-53052-3

A Review of Algebra

- By Heidi Howard, *Florida Community College at Jacksonville*
- Provides additional support for students needing further algebra review
  ISBN: 0-201-77347-3 & 978-0-201-77347-7

Most of the teacher supplements and resources for this book are available electronically for preview and download on the Instructor Resource Center (IRC). Please go to pearsonschool.com/advanced and click "Online Teacher Supplements" for directions. You will be required to complete a one time registration subject to verification before being emailed access information to download materials.

Annotated Instructor's Edition

- Special edition of the text
- Provides answers in the margins to almost all text exercises, plus helpful Teaching Tips and all-NEW! Classroom Examples
  ISBN: 0-321-53043-8 & 978-0-321-53043-1

Instructor's Solutions Manual

- By Beverly Fusfield
- Provides complete solutions to all text exercises
  ISBN: 0-321-53041-1 & 978-0-321-53041-7

Instructor's Testing Manual

- By Christopher Mason, *Community College of Vermont*
- Includes diagnostic pretests, chapter tests, and additional test items, grouped by section, with answers provided
  ISBN: 0-321-53051-9 & 978-0-321-53051-6

TestGen®

- Enables instructors to build, edit, print, and administer tests
- Features a computerized bank of questions developed to cover all text objectives
- Available from the Instructor Resource Center for download only

Insider's Guide

- Includes resources to assist faculty with course preparation and classroom management
- Provides helpful teaching tips correlated to the sections of the text, as well as general teaching advice
  ISBN: 0-321-53050-0 & 978-0-321-53050-9

(continued)

Instructor Supplements

PowerPoint Lecture Presentations and Active Learning Questions
- Features presentations written and designed specifically for this text, including figures and examples from the text
- Provides active learning questions for use with classroom response systems, including multiple choice questions to review lecture material
- Available from the Instructor Resource Center for download only

NEW! Classroom Example PowerPoints
- Include full worked-out solutions to all Classroom Examples
- Abbreviated PDF also available for download
- Available from the Instructor Resource Center

Media Resources

MathXL®

MathXL is a powerful online homework, tutorial, and assessment system that accompanies Pearson textbooks in mathematics or statistics. With MathXL, instructors can create, edit, and assign online homework and tests using algorithmically generated exercises correlated with the textbook. They can also create and assign their own online exercises and import TestGen tests for added flexibility. All student work is tracked in MathXL's online gradebook. Students can take chapter tests in MathXL and receive personalized study plans based on their test results. The study plan diagnoses weaknesses and links students directly to tutorial exercises for the material they need to study and on which they need to be retested. Students can also access supplemental animations and video clips directly from selected exercises. MathXL is available for purchase only. For more information, visit our Web site at www.mathxl.com, or contact your Pearson sales representative.

MathXL Tutorials on CD (ISBN 0-321-53059-4 & 978-0-321-53059-2)

This interactive tutorial CD-ROM provides algorithmically generated practice exercises that are correlated with the exercises in the textbook. Every practice exercise is accompanied by an example and a guided solution designed to involve students in the solution process. Selected exercises may also include a video clip to help students visualize concepts. The software provides helpful feedback for incorrect answers and can generate printed summaries of students' progress. Available for purchase only.

InterAct Math Tutorial Web site: www.interactmath.com

Get practice and tutorial help online! This interactive tutorial Web site provides algorithmically generated practice exercises that correlate directly with the exercises in the textbook. Students can retry an exercise as many times as they like, with new values each time, for unlimited practice and mastery. Every exercise is accompanied by an interactive guided solution that provides helpful feedback for incorrect answers, and students can also view a worked-out sample problem that steps them through an exercise similar to the one they're working on.

Video Lectures on CD with Optional Captioning (ISBN: 0-321-53053-5 & 978-0-321-53053-0)

The video lectures for this text are available on CD-ROM, making it easy and convenient for students to watch the videos from a computer at home or on campus. Quick Reviews cover key definitions and procedures from each section. Example Solutions walk students through the detailed solution process for every example in the textbook. The format provides distance-learning students with comprehensive video instruction for each section in the book, but also allows students needing only small amounts of review to watch instruction on a specific skill or procedure. The videos have an optional text captioning window; the captions can be easily turned on or off for individual student needs. Available for purchase.

1 Trigonometric Functions

Some of the most important concepts in the study of trigonometry deal with the idea of similar right triangles. Informally speaking, two figures in the plane are *similar* if they have the same shape but not necessarily the same size. In the case of two similar right triangles, each has a right angle and two acute angles of corresponding measures. Furthermore, their corresponding sides are proportional. These facts allow us to solve many types of problems in astronomy, navigation, architecture, and other fields.

The depiction of Oliver Wendell Holmes' "Chambered Nautilus" shell can be approximated geometrically by a sequence of similar triangles. It can be found at http://www.mathwright.com/librarya/spiral/spiral5.htm.

In Section 1.2 we investigate *similar triangles* and their properties.

1.1 Angles

Basic Terminology ▪ **Degree Measure** ▪ **Standard Position** ▪ **Coterminal Angles**

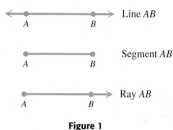

Line *AB*

Segment *AB*

Ray *AB*

Figure 1

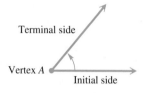

Figure 2

Basic Terminology   Two distinct points A and B determine a line called **line *AB*.** The portion of the line between A and B, including points A and B themselves, is **line segment *AB*,** or simply **segment *AB*.** The portion of line AB that starts at A and continues through B, and on past B, is called **ray *AB*.** Point A is the **endpoint of the ray.** See Figure 1.

In trigonometry, an **angle** consists of two rays in a plane with a common endpoint, or two line segments with a common endpoint. These two rays (or segments) are called the **sides** of the angle, and the common endpoint is called the **vertex** of the angle. Associated with an angle is its measure, generated by a rotation about the vertex. See Figure 2. This measure is determined by rotating a ray starting at one side of the angle, called the **initial side,** to the position of the other side, called the **terminal side.** *A counterclockwise rotation generates a positive measure, while a clockwise rotation generates a negative measure.* The rotation can consist of more than one complete revolution.

Figure 3 shows two angles, one **positive** and one **negative.**

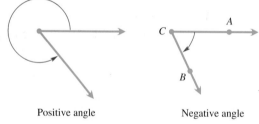

Positive angle   Negative angle

Figure 3

An angle can be named by using the name of its vertex. For example, the angle on the right in Figure 3 can be called angle C. Alternatively, an angle can be named using three letters, with the vertex letter in the middle. Thus, the angle on the right also could be named angle ACB or angle BCA.

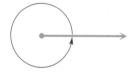

A complete rotation of a ray gives an angle whose measure is 360°. $\frac{1}{360}$ of a complete rotation gives an angle whose measure is 1°.

Figure 4

Degree Measure   The most common unit for measuring angles is the **degree.** Degree measure was developed by the Babylonians, 4000 yr ago. To use degree measure, we assign 360 degrees to a complete rotation of a ray.* In Figure 4, notice that the terminal side of the angle corresponds to its initial side when it makes a complete rotation. One degree, written 1°, represents $\frac{1}{360}$ of a rotation. Therefore, 90° represents $\frac{90}{360} = \frac{1}{4}$ of a complete rotation, and 180° represents $\frac{180}{360} = \frac{1}{2}$ of a complete rotation. An angle measuring between 0° and 90° is called an **acute angle.** An angle measuring exactly 90° is a **right angle.** The symbol ⌐ is often used at the vertex of a right angle to denote the 90° measure. An angle measuring more than 90° but less than 180° is an **obtuse angle,** and an angle of exactly 180° is a **straight angle.**

---

*The Babylonians were the first to subdivide the circumference of a circle into 360 parts. There are various theories as to why the number 360 was chosen. One is that it is approximately the number of days in a year, and it has many divisors, which makes it convenient to work with.

In Figure 5, we use the **Greek letter** θ **(theta)*** to name each angle.

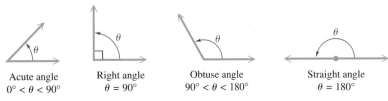

Acute angle
$0° < \theta < 90°$

Right angle
$\theta = 90°$

Obtuse angle
$90° < \theta < 180°$

Straight angle
$\theta = 180°$

Figure 5

If the sum of the measures of two positive angles is 90°, the angles are called **complementary** and the angles are **complements** of each other. Two positive angles with measures whose sum is 180° are **supplementary,** and the angles are **supplements.**

▶ EXAMPLE 1   FINDING THE COMPLEMENT AND THE SUPPLEMENT OF AN ANGLE

For an angle measuring 40°, find the measure of its **(a)** complement and **(b)** supplement.

Solution

(a) To find the measure of its complement, subtract the measure of the angle from 90°.

$$90° - 40° = 50° \quad \text{Complement of } 40°$$

(b) To find the measure of its supplement, subtract the measure of the angle from 180°.

$$180° - 40° = 140° \quad \text{Supplement of } 40°$$

NOW TRY EXERCISE 1. ◀

▶ EXAMPLE 2   FINDING MEASURES OF COMPLEMENTARY AND SUPPLEMENTARY ANGLES

Find the measure of each marked angle in Figure 6.

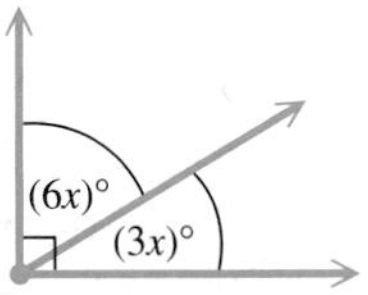

Solution

(a) In Figure 6(a), since the two angles form a right angle, they are complementary angles. Thus,

$$6x + 3x = 90$$
$$9x = 90 \quad \text{Combine terms.}$$

Don't stop here. ⟶ $x = 10.$ Divide by 9. **(Appendix A)**

Be sure to determine the measure of each angle by substituting 10 for x. The two angles have measures of $6(10) = 60°$ and $3(10) = 30°$.

$(6x)°$
$(3x)°$

(a)
Figure 6

----

*In addition to θ (theta), other Greek letters such as α (alpha) and β (beta) are often used.

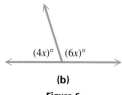

(b)

Figure 6

(b) The angles in Figure 6(b) are supplementary, so their sum must be 180°. Therefore,

$$4x + 6x = 180$$
$$10x = 180$$
$$x = 18.$$

These angle measures are $4(18) = 72°$ and $6(18) = 108°$.

NOW TRY EXERCISES 13 AND 15. ◀

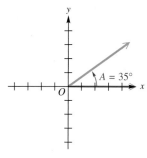

Figure 7

The measure of angle A in Figure 7 is 35°. This measure is often expressed by saying that **m(angle A)** is 35°, where m(angle A) is read **"the measure of angle A."** It is convenient, however, to abbreviate the symbolism m(angle A) = 35° as $A = 35°$.

Traditionally, portions of a degree have been measured with minutes and seconds. One **minute,** written **1′,** is $\frac{1}{60}$ of a degree.

$$1' = \frac{1}{60}^{\circ} \qquad \text{or} \qquad 60' = 1°$$

One **second, 1″,** is $\frac{1}{60}$ of a minute.

$$1'' = \frac{1}{60}' = \frac{1}{3600}^{\circ} \qquad \text{or} \qquad 60'' = 1'$$

The measure $12° \, 42' \, 38''$ represents 12 degrees, 42 minutes, 38 seconds.

▶ **EXAMPLE 3**   **CALCULATING WITH DEGREES, MINUTES, AND SECONDS**

Perform each calculation.

(a) $51° \, 29' + 32° \, 46'$

(b) $90° - 73° \, 12'$

Solution

(a) $51° \, 29'$ Add degrees and minutes separately.
      $+ \, 32° \, 46'$
      $\overline{ 83° \, 75'}$

The sum $83° \, 75'$ can be rewritten as

$$83° \, 75' = 83° + 1° \, 15' = 84° \, 15'. \quad 75' = 60' + 15' = 1° \, 15'$$

(b) $89° \, 60'$ Write 90° as 89° 60′.
      $- \, 73° \, 12'$
      $\overline{ 16° \, 48'}$

NOW TRY EXERCISES 33 AND 37. ◀

Because calculators are now so prevalent, angles are commonly measured in decimal degrees. For example, 12.4238° represents

$$12.4238° = 12\frac{4238}{10{,}000}°.$$

▶ EXAMPLE 4   CONVERTING BETWEEN DECIMAL DEGREES AND DEGREES, MINUTES, AND SECONDS

(a) Convert 74° 8′ 14″ to decimal degrees to the nearest thousandth.

(b) Convert 34.817° to degrees, minutes, and seconds.

Solution

(a) $74° \, 8′ \, 14″ = 74° + \dfrac{8}{60}° + \dfrac{14}{3600}°$ $1′ = \frac{1}{60}°$ and $1″ = \frac{1}{3600}°$.

$\approx 74° + .1333° + .0039°$

$\approx 74.137°$ Add; round to the nearest thousandth.

(b) $34.817° = 34° + .817°$

$= 34° + .817(60′)$ $1° = 60′$

$= 34° + 49.02′$

$= 34° + 49′ + .02′$

$= 34° + 49′ + .02(60″)$ $1′ = 60″$

$= 34° + 49′ + 1.2″$

$= 34° \, 49′ \, 1.2″$

NOW TRY EXERCISES 53 AND 63. ◀

74°8'14"
            74.13722222
74°8'14"
                 74.137
34.817▶DMS
            34°49'1.2"

A graphing calculator performs the conversions in Example 4 as shown above. The ▶DMS option is found in the ANGLE Menu of the TI-83/84 Plus calculator.

Standard Position   An angle is in **standard position** if its vertex is at the origin and its initial side lies on the positive x-axis. The angles in Figures 8(a) and 8(b) are in standard position. An angle in standard position is said to lie in the quadrant in which its terminal side lies. An acute angle is in quadrant I (Figure 8(a)) and an obtuse angle is in quadrant II (Figure 8(b)). Figure 8(c) shows ranges of angle measures for each quadrant when $0° < \theta < 360°$.

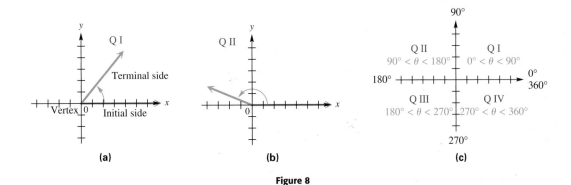

Figure 8

> QUADRANTAL ANGLES
>
> Angles in standard position whose terminal sides lie on the *x*-axis or *y*-axis, such as angles with measures 90°, 180°, 270°, and so on, are called **quadrantal angles.**

Coterminal Angles   A complete rotation of a ray results in an angle measuring 360°. By continuing the rotation, angles of measure larger than 360° can be produced. The angles in Figure 9 with measures 60° and 420° have the same initial side and the same terminal side, but different amounts of rotation. Such angles are called **coterminal angles;** *their measures differ by a multiple of* **360°.** As shown in Figure 10, angles with measures 110° and 830° are coterminal.

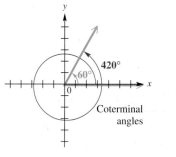

Figure 9

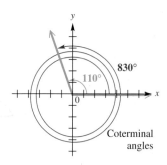

Figure 10

▶ **EXAMPLE 5**   **FINDING MEASURES OF COTERMINAL ANGLES**

Find the angles of least possible positive measure coterminal with each angle.

(a) 908°   **(b)** −75°   **(c)** −800°

Solution

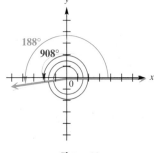

Figure 11

(a) Add or subtract 360° as many times as needed to obtain an angle with measure greater than 0° but less than 360°. Since

$$908° - 2 \cdot 360° = 188°,$$

an angle of 188° is coterminal with an angle of 908°. See Figure 11.

(b) Use a rotation of 360° + (−75°) = 285°. See Figure 12.

(c) The least integer multiple of 360° greater than 800° is

$$360° \cdot 3 = 1080°.$$

Add 1080° to −800° to obtain

$$1080° + (-800°) = 280°.$$

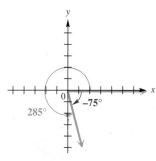

Figure 12

NOW TRY EXERCISES 73, 83, AND 87. ◀

Sometimes it is necessary to find an expression that will generate all angles coterminal with a given angle. For example, we can obtain any angle coterminal with 60° by adding an appropriate integer multiple of 360° to 60°. Let n represent any integer; then the expression

$$60° + n \cdot 360°$$ Angles coterminal with 60°

represents all such coterminal angles. The table shows a few possibilities.

Value of n	Angle Coterminal with 60°
2	$60° + 2 \cdot 360° = 780°$
1	$60° + 1 \cdot 360° = 420°$
0	$60° + 0 \cdot 360° = 60°$ (the angle itself)
-1	$60° + (-1) \cdot 360° = -300°$

The table in the margin shows some examples of coterminal quadrantal angles.

Examples of Coterminal Quadrantal Angles

Quadrantal Angle θ	Coterminal with θ
0°	$\pm 360°, \pm 720°$
90°	$-630°, -270°, 450°$
180°	$-180°, 540°, 900°$
270°	$-450°, -90°, 630°$

▶ **EXAMPLE 6** ANALYZING THE REVOLUTIONS OF A CD PLAYER

CAV (Constant Angular Velocity) CD players always spin at the same speed. Suppose a CAV player makes 480 revolutions per min. Through how many degrees will a point on the edge of a CD move in 2 sec?

Solution The player revolves 480 times in 1 min or $\frac{480}{60}$ times = 8 times per sec (since 60 sec = 1 min). In 2 sec, the player will revolve $2 \cdot 8 = 16$ times. Each revolution is 360°, so a point on the edge of the CD will revolve $16 \cdot 360° = 5760°$ in 2 sec.

NOW TRY EXERCISE 127. ◀

1.1 **Exercises**

Find **(a)** *the complement and* **(b)** *the supplement of an angle with the given measure. See Examples 1 and 3.*

1. 30°	**2.** 60°	**3.** 45°	**4.** 18°
5. 54°	**6.** 89°	**7.** 1°	**8.** 10°
9. 14° 20′	**10.** 39° 50′	**11.** 20° 10′ 30″	**12.** 50° 40′ 50″

Find the measure of each unknown angle in Exercises 13–22. See Example 2.

13.
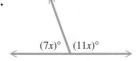
$(7x)° \quad (11x)°$

14.
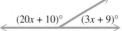
$(20x + 10)° \quad (3x + 9)°$

15.
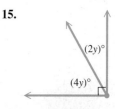
$(2y)°$
$(4y)°$

16.

$(5k + 5)°$

$(3k + 5)°$

17.

$(-4x)°$

$(-14x)°$

18.

$(9x)°$

$(9x)°$

19. supplementary angles with measures $10x + 7$ and $7x + 3$ degrees

20. supplementary angles with measures $6x - 4$ and $8x - 12$ degrees

21. complementary angles with measures $9x + 6$ and $3x$ degrees

22. complementary angles with measures $3x - 5$ and $6x - 40$ degrees

23. *Concept Check*   What is the measure of an angle that is its own complement?

24. *Concept Check*   What is the measure of an angle that is its own supplement?

Find the measure of the smaller angle formed by the hands of a clock at the following times.

25.

26.

27. 3:15

28. 9:00

Concept Check   *Answer each question.*

29. If an angle measures $x°$, how can we represent its complement?

30. If an angle measures $x°$, how can we represent its supplement?

31. If a positive angle has measure $x°$ between $0°$ and $60°$, how can we represent the first negative angle coterminal with it?

32. If a negative angle has measure $x°$ between $0°$ and $-60°$, how can we represent the first positive angle coterminal with it?

Perform each calculation. See Example 3.

33. $62° 18' + 21° 41'$ **34.** $75° 15' + 83° 32'$ **35.** $71° 18' - 47° 29'$

36. $47° 23' - 73° 48'$ **37.** $90° - 51° 28'$ **38.** $90° - 17° 13'$

39. $180° - 119° 26'$ **40.** $180° - 124° 51'$

41. $26° 20' + 18° 17' - 14° 10'$ **42.** $55° 30' + 12° 44' - 8° 15'$

43. $90° - 72° 58' 11''$ **44.** $90° - 36° 18' 47''$

Convert each angle measure to decimal degrees. If applicable, round to the nearest thousandth of a degree. See Example 4(a).

45. $35° 30'$ **46.** $82° 30'$ **47.** $112° 15'$

48. $133° 45'$ **49.** $-60° 12'$ **50.** $-70° 48'$

51. $20° 54' 00''$ **52.** $38° 42' 00''$ **53.** $91° 35' 54''$

54. $34° 51' 35''$ **55.** $274° 18' 59''$ **56.** $165° 51' 9''$

Convert each angle measure to degrees, minutes, and seconds. See Example 4(b).

57. 39.25° **58.** 46.75° **59.** 126.76° **60.** 174.255°

61. −18.515° **62.** −25.485° **63.** 31.4296° **64.** 59.0854°

65. 89.9004° **66.** 102.3771° **67.** 178.5994° **68.** 122.6853°

Find the angle of least positive measure (not equal to the given measure) coterminal with each angle. See Example 5.

69. 32° **70.** 86° **71.** 26° 30′ **72.** 58° 40′

73. −40° **74.** −98° **75.** −125° **76.** −203°

77. 361° **78.** 541° **79.** −361° **80.** −541°

81. 539° **82.** 699° **83.** 850° **84.** 1000°

85. 5280° **86.** 8440° **87.** −5280° **88.** −8440°

Give two positive and two negative angles that are coterminal with the given quadrantal angle.

89. 90° **90.** 180° **91.** 0° **92.** 270°

Give an expression that generates all angles coterminal with each angle. Let n represent any integer.

93. 30° **94.** 45° **95.** 135° **96.** 225°

97. −90° **98.** −180° **99.** 0° **100.** 360°

101. Explain why the answers to Exercises 99 and 100 give the same set of angles.

102. *Concept Check* Which two of the following are not coterminal with $r°$?

 A. $360° + r°$ **B.** $r° − 360°$ **C.** $360° − r°$ **D.** $r° + 180°$

Concept Check Sketch each angle in standard position. Draw an arrow representing the correct amount of rotation. Find the measure of two other angles, one positive and one negative, that are coterminal with the given angle. Give the quadrant of each angle, if applicable.

103. 75° **104.** 89° **105.** 174° **106.** 234°

107. 300° **108.** 512° **109.** −61° **110.** −159°

111. 90° **112.** 180° **113.** −90° **114.** −180°

*Concept Check Locate each point in a coordinate system. Draw a ray from the origin through the given point. Indicate with an arrow the angle in standard position having least positive measure. Then find the distance r from the origin to the point, using the distance formula of **Appendix B.***

115. $(−3, −3)$ **116.** $(4, −4)$ **117.** $(−3, −5)$ **118.** $(−5, 2)$

119. $(\sqrt{2}, −\sqrt{2})$ **120.** $(−2\sqrt{2}, 2\sqrt{2})$ **121.** $(−1, \sqrt{3})$ **122.** $(\sqrt{3}, 1)$

123. $(−2, 2\sqrt{3})$ **124.** $(4\sqrt{3}, −4)$ **125.** $(0, −4)$ **126.** $(0, 2)$

Solve each problem. See Example 6.

127. *Revolutions of a Turntable* A turntable in a shop makes 45 revolutions per min. How many revolutions does it make per second?

128. *Revolutions of a Windmill* A windmill makes 90 revolutions per min. How many revolutions does it make per second?

129. *Rotating Tire* A tire is rotating 600 times per min. Through how many degrees does a point on the edge of the tire move in $\frac{1}{2}$ sec?

130. *Rotating Airplane Propeller* An airplane propeller rotates 1000 times per min. Find the number of degrees that a point on the edge of the propeller will rotate in 1 sec.

131. *Rotating Pulley* A pulley rotates through 75° in 1 min. How many rotations does the pulley make in an hour?

132. *Surveying* One student in a surveying class measures an angle as 74.25°, while another student measures the same angle as 74° 20′. Find the difference between these measurements, both to the nearest minute and to the nearest hundredth of a degree.

133. *Viewing Field of a Telescope* Due to Earth's rotation, celestial objects like the moon and the stars appear to move across the sky, rising in the east and setting in the west. As a result, if a telescope on Earth remains stationary while viewing a celestial object, the object will slowly move outside the viewing field of the telescope. For this reason, a motor is often attached to telescopes so that the telescope rotates at the same rate as Earth. Determine how long it should take the motor to turn the telescope through an angle of 1 min in a direction perpendicular to Earth's axis.

134. *Angle Measure of a Star on the American Flag* Determine the measure of the angle in each point of the five-pointed star appearing on the American flag. (*Hint:* Inscribe the star in a circle, and use the following theorem from geometry: *An angle whose vertex lies on the circumference of a circle is equal to half the central angle that cuts off the same arc.* See the figure.)

1.2 Angle Relationships and Similar Triangles

Geometric Properties ▪ **Triangles**

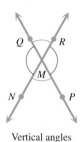

Vertical angles

Figure 13

Geometric Properties In Figure 13, we extended the sides of angle *NMP* to form another angle, *RMQ*. The pair of angles *NMP* and *RMQ* are called **vertical angles.** Another pair of vertical angles, *NMQ* and *PMR*, are also formed. Vertical angles have the following important property.

VERTICAL ANGLES

Vertical angles have equal measures.

Parallel lines are lines that lie in the same plane and do not intersect. Figure 14 shows parallel lines *m* and *n*. When a line *q* intersects two parallel lines, *q* is called a **transversal.** In Figure 14, the transversal intersecting the parallel lines forms eight angles, indicated by numbers.

We learn in geometry that the degree measures of angles 1 through 8 in Figure 14 possess some special properties. The following chart gives the names of these angles and rules about their measures.

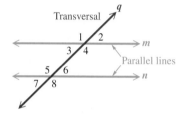

Figure 14

Name	Sketch	Rule
Alternate interior angles	*q* 5 4 *m* *n* (also 3 and 6)	Angle measures are equal.
Alternate exterior angles	*q* 1 *m* *n* 8 (also 2 and 7)	Angle measures are equal.
Interior angles on same side of transversal	*q* 6 4 *m* *n* (also 3 and 5)	Angle measures add to 180°.
Corresponding angles	*q* 2 *m* 6 *n* (also 1 and 5, 3 and 7, 4 and 8)	Angle measures are equal.

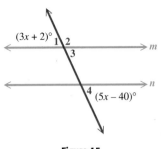

Figure 15

> ▶ **EXAMPLE 1**   FINDING ANGLE MEASURES

Find the measures of angles 1, 2, 3, and 4 in Figure 15, given that lines m and n are parallel.

Solution   Angles 1 and 4 are alternate exterior angles, so they are equal.

$$3x + 2 = 5x - 40$$
$$42 = 2x \qquad \text{Subtract } 3x; \text{ add } 40. \textbf{ (Appendix A)}$$
$$21 = x \qquad \text{Divide by 2.}$$

Angle 1 has measure

$$3x + 2 = 3 \cdot 21 + 2 = 65°, \quad \text{Substitute 21 for } x.$$

and angle 4 has measure

$$5x - 40 = 5 \cdot 21 - 40 = 65°. \quad \text{Substitute 21 for } x.$$

Angle 2 is the supplement of a 65° angle, so it has measure

$$180° - 65° = 115°.$$

Angle 3 is a vertical angle to angle 1, so its measure is 65°. (There are other ways to determine these measures.)

> NOW TRY EXERCISES 3 AND 11. ◀

Triangles   An important property of triangles, first proved by Greek geometers, deals with the sum of the measures of the angles of any triangle.

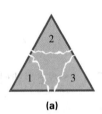

(a)

> ### ANGLE SUM OF A TRIANGLE
>
> ***The sum of the measures of the angles of any triangle is 180°.***

(b)

Figure 16

While it is not an actual proof, we give a rather convincing argument for the truth of this statement, using any size triangle cut from a piece of paper. Tear each corner from the triangle, as suggested in Figure 16(a). You should be able to rearrange the pieces so that the three angles form a straight angle, which has measure 180°, as shown in Figure 16(b).

> **CONNECTIONS**   Use this figure to discuss why the measures of the angles of a triangle must add up to the same sum as the measure of a straight angle.
>
>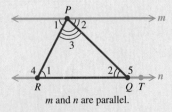
>
> m and n are parallel.

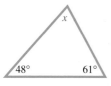

Figure 17

▶ EXAMPLE 2   APPLYING THE ANGLE SUM OF A TRIANGLE PROPERTY

The measures of two of the angles of a triangle are 48° and 61°. (See Figure 17.)
Find the measure of the third angle, x.

Solution $48° + 61° + x = 180°$ The sum of the angles is 180°.

$109° + x = 180°$ Add.

$x = 71°$ Subtract 109°.

The third angle of the triangle measures 71°.

NOW TRY EXERCISES 5 AND 15. ◀

We classify triangles according to angles and sides, as shown below.

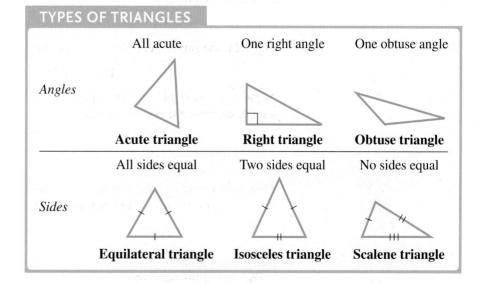

TYPES OF TRIANGLES

Angles	All acute	One right angle	One obtuse angle
	Acute triangle	**Right triangle**	**Obtuse triangle**
Sides	All sides equal	Two sides equal	No sides equal
	Equilateral triangle	**Isosceles triangle**	**Scalene triangle**

NOW TRY EXERCISES 25, 27, AND 31. ◀

Similar triangles are triangles of exactly the same shape but not necessarily
the same size. Figure 18 shows three pairs of similar triangles. The two triangles
in Figure 18(c) not only have the same shape but also the same size. Triangles
that are both the same size and the same shape are called **congruent triangles.** If
two triangles are congruent, then it is possible to pick one of them up and place it
on top of the other so that they coincide. *If two triangles are congruent, then
they must be similar. However, two similar triangles need not be congruent.*

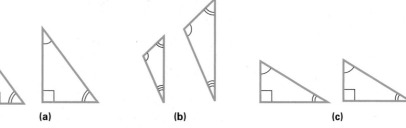

(a)   (b)   (c)

Figure 18

As shown in the figure, the triangular supports for a child's swing set are congruent (and thus similar) triangles, machine-produced with exactly the same dimensions each time. These supports are just one example of similar triangles. The supports of a long bridge, all the same shape but decreasing in size toward the center of the bridge, are examples of similar (but not congruent) triangles.

Suppose a correspondence between two triangles *ABC* and *DEF* is set up as shown in Figure 19.

Angle *A* corresponds to angle *D*.
Angle *B* corresponds to angle *E*.
Angle *C* corresponds to angle *F*.
Side *AB* corresponds to side *DE*.
Side *BC* corresponds to side *EF*.
Side *AC* corresponds to side *DF*.

The small arcs found at the angles in Figure 19 denote the corresponding angles in the triangles.

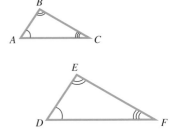

Figure 19

CONDITIONS FOR SIMILAR TRIANGLES

For triangle *ABC* to be similar to triangle *DEF*, the following conditions must hold.

1. Corresponding angles must have the same measure.
2. Corresponding sides must be proportional. (That is, the ratios of their corresponding sides must be equal.)

NOW TRY EXERCISE 41. ◀

▶ **EXAMPLE 3** FINDING ANGLE MEASURES IN SIMILAR TRIANGLES

In Figure 20, triangles *ABC* and *NMP* are similar. Find the measures of angles *B* and *C*.

Solution Since the triangles are similar, corresponding angles have the same measure. Since *C* corresponds to *P* and *P* measures 104°, angle *C* also measures 104°. Since angles *B* and *M* correspond, *B* measures 31°.

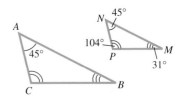

Figure 20

NOW TRY EXERCISE 47. ◀

▶ **EXAMPLE 4** FINDING SIDE LENGTHS IN SIMILAR TRIANGLES

Given that triangle *ABC* and triangle *DFE* in Figure 21 are similar, find the lengths of the unknown sides of triangle *DFE*.

Solution Similar triangles have corresponding sides in proportion. Use this fact to find the unknown side lengths in triangle *DFE*.

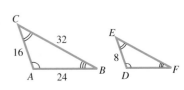

Figure 21

Side *DF* of triangle *DFE* corresponds to side *AB* of triangle *ABC*, and sides *DE* and *AC* correspond. This leads to the proportion

$$\frac{8}{16} = \frac{DF}{24}.$$

Recall this property of proportions from algebra.

If $\dfrac{a}{b} = \dfrac{c}{d}$,   **then** $ad = bc$.

We use this property to solve the equation for *DF*.

$$\frac{8}{16} = \frac{DF}{24}$$

$$8 \cdot 24 = 16 \cdot DF$$

$$192 = 16 \cdot DF \quad \text{Multiply.}$$

$$12 = DF \quad \text{Divide by 16.}$$

Side *DF* has length 12.

Side *EF* corresponds to *CB*. This leads to another proportion.

$$\frac{8}{16} = \frac{EF}{32}$$

$$8 \cdot 32 = 16 \cdot EF$$

$$16 = EF \quad \text{Solve for } EF.$$

Side *EF* has length 16.

NOW TRY EXERCISE 53. ◀

▶ **EXAMPLE 5**   **FINDING THE HEIGHT OF A FLAGPOLE**

Firefighters at the Morganza Fire Station need to measure the height of the station flagpole. They find that at the instant when the shadow of the station is 18 m long, the shadow of the flagpole is 99 ft long. The station is 10 m high. Find the height of the flagpole.

Solution   Figure 22 shows the information given in the problem. The two triangles are similar, so corresponding sides are in proportion.

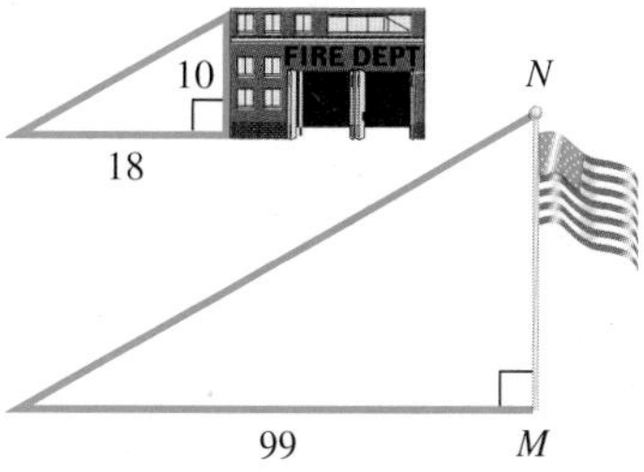

$$\frac{MN}{10} = \frac{99}{18}$$

$$\frac{MN}{10} = \frac{11}{2} \quad \text{Lowest terms}$$

$$MN \cdot 2 = 10 \cdot 11$$

$$MN = 55 \quad \text{Solve for } MN.$$

Figure 22

The flagpole is 55 ft high.

NOW TRY EXERCISE 57. ◀

1.2 **Exercises**

1. *Concept Check* Use the given figure to find the measures of the numbered angles, given that lines *m* and *n* are parallel.

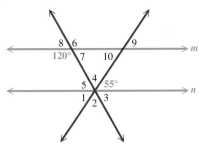

2. Consider Figure 14. If the measure of one of the angles is known, can the measures of the remaining seven angles be determined? Explain.

Find the measure of each marked angle. In Exercises 11–14, m and n are parallel. See Examples 1 and 2.

3.

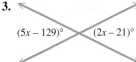

$(5x - 129)°$ $(2x - 21)°$

4.

$(11x - 37)°$ $(7x + 27)°$

5.

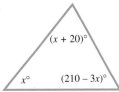

$(x + 20)°$

$x°$ $(210 - 3x)°$

6. $(x + 15)°$

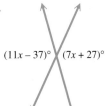

$(x + 5)°$

$(10x - 20)°$

7.

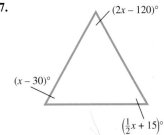

$(2x - 120)°$

$(x - 30)°$

$\left(\frac{1}{2}x + 15\right)°$

8.

$(2x + 16)°$

$(3x - 6)°$

$(5x - 50)°$

9.

$(6x + 3)°$

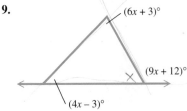

$(9x + 12)°$

$(4x - 3)°$

10.

$(-5x)°$

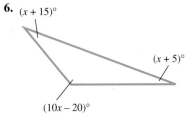

$(-8x + 3)°$

$(7 - 12x)°$

11.

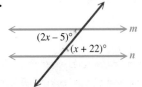

$(2x - 5)°$
$(x + 22)°$

12.

$(2x + 61)°$

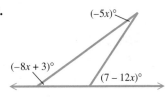

$(6x - 51)°$

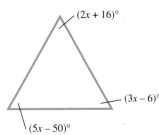

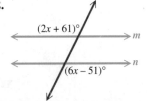

13.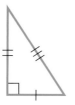

14.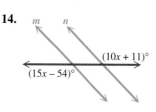

The measures of two angles of a triangle are given. Find the measure of the third angle. See Example 2.

15. $37°, 52°$

16. $29°, 104°$

17. $147° \, 12', 30° \, 19'$

18. $136° \, 50', 41° \, 38'$

19. $74.2°, 80.4°$

20. $29.6°, 49.7°$

21. $51° \, 20' \, 14'', 106° \, 10' \, 12''$

22. $17° \, 41' \, 13'', 96° \, 12' \, 10''$

23. Can a triangle have angles of measures $85°$ and $100°$? Explain.

24. Can a triangle have two obtuse angles? Explain.

Concept Check   *Classify each triangle in Exercises 25–36 as* acute, right, *or* obtuse. *Also classify each as* equilateral, isosceles, *or* scalene.

25.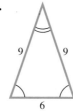

26.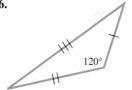

27.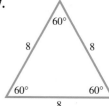

28.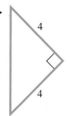

29.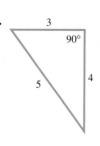

30.

31.

32.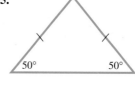

33.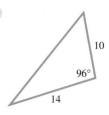

34.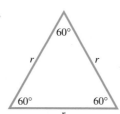

35.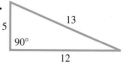

36.

37. Write a definition of *isosceles right triangle*.

38. Explain why the sum of the lengths of any two sides of a triangle must be greater than the length of the third side.

39. Must all equilateral triangles be similar? Explain.

40. *Carpentry Technique*   The following technique is used by carpenters to draw a 60° angle with a straightedge and compass. Explain why this technique works. (*Source:* Hamilton, J. E. and M. S. Hamilton, *Math to Build On,* Construction Trades Press, 1993.)

Step 1   Draw a straight line segment, and mark a point near the center of the line.

Step 2   Place the compass tip on the marked point, and draw a semicircle.

Step 3   Without changing the setting of the compass, place the tip of the compass at the right intersection of the line and the semicircle and then mark a small arc across the semicircle.

Step 4   Draw a line segment from the marked point on the line to the point where the arc crosses the semicircle. This line will make a 60° angle with the original line.

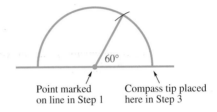

Point marked
on line in Step 1

Compass tip placed
here in Step 3

Concept Check   *Name the corresponding angles and the corresponding sides of each pair of similar triangles.*

41.

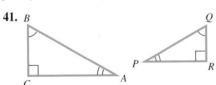

42.

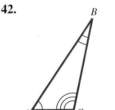

43. (*EA* is parallel to *CD*.)

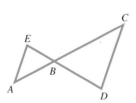

44. (*HK* is parallel to *EF*.)

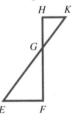

Find all unknown angle measures in each pair of similar triangles. See Example 3.

45.

46.

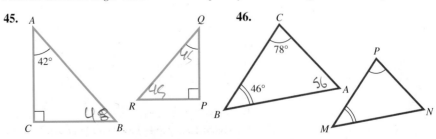

47.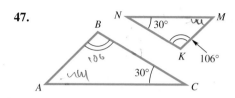

48.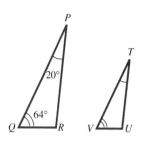

49.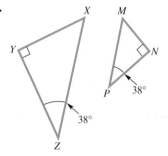

50.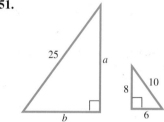

Find the unknown side lengths labeled with a variable in each pair of similar triangles. See Example 4.

51.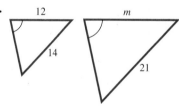

52.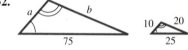

53.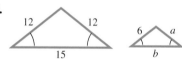

54.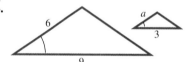

55.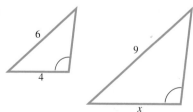

56.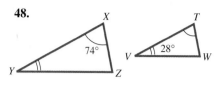

Solve each problem. See Example 5.

57. *Height of a Tree*   A tree casts a shadow 45 m long. At the same time, the shadow cast by a vertical 2-m stick is 3 m long. Find the height of the tree.

58. *Height of a Lookout Tower*   A forest fire lookout tower casts a shadow 180 ft long at the same time that the shadow of a 9-ft truck is 15 ft long. Find the height of the tower.

59. *Lengths of Sides of a Triangle*   On a photograph of a triangular piece of land, the lengths of the three sides are 4 cm, 5 cm, and 7 cm, respectively. The shortest side of the actual piece of land is 400 m long. Find the lengths of the other two sides.

60. *Height of a Lighthouse*
The Biloxi lighthouse in the figure casts a shadow 28 m long at 7 P.M. At the same time, the shadow of the lighthouse keeper, who is 1.75 m tall, is 3.5 m long. How tall is the lighthouse?

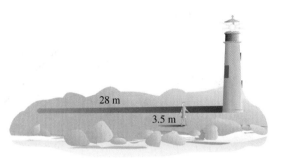

Not to scale

61. *Height of a Building* A house is 15 ft tall. Its shadow is 40 ft long at the same time the shadow of a nearby building is 300 ft long. Find the height of the building.

62. *Height of a Carving of Lincoln* Assume that Lincoln was $6\frac{1}{3}$ ft tall and his head $\frac{3}{4}$ ft long. Knowing that the carved head of Lincoln at Mt. Rushmore is 60 ft tall, find how tall his entire body would be if it were carved into the mountain.

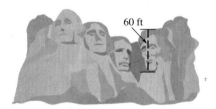

In each diagram, there are two similar triangles. Find the unknown measurement. (Hint: In the sketch for Exercise 63, the side of length 100 in the small triangle corresponds to the side of the length 100 + 120 = 220 in the large triangle.)

63.

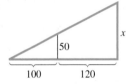

64.

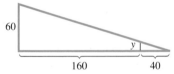

65.

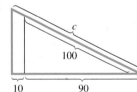

66.

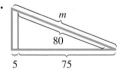

Solve each problem.

67. *Lengths of Sides of a Quadrilateral* Two quadrilaterals (four-sided figures) are similar. The lengths of the three shortest sides of the first quadrilateral are 18 cm, 24 cm, and 32 cm. The lengths of the two longest sides of the second quadrilateral are 48 cm and 60 cm. Find the unknown lengths of the sides of these two figures.

68. *Distance Between Two Cities* By drawing lines on a map, a triangle can be formed by the cities of Phoenix, Tucson, and Yuma. On the map, the distance between Phoenix and Tucson is 8 cm, the distance between Phoenix and Yuma is 12 cm, and the distance between Tucson and Yuma is 17 cm. The actual straight-line distance from Phoenix to Yuma is 230.0 km. Find the distances between the other pairs of cities to the nearest tenth of a kilometer.

69. *Solar Eclipse on Earth*   The sun has a diameter of about 865,000 mi with a maximum distance from Earth's surface of about 94,500,000 mi. The moon has a smaller diameter of 2159 mi. For a total solar eclipse to occur, the moon must pass between Earth and the sun. The moon must also be close

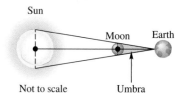

enough to Earth for the moon's **umbra** (shadow) to reach the surface of Earth. (*Source:* Karttunen, H., P. Kröger, H. Oja, M. Putannen, and K. Donners, Editors, *Fundamental Astronomy,* Fourth Edition, Springer-Verlag, 2003.)

    (a) Calculate the maximum distance that the moon can be from Earth and still have a total solar eclipse occur. (*Hint:* Use similar triangles.)

    (b) The closest approach of the moon to Earth's surface was 225,745 mi and the farthest was 251,978 mi. (*Source: World Almanac and Book of Facts.*) Can a total solar eclipse occur every time the moon is between Earth and the sun?

70. *Solar Eclipse on Neptune*   (Refer to Exercise 69.) The sun's distance from Neptune is approximately 2,800,000,000 mi (2.8 billion mi). The largest moon of Neptune is Triton, with a diameter of approximately 1680 mi. (*Source: World Almanac and Book of Facts.*)

    (a) Calculate the maximum distance that Triton can be from Neptune for a total eclipse of the sun to occur on Neptune. (*Hint:* Use similar triangles.)

    (b) Triton is approximately 220,000 mi from Neptune. Is it possible for Triton to cause a total eclipse on Neptune?

71. *Solar Eclipse on Mars*   (Refer to Exercise 69.) The sun's distance from the surface of Mars is approximately 142,000,000 mi. One of Mars' two moons, Phobos, has a maximum diameter of 17.4 mi. (*Source:* Zeilik, M., S. Gregory, and E. Smith, *Introductory Astronomy and Astrophysics,* Second Edition, Saunders College Publishers, 1998.)

    (a) Calculate the maximum distance that the moon Phobos can be from Mars for a total eclipse of the sun to occur on Mars.

    (b) Phobos is approximately 5800 mi from Mars. Is it possible for Phobos to cause a total eclipse on Mars?

72. *Solar Eclipse on Jupiter*   (Refer to Exercise 69.) The sun's distance from the surface of Jupiter is approximately 484,000,000 mi. One of Jupiter's moons, Ganymede, has a diameter of 3270 mi. (*Source:* Wright, J. W., General Editor, *The Universal Almanac,* Andrews and McMeel, 1997.)

    (a) Calculate the maximum distance that the moon Ganymede can be from Jupiter for a total eclipse of the sun to occur on Jupiter.

    (b) Ganymede is approximately 665,000 mi from Jupiter. Is it possible for Ganymede to cause a total eclipse on Jupiter?

73. *Sizes and Distances in the Sky*   Astronomers use degrees, minutes, and seconds to measure sizes and distances in the sky along an arc from the horizon to the zenith point directly overhead. An adult observer on Earth can judge distances in the sky using his or her hand at arm's length. An outstretched hand will be about 20 arc degrees wide from the tip of

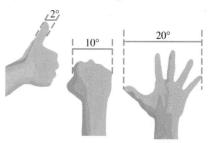

the thumb to the tip of the little finger. A clenched fist at arm's length measures about 10 arc degrees, and a thumb corresponds to about 2 arc degrees. (*Source:* Levy, D. H., *Skywatching,* The Nature Company, 1994.)

    (a) The apparent size of the moon is about 31 arc minutes. What part of your thumb would cover the moon?

(b) If an outstretched hand plus a fist cover the distance between two bright stars, about how far apart in arc degrees are the stars?

In each figure, two similar triangles are present. Find the value of each variable.

74.

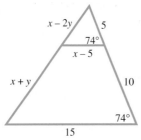

75.

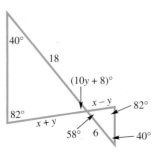

---

1. For an angle measuring $19°$, give the measure of its **(a)** complement and **(b)** supplement.

Find the measure of each unknown angle.

2.

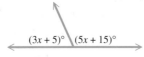

3.

4.

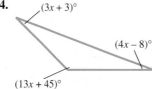

5.

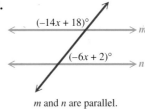

m and n are parallel.

6. Perform each indicated conversion.

    (a) $77° \, 12' \, 09''$ to decimal degrees   **(b)** $22.0250°$ to degrees, minutes, seconds

7. Find the angle of least positive measure (not equal to the given angle) coterminal with each angle.

    (a) $410°$ **(b)** $-60°$ **(c)** $890°$ **(d)** $57°$

8. *Rotating Flywheel* A flywheel rotates 300 times per min. Through how many degrees does a point on the edge of the flywheel move in 1 sec?

9. *Length of a Shadow* If a vertical antenna 45 ft tall casts a shadow 15 ft long, how long would the shadow of a 30-ft pole be at the same time and place?

10. Find the unknown side lengths x and y in this pair of similar triangles.

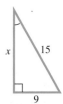

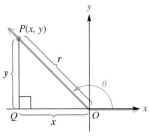

Figure 23

1.3 Trigonometric Functions

Trigonometric Functions ▪ Quadrantal Angles

Trigonometric Functions The study of trigonometry covers the six **trigonometric functions** defined here. To define these functions, we start with an angle θ in standard position, and choose any point P having coordinates (x, y) on the terminal side of angle θ. (The point P must not be the vertex of the angle.) See Figure 23. A perpendicular from P to the x-axis at point Q determines a right triangle, having vertices at O, P, and Q. We find the distance r from $P(x, y)$ to the origin, $(0, 0)$, using the distance formula.

$$r = \sqrt{(x - 0)^2 + (y - 0)^2} = \sqrt{x^2 + y^2} \quad \text{(Appendix B)}$$

Notice that $r > 0$ since this is the undirected distance.

The six trigonometric functions of angle θ are **sine, cosine, tangent, cotangent, secant,** and **cosecant.** In the following definitions, we use the customary abbreviations for the names of these functions: **sin, cos, tan, cot, sec,** and **csc.**

TRIGONOMETRIC FUNCTIONS

Let (x, y) be a point other than the origin on the terminal side of an angle θ in standard position. The distance from the point to the origin is $r = \sqrt{x^2 + y^2}$. The six trigonometric functions of θ are defined as follows.

$$\sin \theta = \frac{y}{r} \qquad\qquad \cos \theta = \frac{x}{r} \qquad\qquad \tan \theta = \frac{y}{x} \quad (x \neq 0)$$

$$\csc \theta = \frac{r}{y} \quad (y \neq 0) \qquad \sec \theta = \frac{r}{x} \quad (x \neq 0) \qquad \cot \theta = \frac{x}{y} \quad (y \neq 0)$$

▶ **EXAMPLE 1**   FINDING FUNCTION VALUES OF AN ANGLE

The terminal side of an angle θ in standard position passes through the point $(8, 15)$. Find the values of the six trigonometric functions of angle θ.

Solution   Figure 24 shows angle θ and the triangle formed by dropping a perpendicular from the point $(8, 15)$ to the x-axis. The point $(8, 15)$ is 8 units to the right of the y-axis and 15 units above the x-axis, so $x = 8$ and $y = 15$. Since $r = \sqrt{x^2 + y^2}$,

$$r = \sqrt{8^2 + 15^2} = \sqrt{64 + 225} = \sqrt{289} = 17.$$

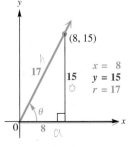

Figure 24

We can now find the values of the six trigonometric functions of angle θ.

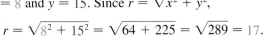

$$\sin \theta = \frac{y}{r} = \frac{15}{17} \qquad\qquad \cos \theta = \frac{x}{r} = \frac{8}{17} \qquad\qquad \tan \theta = \frac{y}{x} = \frac{15}{8}$$

$$\csc \theta = \frac{r}{y} = \frac{17}{15} \qquad\qquad \sec \theta = \frac{r}{x} = \frac{17}{8} \qquad\qquad \cot \theta = \frac{x}{y} = \frac{8}{15}$$

NOW TRY EXERCISE 5. ◀

▶ **EXAMPLE 2**   **FINDING FUNCTION VALUES OF AN ANGLE**

The terminal side of an angle θ in standard position passes through the point $(-3, -4)$. Find the values of the six trigonometric functions of angle θ.

Solution   As shown in Figure 25, $x = -3$ and $y = -4$. The value of r is

$$r = \sqrt{(-3)^2 + (-4)^2} = \sqrt{25} = 5. \quad \text{Remember that } r > 0.$$

Then use the definitions of the trigonometric functions.

$$\sin \theta = \frac{-4}{5} = -\frac{4}{5} \qquad \cos \theta = \frac{-3}{5} = -\frac{3}{5} \qquad \tan \theta = \frac{-4}{-3} = \frac{4}{3}$$

$$\csc \theta = \frac{5}{-4} = -\frac{5}{4} \qquad \sec \theta = \frac{5}{-3} = -\frac{5}{3} \qquad \cot \theta = \frac{-3}{-4} = \frac{3}{4}$$

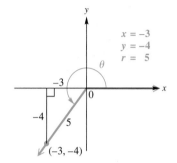

Figure 25

NOW TRY EXERCISE 19. ◀

We can find the six trigonometric functions using *any* point other than the origin on the terminal side of an angle. To see why any point may be used, refer to Figure 26, which shows an angle θ and two distinct points on its terminal side. Point P has coordinates (x, y), and point $\boldsymbol{P'}$ (read **"P-prime"**) has coordinates (x', y'). Let r be the length of the hypotenuse of triangle OPQ, and let r' be the length of the hypotenuse of triangle $OP'Q'$. Since corresponding sides of similar triangles are proportional,

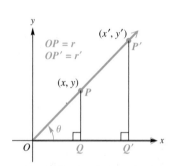

Figure 26

$$\frac{y}{r} = \frac{y'}{r'}, \quad \text{(Section 1.2)}$$

so $\sin \theta = \frac{y}{r}$ is the same no matter which point is used to find it. A similar result holds for the other five trigonometric functions.

We can also find the trigonometric function values of an angle if we know the equation of the line coinciding with the terminal ray. Recall from algebra that the graph of the equation

$$Ax + By = 0 \quad \textbf{(Appendix B)}$$

is a line that passes through the origin. If we restrict x to have only nonpositive or only nonnegative values, we obtain as the graph a ray with endpoint at the origin. For example, the graph of $x + 2y = 0$, $x \geq 0$, shown in Figure 27, is a ray that can serve as the terminal side of an angle θ in standard position. By choosing a point on the ray, we can find the trigonometric function values of the angle.

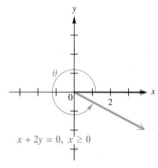

Figure 27

▶ **EXAMPLE 3**   **FINDING FUNCTION VALUES OF AN ANGLE**

Find the six trigonometric function values of the angle θ in standard position, if the terminal side of θ is defined by $x + 2y = 0$, $x \geq 0$.

Solution   The angle is shown in Figure 28 on the next page. We can use *any* point except $(0, 0)$ on the terminal side of θ to find the trigonometric function values. We choose $x = 2$ and find the corresponding y-value.

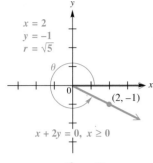

$x = 2$
$y = -1$
$r = \sqrt{5}$

$x + 2y = 0, \; x \geq 0$

Figure 28

$$x + 2y = 0, \; x \geq 0$$
$$2 + 2y = 0 \qquad \text{Let } x = 2.$$
$$2y = -2 \qquad \text{Subtract 2. (Appendix A)}$$
$$y = -1 \qquad \text{Divide by 2.}$$

The point $(2, -1)$ lies on the terminal side, and the corresponding value of r is $r = \sqrt{2^2 + (-1)^2} = \sqrt{5}$. Now we use the definitions of the trigonometric functions.

$$\sin \theta = \frac{y}{r} = \frac{-1}{\sqrt{5}} = \frac{-1}{\sqrt{5}} \cdot \frac{\sqrt{5}}{\sqrt{5}} = -\frac{\sqrt{5}}{5}$$

Multiply by $\frac{\sqrt{5}}{\sqrt{5}}$, which equals 1, to rationalize the denominators.

$$\cos \theta = \frac{x}{r} = \frac{2}{\sqrt{5}} = \frac{2}{\sqrt{5}} \cdot \frac{\sqrt{5}}{\sqrt{5}} = \frac{2\sqrt{5}}{5}$$

$$\tan \theta = \frac{y}{x} = \frac{-1}{2} = -\frac{1}{2}$$

$$\csc \theta = \frac{r}{y} = \frac{\sqrt{5}}{-1} = -\sqrt{5} \qquad \sec \theta = \frac{r}{x} = \frac{\sqrt{5}}{2} \qquad \cot \theta = \frac{x}{y} = \frac{2}{-1} = -2$$

NOW TRY EXERCISE 45. ◀

Recall that when the equation of a line is written in slope-intercept form $y = mx + b$, the coefficient of x is the slope of the line. In Example 3, the equation $x + 2y = 0$ can be written as $y = -\frac{1}{2}x$, so the slope is $-\frac{1}{2}$. Notice that $\tan \theta = -\frac{1}{2}$. *In general, it is true that $m = \tan \theta$.*

> ▶ **Note** The trigonometric function values we found in Examples 1–3 are *exact*. If we were to use a calculator to approximate these values, the decimal results would not be acceptable if exact values were required.

Quadrantal Angles If the terminal side of an angle in standard position lies along the y-axis, any point on this terminal side has x-coordinate 0. Similarly, an angle with terminal side on the x-axis has y-coordinate 0 for any point on the terminal side. Since the values of x and y appear in the denominators of some trigonometric functions, and since a fraction is undefined if its denominator is 0, some trigonometric function values of quadrantal angles (i.e., those with terminal side on an axis) are undefined.

When determining trigonometric function values of quadrantal angles, Figure 29 can help find the ratios. Because *any* point on the terminal side can be used, it is convenient to choose the point one unit from the origin, with $r = 1$. (In **Chapter 3** we extend this idea to the *unit circle*.)

To find the function values of a quadrantal angle, determine the position of the terminal side, choose the one of these four points that lies on this terminal side, and then use the definitions involving x, y, and r.

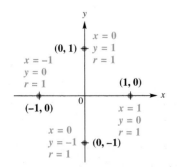

$x = 0$
$(0, 1) \quad y = 1$
$x = -1 \qquad r = 1$
$y = 0$
$r = 1 \qquad\qquad (1, 0)$

$(-1, 0)$
$x = 1$
$y = 0$
$x = 0 \qquad r = 1$
$y = -1 \quad (0, -1)$
$r = 1$

Figure 29

▶ **EXAMPLE 4** **FINDING FUNCTION VALUES OF QUADRANTAL ANGLES**

Find the values of the six trigonometric functions for each angle.

(a) an angle of 90°

(b) an angle θ in standard position with terminal side through $(-3, 0)$

Solution

(a) The sketch in Figure 30 shows that the terminal side passes through $(0, 1)$. So $x = 0$, $y = 1$, and $r = 1$. Thus,

$$\sin 90° = \frac{1}{1} = 1 \qquad \cos 90° = \frac{0}{1} = 0 \qquad \tan 90° = \frac{1}{0} \text{ (undefined)}$$

$$\csc 90° = \frac{1}{1} = 1 \qquad \sec 90° = \frac{1}{0} \text{ (undefined)} \qquad \cot 90° = \frac{0}{1} = 0.$$

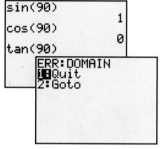

A calculator in degree mode returns the correct values for sin 90° and cos 90°. The second screen shows an ERROR message for tan 90°, because 90° is not in the domain of the tangent function.

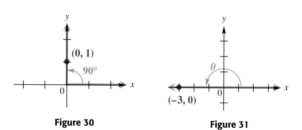

Figure 30   **Figure 31**

(b) Figure 31 shows the angle. Here, $x = -3$, $y = 0$, and $r = 3$, so the trigonometric functions have the following values.

$$\sin \theta = \frac{0}{3} = 0 \qquad \cos \theta = \frac{-3}{3} = -1 \qquad \tan \theta = \frac{0}{-3} = 0$$

$$\csc \theta = \frac{3}{0} \text{ (undefined)} \qquad \sec \theta = \frac{3}{-3} = -1 \qquad \cot \theta = \frac{-3}{0} \text{ (undefined)}$$

(Verify that these values can also be found by using the point $(-1, 0)$.)

NOW TRY EXERCISES 13, 57, 59, 63, AND 65. ◀

The conditions under which the trigonometric function values of quadrantal angles are undefined are summarized here.

UNDEFINED FUNCTION VALUES

If the terminal side of a quadrantal angle lies along the y-axis, then the tangent and secant functions are undefined. If it lies along the x-axis, then the cotangent and cosecant functions are undefined.

The function values of the most commonly used quadrantal angles, 0°, 90°, 180°, 270°, and 360°, are summarized in the table on the next page. They can be determined when needed by using Figure 29 and the method of Example 4(a).

Function Values of Quadrantal Angles

θ	$\sin \theta$	$\cos \theta$	$\tan \theta$	$\cot \theta$	$\sec \theta$	$\csc \theta$
0°	0	1	0	Undefined	1	Undefined
90°	1	0	Undefined	0	Undefined	1
180°	0	-1	0	Undefined	-1	Undefined
270°	-1	0	Undefined	0	Undefined	-1
360°	0	1	0	Undefined	1	Undefined

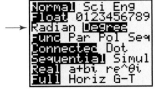

TI-83 Plus

TI-84 Plus

Figure 32

The values given in this table can be found with a calculator that has trigonometric function keys. *Make sure the calculator is set in degree mode.*

▶ **Caution** *One of the most common errors involving calculators in trigonometry occurs when the calculator is set for radian measure, rather than degree measure.* (Radian measure of angles is discussed in **Chapter 3.**) Be sure you know how to set your calculator in degree mode. See Figure 32, which illustrates degree mode for TI-83/84 Plus calculators.

1.3 Exercises

Concept Check   Sketch an angle θ in standard position such that θ has the least possible positive measure, and the given point is on the terminal side of θ. Then find the values of the six trigonometric functions for each angle. Rationalize denominators when applicable. See Examples 1, 2, and 4.

1. $(5, -12)$ **2.** $(-12, -5)$ **3.** $(-3, 4)$ **4.** $(-4, -3)$

5. $(-8, 15)$ **6.** $(15, -8)$ **7.** $(7, -24)$ **8.** $(-24, -7)$

9. $(0, 2)$ **10.** $(0, 5)$ **11.** $(-4, 0)$ **12.** $(-5, 0)$

13. $(0, -4)$ **14.** $(0, -3)$ **15.** $\left(1, \sqrt{3}\right)$ **16.** $\left(-1, \sqrt{3}\right)$

17. $\left(\sqrt{2}, \sqrt{2}\right)$ **18.** $\left(-\sqrt{2}, -\sqrt{2}\right)$ **19.** $\left(-2\sqrt{3}, -2\right)$ **20.** $\left(-2\sqrt{3}, 2\right)$

21. For any nonquadrantal angle θ, $\sin \theta$ and $\csc \theta$ will have the same sign. Explain why.

22. *Concept Check*   How is the value of r interpreted geometrically in the definitions of the sine, cosine, secant, and cosecant functions?

23. *Concept Check*   If $\cot \theta$ is undefined, what is the value of $\tan \theta$?

24. *Concept Check*   If the terminal side of an angle θ is in quadrant III, what is the sign of each of the trigonometric function values of θ?

Concept Check   Suppose that the point (x, y) is in the indicated quadrant. Decide whether the given ratio is positive or negative. Recall that $r = \sqrt{x^2 + y^2}$. (Hint: Drawing a sketch may help.)

25. II, $\dfrac{x}{r}$ **26.** III, $\dfrac{y}{r}$ **27.** IV, $\dfrac{y}{x}$ **28.** IV, $\dfrac{x}{y}$

29. II, $\dfrac{y}{r}$ **30.** III, $\dfrac{x}{r}$ **31.** IV, $\dfrac{x}{r}$ **32.** IV, $\dfrac{y}{r}$

33. II, $\dfrac{x}{y}$ **34.** II, $\dfrac{y}{x}$ **35.** III, $\dfrac{y}{x}$ **36.** III, $\dfrac{x}{y}$

37. IV, $\dfrac{x}{y}$ **38.** IV, $\dfrac{y}{x}$ **39.** I, $\dfrac{x}{y}$ **40.** I, $\dfrac{y}{x}$

41. I, $\dfrac{y}{r}$ **42.** I, $\dfrac{x}{r}$ **43.** I, $\dfrac{r}{x}$ **44.** I, $\dfrac{r}{y}$

In Exercises 45–54, an equation of the terminal side of an angle θ in standard position is given with a restriction on x. Sketch the least positive such angle θ, and find the values of the six trigonometric functions of θ. See Example 3.

45. $2x + y = 0, x \geq 0$ **46.** $3x + 5y = 0, x \geq 0$

47. $-6x - y = 0, x \leq 0$ **48.** $-5x - 3y = 0, x \leq 0$

49. $-4x + 7y = 0, x \leq 0$ **50.** $6x - 5y = 0, x \geq 0$

51. $x + y = 0, x \geq 0$ **52.** $x - y = 0, x \geq 0$

53. $-\sqrt{3}x + y = 0, x \leq 0$ **54.** $\sqrt{3}x + y = 0, x \leq 0$

To work Exercises 55–72, begin by reproducing the graph in Figure 29 on page 25. Keep in mind that for each of the four points labeled in the figure, r = 1. For each quadrantal angle, identify the appropriate values of x, y, and r to find the indicated function value. If it is undefined, say so. See Example 4.

55. $\cos 90°$ **56.** $\sin 90°$ **57.** $\tan 180°$

58. $\cot 90°$ **59.** $\sec 180°$ **60.** $\csc 270°$

61. $\sin(-270°)$ **62.** $\cos(-90°)$ **63.** $\cot 540°$

64. $\tan 450°$ **65.** $\csc(-450°)$ **66.** $\sec(-540°)$

67. $\sin 1800°$ **68.** $\cos 1800°$ **69.** $\csc 1800°$

70. $\cot 1800°$ **71.** $\sec 1800°$ **72.** $\tan 1800°$

Use the trigonometric function values of quadrantal angles given in this section to evaluate each expression. An expression such as $\cot^2 90°$ means $(\cot 90°)^2$, which is equal to $0^2 = 0$.

73. $\cos 90° + 3 \sin 270°$ **74.** $\tan 0° - 6 \sin 90°$

75. $3 \sec 180° - 5 \tan 360°$ **76.** $4 \csc 270° + 3 \cos 180°$

77. $\tan 360° + 4 \sin 180° + 5 \cos^2 180°$ **78.** $2 \sec 0° + 4 \cot^2 90° + \cos 360°$

79. $\sin^2 180° + \cos^2 180°$ **80.** $\sin^2 360° + \cos^2 360°$

81. $\sec^2 180° - 3 \sin^2 360° + \cos 180°$ **82.** $5 \sin^2 90° + 2 \cos^2 270° - \tan 360°$

83. $-2 \sin^4 0° + 3 \tan^2 0°$ **84.** $-3 \sin^4 90° + 4 \cos^3 180°$

If n is an integer, n · 180° represents an integer multiple of 180°, (2n + 1) · 90° represents an odd integer multiple of 90°, and so on. Decide whether each expression is equal to 0, 1, −1, or is undefined.

85. $\cos[(2n + 1) \cdot 90°]$ **86.** $\sin[n \cdot 180°]$

87. $\tan[n \cdot 180°]$ **88.** $\tan[(2n + 1) \cdot 90°]$

89. $\sin[270° + n \cdot 360°]$ **90.** $\cot[n \cdot 180°]$

91. $\cot[(2n + 1) \cdot 90°]$

92. $\cos[n \cdot 360°]$

93. $\sec[(2n + 1) \cdot 90°]$

94. $\csc[n \cdot 180°]$

Concept Check   *In later chapters we will study trigonometric functions of angles other than quadrantal angles, such as 15°, 30°, 60°, 75°, and so on. To prepare for some important concepts, provide conjectures in Exercises 95–98. Be sure that your calculator is in degree mode.*

95. The angles 15° and 75° are complementary. With your calculator determine sin 15° and cos 75°. Make a conjecture about the sines and cosines of complementary angles, and test your hypothesis with other pairs of complementary angles. (*Note:* This relationship will be discussed in detail in **Section 2.1.**)

96. The angles 25° and 65° are complementary. With your calculator determine tan 25° and cot 65°. Make a conjecture about the tangents and cotangents of complementary angles, and test your hypothesis with other pairs of complementary angles. (*Note:* This relationship will be discussed in detail in **Section 2.1.**)

97. With your calculator determine sin 10° and sin(−10°). Make a conjecture about the sines of an angle and its negative, and test your hypothesis with other angles. (*Note:* This relationship will be discussed in detail in **Section 5.1.**)

98. With your calculator determine cos 20° and cos(−20°). Make a conjecture about the cosines of an angle and its negative, and test your hypothesis with other angles. (*Note:* This relationship will be discussed in detail in **Section 5.1.**)

In Exercises 99–104, set your graphing calculator in parametric and degree modes. Set the window and functions (see the third screen) as shown here, and graph. A circle of radius 1 will appear on the screen. Trace to move a short distance around the circle. In the screen, the point on the circle corresponds to an angle T = 25°. *Since r* = 1, cos 25° *is* X = .90630779, *and* sin 25° *is* Y = .42261826.

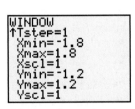

This screen is a continuation of the previous one.

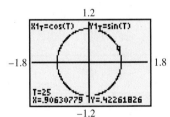

99. Use the right- and left-arrow keys to move to the point corresponding to 20°. What are cos 20° and sin 20°?

100. For what angle T, 0° ≤ T ≤ 90°, is cos T ≈ .766?

101. For what angle T, 0° ≤ T ≤ 90°, is sin T ≈ .574?

102. For what angle T, 0° ≤ T ≤ 90°, does cos T = sin T?

103. As T increases from 0° to 90°, does the cosine increase or decrease? What about the sine?

104. As T increases from 90° to 180°, does the cosine increase or decrease? What about the sine?

1.4 Using the Definitions of the Trigonometric Functions

Reciprocal Identities ▪ **Signs and Ranges of Function Values** ▪ **Pythagorean Identities** ▪ **Quotient Identities**

Identities are equations that are true for all values of the variables for which all expressions are defined. Identities are studied in more detail in **Chapter 5.**

$$(x + y)^2 = x^2 + 2xy + y^2 \qquad 2(x + 3) = 2x + 6 \quad \text{Identities (Appendix A)}$$

Reciprocal Identities   Recall the definition of a reciprocal: the **reciprocal** of the nonzero number x is $\frac{1}{x}$. For example, the reciprocal of 2 is $\frac{1}{2}$, and the reciprocal of $\frac{8}{11}$ is $\frac{11}{8}$. There is no reciprocal for 0. Scientific calculators have a reciprocal key, usually labeled $\boxed{1/x}$ or $\boxed{x^{-1}}$. Using this key gives the reciprocal of any nonzero number entered in the display.

The definitions of the trigonometric functions in the previous section on page 23 were written so that functions in the same column are reciprocals of each other. Since $\sin \theta = \frac{y}{r}$ and $\csc \theta = \frac{r}{y}$,

$$\sin \theta = \frac{1}{\csc \theta} \qquad \text{and} \qquad \csc \theta = \frac{1}{\sin \theta}, \quad \text{provided } \sin \theta \neq 0.$$

Also, $\cos \theta$ and $\sec \theta$ are reciprocals, as are $\tan \theta$ and $\cot \theta$. The **reciprocal identities** hold for any angle θ that does not lead to a 0 denominator.

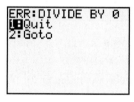

(a)

RECIPROCAL IDENTITIES

For all angles θ for which both functions are defined,

$$\sin \theta = \frac{1}{\csc \theta} \qquad \cos \theta = \frac{1}{\sec \theta} \qquad \tan \theta = \frac{1}{\cot \theta}$$

$$\csc \theta = \frac{1}{\sin \theta} \qquad \sec \theta = \frac{1}{\cos \theta} \qquad \cot \theta = \frac{1}{\tan \theta}.$$

```
ERR:DIVIDE BY 0
1∎Quit
2:Goto
```

(b)

Figure 33

⊟ The screen in Figure 33(a) shows how to find csc 90°, sec 180°, and csc(−270°), using the appropriate reciprocal identities and the reciprocal key of a graphing calculator in degree mode. Attempting to find sec 90° by entering $\frac{1}{\cos 90°}$ produces an ERROR message, indicating the reciprocal is undefined. See Figure 33(b). Compare these results with the ones found in the table of quadrantal angle function values in **Section 1.3.** ▪

▶ **Caution** *Be sure not to use the inverse trigonometric function keys to find reciprocal function values.* For example,

$$\sin^{-1}(90°) \neq \frac{1}{\sin(90°)}.$$

Inverse trigonometric functions are covered in **Section 2.3.**

> ▶ **Note** Identities can be written in different forms. For example,
>
> $$\sin \theta = \frac{1}{\csc \theta} \quad \text{can be written} \quad \csc \theta = \frac{1}{\sin \theta}, \quad \text{or} \quad (\sin \theta)(\csc \theta) = 1.$$

▶ **EXAMPLE 1**   USING THE RECIPROCAL IDENTITIES

Find each function value.

(a) $\cos \theta$, given that $\sec \theta = \frac{5}{3}$ **(b)** $\sin \theta$, given that $\csc \theta = -\frac{\sqrt{12}}{2}$

Solution

(a) Since $\cos \theta$ is the reciprocal of $\sec \theta$,

$$\cos \theta = \frac{1}{\sec \theta} = \frac{1}{\frac{5}{3}} = 1 \div \frac{5}{3} = 1 \cdot \frac{3}{5} = \frac{3}{5}.$$

Simplify the complex fraction.

(b) $\sin \theta = \dfrac{1}{-\dfrac{\sqrt{12}}{2}}$ $\sin \theta = \frac{1}{\csc \theta}$

$\quad = -\dfrac{2}{\sqrt{12}}$ Simplify the complex fraction as in part (a).

$\quad = -\dfrac{2}{2\sqrt{3}}$ $\sqrt{12} = \sqrt{4 \cdot 3} = 2\sqrt{3}$

$\quad = -\dfrac{1}{\sqrt{3}}$ We are multiplying by $1 = \frac{\sqrt{3}}{\sqrt{3}}$.

$\quad = -\dfrac{1}{\sqrt{3}} \cdot \dfrac{\sqrt{3}}{\sqrt{3}} = -\dfrac{\sqrt{3}}{3}$ Rationalize the denominator.

NOW TRY EXERCISES 1 AND 9. ◀

Signs and Ranges of Function Values

In the definitions of the trigonometric functions, r is the distance from the origin to the point (x, y). This distance is undirected, so $r > 0$. If we choose a point (x, y) in quadrant I, then both x and y will be positive, and the values of all six functions will be positive.

A point (x, y) in quadrant II has $x < 0$ and $y > 0$. This makes the values of sine and cosecant positive for quadrant II angles, while the other four functions take on negative values. Similar results can be obtained for the other quadrants.

This important information is summarized here.

Signs of Function Values

θ in Quadrant	$\sin \theta$	$\cos \theta$	$\tan \theta$	$\cot \theta$	$\sec \theta$	$\csc \theta$
I	+	+	+	+	+	+
II	+	−	−	−	−	+
III	−	−	+	+	−	−
IV	−	+	−	−	+	−

$x < 0, y > 0, r > 0$ | $x > 0, y > 0, r > 0$

II Sine and cosecant positive | **I** All functions positive

$x < 0, y < 0, r > 0$ | $x > 0, y < 0, r > 0$

III Tangent and cotangent positive | **IV** Cosine and secant positive

> ► **EXAMPLE 2** DETERMINING SIGNS OF FUNCTIONS OF NONQUADRANTAL ANGLES

Determine the signs of the trigonometric functions of an angle in standard position with the given measure.

(a) 87°   **(b)** 300°   **(c)** −200°

Solution

(a) An angle of 87° is in the first quadrant, with x, y, and r all positive, so all of its trigonometric function values are positive.

(b) A 300° angle is in quadrant IV, so the cosine and secant are positive, while the sine, cosecant, tangent, and cotangent are negative.

(c) A −200° angle is in quadrant II. The sine and cosecant are positive, and all other function values are negative.

> NOW TRY EXERCISES 19, 21, AND 25. ◄

> ► **Note** Because numbers that are reciprocals will always have the same sign, knowing the sign of a function value will automatically determine the sign of the reciprocal function value.

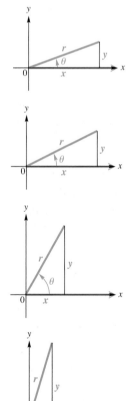

> ► **EXAMPLE 3** IDENTIFYING THE QUADRANT OF AN ANGLE

Identify the quadrant (or possible quadrants) of an angle θ that satisfies the given conditions.

(a) $\sin \theta > 0$, $\tan \theta < 0$ **(b)** $\cos \theta < 0$, $\sec \theta < 0$

Solution

(a) Since $\sin \theta > 0$ in quadrants I and II and $\tan \theta < 0$ in quadrants II and IV, both conditions are met only in quadrant II.

(b) The cosine and secant functions are both negative in quadrants II and III, so in this case θ could be in either of these two quadrants.

> NOW TRY EXERCISES 35 AND 41. ◄

Figure 34 shows an angle θ as it increases in measure from near 0° toward 90°. In each case, the value of r is the same. As the measure of the angle increases, y increases but never exceeds r, so $y \leq r$. Dividing both sides by the positive number r gives $\frac{y}{r} \leq 1$.

In a similar way, angles in quadrant IV suggest that

$$-1 \leq \frac{y}{r},$$

so

$$-1 \leq \frac{y}{r} \leq 1$$

and

$$-1 \leq \sin \theta \leq 1. \quad \text{$\frac{y}{r} = \sin \theta$ for any angle θ. (Section 1.3)}$$

Figure 34

Similarly, $-1 \leq \cos \theta \leq 1.$

The tangent of an angle is defined as $\frac{y}{x}$. It is possible that $x < y$, $x = y$, or $x > y$. Thus, $\frac{y}{x}$ can take any value, so **tan θ can be any real number, as can cot θ.**

The functions sec θ and csc θ are reciprocals of the functions cos θ and sin θ, respectively, making

$$\textbf{sec } \theta \leq -1 \quad \text{or} \quad \textbf{sec } \theta \geq 1 \quad \text{and} \quad \textbf{csc } \theta \leq -1 \quad \text{or} \quad \textbf{csc } \theta \geq 1.$$

In summary, the ranges of the trigonometric functions are as follows.

RANGES OF TRIGONOMETRIC FUNCTIONS

Trigonometric Function of θ	Range (Set-Builder Notation)	Range (Interval Notation)
sin θ, cos θ	$\{y \mid |y| \leq 1\}$	$[-1, 1]$
tan θ, cot θ	$\{y \mid y \text{ is a real number}\}$	$(-\infty, \infty)$
sec θ, csc θ	$\{y \mid |y| \geq 1\}$	$(-\infty, -1] \cup [1, \infty)$

▶ **EXAMPLE 4**   **DECIDING WHETHER A VALUE IS IN THE RANGE OF A TRIGONOMETRIC FUNCTION**

Decide whether each statement is *possible* or *impossible.*

(a) $\sin \theta = 2.5$ **(b)** $\tan \theta = 110.47$ **(c)** $\sec \theta = .6$

Solution

(a) For any value of θ, $-1 \leq \sin \theta \leq 1$. Since $2.5 > 1$, it is impossible to find a value of θ with $\sin \theta = 2.5$.

(b) Tangent can take on any real number value. Thus, $\tan \theta = 110.47$ is possible.

(c) Since $|\sec \theta| \geq 1$ for all θ for which the secant is defined, the statement $\sec \theta = .6$ is impossible.

NOW TRY EXERCISES 45, 49, AND 51. ◀

The six trigonometric functions are defined in terms of x, y, and r, where the Pythagorean theorem shows that $r^2 = x^2 + y^2$ and $r > 0$. With these relationships, knowing the value of only one function and the quadrant in which the angle lies makes it possible to find the values of the other trigonometric functions.

▶ **EXAMPLE 5**   **FINDING ALL FUNCTION VALUES GIVEN ONE VALUE AND THE QUADRANT**

Suppose that angle θ is in quadrant II and $\sin \theta = \frac{2}{3}$. Find the values of the other five trigonometric functions.

Solution   Choose any point on the terminal side of angle θ. For simplicity, since $\sin \theta = \frac{y}{r}$, choose the point with $r = 3$.

$$\sin \theta = \frac{2}{3} \quad \text{Given value}$$

$$\frac{y}{r} = \frac{2}{3} \quad \text{Substitute } \tfrac{y}{r} \text{ for sin } \theta.$$

Since $\frac{y}{r} = \frac{2}{3}$ and $r = 3$, then $y = 2$. To find x, use the equation $x^2 + y^2 = r^2$.

$$x^2 + y^2 = r^2$$

$$x^2 + 2^2 = 3^2 \qquad \text{Substitute.}$$

$$x^2 + 4 = 9 \qquad \text{Apply exponents.}$$

$$x^2 = 5 \qquad \text{Subtract 4. (Appendix A)}$$

Remember **both roots.**

$$x = \sqrt{5} \qquad \text{or} \qquad x = -\sqrt{5} \qquad \text{Square root property (Appendix A)}$$

Since θ is in quadrant II, x must be negative, as shown in Figure 35, so $x = -\sqrt{5}$, and the point $(-\sqrt{5}, 2)$ is on the terminal side of θ. Now we can find the values of the remaining trigonometric functions.

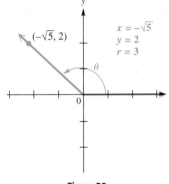

$x = -\sqrt{5}$
$y = 2$
$r = 3$

Figure 35

$$\cos \theta = \frac{x}{r} = \frac{-\sqrt{5}}{3} = -\frac{\sqrt{5}}{3}$$

$$\sec \theta = \frac{r}{x} = \frac{3}{-\sqrt{5}} = -\frac{3}{\sqrt{5}} \cdot \frac{\sqrt{5}}{\sqrt{5}} = -\frac{3\sqrt{5}}{5}$$

Remember to rationalize denominators.

$$\tan \theta = \frac{y}{x} = \frac{2}{-\sqrt{5}} = -\frac{2}{\sqrt{5}} \cdot \frac{\sqrt{5}}{\sqrt{5}} = -\frac{2\sqrt{5}}{5}$$

$$\cot \theta = \frac{x}{y} = \frac{-\sqrt{5}}{2} = -\frac{\sqrt{5}}{2}$$

$$\csc \theta = \frac{r}{y} = \frac{3}{2}$$

NOW TRY EXERCISE 71. ◄

Pythagorean Identities We derive three new identities from the relationship $x^2 + y^2 = r^2$.

$$\frac{x^2}{r^2} + \frac{y^2}{r^2} = \frac{r^2}{r^2} \qquad \text{Divide by } r^2.$$

$$\left(\frac{x}{r}\right)^2 + \left(\frac{y}{r}\right)^2 = 1 \qquad \text{Power rule for exponents; } \frac{a^m}{b^m} = \left(\frac{a}{b}\right)^m$$

$$(\cos \theta)^2 + (\sin \theta)^2 = 1 \qquad \cos \theta = \frac{x}{r}, \sin \theta = \frac{y}{r} \textbf{ (Section 1.3)}$$

or $$\mathbf{\sin^2 \theta + \cos^2 \theta = 1}$$

Starting again with $x^2 + y^2 = r^2$ and dividing through by x^2 gives

$$\frac{x^2}{x^2} + \frac{y^2}{x^2} = \frac{r^2}{x^2} \qquad \text{Divide by } x^2.$$

$$1 + \left(\frac{y}{x}\right)^2 = \left(\frac{r}{x}\right)^2 \qquad \text{Power rule for exponents}$$

$$1 + (\tan \theta)^2 = (\sec \theta)^2 \qquad \tan \theta = \frac{y}{x}, \sec \theta = \frac{r}{x} \textbf{ (Section 1.3)}$$

or $$\mathbf{\tan^2 \theta + 1 = \sec^2 \theta.}$$

Similarly, dividing through by y^2 leads to

$$\mathbf{1 + \cot^2 \theta = \csc^2 \theta.}$$

These three identities are called the **Pythagorean identities** since the original equation that led to them, $x^2 + y^2 = r^2$, comes from the Pythagorean theorem.

PYTHAGOREAN IDENTITIES

For all angles θ for which the function values are defined,

$$\sin^2 \theta + \cos^2 \theta = 1 \qquad \tan^2 \theta + 1 = \sec^2 \theta \qquad 1 + \cot^2 \theta = \csc^2 \theta.$$

As before, we have given only one form of each identity. However, algebraic transformations produce equivalent identities. For example, by subtracting $\sin^2 \theta$ from both sides of $\sin^2 \theta + \cos^2 \theta = 1$, we get the equivalent identity

$$\cos^2 \theta = 1 - \sin^2 \theta. \qquad \text{Alternative form}$$

You should be able to transform these identities quickly and also recognize their equivalent forms.

▼ **LOOKING AHEAD TO CALCULUS**
The reciprocal, Pythagorean, and quotient identities are used in calculus to find derivatives and integrals of trigonometric functions. A standard technique of integration called **trigonometric substitution** relies on the Pythagorean identities.

Quotient Identities   Consider the quotient of $\sin \theta$ and $\cos \theta$, for $\cos \theta \neq 0$.

$$\frac{\sin \theta}{\cos \theta} = \frac{\frac{y}{r}}{\frac{x}{r}} = \frac{y}{r} \div \frac{x}{r} = \frac{y}{r} \cdot \frac{r}{x} = \frac{y}{x} = \tan \theta$$

Similarly, $\frac{\cos \theta}{\sin \theta} = \cot \theta$, for $\sin \theta \neq 0$. Thus, we have the **quotient identities.**

QUOTIENT IDENTITIES

For all angles θ for which the denominators are not zero,

$$\frac{\sin \theta}{\cos \theta} = \tan \theta \qquad\qquad \frac{\cos \theta}{\sin \theta} = \cot \theta.$$

▶ **EXAMPLE 6**   FINDING OTHER FUNCTION VALUES GIVEN ONE VALUE AND THE QUADRANT

Find $\sin \theta$ and $\tan \theta$, given that $\cos \theta = -\frac{\sqrt{3}}{4}$ and $\sin \theta > 0$.

Solution   Start with $\sin^2 \theta + \cos^2 \theta = 1$.

$$\sin^2 \theta + \left(-\frac{\sqrt{3}}{4}\right)^2 = 1 \qquad \text{Replace } \cos \theta \text{ with } -\frac{\sqrt{3}}{4}.$$

$$\sin^2 \theta + \frac{3}{16} = 1 \qquad \text{Square } -\frac{\sqrt{3}}{4}.$$

$$\sin^2 \theta = \frac{13}{16} \qquad \text{Subtract } \frac{3}{16}.$$

$$\sin \theta = \pm \frac{\sqrt{13}}{4} \qquad \text{Take square roots.}$$

Choose the correct sign here.

$$\sin \theta = \frac{\sqrt{13}}{4} \qquad \text{Choose the positive square root since } \sin \theta \text{ is positive.}$$

To find $\tan\theta$, use the quotient identity $\tan\theta = \frac{\sin\theta}{\cos\theta}$.

$$\tan\theta = \frac{\sin\theta}{\cos\theta} = \frac{\frac{\sqrt{13}}{4}}{-\frac{\sqrt{3}}{4}} = \frac{\sqrt{13}}{4}\left(-\frac{4}{\sqrt{3}}\right) = -\frac{\sqrt{13}}{\sqrt{3}}$$

$$= -\frac{\sqrt{13}}{\sqrt{3}}\cdot\frac{\sqrt{3}}{\sqrt{3}} = -\frac{\sqrt{39}}{3} \qquad \text{Rationalize the denominator.}$$

NOW TRY EXERCISE 75. ◄

▶ **Caution** *In problems like those in Examples 5 and 6, be careful to choose the correct sign when square roots are taken.*

▶ **EXAMPLE 7** FINDING OTHER FUNCTION VALUES GIVEN ONE VALUE AND THE QUADRANT

Find $\sin\theta$ and $\cos\theta$, given that $\tan\theta = \frac{4}{3}$ and θ is in quadrant III.

Solution Since θ is in quadrant III, $\sin\theta$ and $\cos\theta$ will both be negative. It is tempting to say that since $\tan\theta = \frac{\sin\theta}{\cos\theta}$ and $\tan\theta = \frac{4}{3}$, then $\sin\theta = -4$ and $\cos\theta = -3$. This is *incorrect*, however, since both $\sin\theta$ and $\cos\theta$ must be in the interval $[-1, 1]$.

We use the Pythagorean identity $\tan^2\theta + 1 = \sec^2\theta$ to find $\sec\theta$, and then the reciprocal identity $\cos\theta = \frac{1}{\sec\theta}$ to find $\cos\theta$.

$$\tan^2\theta + 1 = \sec^2\theta$$

$$\left(\frac{4}{3}\right)^2 + 1 = \sec^2\theta \qquad \tan\theta = \frac{4}{3}$$

$$\frac{16}{9} + 1 = \sec^2\theta$$

$$\frac{25}{9} = \sec^2\theta$$

Be careful to choose the correct sign here.

$$-\frac{5}{3} = \sec\theta \qquad \text{Choose the negative square root since } \sec\theta \text{ is negative when } \theta \text{ is in quadrant III.}$$

$$-\frac{3}{5} = \cos\theta \qquad \text{Secant and cosine are reciprocals.}$$

Since $\sin^2\theta = 1 - \cos^2\theta$,

$$\sin^2\theta = 1 - \left(-\frac{3}{5}\right)^2 \qquad \cos\theta = -\frac{3}{5}$$

$$\sin^2\theta = 1 - \frac{9}{25}$$

$$\sin^2\theta = \frac{16}{25}$$

Again, be careful.

$$\sin\theta = -\frac{4}{5}. \qquad \text{Choose the negative square root.}$$

NOW TRY EXERCISE 73. ◄

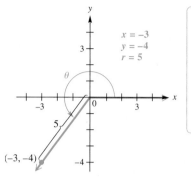

Figure 36

▶ **Note** Example 7 can also be worked by drawing θ in standard position in quadrant III, finding r to be 5, and then using the definitions of $\sin \theta$ and $\cos \theta$ in terms of x, y, and r. See Figure 36.

When using this method, be sure to choose the correct signs for x and y. This is analogous to choosing the correct signs after applying the Pythagorean identities. Always check to be sure that the signs of the functions correspond to those found in the table at the bottom of page 31.

1.4   Exercises

Use the appropriate reciprocal identity to find each function value. Rationalize denominators when applicable. See Example 1.

1. $\sec \theta$, given that $\cos \theta = \frac{2}{3}$

2. $\sec \theta$, given that $\cos \theta = \frac{5}{8}$

3. $\csc \theta$, given that $\sin \theta = -\frac{3}{7}$

4. $\csc \theta$, given that $\sin \theta = -\frac{8}{43}$

5. $\cot \theta$, given that $\tan \theta = 5$

6. $\cot \theta$, given that $\tan \theta = 18$

7. $\cos \theta$, given that $\sec \theta = -\frac{5}{2}$

8. $\cos \theta$, given that $\sec \theta = -\frac{11}{7}$

9. $\sin \theta$, given that $\csc \theta = \frac{\sqrt{8}}{2}$

10. $\sin \theta$, given that $\csc \theta = \frac{\sqrt{24}}{3}$

11. $\tan \theta$, given that $\cot \theta = -2.5$

12. $\tan \theta$, given that $\cot \theta = -.01$

13. $\sin \theta$, given that $\csc \theta = 1.42716321$

14. $\cos \theta$, given that $\sec \theta = 9.80425133$

15. Can a given angle θ satisfy both $\sin \theta > 0$ and $\csc \theta < 0$? Explain.

16. Explain what is wrong with the following item that appears on a trigonometry test:

   "*Find* $\sec \theta$, *given that* $\cos \theta = \dfrac{3}{2}$."

17. *Concept Check*   What is wrong with the following statement? $\tan 90° = \frac{1}{\cot 90°}$.

18. *Concept Check*   One form of a particular reciprocal identity is $\tan \theta = \frac{1}{\cot \theta}$. Give two other equivalent forms of this identity.

Determine the signs of the trigonometric functions of an angle in standard position with the given measure. See Example 2.

19. 74°	**20.** 84°	**21.** 218°	**22.** 195°
23. 178°	**24.** 125°	**25.** −80°	**26.** −15°
27. 845°	**28.** 1005°	**29.** −345°	**30.** −705°

Identify the quadrant (or possible quadrants) of an angle θ that satisfies the given conditions. See Example 3.

31. $\sin \theta > 0$, $\csc \theta > 0$	**32.** $\cos \theta > 0$, $\sec \theta > 0$	**33.** $\cos \theta > 0$, $\sin \theta > 0$
34. $\sin \theta > 0$, $\tan \theta > 0$	**35.** $\tan \theta < 0$, $\cos \theta < 0$	**36.** $\cos \theta < 0$, $\sin \theta < 0$
37. $\sec \theta > 0$, $\csc \theta > 0$	**38.** $\csc \theta > 0$, $\cot \theta > 0$	**39.** $\sec \theta < 0$, $\csc \theta < 0$
40. $\cot \theta < 0$, $\sec \theta < 0$	**41.** $\sin \theta < 0$, $\csc \theta < 0$	**42.** $\tan \theta < 0$, $\cot \theta < 0$

43. Explain why the answers to Exercises 33 and 37 are the same.

44. Explain why there is no angle θ that satisfies $\tan \theta > 0$, $\cot \theta < 0$.

Decide whether each statement is possible *or* impossible *for an angle θ. See Example 4.*

45. $\sin \theta = 2$ **46.** $\sin \theta = 3$ **47.** $\cos \theta = -.96$

48. $\cos \theta = -.56$ **49.** $\tan \theta = .93$ **50.** $\cot \theta = .93$

51. $\sec \theta = -.3$ **52.** $\sec \theta = -.9$ **53.** $\csc \theta = 100$

54. $\csc \theta = -100$ **55.** $\cot \theta = -4$

56. $\cot \theta = -6$ **57.** $\sin \theta = \dfrac{1}{2}$ and $\csc \theta = 2$

58. $\tan \theta = 2$ and $\cot \theta = -2$ **59.** $\cos \theta = -2$ and $\sec \theta = \dfrac{1}{2}$

60. Explain why there is no angle θ that satisfies $\cos \theta = \frac{1}{2}$ and $\sec \theta = -2$.

Use identities to solve each of the following. See Examples 5–7.

61. Find $\cos \theta$, given that $\sin \theta = \frac{3}{5}$ and θ is in quadrant II.

62. Find $\sin \theta$, given that $\cos \theta = \frac{4}{5}$ and θ is in quadrant IV.

63. Find $\csc \theta$, given that $\cot \theta = -\frac{1}{2}$ and θ is in quadrant IV.

64. Find $\sec \theta$, given that $\tan \theta = \frac{\sqrt{7}}{3}$ and θ is in quadrant III.

65. Find $\tan \theta$, given that $\sin \theta = \frac{1}{2}$ and θ is in quadrant II.

66. Find $\cot \theta$, given that $\csc \theta = -2$ and θ is in quadrant III.

67. Find $\cot \theta$, given that $\csc \theta = -3.5891420$ and θ is in quadrant III.

68. Find $\tan \theta$, given that $\sin \theta = .49268329$ and θ is in quadrant II.

Find the five remaining trigonometric function values for each angle θ. See Examples 5–7.

69. $\tan \theta = -\frac{15}{8}$, given that θ is in quadrant II

70. $\cos \theta = -\frac{3}{5}$, given that θ is in quadrant III

71. $\sin \theta = \frac{\sqrt{5}}{7}$, given that θ is in quadrant I

72. $\tan \theta = \sqrt{3}$, given that θ is in quadrant III

73. $\cot \theta = \frac{\sqrt{3}}{8}$, given that θ is in quadrant I

74. $\csc \theta = 2$, given that θ is in quadrant II

75. $\sin \theta = \frac{\sqrt{2}}{6}$, given that $\cos \theta < 0$

76. $\cos \theta = \frac{\sqrt{5}}{8}$, given that $\tan \theta < 0$

77. $\sec \theta = -4$, given that $\sin \theta > 0$

78. $\csc \theta = -3$, given that $\cos \theta > 0$

79. $\sin \theta = .164215$, given that θ is in quadrant II

80. $\cot \theta = -1.49586$, given that θ is in quadrant IV

Work each problem.

81. Derive the identity $1 + \cot^2 \theta = \csc^2 \theta$ by dividing $x^2 + y^2 = r^2$ by y^2.

82. Using a method similar to the one given in this section showing that $\frac{\sin \theta}{\cos \theta} = \tan \theta$, show that $\frac{\cos \theta}{\sin \theta} = \cot \theta$.

83. *Concept Check*   *True* or *false*: For all angles θ, $\sin \theta + \cos \theta = 1$. If false, give an example showing why it is false.

84. *Concept Check*   *True* or *false*: Since $\cot \theta = \frac{\cos \theta}{\sin \theta}$, if $\cot \theta = \frac{1}{2}$ with θ in quadrant I, then $\cos \theta = 1$ and $\sin \theta = 2$. If false, explain why.

Concept Check   *Suppose that* $90° < \theta < 180°$. *Find the sign of each function value.*

85. $\sin 2\theta$

86. $\tan \dfrac{\theta}{2}$

87. $\cot(\theta + 180°)$

88. $\cos(-\theta)$

Concept Check   *Suppose that* $-90° < \theta < 90°$. *Find the sign of each function value.*

89. $\cos \dfrac{\theta}{2}$

90. $\cos(\theta + 180°)$

91. $\sec(-\theta)$

92. $\sec(\theta - 180°)$

Find a value of each variable.

93. $\tan(3\theta - 4°) = \dfrac{1}{\cot(5\theta - 8°)}$

94. $\sec(2\theta + 6°)\cos(5\theta + 3°) = 1$

95. $\sin(4\theta + 2°)\csc(3\theta + 5°) = 1$

96. $\cos(6A + 5°) = \dfrac{1}{\sec(4A + 15°)}$

97. *Concept Check*   The screen below was obtained with the calculator in degree mode. How can we use it to justify that an angle of 14,879° is a quadrant II angle?

```
cos(14879)
        -.4848096202
sin(14879)
         .8746197071
```

98. *Concept Check*   The screen below was obtained with the calculator in degree mode. In which quadrant does a 1294° angle lie?

```
tan(1294)
         .6745085168
sin(1294)
        -.5591929035
```

Chapter 1 Summary

NEW SYMBOLS

⌐ right angle symbol (for a right triangle)	′ minute
θ Greek letter theta	″ second
° degree	

QUICK REVIEW

CONCEPTS	EXAMPLES

1.1 Angles

Types of Angles

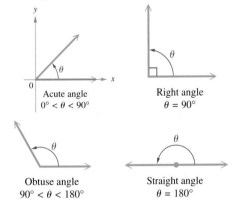

Acute angle
$0° < \theta < 90°$

Right angle
$\theta = 90°$

Obtuse angle
$90° < \theta < 180°$

Straight angle
$\theta = 180°$

If $\theta = 46°$, then angle θ is an acute angle.

If $\theta = 90°$, then angle θ is a right angle.

If $\theta = 148°$, then angle θ is an obtuse angle.

If $\theta = 180°$, then angle θ is a straight angle.

The acute angle θ in the figure at the left is in standard position. If θ measures 46°, find the measure of a negative coterminal angle.

$$46° - 360° = -314°$$

1.2 Angle Relationships and Similar Triangles

Vertical angles have equal measures.

The sum of the measures of the angles of any triangle is 180°.

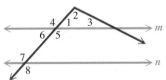

m and n are parallel lines.

Vertical angles 4 and 5 are equal.

The sum of angles 1, 2, and 3 is 180°.

CONCEPTS	EXAMPLES
When a transversal intersects parallel lines, the following angles formed have equal measure: alternate interior, alternate exterior, and corresponding. Interior angles on the same side of the transversal are supplementary.	Refer to the diagram at the bottom of the previous page. Angles 5 and 7 are alternate interior angles, so they are equal. Angles 4 and 8 are alternate exterior angles, so they are equal. Angles 4 and 7 are corresponding angles, so they are equal. Angles 6 and 7 are interior angles on the same side of the transversal, so they are supplementary.

Similar triangles have corresponding angles with the same measures, and corresponding sides proportional.

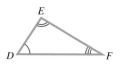

Pairs of corresponding angles as marked in triangles ABC and DEF are equal.

Also, $\dfrac{AB}{DE} = \dfrac{BC}{EF} = \dfrac{AC}{DF}$.

Congruent triangles are the same size and the same shape.

Corresponding angles are equal, and corresponding sides are equal.

1.3 Trigonometric Functions

Definitions of the Trigonometric Functions
Let (x, y) be a point other than the origin on the terminal side of an angle θ in standard position. Let $r = \sqrt{x^2 + y^2}$ represent the distance from the origin to (x, y). Then

$$\sin \theta = \frac{y}{r} \qquad \cos \theta = \frac{x}{r} \qquad \tan \theta = \frac{y}{x} \ (x \neq 0)$$

$$\csc \theta = \frac{r}{y} \ (y \neq 0) \ \sec \theta = \frac{r}{x} \ (x \neq 0) \ \cot \theta = \frac{x}{y} \ (y \neq 0).$$

See the summary table of trigonometric function values for quadrantal angles on page 27.

If the point $(-2, 3)$ is on the terminal side of angle θ in standard position, then $x = -2$, $y = 3$, and $r = \sqrt{(-2)^2 + 3^2} = \sqrt{4 + 9} = \sqrt{13}$. Then

$$\sin \theta = \frac{3\sqrt{13}}{13}, \qquad \cos \theta = -\frac{2\sqrt{13}}{13}, \qquad \tan \theta = -\frac{3}{2},$$

$$\csc \theta = \frac{\sqrt{13}}{3}, \qquad \sec \theta = -\frac{\sqrt{13}}{2}, \qquad \cot \theta = -\frac{2}{3}.$$

1.4 Using the Definitions of the Trigonometric Functions

Reciprocal Identities

$$\sin \theta = \frac{1}{\csc \theta} \qquad \cos \theta = \frac{1}{\sec \theta} \qquad \tan \theta = \frac{1}{\cot \theta}$$

$$\csc \theta = \frac{1}{\sin \theta} \qquad \sec \theta = \frac{1}{\cos \theta} \qquad \cot \theta = \frac{1}{\tan \theta}$$

If $\cot \theta = -\frac{2}{3}$, find $\tan \theta$.

$$\tan \theta = \frac{1}{\cot \theta} = \frac{1}{-\frac{2}{3}} = -\frac{3}{2}$$

(continued)

CONCEPTS	EXAMPLES

Pythagorean Identities

$$\sin^2 \theta + \cos^2 \theta = 1 \qquad \tan^2 \theta + 1 = \sec^2 \theta$$

$$1 + \cot^2 \theta = \csc^2 \theta$$

Use the function values for the example from **Section 1.3** to illustrate the Pythagorean identities.

$$\sin^2 \theta + \cos^2 \theta = \left(\frac{3\sqrt{13}}{13}\right)^2 + \left(-\frac{2\sqrt{13}}{13}\right)^2$$

$$= \frac{9}{13} + \frac{4}{13} = 1,$$

$$\tan^2 \theta + 1 = \left(-\frac{3}{2}\right)^2 + 1 = \frac{13}{4} = \left(-\frac{\sqrt{13}}{2}\right)^2 = \sec^2 \theta,$$

$$1 + \cot^2 \theta = 1 + \left(-\frac{2}{3}\right)^2 = \frac{13}{9} = \left(\frac{\sqrt{13}}{3}\right)^2 = \csc^2 \theta.$$

Quotient Identities

$$\frac{\sin \theta}{\cos \theta} = \tan \theta \qquad \frac{\cos \theta}{\sin \theta} = \cot \theta$$

Use the function values for the example from **Section 1.3** to illustrate $\frac{\sin \theta}{\cos \theta} = \tan \theta$.

$$\frac{\sin \theta}{\cos \theta} = \frac{\frac{3\sqrt{13}}{13}}{-\frac{2\sqrt{13}}{13}} = \frac{3\sqrt{13}}{13}\left(-\frac{13}{2\sqrt{13}}\right) = -\frac{3}{2} = \tan \theta$$

Signs of the Trigonometric Functions

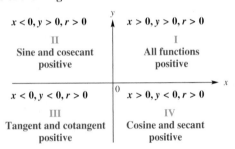

$x < 0, y > 0, r > 0$
II
Sine and cosecant positive

$x > 0, y > 0, r > 0$
I
All functions positive

$x < 0, y < 0, r > 0$
III
Tangent and cotangent positive

$x > 0, y < 0, r > 0$
IV
Cosine and secant positive

Identify the quadrant(s) of any angle θ that satisfies $\sin \theta < 0$, $\tan \theta > 0$.

Since $\sin \theta < 0$ in quadrants III and IV, while $\tan \theta > 0$ in quadrants I and III, both conditions are met only in quadrant III.

CHAPTER 1 ▶ **Review Exercises**

1. Give the measures of the complement and the supplement of an angle measuring $35°$.

Find the angle of least possible positive measure coterminal with each angle.

2. $-51°$ **3.** $-174°$ **4.** $792°$

5. Find the measure of each marked angle.

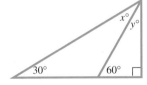

Work each problem.

6. *Rotating Pulley* A pulley is rotating 320 times per min. Through how many degrees does a point on the edge of the pulley move in $\frac{2}{3}$ sec?

7. *Rotating Propeller* The propeller of a speedboat rotates 650 times per min. Through how many degrees will a point on the edge of the propeller rotate in 2.4 sec?

Convert decimal degrees to degrees, minutes, seconds, and convert degrees, minutes, seconds to decimal degrees. Round to the nearest second or the nearest thousandth of a degree, as appropriate. Use a calculator as necessary.

8. $47° \, 25' \, 11''$ **9.** $119° \, 8' \, 3''$ **10.** $-61.5034°$ **11.** $275.1005°$

Find the measure of each marked angle.

12.

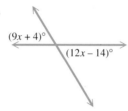

13.

14. Express θ in terms of α and β.

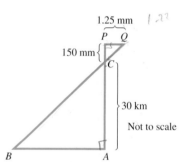

15. *Length of a Road* The flight path CP of a satellite carrying a camera with its lens at C is shown in the figure. Length PC represents the distance from the lens to the film PQ, and BA represents a straight road on the ground. Use the measurements given in the figure to find the length of the road. (*Source:* Kastner, B., *Space Mathematics,* NASA, 1985.)

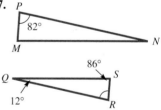

Find all unknown angle measures in each pair of similar triangles.

16.

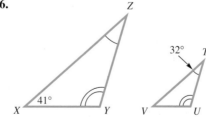

17.

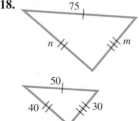

Find the unknown side lengths in each pair of similar triangles.

18.

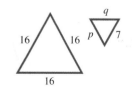

19.

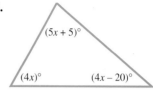

Find the unknown measurement. There are two similar triangles in each figure.

20.

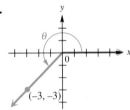

21.

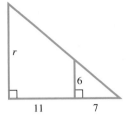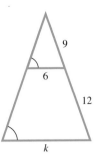

22. *Concept Check* Complete the following statement: If two triangles are similar, their corresponding sides are _____ and the measures of their corresponding angles are _____ .

23. *Length of a Shadow* If a tree 20 ft tall casts a shadow 8 ft long, how long would the shadow of a 30-ft tree be at the same time and place?

Find the six trigonometric function values for each angle. If a value is undefined, say so.

24.

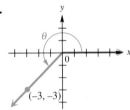

25.

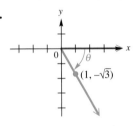

26. 180°

Find the values of the six trigonometric functions for an angle in standard position having each point on its terminal side.

27. $(3, -4)$

28. $(9, -2)$

29. $(-8, 15)$

30. $(1, -5)$

31. $(6\sqrt{3}, -6)$

32. $(-2\sqrt{2}, 2\sqrt{2})$

33. *Concept Check* If the terminal side of a quadrantal angle lies along the y-axis, which of its trigonometric functions are undefined?

34. Find the values of all six trigonometric functions for an angle in standard position having its terminal side defined by the equation $5x - 3y = 0, x \geq 0$.

In Exercises 35 and 36, consider an angle θ in standard position whose terminal side has the equation $y = -5x$, with $x \leq 0$.

35. Sketch θ and use an arrow to show the rotation if $0° \leq \theta < 360°$.

36. Find the exact values of $\sin \theta$, $\cos \theta$, $\tan \theta$, $\cot \theta$, $\sec \theta$, and $\csc \theta$.

Complete the table with the appropriate function values of the given quadrantal angles. If the value is undefined, say so.

	θ	$\sin \theta$	$\cos \theta$	$\tan \theta$	$\cot \theta$	$\sec \theta$	$\csc \theta$
37.	180°						
38.	$-90°$						

39. Decide whether each statement is *possible* or *impossible*.

(a) $\sec \theta = -\dfrac{2}{3}$ 　　　　(b) $\tan \theta = 1.4$ 　　　　(c) $\csc \theta = 5$

Find all six trigonometric function values for each angle. Rationalize denominators when applicable.

40. $\sin \theta = \dfrac{\sqrt{3}}{5}$, given that $\cos \theta < 0$

41. $\cos \theta = -\dfrac{5}{8}$, given that θ is in quadrant III

42. $\tan \theta = 2$, given that θ is in quadrant III

43. $\sec \theta = -\sqrt{5}$, given that θ is in quadrant II

44. $\sin \theta = -\dfrac{2}{5}$, given that θ is in quadrant III

45. $\sec \theta = \dfrac{5}{4}$, given that θ is in quadrant IV

46. *Concept Check* If, for some particular angle θ, $\sin \theta < 0$ and $\cos \theta > 0$, in what quadrant must θ lie? What is the sign of $\tan \theta$?

Solve each problem.

47. *Swimmer in Distress* A lifeguard located 20 yd from the water spots a swimmer in distress. The swimmer is 30 yd from shore and 100 yd east of the lifeguard. Suppose the lifeguard runs, then swims to the swimmer in a direct line, as shown in the figure. How far east from his original position will he enter the water? (*Hint:* Find the value of x in the sketch.)

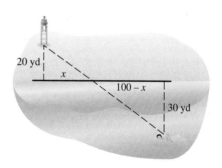

48. *Angle the Celestial North Pole Moves* At present, the north star Polaris is located very near the celestial north pole. However, because Earth is inclined 23.5°, the moon's gravitational pull on Earth is uneven. As a result, Earth slowly precesses (moves in) like a spinning top and the direction of the celestial north pole traces out a circular path once every 26,000 yr. See the figure. For example, in approximately A.D. 14,000 the star Vega will be located at the celestial north pole—and not the star Polaris. As viewed from the center C of this circular path, calculate the angle (to the nearest second) that the celestial north pole moves each year. (*Source:* Zeilik, M., S. Gregory, and E. Smith, *Introductory Astronomy and Astrophysics,* Second Edition, Saunders College Publishers, 1998.)

49. *Depth of a Crater on the Moon*   The depths of unknown craters on the moon can be approximated by comparing the lengths of their shadows to the shadows of nearby craters with known depths. The crater Aristillus is 11,000 ft deep, and its shadow was measured as 1.5 mm on a photograph. Its companion crater, Autolycus, had a shadow of 1.3 mm on the same photograph. Use similar triangles to determine the depth of the crater Autolycus. (*Source:* Webb, T., *Celestial Objects for Common Telescopes*, Dover Publications, 1962.)

50. *Height of a Lunar Peak*   The lunar mountain peak Huygens has a height of 21,000 ft. The shadow of Huygens on a photograph was 2.8 mm, while the nearby mountain Bradley had a shadow of 1.8 mm on the same photograph. Calculate the height of Bradley. (*Source:* Webb, T., *Celestial Objects for Common Telescopes*, Dover Publications, 1962.)

CHAPTER 1 ▶ Test

1. For an angle measuring $67°$, give the measure of its **(a)** complement and **(b)** supplement.

Find the measure of each unknown angle.

2.

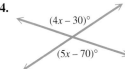

$(7x + 19)°$ $(2x - 1)°$

3.

$(-3x + 5)°$

$(-8x + 30)°$

4.

$(4x - 30)°$

$(5x - 70)°$

5.

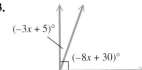

$(10x - 10)°$

$(8x + 14)°$

m and *n* are parallel.

6.

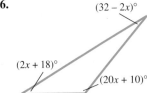

$(32 - 2x)°$

$(2x + 18)°$

$(20x + 10)°$

7. Perform the indicated conversion.

   (a) $74° \, 18' \, 36''$ to decimal degrees   **(b)** $45.2025°$ to degrees, minutes, seconds

8. Find the least positive measure of an angle coterminal with an angle of the given measure.

    (a) 390°   **(b)** −80°   **(c)** 810°

9. *Rotating Tire*   A tire rotates 450 times per min. Through how many degrees does a point on the edge of the tire move in 1 sec?

10. *Length of a Shadow*   If a vertical pole 30 ft tall casts a shadow 8 ft long, how long would the shadow of a 40-ft pole be at the same time and place?

11. Find the unknown side lengths x and y in this pair of similar triangles.

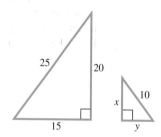

Draw a sketch of an angle in standard position having the given point on its terminal side. Indicate the angle of least positive measure θ, and give the values of sin θ, cos θ, tan θ, cot θ, sec θ, *and* csc θ. *If any of these are undefined, say so.*

12. $(2, -7)$ **13.** $(0, -2)$

14. Draw a sketch of an angle in standard position having the equation $3x - 4y = 0$, $x \le 0$, as its terminal side. Indicate the angle of least positive measure θ, and give the values of sin θ, cos θ, tan θ, cot θ, sec θ, and csc θ.

15. Complete the table with the appropriate function values of the given quadrantal angles. If the value is undefined, say so.

θ	sin θ	cos θ	tan θ	cot θ	sec θ	csc θ
90°						
−360°						
630°						

16. If the terminal side of a quadrantal angle lies along the negative x-axis, which two of its trigonometric function values are undefined?

17. Identify the possible quadrant or quadrants in which θ must lie under the given conditions.

    (a) cos θ > 0, tan θ > 0   **(b)** sin θ < 0, csc θ < 0   **(c)** cot θ > 0, cos θ < 0

18. Decide whether each statement is *possible* or *impossible* for some angle θ.

    (a) sin θ = 1.5   **(b)** sec θ = 4   **(c)** tan θ = 10,000

19. Find the value of sec θ if cos θ $= -\frac{7}{12}$.

20. Find the five remaining trigonometric function values of θ if sin θ $= \frac{3}{7}$ and θ is in quadrant II.

CHAPTER 1 ▶ Quantitative Reasoning

Have you ever gazed up at a redwood tree, a skyscraper, a public monument, or perhaps a dinosaur in a museum and wondered how tall it was?

There is a relatively simple way to make a reasonable estimate. All you need is a 1-ft ruler. Hold the ruler vertically at arm's length as you approach the object to be measured. Stop when one end of the ruler lines up with the top of the object and the other end with its base. Now pace off the distance to the object, taking normal strides. The number of paces will be the approximate height of the object in feet.

The reason this method works depends on a concept presented in this chapter. Furnish the reasons for the following steps, which refer to the figure. (Assume that the length of one pace is *EF*.) Then answer the question.

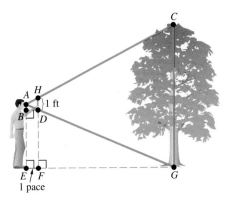

Reasons

Step 1 $CG = \dfrac{CG}{1} = \dfrac{AG}{AD}$ _____

Step 2 $\dfrac{AG}{AD} = \dfrac{EG}{EF}$ _____

Step 3 $\dfrac{EG}{EF} = \dfrac{EG}{BD} = \dfrac{EG}{1}$ _____

Step 4 CG ft $= EG$ paces _____

What is the height of the tree?

2 | Acute Angles and Right Triangles

Highway transportation is critical to the economy of the United States. In 1970 there were 1150 billion miles traveled, and by the year 2000 this increased to approximately 2500 billion miles. When an automobile travels around a curve, objects like trees, buildings, and fences situated on the curve may obstruct a driver's vision. Trigonometry is used to determine how far inside the curve land must be cleared to provide visibility for a safe stopping distance. (*Source:* Mannering, F. and W. Kilareski, *Principles of Highway Engineering and Traffic Analysis,* Second Edition, John Wiley and Sons, 1998.)

A problem like this is presented in Exercise 39 of Section 2.5.

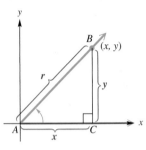

Figure 1

2.1 Trigonometric Functions of Acute Angles

Right-Triangle-Based Definitions of the Trigonometric Functions ▪ Cofunctions ▪
Trigonometric Function Values of Special Angles

Right-Triangle-Based Definitions of the Trigonometric Functions   In **Section 1.3** we used angles in standard position to define the trigonometric functions. There is another way to approach them: as ratios of the lengths of the sides of right triangles. Figure 1 shows an acute angle A in standard position. The definitions of the trigonometric function values of angle A require x, y, and r. As drawn in Figure 1, x and y are the lengths of the two legs of the right triangle ABC, and r is the length of the hypotenuse.

The side of length y is called the **side opposite** angle A, and the side of length x is called the **side adjacent** to angle A. We use the lengths of these sides to replace x and y in the definitions of the trigonometric functions, and the length of the hypotenuse to replace r, to get the following right-triangle-based definitions.

RIGHT-TRIANGLE-BASED DEFINITIONS OF TRIGONOMETRIC FUNCTIONS

For any acute angle A in standard position,

$$\sin A = \frac{y}{r} = \frac{\text{side opposite}}{\text{hypotenuse}} \qquad \csc A = \frac{r}{y} = \frac{\text{hypotenuse}}{\text{side opposite}}$$

$$\cos A = \frac{x}{r} = \frac{\text{side adjacent}}{\text{hypotenuse}} \qquad \sec A = \frac{r}{x} = \frac{\text{hypotenuse}}{\text{side adjacent}}$$

$$\tan A = \frac{y}{x} = \frac{\text{side opposite}}{\text{side adjacent}} \qquad \cot A = \frac{x}{y} = \frac{\text{side adjacent}}{\text{side opposite}}.$$

▶ **EXAMPLE 1**   FINDING TRIGONOMETRIC FUNCTION VALUES OF AN ACUTE ANGLE

Find the sine, cosine, and tangent values for angles A and B in the right triangle in Figure 2.

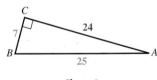

Figure 2

Solution   The length of the side opposite angle A is 7, the length of the side adjacent to angle A is 24, and the length of the hypotenuse is 25. Use the relationships given in the box.

$$\sin A = \frac{\text{side opposite}}{\text{hypotenuse}} = \frac{7}{25} \qquad \cos A = \frac{\text{side adjacent}}{\text{hypotenuse}} = \frac{24}{25} \qquad \tan A = \frac{\text{side opposite}}{\text{side adjacent}} = \frac{7}{24}$$

The length of the side opposite angle B is 24, while the length of the side adjacent to B is 7, so

$$\sin B = \frac{24}{25} \qquad \cos B = \frac{7}{25} \qquad \tan B = \frac{24}{7}.$$

NOW TRY EXERCISE 1. ◀

> ▶ **Note** Because the cosecant, secant, and cotangent ratios are the reciprocals of the sine, cosine, and tangent values, respectively, in Example 1 we can conclude that $\csc A = \frac{25}{7}$, $\sec A = \frac{25}{24}$, $\cot A = \frac{24}{7}$, $\csc B = \frac{25}{24}$, $\sec B = \frac{25}{7}$, and $\cot B = \frac{7}{24}$.

Cofunctions   In Example 1, you may have noticed that $\sin A = \cos B$ and $\cos A = \sin B$. Such relationships are always true for the two acute angles of a right triangle. Figure 3 shows a right triangle with acute angles A and B and a right angle at C. (Whenever we use A, B, and C to name angles in a right triangle, C will be the right angle.) The length of the side opposite A is a, and the length of the side opposite angle B is b. The length of the hypotenuse is c.

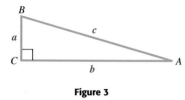

Figure 3

By the preceding definitions, $\sin A = \frac{a}{c}$. Since $\cos B$ is also equal to $\frac{a}{c}$,

$$\sin A = \frac{a}{c} = \cos B.$$

Similarly, $\tan A = \dfrac{a}{b} = \cot B$ and $\sec A = \dfrac{c}{b} = \csc B.$

Since the sum of the three angles in any triangle is $180°$ and angle C equals $90°$, angles A and B must have a sum of $180° - 90° = 90°$. As mentioned in **Section 1.1,** angles with a sum of $90°$ are complementary angles. Since angles A and B are complementary and $\sin A = \cos B$, the functions sine and cosine are called **cofunctions.** Tangent and cotangent are also cofunctions, as are secant and cosecant. And since the angles A and B are complementary, $A + B = 90°$, or $B = 90° - A$, giving

$$\sin A = \cos B = \cos(90° - A).$$

Similar results, called the **cofunction identities,** are true for the other trigonometric functions.

COFUNCTION IDENTITIES

For any acute angle A,

$\sin A = \cos(90° - A)$ $\sec A = \csc(90° - A)$ $\tan A = \cot(90° - A)$

$\cos A = \sin(90° - A)$ $\csc A = \sec(90° - A)$ $\cot A = \tan(90° - A).$

▶ **EXAMPLE 2**   **WRITING FUNCTIONS IN TERMS OF COFUNCTIONS**

Write each function in terms of its cofunction.

(a) $\cos 52°$ **(b)** $\tan 71°$ **(c)** $\sec 24°$

Solution

(a) Since $\cos A = \sin(90° - A)$,

$$\cos 52° = \sin(90° - 52°) = \sin 38°.$$

(b) $\tan 71° = \cot(90° - 71°) = \cot 19°$ **(c)** $\sec 24° = \csc 66°$

NOW TRY EXERCISES 19 AND 25. ◀

▶ **EXAMPLE 3**   SOLVING EQUATIONS USING THE COFUNCTION
IDENTITIES

Find one solution for each equation. Assume all angles involved are acute
angles.

(a) $\cos(\theta + 4°) = \sin(3\theta + 2°)$ **(b)** $\tan(2\theta - 18°) = \cot(\theta + 18°)$

Solution

(a) Since sine and cosine are cofunctions, $\cos(\theta + 4°) = \sin(3\theta + 2°)$ is true if
the sum of the angles is 90°.

$$(\theta + 4°) + (3\theta + 2°) = 90°$$
$$4\theta + 6° = 90° \quad \text{Combine terms.}$$
$$4\theta = 84° \quad \text{Subtract 6°. (Appendix A)}$$
$$\theta = 21° \quad \text{Divide by 4.}$$

(b) Tangent and cotangent are cofunctions, so

$$(2\theta - 18°) + (\theta + 18°) = 90°$$
$$3\theta = 90°$$
$$\theta = 30°.$$

NOW TRY EXERCISES 27 AND 29. ◀

Figure 4 shows three right triangles. From left to right, the length of each
hypotenuse is the same, but angle A increases in measure. As angle A increases
in measure from 0° to 90°, the length of the side opposite angle A also increases.

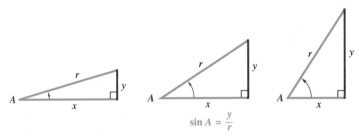

$$\sin A = \frac{y}{r}$$

As A increases, y increases. Since r is fixed, $\sin A$ increases.

Figure 4

Since $\sin A = \dfrac{\text{side opposite}}{\text{hypotenuse}}$,

as angle A increases, the numerator of this fraction also increases, while the
denominator is fixed. This means that $\sin A$ *increases* as A increases from 0°
to 90°.

As angle A increases from 0° to 90°, the length of the side adjacent to A
decreases. Since r is fixed, the ratio $\frac{x}{r}$ will decrease. This ratio gives $\cos A$, showing that the values of cosine *decrease* as the angle measure changes from 0° to
90°. Finally, increasing A from 0° to 90° causes y to increase and x to decrease,
making the values of $\frac{y}{x} = \tan A$ increase.

A similar discussion shows that as A increases from 0° to 90°, the values of
$\sec A$ increase, while the values of $\cot A$ and $\csc A$ decrease.

▶ **EXAMPLE 4**   **COMPARING FUNCTION VALUES OF ACUTE ANGLES**

Determine whether each statement is *true* or *false*.

(a) $\sin 21° > \sin 18°$ **(b)** $\cos 49° \leq \cos 56°$

Solution

(a) In the interval from 0° to 90°, as the angle increases, so does the sine of the angle, which makes $\sin 21° > \sin 18°$ a true statement.

(b) In the interval from 0° to 90°, as the angle increases, the cosine of the angle decreases. The given statement $\cos 49° \leq \cos 56°$ is false.

NOW TRY EXERCISE 37. ◀

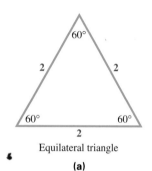

Equilateral triangle

(a)

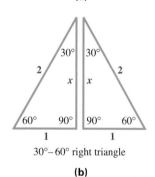

30°–60° right triangle

(b)

Figure 5

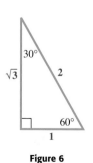

Figure 6

Trigonometric Function Values of Special Angles

Certain special angles, such as 30°, 45°, and 60°, occur so often in trigonometry and in more advanced mathematics that they deserve special study. We start with an equilateral triangle, a triangle with all sides of equal length. Each angle of such a triangle measures 60°. While the results we will obtain are independent of the length, for convenience we choose the length of each side to be 2 units. See Figure 5(a).

Bisecting one angle of this equilateral triangle leads to two right triangles, each of which has angles of 30°, 60°, and 90°, as shown in Figure 5(b). Since the hypotenuse of each right triangle has length 2, the shorter leg will have length 1. (Why?) If x represents the length of the longer leg then,

$$2^2 = 1^2 + x^2 \qquad \text{Pythagorean theorem (\textbf{Appendix B})}$$
$$4 = 1 + x^2$$
$$3 = x^2 \qquad \text{Subtract 1.}$$
$$\sqrt{3} = x. \qquad \text{Square root property (\textbf{Appendix A});}$$
$$\text{choose the positive square root.}$$

Figure 6 summarizes our results using a 30°–60° right triangle. As shown in the figure, the side opposite the 30° angle has length 1; that is, for the 30° angle,

$$\text{hypotenuse} = 2, \quad \text{side opposite} = 1, \quad \text{side adjacent} = \sqrt{3}.$$

Now we use the definitions of the trigonometric functions.

$$\sin 30° = \frac{\text{side opposite}}{\text{hypotenuse}} = \frac{1}{2}$$

$$\cos 30° = \frac{\text{side adjacent}}{\text{hypotenuse}} = \frac{\sqrt{3}}{2}$$

$$\tan 30° = \frac{\text{side opposite}}{\text{side adjacent}} = \frac{1}{\sqrt{3}} = \frac{1}{\sqrt{3}} \cdot \frac{\sqrt{3}}{\sqrt{3}} = \frac{\sqrt{3}}{3}$$

$$\csc 30° = \frac{2}{1} = 2$$

$$\sec 30° = \frac{2}{\sqrt{3}} = \frac{2}{\sqrt{3}} \cdot \frac{\sqrt{3}}{\sqrt{3}} = \frac{2\sqrt{3}}{3}$$

$$\cot 30° = \frac{\sqrt{3}}{1} = \sqrt{3}$$

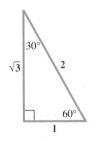

Figure 6 (repeated)

▶ **EXAMPLE 5**   FINDING TRIGONOMETRIC FUNCTION VALUES FOR 60°

Find the six trigonometric function values for a 60° angle.

Solution   Refer to Figure 6 to find the following ratios.

$$\sin 60° = \frac{\sqrt{3}}{2} \qquad \cos 60° = \frac{1}{2} \qquad \tan 60° = \frac{\sqrt{3}}{1} = \sqrt{3}$$

$$\csc 60° = \frac{2}{\sqrt{3}} = \frac{2\sqrt{3}}{3} \qquad \sec 60° = \frac{2}{1} = 2 \qquad \cot 60° = \frac{1}{\sqrt{3}} = \frac{\sqrt{3}}{3}$$

NOW TRY EXERCISES 45, 47, AND 49. ◀

▶ **Note**   The results in Example 5 can also be found using the fact that co-functions of complementary angles are equal.

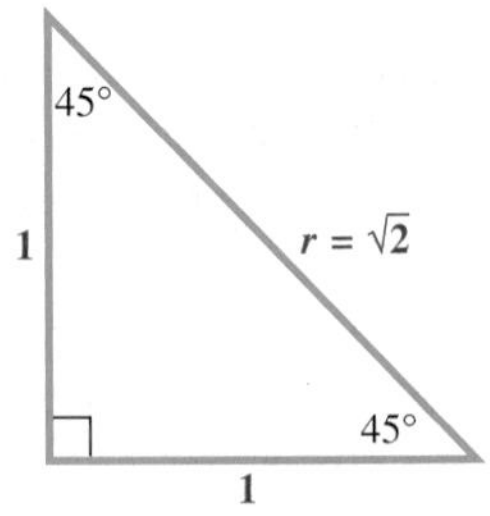

We find the values of the trigonometric functions for 45° by starting with a 45°–45° right triangle, as shown in Figure 7. This triangle is isosceles; we choose the lengths of the equal sides to be 1 unit. (As before, the results are independent of the length of the equal sides.) Since the shorter sides each have length 1, if r represents the length of the hypotenuse, then

$$1^2 + 1^2 = r^2 \quad \text{Pythagorean theorem}$$
$$2 = r^2$$
$$\sqrt{2} = r. \quad \text{Choose the positive square root.}$$

45°–45° right triangle

Figure 7

Now we use the measures indicated on the 45°–45° right triangle in Figure 7.

$$\sin 45° = \frac{1}{\sqrt{2}} = \frac{\sqrt{2}}{2} \qquad \cos 45° = \frac{1}{\sqrt{2}} = \frac{\sqrt{2}}{2} \qquad \tan 45° = \frac{1}{1} = 1$$

$$\csc 45° = \frac{\sqrt{2}}{1} = \sqrt{2} \qquad \sec 45° = \frac{\sqrt{2}}{1} = \sqrt{2} \qquad \cot 45° = \frac{1}{1} = 1$$

Function values for 30°, 45°, and 60° are summarized in the table that follows.

Function Values of Special Angles

θ	$\sin \theta$	$\cos \theta$	$\tan \theta$	$\cot \theta$	$\sec \theta$	$\csc \theta$
30°	$\frac{1}{2}$	$\frac{\sqrt{3}}{2}$	$\frac{\sqrt{3}}{3}$	$\sqrt{3}$	$\frac{2\sqrt{3}}{3}$	2
45°	$\frac{\sqrt{2}}{2}$	$\frac{\sqrt{2}}{2}$	1	1	$\sqrt{2}$	$\sqrt{2}$
60°	$\frac{\sqrt{3}}{2}$	$\frac{1}{2}$	$\sqrt{3}$	$\frac{\sqrt{3}}{3}$	2	$\frac{2\sqrt{3}}{3}$

NOW TRY EXERCISES 5, 7, AND 9. ◀

TI-83

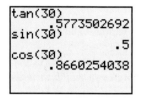

TI-84

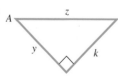

Figure 8

▶ **Note** You will be able to reproduce this table quickly if you learn the values of sin 30°, sin 45°, and sin 60°. Then you can complete the rest of the table using the reciprocal, cofunction, and quotient identities.

Since a calculator finds trigonometric function values at the touch of a key, you may wonder why we spend so much time finding values for special angles. We do this because a calculator gives only *approximate* values in most cases, while we often need *exact* values. For example, tan 30° can be found on a scientific calculator by first setting it in *degree mode,* then entering 30 and pressing the tan key to get

$$\tan 30° \approx .57735027. \quad \approx \text{ means "is approximately equal to."}$$

Earlier, however, we found the exact value:

$$\tan 30° = \frac{\sqrt{3}}{3}.$$

To use a graphing calculator to approximate sine, cosine, or tangent function values, press the appropriate function key *first,* and then enter the angle measure. (The calculator must be in degree mode to enter the angle measure in degrees.) See Figure 8. ■

2.1 Exercises

Find exact values or expressions for sin A, cos A, *and* tan A. *See Example 1.*

1.

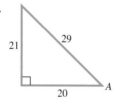

2.

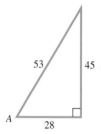

3.

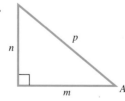

4.

Concept Check For each trigonometric function in Column I, choose its value from Column II.

I		II

5. sin 30°   **6.** cos 45°

7. tan 45°   **8.** sec 60°

9. csc 60°   **10.** cot 30°

A. $\sqrt{3}$ **B.** 1   **C.** $\frac{1}{2}$

D. $\frac{\sqrt{3}}{2}$ **E.** $\frac{2\sqrt{3}}{3}$ **F.** $\frac{\sqrt{3}}{3}$

G. 2   **H.** $\frac{\sqrt{2}}{2}$ **I.** $\sqrt{2}$

*Suppose ABC is a right triangle with sides of lengths a, b, and c and right angle at C. (See Figure 3.) Find the unknown side length using the Pythagorean theorem (**Appendix B**), and then find the values of the six trigonometric functions for angle B. Rationalize denominators when applicable.*

11. $a = 5, b = 12$ **12.** $a = 3, b = 5$

13. $a = 6, c = 7$ **14.** $b = 7, c = 12$

15. $a = 3, c = 5$ **16.** $b = 8, c = 11$

17. *Concept Check*   Give a summary of the six cofunction relationships.

Write each function in terms of its cofunction. Assume that all angles in which an unknown appears are acute angles. See Example 2.

18. $\cot 73°$ **19.** $\sec 39°$ **20.** $\sin 27°$

21. $\sec(\theta + 15°)$ **22.** $\cos(\alpha + 20°)$ **23.** $\cot(\theta - 10°)$

24. $\tan 25.4°$ **25.** $\sin 38.7°$

26. With a calculator, evaluate $\sin(90° - A)$ and $\cos A$ for various values of A. (Include values greater than 90° and less than 0°.) What do you find?

Find one solution for each equation. Assume that all angles in which an unknown appears are acute angles. See Example 3.

27. $\tan \alpha = \cot(\alpha + 10°)$ **28.** $\cos \theta = \sin 2\theta$

29. $\sin(2\theta + 10°) = \cos(3\theta - 20°)$ **30.** $\sec(\beta + 10°) = \csc(2\beta + 20°)$

31. $\tan(3B + 4°) = \cot(5B - 10°)$ **32.** $\cot(5\theta + 2°) = \tan(2\theta + 4°)$

33. $\sin(\theta - 20°) = \cos(2\theta + 5°)$ **34.** $\cos(2\theta + 50°) = \sin(2\theta - 20°)$

35. $\sec(3\beta + 10°) = \csc(\beta + 8°)$ **36.** $\csc(\beta + 40°) = \sec(\beta - 20°)$

Determine whether each statement is true *or* false. *See Example 4.*

37. $\sin 50° > \sin 40°$ **38.** $\tan 28° \leq \tan 40°$

39. $\sin 46° < \cos 46°$ **40.** $\cos 28° < \sin 28°$
      (*Hint:* $\cos 46° = \sin 44°$)   (*Hint:* $\sin 28° = \cos 62°$)

41. $\tan 41° < \cot 41°$ **42.** $\cot 30° < \tan 40°$

43. $\sec 60° > \sec 30°$ **44.** $\csc 20° < \csc 30°$

For each expression, give the exact value. See Example 5.

45. $\tan 30°$ **46.** $\cot 30°$ **47.** $\sin 30°$ **48.** $\cos 30°$

49. $\sec 30°$ **50.** $\csc 30°$ **51.** $\csc 45°$ **52.** $\sec 45°$

53. $\cos 45°$ **54.** $\cot 45°$ **55.** $\tan 45°$ **56.** $\sin 45°$

57. $\sin 60°$ **58.** $\cos 60°$ **59.** $\tan 60°$ **60.** $\csc 60°$

RELATING CONCEPTS

For individual or collaborative investigation
(Exercises 61–64)

The figure shows a 45° central angle in a circle with radius 4 units. To find the coordinates of point P on the circle, work Exercises 61–64 in order.

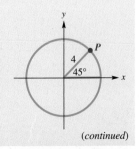

61. Sketch a line segment from P perpendicular to the x-axis.

(continued)

62. Use the trigonometric ratios for a 45° angle to label the sides of the right triangle you sketched in Exercise 61.

63. Which sides of the right triangle give the coordinates of point P? What are the coordinates of P?

64. The figure at the right shows a 60° central angle in a circle of radius 2 units. Follow the same procedure as in Exercises 61–63 to find the coordinates of P in the figure.

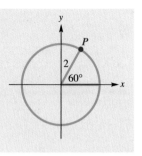

65. *Concept Check*   Refer to the table. What trigonometric functions are y_1 and y_2?

$x°$	y_1	y_2
0	0	0
15	.25882	.26795
30	.5	.57735
45	.70711	1
60	.86603	1.7321
75	.96593	3.7321
90	1	undefined

66. *Concept Check*   Refer to the table. What trigonometric functions are y_1 and y_2?

$x°$	y_1	y_2
0	1	undefined
15	.96593	3.8637
30	.86603	2
45	.70711	1.4142
60	.5	1.1547
75	.25882	1.0353
90	0	1

67. *Concept Check*   What value of A between 0° and 90° will produce the output shown on the graphing calculator screen?

```
√(3)/2
            .8660254038
sin(A)
            .8660254038
```

68. A student was asked to give the exact value of sin 45°. Using a calculator, he gave the answer .7071067812. The teacher did not give him credit. What was the teacher's reason for this?

69. With a graphing calculator, find the coordinates of the point of intersection of $y = x$ and $y = \sqrt{1 - x^2}$. These coordinates are the cosine and sine of what angle between 0° and 90°?

Concept Check   *Work each problem.*

70. Find the equation of the line passing through the origin and making a 60° angle with the x-axis.

71. Find the equation of the line passing through the origin and making a 30° angle with the x-axis.

72. What angle does the line $y = \dfrac{\sqrt{3}}{3}x$ make with the positive x-axis?

73. What angle does the line $y = \sqrt{3}x$ make with the positive x-axis?

74. Construct a square with each side of length k.

    (a) Draw a diagonal of the square. What is the measure of each angle formed by a side of the square and this diagonal?

    (b) What is the length of the diagonal?

    (c) From the results of parts (a) and (b), complete the following statement: In a $45°$–$45°$ right triangle, the hypotenuse has a length that is _____ times as long as either leg.

75. Construct an equilateral triangle with each side having length $2k$.

    (a) What is the measure of each angle?

    (b) Label one angle A. Drop a perpendicular from A to the side opposite A. Two $30°$ angles are formed at A, and two right triangles are formed. What is the length of the sides opposite the $30°$ angles?

    (c) What is the length of the perpendicular constructed in part (b)?

    (d) From the results of parts (a)–(c), complete the following statement: In a $30°$–$60°$ right triangle, the hypotenuse is always _____ times as long as the shorter leg, and the longer leg has a length that is _____ times as long as that of the shorter leg. Also, the shorter leg is opposite the _____ angle, and the longer leg is opposite the _____ angle.

Find the exact value of each part labeled with a variable in each figure.

76.

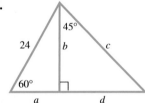

77.

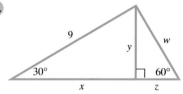

78.

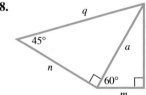

79.

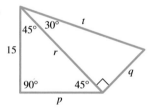

Find a formula for the area of each figure in terms of s.

80.

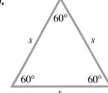

81.

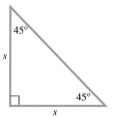

82. Suppose you know the length of one side and one acute angle of a right triangle. Can you determine the measures of all the sides and angles of the triangle?

83. Why is it important to be able to find trigonometric values for the special angles without using a calculator?

Reference Angles ▪ Special Angles as Reference Angles ▪ Finding Angle Measures with Special Angles

Reference Angles   Associated with every nonquadrantal angle in standard position is a positive acute angle called its *reference angle*. A **reference angle** for an angle θ, written θ', is the positive acute angle made by the terminal side of angle θ and the *x*-axis. Figure 9 shows several angles θ (each less than one complete counterclockwise revolution) in quadrants II, III, and IV, respectively, with the reference angle θ' also shown. In quadrant I, θ and θ' are the same. If an angle θ is negative or has measure greater than 360°, its reference angle is found by first finding its coterminal angle that is between 0° and 360°, and then using the diagrams in Figure 9.

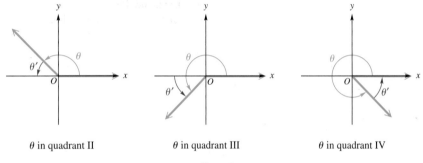

| θ in quadrant II | θ in quadrant III | θ in quadrant IV |

Figure 9

> ▶ **Caution** A common error is to find the reference angle by using the terminal side of θ and the *y*-axis. *The reference angle is always found with reference to the x-axis.*

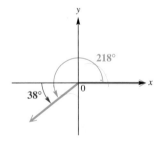

$218° - 180° = 38°$

Figure 10

▶ **EXAMPLE 1**   FINDING REFERENCE ANGLES

Find the reference angle for each angle.

(a) 218°   **(b)** 1387°

Solution

(a) As shown in Figure 10, the positive acute angle made by the terminal side of this angle and the *x*-axis is $218° - 180° = 38°$. For $\theta = 218°$, the reference angle $\theta' = 38°$.

(b) First find a coterminal angle between 0° and 360°. Divide 1387° by 360° to get a quotient of about 3.9. Begin by subtracting 360° three times (because of the whole number 3 in 3.9):

$$1387° - 3 \cdot 360° = 1387° - 1080° = 307°. \quad \text{(Section 1.1)}$$

The reference angle for 307° (and thus for 1387°) is $360° - 307° = 53°$. See Figure 11.

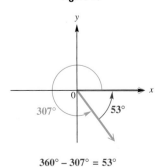

$360° - 307° = 53°$

Figure 11

NOW TRY EXERCISES 1 AND 5. ◀

The preceding example suggests the following table for finding the reference angle θ' for any angle θ between $0°$ and $360°$.

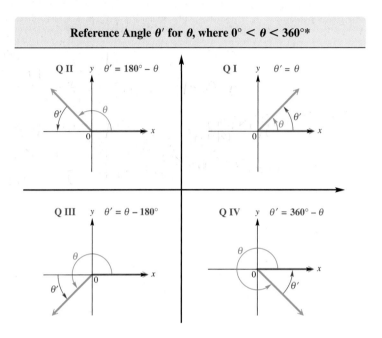

Reference Angle θ' for θ, where $0° < \theta < 360°$*

Special Angles as Reference Angles

We can now find exact trigonometric function values of angles with reference angles of $30°$, $45°$, or $60°$.

▶ **EXAMPLE 2** FINDING TRIGONOMETRIC FUNCTION VALUES OF A QUADRANT III ANGLE

Find the values of the six trigonometric functions for $210°$.

Solution An angle of $210°$ is shown in Figure 12. The reference angle is $210° - 180° = 30°$. To find the trigonometric function values of $210°$, choose point P on the terminal side of the angle so that the distance from the origin O to P is 2. By the results from $30°$–$60°$ right triangles, the coordinates of point P become $\left(-\sqrt{3}, -1\right)$, with $x = -\sqrt{3}$, $y = -1$, and $r = 2$. Then, by the definitions of the trigonometric functions in **Section 2.1,**

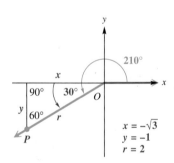

Figure 12

$$\sin 210° = \frac{-1}{2} = -\frac{1}{2} \qquad \csc 210° = \frac{2}{-1} = -2$$

$$\cos 210° = \frac{-\sqrt{3}}{2} = -\frac{\sqrt{3}}{2} \qquad \sec 210° = \frac{2}{-\sqrt{3}} = -\frac{2\sqrt{3}}{3}$$

$$\tan 210° = \frac{-1}{-\sqrt{3}} = \frac{\sqrt{3}}{3} \qquad \cot 210° = \frac{-\sqrt{3}}{-1} = \sqrt{3}.$$

NOW TRY EXERCISE 19. ◀

---

*The authors would like to thank Bethany Vaughn and Theresa Matick, of Vincennes Lincoln High School, for their suggestions concerning this table.

Notice in Example 2 that the trigonometric function values of 210° correspond in absolute value to those of its reference angle 30°. The signs are different for the sine, cosine, secant, and cosecant functions because 210° is a quadrant III angle. These results suggest a shortcut for finding the trigonometric function values of a non-acute angle, using the reference angle. In Example 2, the reference angle for 210° is 30°. Using the trigonometric function values of 30°, and choosing the correct signs for a quadrant III angle, we obtain the same results.

We determine the values of the trigonometric functions for any nonquadrantal angle θ as follows.

FINDING TRIGONOMETRIC FUNCTION VALUES FOR ANY NONQUADRANTAL ANGLE θ

Step 1 If θ > 360°, or if θ < 0°, then find a coterminal angle by adding or subtracting 360° as many times as needed to get an angle greater than 0° but less than 360°.

Step 2 Find the reference angle θ′.

Step 3 Find the trigonometric function values for reference angle θ′.

Step 4 Determine the correct signs for the values found in Step 3. (Use the table of signs in **Section 1.4,** if necessary.) This gives the values of the trigonometric functions for angle θ.

▶ **EXAMPLE 3** FINDING TRIGONOMETRIC FUNCTION VALUES USING REFERENCE ANGLES

Find the exact value of each expression.

(a) cos(−240°)   **(b)** tan 675°

Solution

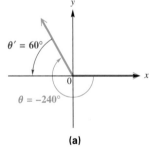

(a)

(a) Since an angle of −240° is coterminal with an angle of

$$-240° + 360° = 120°, \quad \text{(Section 1.1)}$$

the reference angle is 180° − 120° = 60°, as shown in Figure 13(a). Since the cosine is negative in quadrant II,

$$\cos(-240°) = \cos 120° = -\cos 60° = -\frac{1}{2}.$$

                 Coterminal   Reference
                 angle   angle

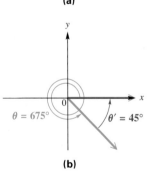

(b)

Figure 13

(b) Begin by subtracting 360° to get a coterminal angle between 0° and 360°.

$$675° - 360° = 315°$$

As shown in Figure 13(b), the reference angle is 360° − 315° = 45°. An angle of 315° is in quadrant IV, so the tangent will be negative, and

$$\tan 675° = \tan 315° = -\tan 45° = -1.$$

NOW TRY EXERCISES 37 AND 39. ◀

▶ **EXAMPLE 4** EVALUATING AN EXPRESSION WITH FUNCTION VALUES OF SPECIAL ANGLES

Evaluate $\cos 120° + 2 \sin^2 60° - \tan^2 30°$.

Solution Since $\cos 120° = -\frac{1}{2}$, $\sin 60° = \frac{\sqrt{3}}{2}$, and $\tan 30° = \frac{\sqrt{3}}{3}$,

$$\cos 120° + 2 \sin^2 60° - \tan^2 30° = -\frac{1}{2} + 2\left(\frac{\sqrt{3}}{2}\right)^2 - \left(\frac{\sqrt{3}}{3}\right)^2$$

$$= -\frac{1}{2} + 2\left(\frac{3}{4}\right) - \frac{3}{9}$$

$$= \frac{2}{3}.$$

NOW TRY EXERCISE 45. ◀

Recall that trigonometric function values of coterminal angles are the same.

▶ **EXAMPLE 5** USING COTERMINAL ANGLES TO FIND FUNCTION VALUES

Evaluate each function by first expressing the function in terms of an angle between 0° and 360°.

(a) $\cos 780°$ **(b)** $\tan(-405°)$

Solution

(a) Add or subtract 360° as many times as necessary to get an angle between 0° and 360°. Subtracting 720°, which is $2 \cdot 360°$, gives

$$\cos 780° = \cos(780° - 2 \cdot 360°)$$

$$= \cos 60°$$

$$= \frac{1}{2}.$$

> Multiply first, and then subtract.

(b) Add 360° twice to get $-405° + 2(360°) = 315°$. This angle is located in quadrant IV and its reference angle is 45°. The tangent function is negative in quadrant IV; thus

$$\tan(-405°) = \tan 315° = -\tan 45° = -1.$$

NOW TRY EXERCISES 27 AND 31. ◀

Finding Angle Measures with Special Angles
The ideas discussed in this section can also be used to find the measures of certain angles, given a trigonometric function value and an interval in which the angle must lie. We are most often interested in the interval $[0°, 360°)$.

▶ **EXAMPLE 6**   FINDING ANGLE MEASURES GIVEN AN INTERVAL AND A FUNCTION VALUE

Find all values of θ, if θ is in the interval $[0°, 360°)$ and $\cos \theta = -\frac{\sqrt{2}}{2}$.

Solution   Since $\cos \theta$ is negative, θ must lie in quadrant II or III. Since the absolute value of $\cos \theta$ is $\frac{\sqrt{2}}{2}$, the reference angle θ' must be $45°$. The two possible angles θ are sketched in Figure 14. The quadrant II angle θ equals

$$180° - 45° = 135°,$$

and the quadrant III angle θ equals

$$180° + 45° = 225°.$$

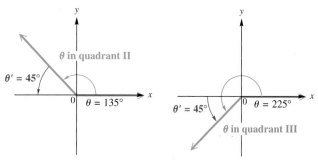

Figure 14

NOW TRY EXERCISE 71. ◀

<h1>2.2 Exercises</h1>

Match each angle in Column I with its reference angle in Column II. Choices may be used once, more than once, or not at all. See Example 1.

I

1. 98°   **2.** 212°

3. −135°   **4.** −60°

5. 750°   **6.** 480°

II

3 **A.** 45°   **B.** 60°

1 **C.** 82°   *5* **D.** 30°

E. 38°   **F.** 32°

Give a short explanation in Exercises 7–9.

7. In Example 2, why was 2 a good choice for *r*? Could any other positive number have been used?

8. Explain how the reference angle is used to find values of the trigonometric functions for an angle in quadrant III.

9. Explain why two coterminal angles have the same values for their trigonometric functions.

Complete the table with exact trigonometric function values. Do not use a calculator. See Examples 2 and 3.

	θ	$\sin \theta$	$\cos \theta$	$\tan \theta$	$\cot \theta$	$\sec \theta$	$\csc \theta$
10.	30°	$\dfrac{1}{2}$	$\dfrac{\sqrt{3}}{2}$			$\dfrac{2\sqrt{3}}{3}$	2
11.	45°	$\dfrac{\sqrt{2}}{2}$	$\dfrac{\sqrt{2}}{2}$	1	1	$\sqrt{2}$	$\sqrt{2}$
12.	60°		$\dfrac{1}{2}$	$\sqrt{3}$		2	
13.	120°	$\dfrac{\sqrt{3}}{2}$		$-\sqrt{3}$			$\dfrac{2\sqrt{3}}{3}$
14.	135°	$\dfrac{\sqrt{2}}{2}$	$-\dfrac{\sqrt{2}}{2}$			$-\sqrt{2}$	$\sqrt{2}$
15.	150°	$\dfrac{1}{2}$	$-\dfrac{\sqrt{3}}{2}$	$-\dfrac{\sqrt{3}}{3}$	$-\sqrt{3}$	$-\dfrac{2\sqrt{3}}{3}$	2
16.	210°	$-\dfrac{1}{2}$		$\dfrac{\sqrt{3}}{3}$	$\sqrt{3}$		-2
17.	240°	$-\dfrac{\sqrt{3}}{2}$	$-\dfrac{1}{2}$			-2	$-\dfrac{2\sqrt{3}}{3}$

Find exact values of the six trigonometric functions for each angle. Rationalize denominators when applicable. See Examples 2, 3, and 5.

18. 300°	**19.** 315°	**20.** 405°	**21.** $-300°$	**22.** 420°	**23.** 480°
24. 495°	**25.** 570°	**26.** 750°	**27.** 1305°	**28.** 1500°	**29.** 2670°
30. $-390°$	**31.** $-510°$	**32.** $-1020°$	**33.** $-1290°$	**34.** $-855°$	**35.** $-1860°$

Find the exact value of each expression. See Example 3.

36. $\sin 1305°$ **37.** $\cos(-510°)$ **38.** $\tan(-1020°)$ **39.** $\sin 1500°$

40. $\sec(-495°)$ **41.** $\csc(-855°)$ **42.** $\cot 2280°$ **43.** $\tan 3015°$

Determine whether each statement is true *or* false. *If false, tell why. See Example 4.*

44. $\sin 30° + \sin 60° = \sin(30° + 60°)$

45. $\sin(30° + 60°) = \sin 30° \cdot \cos 60° + \sin 60° \cdot \cos 30°$

46. $\cos 60° = 2 \cos^2 30° - 1$

47. $\cos 60° = 2 \cos 30°$

48. $\sin 120° = \sin 150° - \sin 30°$

49. $\sin 120° = \sin 180° \cdot \cos 60° - \sin 60° \cdot \cos 180°$

50. $\sin(2 \cdot 30°) = 2 \sin 30° \cdot \cos 30°$

51. $\sin^2 45° + \cos^2 45° = 1$

52. $\tan^2 60° + 1 = \sec^2 60°$

53. $\cos(30° + 60°) = \cos 30° + \cos 60°$

Concept Check   *Find the coordinates of the point P on the circumference of each circle. (Hint: Add x- and y-axes, assuming that the angle is in standard position.)*

54.

225°

10

P

55.

P

150°

6

56. *Concept Check*   Does there exist an angle θ with the function values $\cos \theta = .6$ and $\sin \theta = -.8$?

57. *Concept Check*   Does there exist an angle θ with the function values $\cos \theta = \frac{2}{3}$ and $\sin \theta = \frac{3}{4}$?

Suppose θ is in the interval $(90°, 180°)$. Find the sign of each of the following.

58. $\sin \dfrac{\theta}{2}$ **59.** $\cos \dfrac{\theta}{2}$ **60.** $\cot(\theta + 180°)$

61. $\sec(\theta + 180°)$ **62.** $\cos(-\theta)$ **63.** $\sin(-\theta)$

64. Explain why $\sin \theta = \sin(\theta + n \cdot 360°)$ for any angle θ and any integer n.

65. Explain why $\cos \theta = \cos(\theta + n \cdot 360°)$ for any angle θ and any integer n.

Concept Check   *Work Exercises 66–69.*

66. Without using a calculator, determine which of the following numbers is closest to $\cos 115°$: .4, .6, 0, −.4, or −.6.

67. Without using a calculator, determine which of the following numbers is closest to $\sin 115°$: .9, .1, 0, −.9, or −.1.

68. For what angles θ between 0° and 360° does $\cos \theta = -\sin \theta$?

69. For what angles θ between 0° and 360° does $\cos \theta = \sin \theta$?

70. *(Modeling) Length of a Sag Curve*   When a highway goes downhill and then uphill, it is said to have a **sag curve**. Sag curves are designed so that at night, headlights shine sufficiently far down the road to allow a safe stopping distance. See the figure.

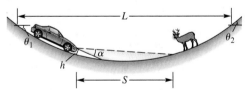

The minimum length L of a sag curve is determined by the height h of the car's headlights above the pavement, the downhill grade $\theta_1 < 0°$, the uphill grade $\theta_2 > 0°$, and the safe stopping distance S for a given speed limit. In addition, L is dependent on the vertical alignment of the headlights. Headlights are usually pointed upward at a slight angle α above the horizontal of the car. Using these quantities, for a 55 mph speed limit, L can be modeled by the formula

$$L = \frac{(\theta_2 - \theta_1)S^2}{200(h + S \tan \alpha)},$$

where $S < L$. (*Source:* Mannering, F. and W. Kilareski, *Principles of Highway Engineering and Traffic Analysis,* Second Edition, John Wiley and Sons, 1998.)

(a) Compute L if $h = 1.9$ ft, $\alpha = .9°$, $\theta_1 = -3°$, $\theta_2 = 4°$, and $S = 336$ ft.

(b) Repeat part (a) with $\alpha = 1.5°$.

(c) How does the alignment of the headlights affect the value of L?

Find all values of θ, if θ is in the interval $[0°, 360°)$ and has the given function value. See Example 6.

71. $\sin \theta = \dfrac{1}{2}$ **72.** $\cos \theta = \dfrac{\sqrt{3}}{2}$ **73.** $\tan \theta = -\sqrt{3}$

74. $\sec \theta = -\sqrt{2}$ **75.** $\cos \theta = \dfrac{\sqrt{2}}{2}$ **76.** $\cot \theta = -\dfrac{\sqrt{3}}{3}$

77. $\csc \theta = -2$ **78.** $\sin \theta = -\dfrac{\sqrt{3}}{2}$ **79.** $\tan \theta = \dfrac{\sqrt{3}}{3}$

80. $\cos \theta = -\dfrac{1}{2}$ **81.** $\csc \theta = -\sqrt{2}$ **82.** $\cot \theta = -1$

2.3 Finding Trigonometric Function Values Using a Calculator

Finding Function Values Using a Calculator ▪ Finding Angle Measures Using a Calculator

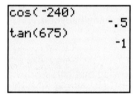

Degree mode

Figure 15

Finding Function Values Using a Calculator Calculators are capable of finding trigonometric function values. For example, the values of $\cos(-240°)$ and $\tan 675°$ in Example 3 of **Section 2.2** are found with a calculator as shown in Figure 15.

> ▶ **Caution** *When evaluating trigonometric functions of angles given in degrees, remember that the calculator must be set in degree mode.* Get in the habit of always starting work by entering sin 90. If the displayed answer is 1, then the calculator is set for degree measure. Remember that most calculator values of trigonometric functions are *approximations*.

> ▶ **EXAMPLE 1** FINDING FUNCTION VALUES WITH A CALCULATOR

Approximate the value of each expression.

(a) $\sin 49° 12'$ **(b)** $\sec 97.977°$ **(c)** $\dfrac{1}{\cot 51.4283°}$ **(d)** $\sin(-246°)$

Solution

(a) $49° \, 12' = 49\dfrac{12}{60}^{\circ} = 49.2°$ Convert $49° \, 12'$ to decimal degrees. **(Section 1.1)**

   $\sin 49° \, 12' = \sin 49.2° \approx .75699506$ To eight decimal places

(b) Calculators do not have secant keys. However, $\sec \theta = \frac{1}{\cos \theta}$ for all angles θ where $\cos \theta \neq 0$. First find $\cos 97.977°$, and then take the reciprocal to get

$$\sec 97.977° \approx -7.20587921.$$

(c) Use the reciprocal identity $\tan \theta = \frac{1}{\cot \theta}$ from **Section 1.4** to get

$$\frac{1}{\cot 51.4283°} = \tan 51.4283° \approx 1.25394815.$$

(d) $\sin(-246°) \approx .91354546$

These screens support the results of Example 7. We entered the angle measure in degrees and minutes for part (a). In the fifth line of the first screen, Ans^{-1} tells the calculator to find the reciprocal of the answer given in the previous line.

NOW TRY EXERCISES 5, 7, 13, AND 17. ◀

Finding Angle Measures Using a Calculator   Sometimes we need to find the measure of an angle having a certain trigonometric function value. Graphing calculators have three *inverse functions* (denoted **sin⁻¹, cos⁻¹,** and **tan⁻¹**) that do just that. If x is an appropriate number, then $\sin^{-1}x$, $\cos^{-1}x$, or $\tan^{-1}x$ gives the measure of an angle whose sine, cosine, or tangent is x. For the applications in this chapter, these functions will return values of x in quadrant I.

▶ EXAMPLE 2   USING INVERSE TRIGONOMETRIC FUNCTIONS TO FIND ANGLES

Use a calculator to find an angle θ in the interval $[0°, 90°]$ that satisfies each condition.

(a) $\sin \theta \approx .9677091705$ **(b)** $\sec \theta \approx 1.0545829$

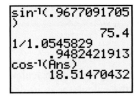

Degree mode

Figure 16

Solution

(a) Using degree mode and the inverse sine function, we find that an angle θ having sine value .9677091705 is 75.4°. (While there are infinitely many such angles, the calculator gives only this one.) We write this result as

$$\theta \approx \sin^{-1} .9677091705 \approx 75.4°.$$

See Figure 16.

(b) Use the identity $\cos \theta = \frac{1}{\sec \theta}$. Find the reciprocal of 1.0545829 to get $\cos \theta \approx .9482421913$. Now find θ as shown in part (a), using the inverse cosine function. The result is

$$\theta \approx \cos^{-1} (.9482421913) \approx 18.514704°.$$

See Figure 16.

NOW TRY EXERCISES 25 AND 29. ◀

▶ **Caution** Compare Examples 1(b) and 2(b). Note that the reciprocal is used *before* the inverse cosine key when finding the angle, but *after* the cosine key when finding the trigonometric function value.

▶ EXAMPLE 3   FINDING GRADE RESISTANCE

When an automobile travels uphill or downhill on a highway, it experiences a force due to gravity. This force F in pounds is called **grade resistance** and is modeled by the equation $F = W \sin \theta$, where θ is the grade and W is the weight of the automobile. If the automobile is moving uphill, then $\theta > 0°$; if downhill, then $\theta < 0°$. See Figure 17. (*Source:* Mannering, F. and W. Kilareski, *Principles of Highway Engineering and Traffic Analysis,* Second Edition, John Wiley and Sons, 1998.)

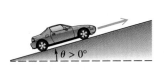
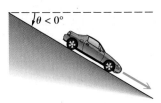

Figure 17

(a) Calculate F to the nearest 10 lb for a 2500-lb car traveling an uphill grade with $\theta = 2.5°$.

(b) Calculate F to the nearest 10 lb for a 5000-lb truck traveling a downhill grade with $\theta = -6.1°$.

(c) Calculate F for $\theta = 0°$ and $\theta = 90°$. Do these answers agree with your intuition?

Solution

(a) $F = W \sin \theta = 2500 \sin 2.5° \approx 110$ lb

(b) $F = W \sin \theta = 5000 \sin(-6.1°) \approx -530$ lb
F is negative because the truck is moving downhill.

(c) $F = W \sin \theta = W \sin 0° = W(0) = 0$ lb
$F = W \sin \theta = W \sin 90° = W(1) = W$ lb

This agrees with intuition because if $\theta = 0°$, then there is level ground and gravity does not cause the vehicle to roll. If $\theta = 90°$, the road would be vertical and the full weight of the vehicle would be pulled downward by gravity, so $F = W$.

> NOW TRY EXERCISES 57 AND 59. ◀

2.3 Exercises

Concept Check *Fill in the blanks to complete each statement.*

1. The CAUTION at the beginning of this section suggests verifying that a calculator is in degree mode by finding _____ 90°. If the calculator is in degree
(sin/cos/tan)

   mode, the display should be _____.

2. When a scientific or graphing calculator is used to find a trigonometric function value, in most cases the result is an _____ value.
(exact/approximate)

3. To find values of the cotangent, secant, and cosecant functions with a calculator, it is necessary to find the _____ of the _____ function value.

4. The reciprocal is used _____ the inverse function key when finding the
(before/after)

   angle, but _____ the function key when finding the trigonometric function
(before/after)

   value.

Use a calculator to find a decimal approximation for each value. Give as many digits as your calculator displays. In Exercises 16–22, simplify the expression before using the calculator. See Example 1.

5. $\sin 38° 42'$	**6.** $\cot 41° 24'$	**7.** $\sec 13° 15'$
8. $\csc 145° 45'$	**9.** $\cot 183° 48'$	**10.** $\cos 421° 30'$
11. $\sec 312° 12'$	**12.** $\tan(-80° 6')$	**13.** $\sin(-317° 36')$
14. $\cot(-512° 20')$	**15.** $\cos(-15')$	**16.** $\dfrac{1}{\sec 14.8°}$
17. $\dfrac{1}{\cot 23.4°}$	**18.** $\dfrac{\sin 33°}{\cos 33°}$	**19.** $\dfrac{\cos 77°}{\sin 77°}$
20. $\cos(90° - 3.69°)$	**21.** $\cot(90° - 4.72°)$	**22.** $\dfrac{1}{\tan(90° - 22°)}$

Find a value of θ in the interval $[0°, 90°]$ that satisfies each statement. Leave answers in decimal degrees. See Example 2.

23. $\tan \theta = 1.4739716$	**24.** $\tan \theta = 6.4358841$	**25.** $\sin \theta = .27843196$
26. $\sec \theta = 1.1606249$	**27.** $\cot \theta = 1.2575516$	**28.** $\csc \theta = 1.3861147$
29. $\sec \theta = 2.7496222$	**30.** $\sin \theta = .84802194$	**31.** $\cos \theta = .70058013$

32. A student, wishing to use a calculator to verify the value of sin 30°, enters the information correctly but gets a display of −.98803162. He knows that the display should be .5, and he also knows that his calculator is in good working order. What do you think is the problem?

33. At one time, a certain make of calculator did not allow the input of angles outside of a particular interval when finding trigonometric function values. For example, trying to find cos 2000° using the methods of this section gave an error message, despite the fact that cos 2000° can be evaluated. Explain how you would find cos 2000° using this calculator.

34. What value of A between 0° and 90° will produce the output in the graphing calculator screen?

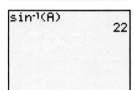

```
tan(A)
        1.482560969
```

35. What value of A will produce the output (in degrees) in the graphing calculator screen?

```
sin⁻¹(A)
                22
```

Use a calculator to evaluate each expression.

36. sin 35° cos 55° + cos 35° sin 55°   **37.** cos 100° cos 80° − sin 100° sin 80°

38. cos 75° 29′ cos 14° 31′ − sin 75° 29′ sin 14° 31′

39. sin 28° 14′ cos 61° 46′ + cos 28° 14′ sin 61° 46′

40. $\sin^2 36° + \cos^2 36°$ **41.** 2 sin 25° 13′ cos 25° 13′ − sin 50° 26′

Work each problem.

42. *Measuring Speed by Radar*   Any offset between a stationary radar gun and a moving target creates a "cosine effect" that reduces the radar mileage reading by the cosine of the angle between the gun and the vehicle. That is, the radar speed reading is the product of the actual reading and the cosine of the angle. Find the radar readings to the nearest hundredth for Auto A and Auto B shown in the figure. (*Source:* Fischetti, M., "Working Knowledge," *Scientific American,* March 2001.)

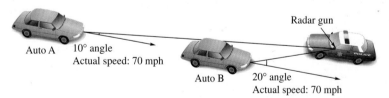

Radar gun

Auto A   10° angle
Actual speed: 70 mph

Auto B   20° angle
Actual speed: 70 mph

43. *Measuring Speed by Radar*   In Exercise 42, we saw that the mileage reported by a radar gun is reduced by the cosine of angle θ, shown in the figure. In the figure, r represents reduced speed and a represents the actual speed. Use the figure to show why this "cosine effect" occurs.

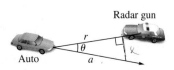

Radar gun

Auto

r

θ

a

Use a calculator to decide whether each statement is true *or* false. *It may be that a true statement will lead to results that differ in the last decimal place due to rounding error.*

44. $\cos 40° = 2 \cos 20°$

45. $\sin 10° + \sin 10° = \sin 20°$

46. $\cos 70° = 2 \cos^2 35° - 1$

47. $\sin 50° = 2 \sin 25° \cos 25°$

48. $2 \cos 38° 22' = \cos 76° 44'$

49. $\cos 40° = 1 - 2 \sin^2 80°$

50. $\dfrac{1}{2} \sin 40° = \sin \dfrac{1}{2}(40°)$

51. $\sin 39° 48' + \cos 39° 48' = 1$

52. $\cos(30° + 20°) = \cos 30° + \cos 20°$

53. $\cos(30° + 20°) = \cos 30° \cos 20° - \sin 30° \sin 20°$

54. $\tan^2 72° 25' + 1 = \sec^2 72° 25'$

55. $1 + \cot^2 42.5° = \csc^2 42.5°$

(Modeling) Grade Resistance *See Example 3 to work Exercises 56–61.*

56. What is the grade resistance of a 2400-lb car traveling on a $-2.4°$ downhill grade?

57. What is the grade resistance of a 2100-lb car traveling on a $1.8°$ uphill grade?

58. A car traveling on a $-3°$ downhill grade has a grade resistance of -145 lb. What is the weight of the car?

59. A 2600-lb car traveling downhill has a grade resistance of -130 lb. What is the angle of the grade?

60. Which has the greater grade resistance: a 2200-lb car on a $2°$ uphill grade or a 2000-lb car on a $2.2°$ uphill grade?

61. A 3000-lb car traveling uphill has a grade resistance of 150 lb. What is the angle of the grade?

62. *Highway Grades* Complete the table for the values of $\sin \theta$, $\tan \theta$, and $\dfrac{\pi\theta}{180}$ to four decimal places.

θ	$\sin \theta$	$\tan \theta$	$\dfrac{\pi\theta}{180}$
0°			
.5°			
1°			
1.5°			
2°			
2.5°			
3°			
3.5°			
4°			

    (a) How do $\sin \theta$, $\tan \theta$, and $\dfrac{\pi\theta}{180}$ compare for small grades θ?

    (b) Highway grades are usually small. Give two approximations of the grade resistance $F = W \sin \theta$ that do not use the sine function.

(c) A stretch of highway has a 4-ft vertical rise for every 100 ft of horizontal run. Use an approximation from part (a) to estimate the grade resistance for a 2000-lb car on this stretch of highway.

(d) Without evaluating a trigonometric function, estimate the grade resistance for an 1800-lb car on a stretch of highway that has a 3.75° grade.

(Modeling) *Solve each problem.*

63. *Design of Highway Curves*   When highway curves are designed, the outside of the curve is often slightly elevated or inclined above the inside of the curve. See the figure. This incli-nation is called **superelevation.** For safety  reasons, it is important that both the curve's radius and superelevation are correct for a given speed limit. If an automobile is traveling at velocity *V* (in feet per second), the safe radius *R* for a curve with superelevation θ is modeled by the formula

$$R = \frac{V^2}{g(f + \tan \theta)},$$

where *f* and *g* are constants. (*Source:* Mannering, F. and W. Kilareski, *Principles of Highway Engineering and Traffic Analysis,* Second Edition, John Wiley and Sons, 1998.)

(a) A roadway is being designed for automobiles traveling at 45 mph. If $\theta = 3°$, $g = 32.2$, and $f = .14$, calculate *R* to the nearest foot. (*Hint:* 45 mph = 66 ft per sec)

(b) Determine the radius of the curve, to the nearest foot, if the speed in part (a) is increased to 70 mph.

(c) How would increasing the angle θ affect the results? Verify your answer by re-peating parts (a) and (b) with $\theta = 4°$.

64. *Speed Limit on a Curve*   Refer to Exercise 63 and use the same values for *f* and *g*. A highway curve has radius $R = 1150$ ft and a superelevation of $\theta = 2.1°$. What should the speed limit (in miles per hour) be for this curve?

(Modeling) Speed of Light   When a light ray travels from one medium, such as air, to another medium, such as water or glass, the speed of the light changes, and the direction in which the ray is traveling changes. (This is why a fish under water is in a different position than it appears to be.) These changes are given by **Snell's** *law*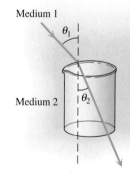

$$\frac{c_1}{c_2} = \frac{\sin \theta_1}{\sin \theta_2},$$

where c_1 is the speed of light in the first medium, c_2 is the speed of light in the second medium, and θ_1 and θ_2 are the angles shown in the figure. (*Source:* The Physics Classroom, www.glenbrook.k12.il.us) *In Exercises 65 and 66, assume that* $c_1 = 3 \times 10^8$ *m per sec.*

65. Find the speed of light in the second medium for each of the following.

(a) $\theta_1 = 46°, \theta_2 = 31°$ **(b)** $\theta_1 = 39°, \theta_2 = 28°$

66. Find θ_2 for each of the following values of θ_1 and c_2. Round to the nearest degree.

(a) $\theta_1 = 40°, c_2 = 1.5 \times 10^8$ m per sec   **(b)** $\theta_1 = 62°, c_2 = 2.6 \times 10^8$ m per sec

(Modeling) Fish's View of the World *The figure shows a fish's view of the world above the surface of the water.* (*Source:* Walker, J., "The Amateur Scientist," *Scientific American,* March 1984.) *Suppose that a light ray comes from the horizon, enters the water, and strikes the fish's eye.*

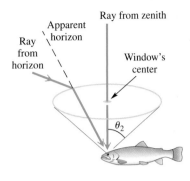

67. Assume that this ray gives a value of 90° for angle θ_1 in the formula for Snell's law. (In a practical situation, this angle would probably be a little less than 90°.) The speed of light in water is about 2.254×10^8 m per sec. Find angle θ_2 to the nearest tenth.

68. Refer to Exercise 67. Suppose an object is located at a true angle of 29.6° above the horizon. Find the apparent angle above the horizon to a fish.

69. *(Modeling) Braking Distance* If aerodynamic resistance is ignored, the braking distance D (in feet) for an automobile to change its velocity from V_1 to V_2 (feet per second) can be modeled using the equation

$$D = \frac{1.05(V_1^2 - V_2^2)}{64.4(K_1 + K_2 + \sin \theta)}.$$

K_1 is a constant determined by the efficiency of the brakes and tires, K_2 is a constant determined by the rolling resistance of the automobile, and θ is the grade of the highway. (*Source:* Mannering, F. and W. Kilareski, *Principles of Highway Engineering and Traffic Analysis,* Second Edition, John Wiley and Sons, 1998.)

(a) Compute the number of feet required to slow a car from 55 mph to 30 mph while traveling uphill with a grade of $\theta = 3.5°$. Let $K_1 = .4$ and $K_2 = .02$. (*Hint:* Change miles per hour to feet per second.)

(b) Repeat part (a) with $\theta = -2°$.

(c) How is braking distance affected by grade θ? Does this agree with your driving experience?

70. *(Modeling) Car's Speed at Collision* Refer to Exercise 69. An automobile is traveling at 90 mph on a highway with a downhill grade of $\theta = -3.5°$. The driver sees a stalled truck in the road 200 ft away and immediately applies the brakes. Assuming that a collision cannot be avoided, how fast (in miles per hour) is the car traveling when it hits the truck? (Use the same values for K_1 and K_2 as in Exercise 69.)

CHAPTER 2 ▶ Quiz (Sections 2.1–2.3)

1. Find the exact values for sin A, cos A, and tan A in the figure.

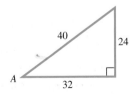

2. Find exact values of the trigonometric functions to complete the table.

θ	sin θ	cos θ	tan θ	cot θ	sec θ	csc θ
30°						
45°						
60°						

3. Find the exact value of each variable in the figure.

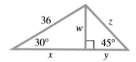

4. *Area of a Solar Cell*   A solar cell converts the energy of sunlight directly into electrical energy. The amount of energy a cell produces depends on its area. Suppose a solar cell is hexagonal, as shown in the figure on the left. Express its area in terms of sin θ and any side x. (*Hint:* Consider one of the six equilateral triangles from the hexagon. See the figure on the right.) (*Source:* Kastner, B., *Space Mathematics,* NASA, 1985.)

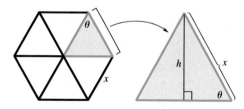

Find exact values of the six trigonometric functions for each angle. Rationalize denominators when applicable.

5. 135°   **6.** −150°   **7.** 1020°

Find all values of θ in the interval [0°, 360°) that have the given function value.

8. $\sin \theta = \dfrac{\sqrt{3}}{2}$ **9.** $\sec \theta = -\sqrt{2}$

Use a calculator to approximate each value.

10. $\sin 42° \, 18'$ **11.** $\sec(-212° \, 12')$

Use a calculator to find the value of θ in the interval [0°, 90°] that satisfies each statement.

12. $\tan \theta = 2.6743210$ **13.** $\csc \theta = 2.3861147$

Determine whether each statement is true *or* false.

14. $\sin(60° + 30°) = \sin 60° + \sin 30°$

15. $\tan(90° − 35°) = \cot 35°$

2.4 Solving Right Triangles

Significant Digits ▪ **Solving Triangles** ▪ **Angles of Elevation or Depression**

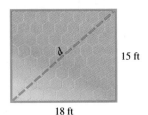

18 ft

Figure 18

Significant Digits   A number that represents the result of counting, or a number that results from theoretical work and is not the result of measurement, is an **exact number.** There are 50 states in the United States, so in that statement, 50 is an exact number.

Most values obtained for trigonometric applications are measured values that are *not* exact. Suppose we quickly measure a room as 15 ft by 18 ft. See Figure 18. To calculate the length of a diagonal of the room, we can use the Pythagorean theorem.

$$d^2 = 15^2 + 18^2 \quad \text{(Appendix B)}$$
$$d^2 = 549$$
$$d = \sqrt{549} \approx 23.430749$$

Should this answer be given as the length of the diagonal of the room? Of course not. The number 23.430749 contains six decimal places, while the original data of 15 ft and 18 ft are only accurate to the nearest foot. In practice, the results of a calculation can be no more accurate than the least accurate number in the calculation. Thus, we should indicate that the diagonal of the 15-by-18-ft room is approximately 23 ft.

If a wall measured to the nearest foot is 18 ft long, this actually means that the wall has length between 17.5 ft and 18.5 ft. If the wall is measured more accurately as 18.3 ft long, then its length is really between 18.25 ft and 18.35 ft. The results of physical measurement are only approximately accurate and depend on the precision of the measuring instrument as well as the aptness of the observer. The digits obtained by actual measurement are called **significant digits.** The measurement 18 ft is said to have two significant digits; 18.3 ft has three significant digits.

The following numbers have their significant digits identified in color.

408   21.5   18.00   6.700   .0025   .09810   7300

Notice that 18.00 has four significant digits. The zeros in this number represent measured digits and are significant. The number .0025 has only two significant digits, 2 and 5, because the zeros here are used only to locate the decimal point. The number 7300 causes some confusion because it is impossible to determine whether the zeros are measured values. The number 7300 may have two, three, or four significant digits. When presented with this situation, we assume that the zeros are not significant, unless the context of the problem indicates otherwise.

To determine the number of significant digits for answers in applications of angle measure, use the following table.

Angle Measure to Nearest	Examples	Answer to Number of Significant Digits
Degree	62°, 36°	2
Ten minutes, or nearest tenth of a degree	52° 30′, 60.4°	3
Minute, or nearest hundredth of a degree	81° 48′, 71.25°	4
Ten seconds, or nearest thousandth of a degree	10° 52′ 20″, 21.264°	5

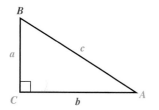

When solving triangles, a labeled sketch is an important aid.

Figure 19

To perform calculations with measured numbers, start by identifying the number with the fewest significant digits. Round your final answer to the same number of significant digits as this number. *Remember that your answer is no more accurate than the least accurate number in your calculation.*

Solving Triangles   To *solve a triangle* means to find the measures of all the angles and sides of the triangle. As shown in Figure 19, we use a to represent the length of the side opposite angle A, b for the length of the side opposite angle B, and so on. In a right triangle, the letter c is reserved for the hypotenuse.

▶ **EXAMPLE 1**   SOLVING A RIGHT TRIANGLE GIVEN AN ANGLE AND A SIDE

Solve right triangle ABC, if $A = 34°\ 30'$ and $c = 12.7$ in. See Figure 20.

Solution   To solve the triangle, find the measures of the remaining sides and angles. To find the value of a, use a trigonometric function involving the known values of angle A and side c. Since the sine of angle A is given by the quotient of the side opposite A and the hypotenuse, use $\sin A$.

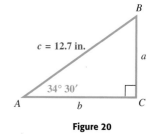

Figure 20

$$\sin A = \frac{a}{c} \qquad \sin A = \frac{\text{side opposite}}{\text{hypotenuse}} \text{ (Section 2.1)}$$

$$\sin 34°\ 30' = \frac{a}{12.7} \qquad A = 34°\ 30',\ c = 12.7$$

$$a = 12.7 \sin 34°\ 30' \qquad \text{Multiply by 12.7; rewrite.}$$

$$a = 12.7 \sin 34.5° \qquad \text{Convert to decimal degrees. (Section 1.1)}$$

$$a = 12.7(.56640624) \qquad \text{Use a calculator.}$$

$$a \approx 7.19 \text{ in.} \qquad \text{Three significant digits}$$

Assuming that $34°\ 30'$ is given to the nearest ten minutes, we rounded the answer to three significant digits.

To find the value of b, we could substitute the value of a just calculated and the given value of c in the Pythagorean theorem. It is better, however, to use the information given in the problem rather than a result just calculated. If a mistake is made in finding a, then b also would be incorrect. Also, rounding more than once may cause the result to be less accurate. Use $\cos A$.

$$\cos A = \frac{b}{c} \qquad \cos A = \frac{\text{side adjacent}}{\text{hypotenuse}} \text{ (Section 2.1)}$$

$$\cos 34°\ 30' = \frac{b}{12.7}$$

$$b = 12.7 \cos 34°\ 30'$$

$$b \approx 10.5 \text{ in.}$$

Once b is found, the Pythagorean theorem can be used as a check.

▼ LOOKING AHEAD TO CALCULUS

The derivatives of the **parametric equations** $x = f(t)$ and $y = g(t)$ often represent the rate of change of physical quantities, such as velocities. When x and y are related by an equation, the derivatives are called **related rates** because a change in one causes a related change in the other. Determining these rates in calculus often requires solving a right triangle.

All that remains to solve triangle ABC is to find the measure of angle B. Since $A + B = 90°$,

$$B = 90° - A \qquad \text{(Section 1.1)}$$
$$B = 89° \, 60' - 34° \, 30' \quad A = 34° \, 30'$$
$$B = 55° \, 30'.$$

NOW TRY EXERCISE 21. ◄

▶ **Note** In Example 1, we could have found the measure of angle B first, and then used the trigonometric function values of B to find the unknown sides. The process of solving a right triangle can usually be done in several ways, each producing the correct answer. *To maintain accuracy, always use given information as much as possible, and avoid rounding off in intermediate steps.*

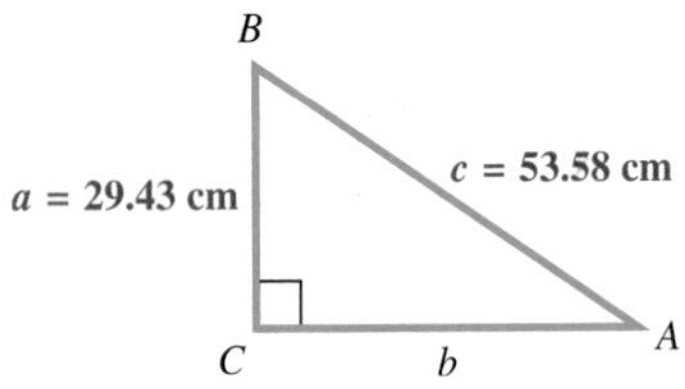

Figure 21

▶ **EXAMPLE 2** SOLVING A RIGHT TRIANGLE GIVEN TWO SIDES

Solve right triangle ABC, if $a = 29.43$ cm and $c = 53.58$ cm.

Solution We draw a sketch showing the given information, as in Figure 21. One way to begin is to find angle A by using the sine function.

$$\sin A = \frac{\text{side opposite}}{\text{hypotenuse}} = \frac{29.43}{53.58} \approx .5492721165 \qquad \text{(Section 2.1)}$$
$$A = \sin^{-1} .5492721165 \approx 33.32° \qquad \text{(Section 2.3)}$$

The measure of B is approximately

$$90° - 33.32° = 56.68°.$$

We now find b from the Pythagorean theorem.

$$b^2 = c^2 - a^2 \qquad \text{Pythagorean theorem solved for } b^2 \text{ (Appendix B)}$$
$$b^2 = 53.58^2 - 29.43^2 \qquad c = 53.58, a = 29.43$$
$$b = \sqrt{2004.6915}$$
$$b \approx 44.77 \text{ cm} \qquad \boxed{\text{Keep only the positive square root.}}$$

NOW TRY EXERCISE 31. ◄

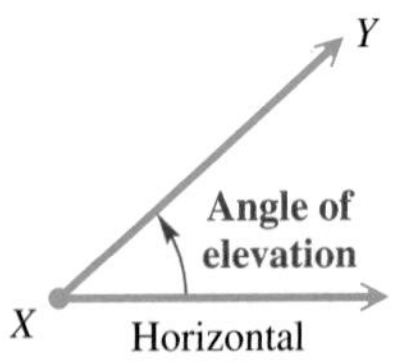

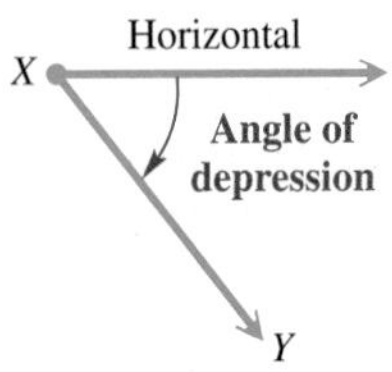

Figure 22

Angles of Elevation or Depression

Many applications of right triangles involve angles of elevation or depression. The **angle of elevation** from point X to point Y (above X) is the acute angle formed by ray XY and a horizontal ray with endpoint at X. See Figure 22(a). The **angle of depression** from point X to point Y (below X) is the acute angle formed by ray XY and a horizontal ray with endpoint X. See Figure 22(b).

> ▶ **Caution** Be careful when interpreting the angle of depression. ***Both the angle of elevation and the angle of depression are measured between the line of sight and a horizontal line.***

To solve applied trigonometry problems, follow the same procedure as solving a triangle.

SOLVING AN APPLIED TRIGONOMETRY PROBLEM

Step 1 Draw a sketch, and label it with the given information. Label the quantity to be found with a variable.

Step 2 Use the sketch to write an equation relating the given quantities to the variable.

Step 3 Solve the equation, and check that your answer makes sense.

Drawing a triangle and labeling it correctly in Step 1 is crucial.

▶ **EXAMPLE 3**   FINDING A LENGTH WHEN THE ANGLE OF ELEVATION IS KNOWN

Shelly McCarthy knows that when she stands 123 ft from the base of a flagpole, the angle of elevation to the top of the flagpole is 26° 40′. If her eyes are 5.30 ft above the ground, find the height of the flagpole.

Solution

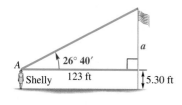

Figure 23

Step 1 The length of the side adjacent to Shelly is known, and the length of the side opposite her must be found. See Figure 23.

Step 2 The tangent ratio involves the given values. Write an equation.

$$\tan A = \frac{\text{side opposite}}{\text{side adjacent}} \qquad \text{Tangent ratio (Section 2.1)}$$

$$\tan 26° 40′ = \frac{a}{123} \qquad A = 26° 40′; \text{ side adjacent} = 123$$

Step 3
$$a = 123 \tan 26° 40′ \qquad \text{Multiply by 123; rewrite.}$$
$$a = 123 \tan 26.66666667° \qquad \text{Convert to decimal degrees.}$$
$$a = 123(.50221888) \qquad \text{Use a calculator.}$$
$$a \approx 61.8 \text{ ft} \qquad \text{Three significant digits}$$

Since Shelly's eyes are 5.30 ft above the ground, the height of the flagpole is

$$61.8 + 5.30 = 67.1 \text{ ft.}$$

NOW TRY EXERCISE 51. ◀

▶ EXAMPLE 4   FINDING THE ANGLE OF ELEVATION WHEN LENGTHS ARE KNOWN

The length of the shadow of a building 34.09 m tall is 37.62 m. Find the angle of elevation of the sun.

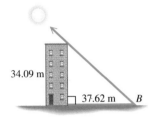

Solution   As shown in Figure 24, the angle of elevation of the sun is angle B. Since the side opposite B and the side adjacent to B are known, use the tangent ratio to find B.

$$\tan B = \frac{34.09}{37.62}, \quad \text{so} \quad B = \tan^{-1}\frac{34.09}{37.62} \approx 42.18°. \quad \text{(Section 2.3)}$$

The angle of elevation of the sun is 42.18°.

34.09 m   37.62 m B

Figure 24

NOW TRY EXERCISE 47. ◀

George Polya (1887–1985)

CONNECTIONS   George Polya proposed an excellent general outline for solving applied problems in his classic book *How to Solve It.*

1. Understand the problem.   **2.** Devise a plan.

3. Carry out the plan.   **4.** Look back.

Polya, a native of Budapest, Hungary, wrote more than 250 papers in many languages, as well as a number of books. He was a brilliant lecturer and teacher, and numerous mathematical properties and theorems bear his name. He once was asked why so many good mathematicians came out of Hungary at the turn of the century. He theorized that it was because mathematics was the cheapest science, requiring no expensive equipment, only pencil and paper.

For Discussion or Writing

Look back at the problem-solving steps given in this section, and compare them to Polya's four steps.

2.4   Exercises

Concept Check   Refer to the discussion of accuracy and significant digits in this section to work Exercises 1–8.

1. *Leading NFL Receiver*   At the end of the 2004 National Football League season, Jerry Rice was the leading career receiver with 22,895 yd. State the range represented by this number. (*Source:* www.nfl.com)

2. *Height of Mt. Everest*   When Mt. Everest was first surveyed, the surveyors obtained a height of 29,000 ft to the nearest foot. State the range represented by this number. (The surveyors thought no one would believe a measurement of 29,000 ft, so they reported it as 29,002.) (*Source:* Dunham, W., *The Mathematical Universe,* John Wiley and Sons, 1994.)

3. *Longest Vehicular Tunnel*   The E. Johnson Memorial Tunnel in Colorado, which measures 8959 ft, is one of the longest land vehicular tunnels in the United States. What is the range of this number? (*Source: The World Almanac and Book of Facts,* 2003.)

4. *Top WNBA Scorer* Women's National Basketball Association player Lauren Jackson of the Seattle Storm received the 2007 award for most points scored, 604. Is it appropriate to consider this number as between 603.5 and 604.5? Why or why not? (*Source:* www.wnba.com)

5. *Circumference of a Circle* The formula for the circumference of a circle is $C = 2\pi r$. Suppose you use the $\boxed{\pi}$ key on your calculator to find the circumference of a circle with radius 54.98 cm, getting 345.44953. Since 2 has only one significant digit, the answer should be given as 3×10^2, or 300 cm. Is this conclusion correct? If not, explain how the answer should be given.

6. Explain the difference between a measurement of 23.0 ft and a measurement of 23.00 ft.

7. If h is the actual height of a building and the height is measured as 58.6 ft, then $|h - 58.6| \leq$ _____.

8. If w is the actual weight of a car and the weight is measured as 1542 lb, then $|w - 1542| \leq$ _____.

Solve each right triangle. When two sides are given, give angles in degrees and minutes. See Example 1.

9.

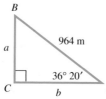

10.

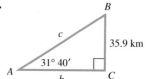

11.

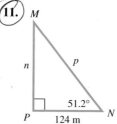

12.

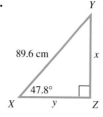

13.

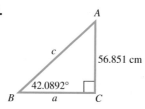

14.

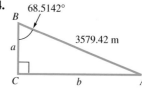

15.

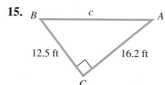

16.

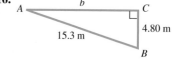

17. Can a right triangle be solved if we are given measures of its two acute angles and no side lengths? Explain.

18. *Concept Check* If we are given an acute angle and a side in a right triangle, what unknown part of the triangle requires the least work to find?

19. Explain why you can always solve a right triangle if you know the measures of one side and one acute angle.

20. Explain why you can always solve a right triangle if you know the lengths of two sides.

Solve each right triangle. In each case, C = 90°. If angle information is given in degrees and minutes, give answers in the same way. If given in decimal degrees, do likewise in answers. When two sides are given, give angles in degrees and minutes. See Examples 1 and 2.

21. $A = 28.0°$, $c = 17.4$ ft   **22.** $B = 46.0°$, $c = 29.7$ m

23. $B = 73.0°$, $b = 128$ in.   **24.** $A = 61.0°$, $b = 39.2$ cm

25. $A = 62.5°$, $a = 12.7$ m   **26.** $B = 51.7°$, $a = 28.1$ ft

27. $a = 13$ m, $c = 22$ m   **28.** $b = 32$ ft, $c = 51$ ft

29. $a = 76.4$ yd, $b = 39.3$ yd   **30.** $a = 958$ m, $b = 489$ m

31. $a = 18.9$ cm, $c = 46.3$ cm   **32.** $b = 219$ m, $c = 647$ m

33. $A = 53° 24'$, $c = 387.1$ ft   **34.** $A = 13° 47'$, $c = 1285$ m

35. $B = 39° 9'$, $c = .6231$ m   **36.** $B = 82° 51'$, $c = 4.825$ cm

37. Explain the meaning of *angle of elevation*.

38. *Concept Check*   Can an angle of elevation be more than 90°?

39. Explain why the angle of depression *DAB* has the same measure as the angle of elevation *ABC* in the figure.

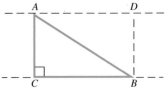

AD is parallel to *BC*.

40. Why is angle *CAB not* an angle of depression in the figure for Exercise 39?

Solve each problem involving triangles. See Examples 1–4.

41. *Height of a Ladder on a Wall*   A 13.5-m fire truck ladder is leaning against a wall. Find the distance *d* the ladder goes up the wall (above the top of the fire truck) if the ladder makes an angle of 43° 50′ with the horizontal.

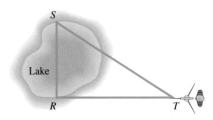

42. *Distance Across a Lake*   To find the distance *RS* across a lake, a surveyor lays off length $RT = 53.1$ m, so that angle $T = 32° 10'$ and angle $S = 57° 50'$. Find length *RS*.

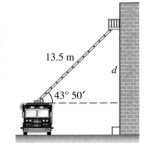

43. *Height of a Building*   From a window 30.0 ft above the street, the angle of elevation to the top of the building across the street is 50.0° and the angle of depression to the base of this building is 20.0°. Find the height of the building across the street.

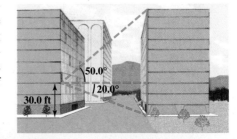

44. *Diameter of the Sun* To determine the diameter of the sun, an astronomer might sight with a **transit** (a device used by surveyors for measuring angles) first to one edge of the sun and then to the other, estimating that the included angle equals 32′. Assuming that the distance d from Earth to the sun is 92,919,800 mi, approximate the diameter of the sun.

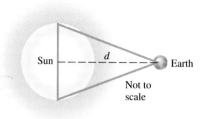

45. *Side Lengths of a Triangle* The length of the base of an isosceles triangle is 42.36 in. Each base angle is 38.12°. Find the length of each of the two equal sides of the triangle. (*Hint:* Divide the triangle into two right triangles.)

46. *Altitude of a Triangle* Find the altitude of an isosceles triangle having base 184.2 cm if the angle opposite the base is 68° 44′.

Solve each problem involving an angle of elevation or depression. See Examples 3 and 4.

47. *Angle of Elevation of Pyramid of the Sun* The Pyramid of the Sun in the ancient Mexican city of Teotihuacan was the largest and most important structure in the city. The base is a square with sides 700 ft long, and the height of the pyramid is 200 ft. Find the angle of elevation of the edge indicated in the figure to two significant digits. (*Hint:* The base of the triangle in the figure is half the diagonal of the square base of the pyramid.)

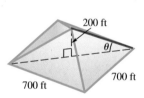

48. *Cloud Ceiling* The U.S. Weather Bureau defines a **cloud ceiling** as the altitude of the lowest clouds that cover more than half the sky. To determine a cloud ceiling, a powerful searchlight projects a circle of light vertically on the bottom of the cloud. An observer sights the circle of light in the crosshairs of a tube called a **clinometer.** A pendant hanging vertically from the tube and resting on a protractor gives the angle of elevation. Find the cloud ceiling if the searchlight is located 1000 ft from the observer and the angle of elevation is 30.0° as measured with a clinometer at eye-height 6 ft. (Assume three significant digits.)

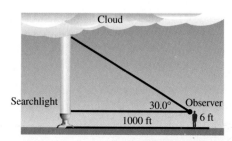

49. *Height of a Tower*   The shadow of a vertical tower is 40.6 m long when the angle of elevation of the sun is 34.6°. Find the height of the tower.

50. *Distance from the Ground to the Top of a Building*   The angle of depression from the top of a building to a point on the ground is 32° 30′. How far is the point on the ground from the top of the building if the building is 252 m high?

51. *Length of a Shadow*   Suppose that the angle of elevation of the sun is 23.4°. Find the length of the shadow cast by Diane Carr, who is 5.75 ft tall.

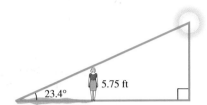

23.4°   5.75 ft

52. *Airplane Distance*   An airplane is flying 10,500 ft above the level ground. The angle of depression from the plane to the base of a tree is 13° 50′. How far horizontally must the plane fly to be directly over the tree?

10,500 ft

53. *Height of a Building*   The angle of elevation from the top of a small building to the top of a nearby taller building is 46° 40′, while the angle of depression to the bottom is 14° 10′. If the shorter building is 28.0 m high, find the height of the taller building.

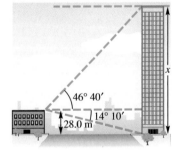

46° 40′
14° 10′
28.0 m
x

54. *Angle of Depression of a Light*   A company safety committee has recommended that a floodlight be mounted in a parking lot so as to illuminate the employee exit. Find the angle of depression of the light.

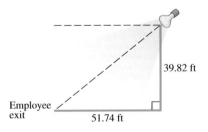

39.82 ft

Employee exit
51.74 ft

55. *Height of Mt. Everest*   The highest mountain peak in the world is Mt. Everest, located in the Himalayas. The height of this enormous mountain was determined in 1856 by surveyors using trigonometry long before it was first climbed in 1953. This difficult measurement had to be done from a great distance. At an altitude of 14,545 ft on a different mountain, the straight line distance to the peak of Mt. Everest is 27.0134 mi and its angle of elevation is $\theta = 5.82°$. (*Source:* Dunham, W., *The Mathematical Universe,* John Wiley and Sons, 1994.)

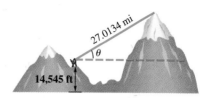

(a) Approximate the height (in feet) of Mt. Everest.

(b) In the actual measurement, Mt. Everest was over 100 mi away and the curvature of Earth had to be taken into account. Would the curvature of Earth make the peak appear taller or shorter than it actually is?

56. *Error in Measurement*   A degree may seem like a very small unit, but an error of one degree in measuring an angle may be very significant. For example, suppose a laser beam directed toward the visible center of the moon misses its assigned target by 30 sec. How far is it (in miles) from its assigned target? Take the distance from the surface of Earth to that of the moon to be 234,000 mi. (*Source: A Sourcebook of Applications of School Mathematics* by Donald Bushaw et al. Copyright © 1980 by The Mathematical Association of America.)

2.5 Further Applications of Right Triangles

Bearing ▪ Further Applications

Bearing   Other applications of right triangles involve **bearing,** an important concept in navigation. There are two methods for expressing bearing. When a single angle is given, such as 164°, it is understood that the bearing is measured in a clockwise direction from due north. Several sample bearings using this first method are shown in Figure 25.

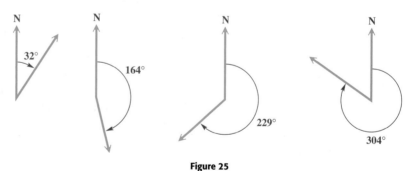

Figure 25

▶ **EXAMPLE 1**   SOLVING A PROBLEM INVOLVING BEARING (FIRST METHOD)

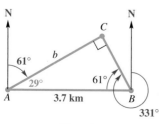

Figure 26

Radar stations A and B are on an east-west line, 3.7 km apart. Station A detects a plane at C, on a bearing of 61°. Station B simultaneously detects the same plane, on a bearing of 331°. Find the distance from A to C.

Solution   Draw a sketch showing the given information, as in Figure 26. Since a line drawn due north is perpendicular to an east-west line, right angles are formed at A and B, so angles CAB and CBA can be found as shown in Figure 26. Angle C is a right angle because angles CAB and CBA are complementary. Find distance b by using the cosine function for angle A.

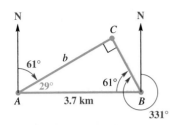

Figure 26 (repeated)

$$\cos 29° = \frac{b}{3.7} \qquad \text{(Section 2.1)}$$

$$3.7 \cos 29° = b \qquad \text{Multiply by 3.7.}$$

$$b \approx 3.2 \text{ km} \qquad \text{Use a calculator; round to the nearest tenth.}$$

NOW TRY EXERCISE 15. ◀

▶ **Caution** *A correctly labeled sketch is crucial* when solving applications like that in Example 1. Some of the necessary information is often not directly stated in the problem and can be determined only from the sketch.

The second method for expressing bearing starts with a north-south line and uses an acute angle to show the direction, either east or west, from this line. Figure 27 shows several sample bearings using this system. Either N or S always comes first, followed by an acute angle, and then E or W.

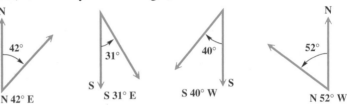

Figure 27

▶ **EXAMPLE 2** SOLVING A PROBLEM INVOLVING BEARING (SECOND METHOD)

The bearing from *A* to *C* is S 52° E. The bearing from *A* to *B* is N 84° E. The bearing from *B* to *C* is S 38° W. A plane flying at 250 mph takes 2.4 hr to go from *A* to *B*. Find the distance from *A* to *C*.

Solution Make a sketch. First draw the two bearings from point *A*. Choose a point *B* on the bearing N 84° E from *A*, and draw the bearing to *C*. Point *C* will be located where the bearing lines from *A* and *B* intersect, as shown in Figure 28.

Since the bearing from *A* to *B* is N 84° E, angle *ABD* is 180° − 84° = 96°. Thus, angle *ABC* is 180° − (96° + 38°) = 46°. Also, angle *BAC* is 180° − (84° + 52°) = 44°. Angle *C* is 180° − (44° + 46°) = 90°. Since a plane flying at 250 mph takes 2.4 hr to go from *A* to *B*, the distance from *A* to *B* is

$$c = \text{rate} \times \text{time} = 250(2.4) = 600 \text{ mi.}$$

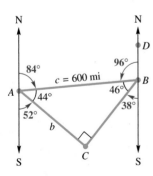

Figure 28

To find *b*, the distance from *A* to *C*, use the sine. (The cosine could also be used.)

$$\sin 46° = \frac{b}{c} \qquad \text{(Section 2.1)}$$

$$\sin 46° = \frac{b}{600} \qquad \text{Let } c = 600.$$

$$600 \sin 46° = b \qquad \text{Multiply by 600.}$$

$$b \approx 430 \text{ mi}$$

NOW TRY EXERCISE 19. ◀

Further Applications

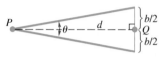

Figure 29

▶ **EXAMPLE 3** USING TRIGONOMETRY TO MEASURE A DISTANCE

A method that surveyors use to determine a small distance d between two points P and Q is called the **subtense bar method.** The subtense bar with length b is centered at Q and situated perpendicular to the line of sight between P and Q. See Figure 29. Angle θ is measured, and then the distance d can be determined.

(a) Find d when $\theta = 1° \, 23' \, 12''$ and $b = 2.0000$ cm.

(b) Angle θ usually cannot be measured more accurately than to the nearest $1''$. How much change would there be in the value of d if θ were measured $1''$ larger?

Solution

(a) From Figure 29, we see that

$$\cot \frac{\theta}{2} = \frac{d}{\frac{b}{2}}$$

$$d = \frac{b}{2} \cot \frac{\theta}{2}. \quad \text{\footnotesize Multiply; rewrite.}$$

Let $b = 2$. To evaluate $\frac{\theta}{2}$, we change θ to decimal degrees.

$$1° \, 23' \, 12'' \approx 1.386667° \quad \text{(Section 1.1)}$$

Then

$$d = \frac{2}{2} \cot \frac{1.386667°}{2} \approx 82.634110 \text{ cm.}$$

(b) Since θ is $1''$ larger, use $\theta = 1° \, 23' \, 13'' \approx 1.386944°$.

$$d = \frac{2}{2} \cot \frac{1.386944°}{2} \approx 82.617558 \text{ cm}$$

The difference is $82.634110 - 82.617558 = .016552$ cm.

NOW TRY EXERCISE 33. ◀

▶ **EXAMPLE 4** SOLVING A PROBLEM INVOLVING ANGLES OF ELEVATION

Francisco needs to know the height of a tree. From a given point on the ground, he finds that the angle of elevation to the top of the tree is $36.7°$. He then moves back 50 ft. From the second point, the angle of elevation to the top of the tree is $22.2°$. See Figure 30. Find the height of the tree to the nearest foot.

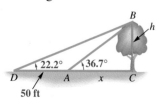

Figure 30

Algebraic Solution

Figure 30 shows two unknowns: x, the distance from the center of the trunk of the tree to the point where the first observation was made, and h, the height of the tree. See Figure 31 in the Graphing Calculator Solution. Since nothing is given about the length of the hypotenuse of either triangle ABC or triangle BCD, use a ratio that does not involve the hypotenuse—namely, the tangent.

In triangle ABC, $\quad \tan 36.7° = \dfrac{h}{x} \quad$ or $\quad h = x \tan 36.7°$.

In triangle BCD, $\quad \tan 22.2° = \dfrac{h}{50 + x} \quad$ or $\quad h = (50 + x) \tan 22.2°$.

Each expression equals h, so the expressions must be equal.

$$x \tan 36.7° = (50 + x) \tan 22.2°$$
Solve for x.

$$x \tan 36.7° = 50 \tan 22.2° + x \tan 22.2°$$
Distributive property

$$x \tan 36.7° - x \tan 22.2° = 50 \tan 22.2°$$
Get x-terms on one side.

$$x(\tan 36.7° - \tan 22.2°) = 50 \tan 22.2°$$
Factor out x.

$$x = \frac{50 \tan 22.2°}{\tan 36.7° - \tan 22.2°}$$
Divide by the coefficient of x.

We saw above that $h = x \tan 36.7°$. Substituting for x,

$$h = \left(\frac{50 \tan 22.2°}{\tan 36.7° - \tan 22.2°}\right) \tan 36.7°.$$

Using a calculator,

$$\tan 36.7° = .74537703 \quad \text{and} \quad \tan 22.2° = .40809244,$$

so $\quad \tan 36.7° - \tan 22.2° = .74537703 - .40809244 = .33728459$

and $\quad h = \left(\dfrac{50(.40809244)}{.33728459}\right).74537703 \approx 45.$

The height of the tree is approximately 45 ft.

Graphing Calculator Solution*

In Figure 31, we superimposed Figure 30 on coordinate axes with the origin at D. By definition, the tangent of the angle between the x-axis and the graph of a line with equation $y = mx + b$ is the slope of the line, m. For line DB, $m = \tan 22.2°$. Since b equals 0, the equation of line DB is $y_1 = (\tan 22.2°)x$. The equation of line AB is $y_2 = (\tan 36.7°)x + b$. Since $b \neq 0$ here, we use the point $A(50, 0)$ and the point-slope form to find the equation.

$$y_2 - y_1 = m(x - x_1)$$
$$y_2 - 0 = m(x - 50) \quad x_1 = 50, y_1 = 0$$
$$y_2 = \tan 36.7°(x - 50)$$

Lines y_1 and y_2 are graphed in Figure 32. The y-coordinate of the point of intersection of the graphs gives the length of BC, or h. Thus, $h \approx 45$.

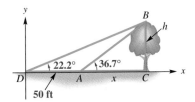

Figure 31

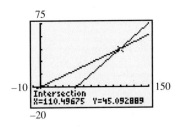

Figure 32

NOW TRY EXERCISE 27. ◀

▶ **Note** In practice, we usually do not write down intermediate calculator approximation steps. We did in Example 4 so you could follow the steps more easily.

Source: Adapted with permission from "Letter to the Editor," by Robert Ruzich (*Mathematics Teacher,* Volume 88, Number 1). Copyright © 1995 by the National Council of Teachers of Mathematics.

2.5 Exercises

Concept Check   Give a short written answer to each question.

1. When bearing is given as a single angle measure, how is the angle represented in a sketch?

2. When bearing is given as N (or S), then the angle measure, and then E (or W), how is the angle represented in a sketch?

3. Why is it important to draw a sketch before solving trigonometric problems like those in the last two sections of this chapter?

4. How should the angle of elevation (or depression) from a point *X* to a point *Y* be represented?

Concept Check   An observer for a radar station is located at the origin of a coordinate system. For each of the points in Exercises 5–12, find the bearing of an airplane located at that point. Express the bearing using both methods.

5. $(-4, 0)$ **6.** $(-3, -3)$ **7.** $(-5, 5)$ **8.** $(0, -2)$

9. $(0, 4)$ **10.** $(2, 2)$ **11.** $(2, -2)$ **12.** $(5, 0)$

13. The ray $y = x$, $x \geq 0$, contains the origin and all points in the coordinate system whose bearing is 45°. Determine the equation of a ray consisting of the origin and all points whose bearing is 240°.

14. Repeat Exercise 13 for a bearing of 150°.

Work each problem. In these exercises, assume the course of a plane or ship is on the indicated bearing. See Examples 1 and 2.

15. *Distance Flown by a Plane*   A plane flies 1.3 hr at 110 mph on a bearing of 38°. It then turns and flies 1.5 hr at the same speed on a bearing of 128°. How far is the plane from its starting point?

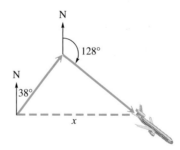

16. *Distance Traveled by a Ship*   A ship travels 55 km on a bearing of 27°, then travels on a bearing of 117° for 140 km. Find the distance traveled from the starting point to the ending point.

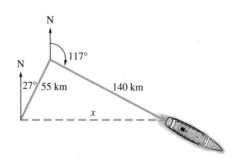

17. *Distance Between Two Ships*   Two ships leave a port at the same time. The first ship sails on a bearing of 40° at 18 knots (nautical miles per hour) and the second at a bearing of 130° at 26 knots. How far apart are they after 1.5 hr?

18. *Distance Between Two Lighthouses*   Two lighthouses are located on a north-south line. From lighthouse *A*, the bearing of a ship 3742 m away is 129° 43′. From lighthouse *B*, the bearing of the ship is 39° 43′. Find the distance between the lighthouses.

19. *Distance Between Two Cities*   The bearing from Winston-Salem, North Carolina, to Danville, Virginia, is N 42° E. The bearing from Danville to Goldsboro, North Carolina, is S 48° E. A car driven by Mark Ferrari, traveling at 65 mph, takes 1.1 hr to go from Winston-Salem to Danville and 1.8 hr to go from Danville to Goldsboro. Find the distance from Winston-Salem to Goldsboro.

20. *Distance Between Two Cities*   The bearing from Atlanta to Macon is S 27° E, and the bearing from Macon to Augusta is N 63° E. An automobile traveling at 62 mph needs $1\frac{1}{4}$ hr to go from Atlanta to Macon and $1\frac{3}{4}$ hr to go from Macon to Augusta. Find the distance from Atlanta to Augusta.

21. *Distance Between Two Ships*   A ship leaves its home port and sails on a bearing of S 61° 50′ E. Another ship leaves the same port at the same time and sails on a bearing of N 28° 10′ E. If the first ship sails at 24.0 mph and the second sails at 28.0 mph, find the distance between the two ships after 4 hr.

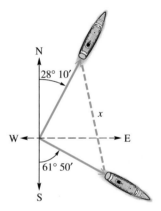

22. *Distance Between Transmitters*   Radio direction finders are set up at two points *A* and *B*, which are 2.50 mi apart on an east-west line. From *A*, it is found that the bearing of a signal from a radio transmitter is N 36° 20′ E, while from *B* the bearing of the same signal is N 53° 40′ W. Find the distance of the transmitter from *B*.

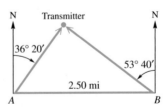

23. Solve the equation $ax = b + cx$ for *x* in terms of *a*, *b*, and *c*. (*Note:* This is in essence the calculation carried out in Example 4.)

24. Explain why the line $y = (\tan \theta)(x - a)$ passes through the point $(a, 0)$ and makes an angle θ with the *x*-axis.

25. Find the equation of the line passing through the point $(25, 0)$ that makes an angle of 35° with the *x*-axis.

26. Find the equation of the line passing through the point $(5, 0)$ that makes an angle of 15° with the *x*-axis.

In Exercises 27–32, use the method of Example 4.

27. Find *h* as indicated in the figure.

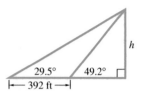

28. Find *h* as indicated in the figure.

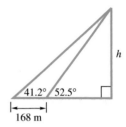

29. *Height of a Pyramid* The angle of elevation from a point on the ground to the top of a pyramid is 35° 30′. The angle of elevation from a point 135 ft farther back to the top of the pyramid is 21° 10′. Find the height of the pyramid.

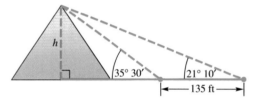

30. *Distance Between a Whale and a Lighthouse* Debbie Glockner-Ferrari, a whale researcher, is watching a whale approach directly toward a lighthouse as she observes from the top of this lighthouse. When she first begins watching the whale, the angle of depression to the whale is 15° 50′. Just as the whale turns away from the lighthouse, the angle of depression is 35° 40′. If the height of the lighthouse is 68.7 m, find the distance traveled by the whale as it approaches the lighthouse.

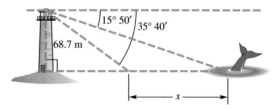

31. *Height of an Antenna* A scanner antenna is on top of the center of a house. The angle of elevation from a point 28.0 m from the center of the house to the top of the antenna is 27° 10′, and the angle of elevation to the bottom of the antenna is 18° 10′. Find the height of the antenna.

32. *Height of Mt. Whitney* The angle of elevation from Lone Pine to the top of Mt. Whitney is 10° 50′. Van Dong Le, traveling 7.00 km from Lone Pine along a straight, level road toward Mt. Whitney, finds the angle of elevation to be 22° 40′. Find the height of the top of Mt. Whitney above the level of the road.

Solve each problem.

33. *(Modeling) Distance Between Two Points* Refer to Example 3. A variation of the subtense bar method that surveyors use to determine larger distances d between two points P and Q is shown in the figure. In this case the subtense bar with length b is placed between the points P and Q so that the bar is centered on and perpendicular to the line of sight connecting P and Q. The angles α and β are measured from points P and Q, respectively. (*Source:* Mueller, I. and K. Ramsayer, *Introduction to Surveying,* Frederick Ungar Publishing Co., 1979.)

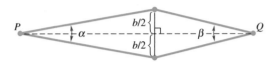

(a) Find a formula for d involving α, β, and b.

(b) Use your formula to determine d if $\alpha = 37'\ 48''$, $\beta = 42'\ 3''$, and $b = 2.000$ cm.

34. *Height of a Plane Above Earth* Find the minimum height h above the surface of Earth so that a pilot at point A in the figure can see an object on the horizon at C, 125 mi away. Assume that the radius of Earth is 4.00×10^3 mi.

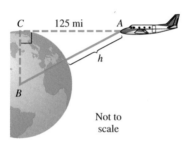

35. *Distance of a Plant from a Fence* In one area, the lowest angle of elevation of the sun in winter is $23°\ 20'$. Find the minimum distance x that a plant needing full sun can be placed from a fence 4.65 ft high.

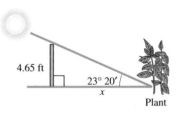

36. *Distance Through a Tunnel* A tunnel is to be dug from A to B. Both A and B are visible from C. If AC is 1.4923 mi and BC is 1.0837 mi, and if C is $90°$, find the measures of angles A and B.

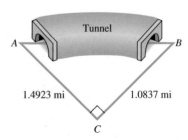

37. *(Modeling) Highway Curves*   A basic highway curve connecting two straight sections of road is often circular. In the figure, the points *P* and *S* mark the beginning and end of the curve. Let *Q* be the point of intersection where the two straight sections of highway leading into the curve would meet if extended. The radius of the curve is *R*, and the central angle θ denotes how many degrees the curve turns. (*Source:* Mannering, F. and W. Kilareski, *Principles of Highway Engineering and Traffic Analysis,* Second Edition, John Wiley and Sons, 1998.)

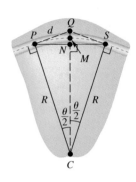

(a) If *R* = 965 ft and $\theta = 37°$, find the distance *d* between *P* and *Q*.

(b) Find an expression in terms of *R* and θ for the distance between points *M* and *N*.

38. *Length of a Side of a Piece of Land*   A piece of land has the shape shown in the figure. Find *x*.

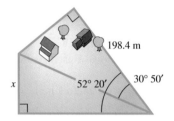

39. *(Modeling) Stopping Distance on a Curve*   Refer to Exercise 37. When an automobile travels along a circular curve, objects like trees and buildings situated on the inside of the curve can obstruct the driver's vision. These obstructions prevent the driver from seeing sufficiently far down the highway to ensure a safe stopping distance. In the figure, the *minimum* distance *d* that should be cleared on the inside of the highway is modeled by the equation

$$d = R\left(1 - \cos\frac{\theta}{2}\right).$$

(*Source:* Mannering, F. and W. Kilareski, *Principles of Highway Engineering and Traffic Analysis,* Second Edition, John Wiley and Sons, 1998.)

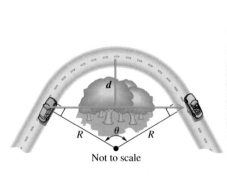

Not to scale

(a) It can be shown that if θ is measured in degrees, then $\theta \approx \frac{57.3S}{R}$, where *S* is the safe stopping distance for the given speed limit. Compute *d* to the nearest foot for a 55 mph speed limit if *S* = 336 ft and *R* = 600 ft.

(b) Compute *d* to the nearest foot for a 65 mph speed limit if *S* = 485 ft and *R* = 600 ft.

(c) How does the speed limit affect the amount of land that should be cleared on the inside of the curve?

Chapter 2 Summary

KEY TERMS

2.1	side opposite side adjacent cofunctions	**2.2** **2.4**	reference angle exact number significant digits	angle of elevation angle of depression	**2.5** bearing

QUICK REVIEW

CONCEPTS	EXAMPLES

2.1 Trigonometric Functions of Acute Angles

Right-Triangle-Based Definitions of the Trigonometric Functions

For any acute angle A in standard position,

$$\sin A = \frac{y}{r} = \frac{\text{side opposite}}{\text{hypotenuse}} \qquad \csc A = \frac{r}{y} = \frac{\text{hypotenuse}}{\text{side opposite}}$$

$$\cos A = \frac{x}{r} = \frac{\text{side adjacent}}{\text{hypotenuse}} \qquad \sec A = \frac{r}{x} = \frac{\text{hypotenuse}}{\text{side adjacent}}$$

$$\tan A = \frac{y}{x} = \frac{\text{side opposite}}{\text{side adjacent}} \qquad \cot A = \frac{x}{y} = \frac{\text{side adjacent}}{\text{side opposite}}.$$

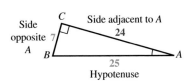

$$\sin A = \frac{7}{25} \qquad \cos A = \frac{24}{25} \qquad \tan A = \frac{7}{24}$$

$$\csc A = \frac{25}{7} \qquad \sec A = \frac{25}{24} \qquad \cot A = \frac{24}{7}$$

Cofunction Identities

For any acute angle A,

$$\sin A = \cos(90° - A) \qquad \cos A = \sin(90° - A)$$

$$\sec A = \csc(90° - A) \qquad \csc A = \sec(90° - A)$$

$$\tan A = \cot(90° - A) \qquad \cot A = \tan(90° - A).$$

$$\sin 55° = \cos(90° - 55°) = \cos 35°$$

$$\sec 48° = \csc(90° - 48°) = \csc 42°$$

$$\tan 72° = \cot(90° - 72°) = \cot 18°$$

Function Values of Special Angles

θ	$\sin \theta$	$\cos \theta$	$\tan \theta$	$\cot \theta$	$\sec \theta$	$\csc \theta$
30°	$\dfrac{1}{2}$	$\dfrac{\sqrt{3}}{2}$	$\dfrac{\sqrt{3}}{3}$	$\sqrt{3}$	$\dfrac{2\sqrt{3}}{3}$	2
45°	$\dfrac{\sqrt{2}}{2}$	$\dfrac{\sqrt{2}}{2}$	1	1	$\sqrt{2}$	$\sqrt{2}$
60°	$\dfrac{\sqrt{3}}{2}$	$\dfrac{1}{2}$	$\sqrt{3}$	$\dfrac{\sqrt{3}}{3}$	2	$\dfrac{2\sqrt{3}}{3}$

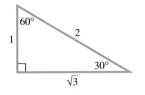

CONCEPTS	EXAMPLES

2.2 Trigonometric Functions of Non-Acute Angles

Reference Angle θ' for θ in $(0°, 360°)$

θ in Quadrant	I	II	III	IV
θ' is	θ	$180° - \theta$	$\theta - 180°$	$360° - \theta$

See the figure on page 60 for illustrations of reference angles.

Finding Trigonometric Function Values for Any Nonquadrantal Angle θ

Step 1   Add or subtract 360° as many times as needed to get an angle greater than 0° but less than 360°.

Step 2   Find the reference angle θ'.

Step 3   Find the trigonometric function values for θ'.

Step 4   Determine the correct signs for the values found in Step 3.

Quadrant I: For $\theta = 25°$, $\theta' = 25°$
Quadrant II: For $\theta = 152°$, $\theta' = 28°$
Quadrant III: For $\theta = 200°$, $\theta' = 20°$
Quadrant IV: For $\theta = 320°$, $\theta' = 40°$

Find sin 1050°.

$$1050° - 2(360°) = 330°$$

Thus, $\theta' = 30°$.

$$\sin 1050° = -\sin 30° = -\frac{1}{2}$$

2.3 Finding Trigonometric Function Values Using a Calculator

To approximate a trigonometric function value of an angle in degrees, make sure your calculator is in degree mode.

Approximate each value.

$$\cos 50° \ 15' = \cos 50.25° \approx .63943900$$

$$\csc 32.5° = \frac{1}{\sin 32.5°} \approx 1.86115900$$

To find the corresponding angle measure given a trigonometric function value, use an appropriate inverse function.

Find an angle θ in the interval $[0°, 90°]$ that satisfies each condition.

$$\cos \theta \approx .73677482$$

$$\theta \approx \cos^{-1}(.73677482)$$

$$\theta \approx 42.542600°$$

$$\csc \theta \approx 1.04766792$$

$$\sin \theta \approx \frac{1}{1.04766792} \qquad \sin \theta = \frac{1}{\csc \theta}$$

$$\theta \approx \sin^{-1}\left(\frac{1}{1.04766792}\right)$$

$$\theta \approx 72.65°$$

(continued)

CONCEPTS	EXAMPLES

2.4 Solving Right Triangles

Solving an Applied Trigonometry Problem

Step 1   Draw a sketch, and label it with the given information. Label the quantity to be found with a variable.

Find the angle of elevation of the sun if a 48.6-ft flagpole casts a shadow 63.1 ft long.

Step 1   See the sketch. We must find θ.

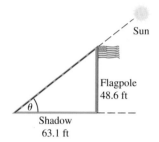

Step 2   Use the sketch to write an equation relating the given quantities to the variable.

Step 3   Solve the equation, and check that your answer makes sense.

Step 2 $\tan\theta = \dfrac{48.6}{63.1} \approx .770206$

Step 3 $\theta = \tan^{-1} .770206 \approx 37.6°$

The angle of elevation rounded to three significant digits is 37.6°, or 37° 40′.

2.5 Further Applications of Right Triangles

Expressing Bearing

Method 1   When a single angle is given, such as 220°, this bearing is measured in a clockwise direction from north.

Method 2   Start with a north-south line and use an acute angle to show direction, either east or west, from this line.

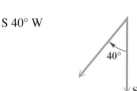

Review Exercises

Find the values of the six trigonometric functions for each angle A.

1.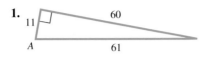

2.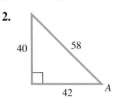

Find one solution for each equation. Assume that all angles are acute angles.

3. $\sin 4\beta = \cos 5\beta$

4. $\sec(2\theta + 10°) = \csc(4\theta + 20°)$

5. $\tan(5x + 11°) = \cot(6x + 2°)$

6. $\cos\left(\dfrac{3\theta}{5} + 11°\right) = \sin\left(\dfrac{7\theta}{10} + 40°\right)$

Tell whether each statement is true *or* false. *If false, tell why.*

7. $\sin 46° < \sin 58°$

8. $\cos 47° < \cos 58°$

9. $\sec 48° \geq \cos 42°$

10. $\sin 22° \geq \csc 68°$

11. Explain why, in the figure, the cosine of angle A is equal to the sine of angle B.

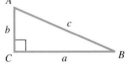

Find exact values of the six trigonometric functions for each angle. Do not use a calculator. Rationalize denominators when applicable.

12. $120°$

13. $1020°$

14. $-225°$

15. $-1470°$

Find all values of θ, if θ is in the interval $[0°, 360°)$ and θ has the given function value.

16. $\sin \theta = -\dfrac{1}{2}$

17. $\cos \theta = -\dfrac{1}{2}$

18. $\cot \theta = -1$

19. $\sec \theta = -\dfrac{2\sqrt{3}}{3}$

Evaluate each expression. Give exact values.

20. $\cos 60° + 2 \sin^2 30°$

21. $\tan^2 120° - 2 \cot 240°$

22. $\sec^2 300° - 2 \cos^2 150° + \tan 45°$

23. Find the sine, cosine, and tangent function values for each angle.

(a)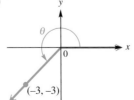

(b)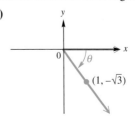

Use a calculator to find each value.

24. sin 72° 30′   **25.** sec 222° 30′   **26.** cot 305.6°

27. csc 78° 21′   **28.** sec 58.9041°   **29.** tan 11.7689°

30. *Concept Check*   Which one of the following cannot be *exactly* determined using the methods of this chapter?

     A. cos 135°   **B.** cot(−45°)   **C.** sin 300°   **D.** tan 140°

Use a calculator to find each value of θ, where θ is in the interval [0°, 90°). *Give answers in decimal degrees.*

31. sin θ = .82584121   **32.** cot θ = 1.1249386   **33.** cos θ = .97540415

34. sec θ = 1.2637891   **35.** tan θ = 1.9633124   **36.** csc θ = 9.5670466

Find two angles in the interval [0°, 360°) *that satisfy each of the following. Leave answers in decimal degrees rounded to the nearest tenth.*

37. sin θ = .73254290   **38.** tan θ = 1.3865342

Tell whether each statement is true *or* false. *If false, tell why. Use a calculator for Exercises 39 and 42.*

39. sin 50° + sin 40° = sin 90°

40. cos 210° = cos 180° · cos 30° − sin 180° · sin 30°

41. sin 240° = 2 sin 120° · cos 120°

42. sin 42° + sin 42° = sin 84°

43. A student wants to use a calculator to find the value of cot 25°. However, instead of entering $\frac{1}{\tan 25}$, he enters $\tan^{-1} 25$. Assuming the calculator is in degree mode, will this produce the correct answer? Explain.

For each angle θ, use a calculator to find cos θ *and* sin θ. *Use your results to decide in which quadrant the angle lies.*

44. θ = 2976°   **45.** θ = 1997°   **46.** θ = 4000°

Solve each right triangle. In Exercise 48, give angles to the nearest minute. In Exercises 49 and 50, label the triangle ABC as in Exercises 47 and 48.

47.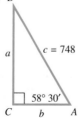

48.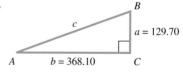

49. *A* = 39.72°, *b* = 38.97 m   **50.** *B* = 47° 53′, *b* = 298.6 m

Solve each problem. (Source for Exercises 51 and 52: Parker, M., Editor, *She Does Math,* Mathematical Association of America, 1995.)

51. *Height of a Tree*   A civil engineer must determine the height of the tree shown in the figure. The given angle was measured with a **clinometer.** Find the height of the tree to the nearest whole number.

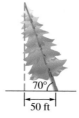

This is a picture of one type of clinometer, called an Abney hand level and clinometer. (Courtesy of Keuffel & Esser Co.)

52. *(Modeling) Double Vision*   To correct mild double vision, a small amount of prism is added to a patient's eyeglasses. The amount of light shift this causes is measured in **prism diopters.** A patient needs 12 prism diopters horizontally and 5 prism diopters vertically. A prism that corrects for both requirements should have length r and be set at angle θ. Find the values of r and θ in the figure.

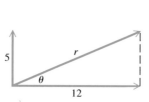

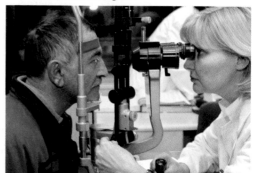

53. *Height of a Tower*   The angle of elevation from a point 93.2 ft from the base of a tower to the top of the tower is 38° 20′. Find the height of the tower.

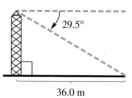

54. *Height of a Tower*   The angle of depression of a television tower to a point on the ground 36.0 m from the bottom of the tower is 29.5°. Find the height of the tower.

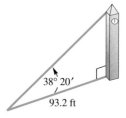

55. *Length of a Diagonal*   One side of a rectangle measures 15.24 cm. The angle between the diagonal and that side is 35.65°. Find the length of the diagonal.

56. *Length of Sides of an Isosceles Triangle*   An isosceles triangle has a base of length 49.28 m. The angle opposite the base is 58.746°. Find the length of each of the two equal sides.

57. *Distance Between Two Points* The bearing of point B from point C is 254°. The bearing of point A from point C is 344°. The bearing of point A from point B is 32°. If the distance from A to C is 780 m, find the distance from A to B.

58. *Distance a Ship Sails* The bearing from point A to point B is S 55° E and from point B to point C is N 35° E. If a ship sails from A to B, a distance of 81 km, and then from B to C, a distance of 74 km, how far is it from A to C?

59. *Distance Between Two Points* Two cars leave an intersection at the same time. One heads due south at 55 mph. The other travels due west. After 2 hr, the bearing of the car headed west from the car headed south is 324°. How far apart are they at that time?

60. Find a formula for h in terms of k, A, and B. Assume $A < B$.

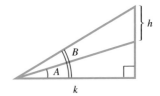

61. Make up a right triangle problem whose solution is 3 tan 25°.

62. Make up a right triangle problem whose solution is found from $\sin \theta = \frac{3}{4}$.

63. *(Modeling) Height of a Satellite* Artificial satellites that orbit Earth often use VHF signals to communicate with the ground. VHF signals travel in straight lines. The height h of the satellite above Earth and the time T that the satellite can communicate with a fixed location on the ground are related by the model

$$h = R\left(\frac{1}{\cos \frac{180T}{P}} - 1\right),$$

where $R = 3955$ mi is the radius of Earth and P is the period for the satellite to orbit Earth. (*Source:* Schlosser, W., T. Schmidt-Kaler, and E. Milone, *Challenges of Astronomy*, Springer-Verlag, 1991.)

(a) Find h to the nearest mile when $T = 25$ min and $P = 140$ min. (Evaluate the cosine function in degree mode.)

(b) What is the value of h to the nearest mile if T is increased to 30 min?

64. *(Modeling) Fundamental Surveying Problem* The first fundamental problem of surveying is to determine the coordinates of a point Q given the coordinates of a point P, the distance between P and Q, and the bearing θ from P to Q. See the figure. (*Source:* Mueller, I. and K. Ramsayer, *Introduction to Surveying*, Frederick Ungar Publishing Co., 1979.)

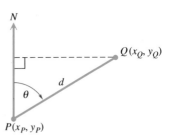

(a) Find a formula for the coordinates (x_Q, y_Q) of the point Q given θ, the coordinates (x_P, y_P) of P, and the distance d between P and Q.

(b) Use your formula to determine (x_Q, y_Q) if $(x_P, y_P) = (123.62, 337.95)$, $\theta = 17° 19' 22''$, and $d = 193.86$ ft.

1. Give the six trigonometric function values of angle A.

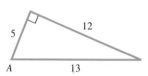

2. Find the exact values of each part labeled with a letter.

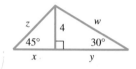

3. Find a solution for $\sin(B + 15°) = \cos(2B + 30°)$.

4. Determine whether each statement is *true* or *false*. If false, tell why.

    (a) $\sin 24° < \sin 48°$ **(b)** $\cos 24° < \cos 48°$

    (c) $\cos(60° + 30°) = \cos 60° \cdot \cos 30° - \sin 60° \cdot \sin 30°$

Find the exact values of the six trigonometric functions for each angle. Rationalize denominators when applicable.

    5. $240°$ **6.** $-135°$ **7.** $990°$

Find all values of θ in the interval $[0°, 360°)$ that have the given function value.

    8. $\cos \theta = -\dfrac{\sqrt{2}}{2}$ **9.** $\csc \theta = -\dfrac{2\sqrt{3}}{3}$ **10.** $\tan \theta = 1$

11. How would you find $\cot \theta$ using a calculator, if $\tan \theta = 1.6778490$? Give $\cot \theta$.

12. Use a calculator to approximate each value.

    (a) $\sin 78° 21'$ **(b)** $\tan 117.689°$ **(c)** $\sec 58.9041°$

13. Find a value of θ in the interval $[0°, 90°)$ in decimal degrees, if $\sin \theta = .27843196$.

14. Solve the triangle.

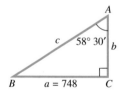

15. *Antenna Mast Guy Wire*   A guy wire 77.4 m long is attached to the top of an antenna mast that is 71.3 m high. Find the angle that the wire makes with the ground.

16. *Height of a Flagpole*   To measure the height of a flagpole, Amado Carillo found that the angle of elevation from a point 24.7 ft from the base to the top is 32° 10′. What is the height of the flagpole?

17. *Altitude of a Mountain*   The highest point in Texas is Guadalupe Peak. The angle of depression from the top of this peak to a small miner's cabin at an approximate elevation of 2000 ft is 26°. The cabin is located 14,000 ft horizontally from a point directly under the top of the mountain. Find the altitude of the top of the mountain to the nearest hundred feet.

18. *Distance Between Two Points* Two ships leave a port at the same time. The first ship sails on a bearing of 32° at 16 knots (nautical miles per hour) and the second on a bearing of 122° at 24 knots. How far apart are they after 2.5 hr?

19. *Distance of a Ship from a Pier* A ship leaves a pier on a bearing of S 62° E and travels for 75 km. It then turns and continues on a bearing of N 28° E for 53 km. How far is the ship from the pier?

20. Find *h* as indicated in the figure.

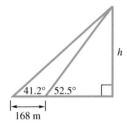

41.2° 52.5°

168 m

CHAPTER 2 ▶ Quantitative Reasoning

Can trigonometry be used to win an Olympic medal?

A shot-putter trying to improve performance may wonder: Is there an optimal angle to aim for, or is the velocity (speed) at which the ball is thrown more important? The figure shows the path of a steel ball thrown by a shot-putter. The distance *D* depends on initial velocity *v*, height *h*, and angle *θ* when the ball is released.

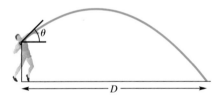

One model developed for this situation gives *D* as

$$D = \frac{v^2 \sin \theta \cos \theta + v \cos \theta \sqrt{(v \sin \theta)^2 + 64h}}{32}.$$

Typical ranges for the variables are *v*: 33–46 ft per sec; *h*: 6–8 ft; and *θ*: 40°–45°. (*Source:* Kreighbaum, E. and K. Barthels, *Biomechanics*, Allyn & Bacon, 1996.)

1. To see how angle *θ* affects distance *D*, let *v* = 44 ft per sec and *h* = 7 ft. Calculate *D* for *θ* = 40°, 42°, and 45°. How does distance *D* change as *θ* increases?

2. To see how velocity *v* affects distance *D*, let *h* = 7 and *θ* = 42°. Calculate *D* for *v* = 43, 44, and 45 ft per sec. How does distance *D* change as *v* increases?

3. Which affects distance *D* more, *v* or *θ*? What should the shot-putter do to improve performance?

3 Radian Measure and Circular Functions

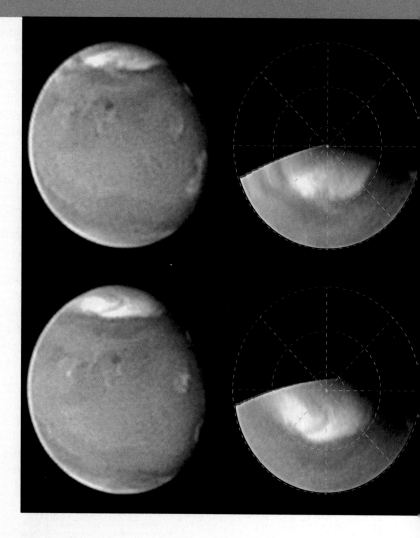

In August 2003, the planet Mars passed closer to Earth than it had in almost 60,000 years. Like Earth, Mars rotates on its axis and thus has days (also called *sols*) and nights. The photo shows a dust cloud/streak in the north polar cap of Mars, taken by the Hubble Space Telescope in 1996. In early 2004, the rovers *Spirit* and *Opportunity* landed on Mars and have provided scientists a wealth of information about the "Red Planet." (*Source:* www.hubblesite.org)

In Exercise 36 of Section 3.4, we examine the length of a Martian *sol* using *radian measure,* an alternative to measuring with degrees.

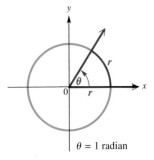

$\theta = 1$ radian

Figure 1

3.1 Radian Measure

Radian Measure ▪ **Converting Between Degrees and Radians** ▪ **Finding Function Values for Angles in Radians**

Radian Measure We have seen that angles can be measured in degrees. In more theoretical work in mathematics, *radian measure* of angles is preferred. Radian measure allows us to treat the trigonometric functions as functions with domains of *real numbers,* rather than angles.

Figure 1 shows an angle θ in standard position along with a circle of radius r. The vertex of θ is at the center of the circle. Because angle θ intercepts an arc on the circle equal in length to the radius of the circle, we say that angle θ has a measure of 1 radian.

> **RADIAN**
>
> An angle with its vertex at the center of a circle that intercepts an arc on the circle equal in length to the radius of the circle has a measure of **1 radian.**

It follows that an angle of measure 2 radians intercepts an arc equal in length to twice the radius of the circle, an angle of measure $\frac{1}{2}$ radian intercepts an arc equal in length to half the radius of the circle, and so on. In general, if θ is a central angle of a circle of radius r and θ intercepts an arc of length s, then the radian measure of θ is $\frac{s}{r}$.

Converting Between Degrees and Radians The **circumference** of a circle—the distance around the circle—is given by $C = 2\pi r$, where r is the radius of the circle. The formula $C = 2\pi r$ shows that the radius can be laid off 2π times around a circle. Therefore, an angle of 360°, which corresponds to a complete circle, intercepts an arc equal in length to 2π times the radius of the circle. Thus, an angle of 360° has a measure of 2π radians:

$$360° = 2\pi \text{ radians.}$$

An angle of 180° is half the size of an angle of 360°, so an angle of 180° has half the radian measure of an angle of 360°.

$$180° = \frac{1}{2}(2\pi) \text{ radians} = \pi \text{ radians} \quad \text{Degree/radian relationship}$$

We can use the relationship $180° = \pi$ radians to develop a method for converting between degrees and radians as follows.

$$180° = \pi \text{ radians}$$

$$1° = \frac{\pi}{180} \text{ radian} \quad \text{Divide by 180.} \qquad \text{or} \qquad 1 \text{ radian} = \frac{180°}{\pi} \quad \text{Divide by } \pi.$$

CONVERTING BETWEEN DEGREES AND RADIANS

1. Multiply a degree measure by $\frac{\pi}{180}$ radian and simplify to convert to radians.

2. Multiply a radian measure by $\frac{180°}{\pi}$ and simplify to convert to degrees.

▶ **EXAMPLE 1** CONVERTING DEGREES TO RADIANS

Convert each degree measure to radians.

(a) 45°   **(b)** −270°   **(c)** 249.8°

Solution

(a) $45° = 45\left(\frac{\pi}{180} \text{ radian}\right) = \frac{\pi}{4} \text{ radian}$ Multiply by $\frac{\pi}{180}$ radian.

(b) $-270° = -270\left(\frac{\pi}{180} \text{ radian}\right) = -\frac{270\pi}{180} \text{ radians}$

$$= -\frac{3\pi}{2} \text{ radians} \qquad \text{Lowest terms}$$

(c) $249.8° = 249.8\left(\frac{\pi}{180} \text{ radian}\right) \approx 4.360 \text{ radians}$ Nearest thousandth

NOW TRY EXERCISES 7, 13, AND 43. ◀

▶ **EXAMPLE 2** CONVERTING RADIANS TO DEGREES

Convert each radian measure to degrees.

(a) $\frac{9\pi}{4}$ **(b)** $-\frac{5\pi}{6}$ **(c)** 4.25

Solution

(a) $\frac{9\pi}{4} \text{ radians} = \frac{9\pi}{4}\left(\frac{180°}{\pi}\right) = 405°$ Multiply by $\frac{180°}{\pi}$.

(b) $-\frac{5\pi}{6} \text{ radians} = -\frac{5\pi}{6}\left(\frac{180°}{\pi}\right) = -150°$

(c) $4.25 \text{ radians} = 4.25\left(\frac{180°}{\pi}\right) \approx 243.5° = 243° \, 30'$ Use a calculator.

NOW TRY EXERCISES 27, 31, AND 55. ◀

▶ **Note** Another way to convert a radian measure that is a rational multiple of π, such as $\frac{9\pi}{4}$, to degrees is to just substitute 180° for π. In Example 2(a), this would be $\frac{9(180°)}{4} = 405°$.

One of the most important facts to remember when working with angles and their measures is summarized in the following statement.

```
45°
        .7853981634
-270°
        -4.71238898
249.8°
        4.359832471
```

A TI-83/84 Plus calculator can convert directly between degrees and radians. This radian mode screen shows the conversions for Example 1. Verify that the first two results are *approximations* for the *exact* values of $\frac{\pi}{4}$ and $-\frac{3\pi}{2}$.

```
(9π/4)ʳ
              405
(-5π/6)ʳ
              -150
4.25ʳ▸DMS
     243°30'25.427"
```

This degree mode screen shows how a TI-83/84 Plus calculator converts the radian measures in Example 2 to degree measures.

AGREEMENT ON ANGLE MEASUREMENT UNITS

If no unit of angle measure is specified, then the angle is understood to be measured in radians.

For example, Figure 2(a) shows an angle of 30°, while Figure 2(b) shows an angle of 30 (which means 30 radians).

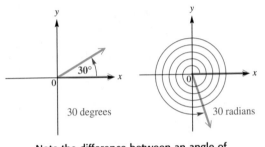

30 degrees

30 radians

Note the difference between an angle of 30 *degrees* and an angle of 30 *radians*.

(a)　　　　　　　**(b)**

Figure 2

The following table and Figure 3 on the next page give some equivalent angle measures in degrees and radians. Keep in mind that **180° = π radians.**

Degrees	Radians		Degrees	Radians	
	Exact	Approximate		Exact	Approximate
0°	0	0	90°	$\frac{\pi}{2}$	1.57
30°	$\frac{\pi}{6}$	.52	180°	π	3.14
45°	$\frac{\pi}{4}$	.79	270°	$\frac{3\pi}{2}$	4.71
60°	$\frac{\pi}{3}$	1.05	360°	2π	6.28

These exact values are *rational multiples of* π.

In calculus, radian measure is much easier to work with than degree measure. If x is measured in radians, then the derivative of $f(x) = \sin x$ is

$$f'(x) = \cos x.$$

However, if x is measured in degrees, then the derivative of $f(x) = \sin x$ is

$$f'(x) = \frac{\pi}{180} \cos x.$$

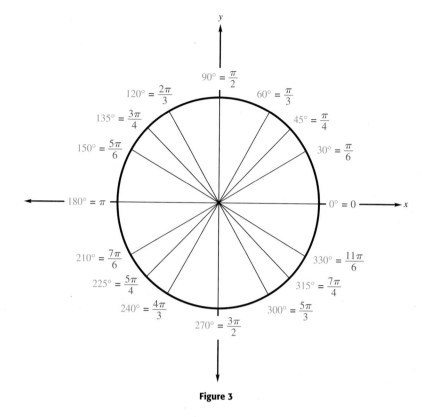

Figure 3

The angles marked in Figure 3 are extremely important in the study of trigonometry. *You should learn these equivalences, as they will appear often in the chapters to follow.*

Finding Function Values for Angles in Radians
Trigonometric function values for angles measured in radians can be found by first converting radian measure to degrees. *(You should try to skip this intermediate step as soon as possible, and find the function values directly from radian measure.)*

▶ **EXAMPLE 3**   FINDING FUNCTION VALUES OF ANGLES IN RADIAN MEASURE

Find each function value.

(a) $\tan \dfrac{2\pi}{3}$ **(b)** $\sin \dfrac{3\pi}{2}$ **(c)** $\cos\left(-\dfrac{4\pi}{3}\right)$

Solution

(a) First convert $\frac{2\pi}{3}$ radians to degrees.

$$\tan \frac{2\pi}{3} = \tan\left(\frac{2\pi}{3} \cdot \frac{180°}{\pi}\right) \qquad \text{Multiply by } \tfrac{180°}{\pi}.$$

$$= \tan 120°$$

$$= -\sqrt{3} \qquad \text{(Section 2.2)}$$

(b) From the table on page 104 and Figure 3 on the preceding page, $\frac{3\pi}{2}$ radians $= 270°$, so

$$\sin\frac{3\pi}{2} = \sin 270° = -1.$$

(c) $\cos\left(-\frac{4\pi}{3}\right) = \cos\left(-\frac{4\pi}{3} \cdot \frac{180°}{\pi}\right)$

$$= -\cos 60° \qquad \text{(Section 2.2)}$$

$$= -\frac{1}{2}$$

NOW TRY EXERCISES 65, 75, AND 79. ◀

3.1 Exercises

Concept Check In Exercises 1–6, each angle θ is an integer when measured in radians. Give the radian measure of the angle.

1.
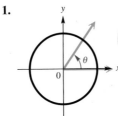

2.
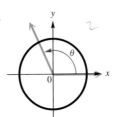

3.
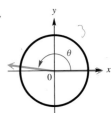

4.
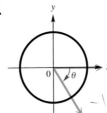

5.
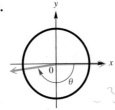

6.
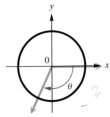

Convert each degree measure to radians. Leave answers as multiples of π. See Examples 1(a) and 1(b).

7. 60°	**8.** 30°	**9.** 90°	**10.** 120°
11. 150°	**12.** 270°	**13.** −300°	**14.** −315°
15. 450°	**16.** 480°	**17.** 1800°	**18.** −3600°

Give a short explanation in Exercises 19–24.

19. In your own words, explain how to convert degree measure to radian measure.

20. In your own words, explain how to convert radian measure to degree measure.

21. In your own words, explain the meaning of radian measure.

22. Explain the difference between degree measure and radian measure.

23. Use an example to show that you can convert from radian measure to degree measure by multiplying by $\frac{180°}{\pi}$.

24. Explain why an angle of radian measure t in standard position intercepts an arc of length t on a circle of radius 1.

Convert each radian measure to degrees. See Examples 2(a) and 2(b).

25. $\frac{\pi}{3}$ **26.** $\frac{8\pi}{3}$ **27.** $\frac{7\pi}{4}$ **28.** $\frac{2\pi}{3}$

29. $\frac{11\pi}{6}$ **30.** $\frac{15\pi}{4}$ **31.** $-\frac{\pi}{6}$ **32.** $-\frac{8\pi}{5}$

33. $\frac{7\pi}{10}$ **34.** $\frac{11\pi}{15}$ **35.** $-\frac{4\pi}{15}$ **36.** $-\frac{7\pi}{20}$

37. $\frac{17\pi}{20}$ **38.** $\frac{11\pi}{30}$ **39.** -5π **40.** 15π

Convert each degree measure to radians. See Example 1(c).

41. $39°$ **42.** $74°$ **43.** $42.5°$ **44.** $264.9°$

45. $139° \, 10'$ 2.43   **46.** $174° \, 50'$ **47.** $64.29°$ **48.** $85.04°$

49. $56° \, 25'$ **50.** $122° \, 37'$ **51.** $47.6925°$ **52.** $23.0143°$

Convert each radian measure to degrees. Write answers to the nearest minute. See Example 2(c).

53. 2 **54.** 5 **55.** 1.74 **56.** 3.06

57. $.3417$ **58.** 9.84763 **59.** -5.01095 **60.** -3.47189

61. *Concept Check*   The value of sin 30 is not $\frac{1}{2}$. Why is this true?

62. Explain in your own words what is meant by an angle of one radian.

Find the exact value of each expression without using a calculator. See Example 3.

63. $\sin\frac{\pi}{3}$ **64.** $\cos\frac{\pi}{6}$ **65.** $\tan\frac{\pi}{4}$ **66.** $\cot\frac{\pi}{3}$

67. $\sec\frac{\pi}{6}$ **68.** $\csc\frac{\pi}{4}$ **69.** $\sin\frac{\pi}{2}$ **70.** $\csc\frac{\pi}{2}$

71. $\tan\frac{5\pi}{3}$ **72.** $\cot\frac{2\pi}{3}$ **73.** $\sin\frac{5\pi}{6}$ **74.** $\tan\frac{5\pi}{6}$

75. $\cos 3\pi$ **76.** $\sec \pi$ **77.** $\sin\left(-\frac{8\pi}{3}\right)$ **78.** $\cot\left(-\frac{2\pi}{3}\right)$

79. $\sin\left(-\frac{7\pi}{6}\right)$ **80.** $\cos\left(-\frac{\pi}{6}\right)$ **81.** $\tan\left(-\frac{14\pi}{3}\right)$ **82.** $\csc\left(-\frac{13\pi}{3}\right)$

83. *Concept Check*   The figure shows the same angles measured in both degrees and radians. Complete the missing measures.

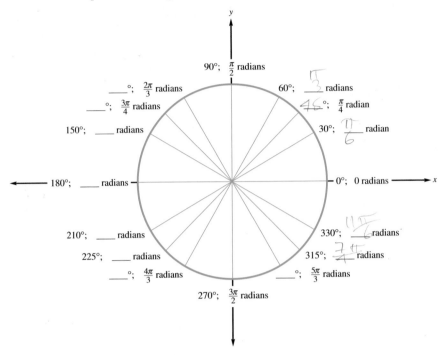

Solve each problem.

84. *Railroad Engineering*   The term **grade** has several different meanings in construction work. Some engineers use the term **grade** to represent $\frac{1}{100}$ of a right angle and express grade as a percent. For instance, an angle of .9° would be referred to as a 1% grade. (*Source:* Hay, W., *Railroad Engineering,* John Wiley and Sons, 1982.)

   (a) By what number should you multiply a grade (disregarding the % symbol) to convert it to radians?

   (b) In a rapid-transit rail system, the maximum grade allowed between two stations is 3.5%. Express this angle in degrees and radians.

85. *Rotating Hour Hand on a Clock*   Through how many radians will the hour hand on a clock rotate in **(a)** 24 hr and **(b)** 4 hr?

86. *Rotating Pulley*   A circular pulley is rotating about its center. Through how many radians would it turn in **(a)** 8 rotations and **(b)** 30 rotations?

87. *Orbits of a Space Vehicle*   A space vehicle is orbiting Earth in a circular orbit. What radian measure corresponds to **(a)** 2.5 orbits and **(b)** $\frac{4}{3}$ orbit?

88. *Revolutions of a Carousel*   A stationary horse on a carousel makes 12 complete revolutions. Through what radian measure angle does the horse revolve?

3.2 Applications of Radian Measure

Arc Length on a Circle ▪ Area of a Sector of a Circle

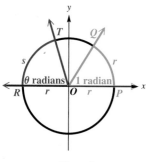

Figure 4

Arc Length on a Circle We use radian measure in the formula to find the length of an arc of a circle. This formula is derived from the fact (proved in geometry) that the length of an arc is proportional to the measure of its central angle. In Figure 4, angle QOP has measure 1 radian and intercepts an arc of length r on the circle. Angle ROT has measure θ radians and intercepts an arc of length s on the circle. Since the lengths of the arcs are proportional to the measures of their central angles,

$$\frac{s}{r} = \frac{\theta}{1}.$$

Multiplying both sides by r gives the following result.

ARC LENGTH

The length s of the arc intercepted on a circle of radius r by a central angle of measure θ radians is given by the product of the radius and the radian measure of the angle, or

$$s = r\theta, \qquad \theta \text{ in radians.}$$

▶ **Caution** *Avoid the common error of applying this formula with θ in degree mode. When applying the formula $s = r\theta$, the value of θ MUST be expressed in radians.*

▶ **EXAMPLE 1**   FINDING ARC LENGTH USING $s = r\theta$

A circle has radius 18.20 cm. Find the length of the arc intercepted by a central angle having each of the following measures.

(a) $\dfrac{3\pi}{8}$ radians

(b) $144°$

Solution

(a) As shown in Figure 5, $r = 18.20$ cm and $\theta = \frac{3\pi}{8}$.

$$s = r\theta \qquad \text{Arc length formula}$$

$$s = 18.20\left(\frac{3\pi}{8}\right) \text{ cm} \quad \text{Substitute for } r \text{ and } \theta.$$

$$s \approx 21.44 \text{ cm}$$

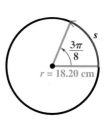

Figure 5

(b) The formula $s = r\theta$ requires that θ be measured in radians. First, convert θ to radians by multiplying $144°$ by $\frac{\pi}{180}$ radian.

$$144° = 144\left(\frac{\pi}{180}\right) = \frac{4\pi}{5} \text{ radians}$$ Convert from degrees to radians. (Section 3.1)

The length s is given by

$$s = r\theta = 18.20\left(\frac{4\pi}{5}\right) \approx 45.74 \text{ cm.}$$

Be sure to use radians for θ in $s = r\theta$.

NOW TRY EXERCISES 11 AND 15. ◀

▶ **EXAMPLE 2**   USING LATITUDES TO FIND THE DISTANCE BETWEEN TWO CITIES

Latitude gives the measure of a central angle with vertex at Earth's center whose initial side goes through the equator and whose terminal side goes through the given location. Reno, Nevada is approximately due north of Los Angeles. The latitude of Reno is $40°$ N, while that of Los Angeles is $34°$ N. (The N in $34°$ N means *north* of the equator.) The radius of Earth is 6400 km. Find the north-south distance between the two cities.

Solution   As shown in Figure 6, the central angle between Reno and Los Angeles is $40° - 34° = 6°$. The distance between the two cities can be found by the formula $s = r\theta$, after $6°$ is first converted to radians.

$$6° = 6\left(\frac{\pi}{180}\right) = \frac{\pi}{30} \text{ radian}$$

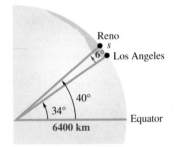

Figure 6

The distance between the two cities is

$$s = r\theta = 6400\left(\frac{\pi}{30}\right) \approx 670 \text{ km.}$$ Let $r = 6400$ and $\theta = \frac{\pi}{30}$.

NOW TRY EXERCISE 21. ◀

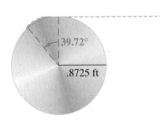

Figure 7

▶ **EXAMPLE 3**   FINDING A LENGTH USING $s = r\theta$

A rope is being wound around a drum with radius .8725 ft. (See Figure 7.) How much rope will be wound around the drum if the drum is rotated through an angle of $39.72°$?

Solution   The length of rope wound around the drum is the arc length for a circle of radius .8725 ft and a central angle of $39.72°$. Use the formula $s = r\theta$, with the angle converted to radian measure. The length of the rope wound around the drum is approximately

$$s = r\theta = .8725\left[39.72\left(\frac{\pi}{180}\right)\right] \approx .6049 \text{ ft.}$$

NOW TRY EXERCISE 33(a). ◀

Figure 8

▶ **EXAMPLE 4** FINDING AN ANGLE MEASURE USING $s = r\theta$

Two gears are adjusted so that the smaller gear drives the larger one, as shown in Figure 8. If the smaller gear rotates through an angle of 225°, through how many degrees will the larger gear rotate?

Solution First find the radian measure of the angle, and then find the arc length on the smaller gear that determines the motion of the larger gear. Since $225° = \frac{5\pi}{4}$ radians, for the smaller gear,

$$s = r\theta = 2.5\left(\frac{5\pi}{4}\right) = \frac{12.5\pi}{4} = \frac{25\pi}{8} \text{ cm.}$$

An arc with this length on the larger gear corresponds to an angle measure θ, in radians, where

$$s = r\theta$$

$$\frac{25\pi}{8} = 4.8\theta \quad \text{Substitute } \tfrac{25\pi}{8} \text{ for } s \text{ and 4.8 for } r.$$

$$\frac{125\pi}{192} = \theta. \quad \begin{array}{l} 4.8 = \tfrac{48}{10} = \tfrac{24}{5}; \text{ multiply by } \tfrac{5}{24} \\ \text{to solve for } \theta. \end{array}$$

Converting θ back to degrees shows that the larger gear rotates through

$$\frac{125\pi}{192}\left(\frac{180°}{\pi}\right) \approx 117°. \quad \text{Convert } \theta = \tfrac{125\pi}{192} \text{ to degrees.}$$

NOW TRY EXERCISE 27. ◀

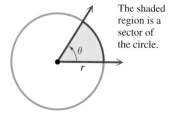

The shaded region is a sector of the circle.

Figure 9

Area of a Sector of a Circle A **sector of a circle** is the portion of the interior of a circle intercepted by a central angle. Think of it as a "piece of pie." See Figure 9. A complete circle can be thought of as an angle with measure 2π radians. If a central angle for a sector has measure θ radians, then the sector makes up the fraction $\frac{\theta}{2\pi}$ of a complete circle. The area of a complete circle with radius r is $A = \pi r^2$. Therefore,

$$\text{area of the sector} = \frac{\theta}{2\pi}(\pi r^2) = \frac{1}{2}r^2\theta, \quad \theta \text{ in radians.}$$

This discussion is summarized as follows.

AREA OF A SECTOR

The area A of a sector of a circle of radius r and central angle θ is given by

$$A = \frac{1}{2}r^2\theta, \quad \theta \text{ in radians.}$$

▶ **Caution** As in the formula for arc length, *the value of θ must be in radians when using this formula for the area of a sector.*

Figure 10

▶ **EXAMPLE 5** FINDING THE AREA OF A SECTOR-SHAPED FIELD

Find the area of the sector-shaped field shown in Figure 10.

Solution First, convert 15° to radians.

$$15° = 15\left(\frac{\pi}{180}\right) = \frac{\pi}{12} \text{ radian}$$

Now use the formula to find the area of a sector of a circle with radius $r = 321$.

$$A = \frac{1}{2}r^2\theta = \frac{1}{2}(321)^2\left(\frac{\pi}{12}\right) \approx 13{,}500 \text{ m}^2$$

NOW TRY EXERCISE 51. ◀

CONNECTIONS **Longitude** is the angular distance (expressed in degrees) East or West of the prime meridian, which goes from the North Pole to the South Pole through Greenwich, England. Arcs of longitude are 110 km apart at the equator. As the figure shows, these sections are similar to those of an orange.

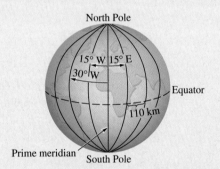

Because Earth revolves 15° per hr, longitude is found by taking the difference between time zones multiplied by 15°. For example, if it is 12 noon where you are (in the United States) and 5 P.M. in Greenwich, you are located at longitude 5(15°) = longitude 75° W. Thus, determining longitude requires only an accurate measure of time. Before 1772, sailors were unable to determine their position at sea because there were no clocks capable of precise measure of time at sea. In 1772, a clock invented by John Harrison, a carpenter's son with no formal education, solved the problem. (*Source:* Ola, P. and E. D'Aulaire, "Taking the Measure of Time," *Smithsonian,* December 1999.)

FOR DISCUSSION OR WRITING
Use time zones to determine the longitude where you live. What would the longitude be at Greenwich, England? Visit the Internet to learn more about longitude.

3.2 Exercises

Concept Check Find the exact length of each arc intercepted by the given central angle.

1.
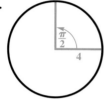

2.
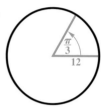

3.
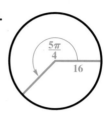

Concept Check Find the radius of each circle.

4.
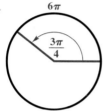

5.
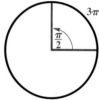

6.
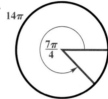

Concept Check Find the measure of each central angle (in radians).

7.
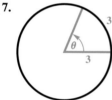

8.
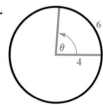

9.
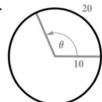

10. Explain in your own words how to find the *degree* measure of a central angle in a circle if both the radius and the length of the intercepted arc are known.

Unless otherwise directed, give calculator approximations in your answers in the rest of this exercise set.

Find the length to three significant digits of each arc intercepted by a central angle θ in a circle of radius r. See Example 1.

11. $r = 12.3$ cm, $\theta = \dfrac{2\pi}{3}$ radians

12. $r = .892$ cm, $\theta = \dfrac{11\pi}{10}$ radians

13. $r = 1.38$ ft, $\theta = \dfrac{5\pi}{6}$ radians

14. $r = 3.24$ mi, $\theta = \dfrac{7\pi}{6}$ radians

15. $r = 4.82$ m, $\theta = 60°$

16. $r = 71.9$ cm, $\theta = 135°$

17. $r = 15.1$ in., $\theta = 210°$

18. $r = 12.4$ ft, $\theta = 330°$

19. *Concept Check* If the radius of a circle is doubled, how is the length of the arc intercepted by a fixed central angle changed?

20. *Concept Check* Radian measure simplifies many formulas, such as the formula for arc length, $s = r\theta$. Give the corresponding formula when θ is measured in degrees instead of radians.

Distance Between Cities Find the distance in kilometers between each pair of cities, assuming they lie on the same north-south line. See Example 2.

21. Panama City, Panama, 9° N, and Pittsburgh, Pennsylvania, 40° N

22. Farmersville, California, 36° N, and Penticton, British Columbia, 49° N

23. New York City, New York, 41° N, and Lima, Peru, 12° S

24. Halifax, Nova Scotia, 45° N, and Buenos Aires, Argentina, 34° S

25. *Latitude of Madison* Madison, South Dakota, and Dallas, Texas, are 1200 km apart and lie on the same north-south line. The latitude of Dallas is 33° N. What is the latitude of Madison?

26. *Latitude of Toronto* Charleston, South Carolina, and Toronto, Canada, are 1100 km apart and lie on the same north-south line. The latitude of Charleston is 33° N. What is the latitude of Toronto?

Work each problem. See Examples 3 and 4.

27. *Gear Movement* Two gears are adjusted so that the smaller gear drives the larger one, as shown in the figure. If the smaller gear rotates through an angle of 300°, through how many degrees will the larger gear rotate?

28. *Gear Movement* Repeat Exercise 27 for gear radii of 4.8 in. and 7.1 in., and for an angle of 315° for the smaller gear.

29. *Rotating Wheels* The rotation of the smaller wheel in the figure causes the larger wheel to rotate. Through how many degrees will the larger wheel rotate if the smaller one rotates through 60.0°?

30. *Rotating Wheels* Repeat Exercise 29 for wheel radii of 6.84 in. and 12.46 in. and an angle of 150° for the smaller wheel.

31. *Rotating Wheels* Find the radius of the larger wheel in the figure if the smaller wheel rotates 80.0° when the larger wheel rotates 50.0°.

32. *Rotating Wheels* Repeat Exercise 31 if the smaller wheel of radius 14.6 in. rotates 120° when the larger wheel rotates 60°.

33. *Pulley Raising a Weight*

(a) How many inches will the weight in the figure rise if the pulley is rotated through an angle of 71° 50′?

(b) Through what angle, to the nearest minute, must the pulley be rotated to raise the weight 6 in.?

34. *Pulley Raising a Weight* Find the radius of the pulley in the figure if a rotation of 51.6° raises the weight 11.4 cm.

35. *Bicycle Chain Drive* The figure shows the chain drive of a bicycle. How far will the bicycle move if the pedals are rotated through 180°? Assume the radius of the bicycle wheel is 13.6 in.

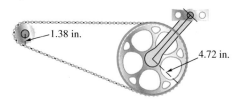
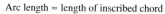
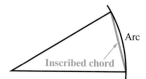

36. *Car Speedometer* The speedometer of Terry's Honda CR-V is designed to be accurate with tires of radius 14 in.

(a) Find the number of rotations of a tire in 1 hr if the car is driven at 55 mph.

(b) Suppose that oversize tires of radius 16 in. are placed on the car. If the car is now driven for 1 hr with the speedometer reading 55 mph, how far has the car gone? If the speed limit is 55 mph, does Terry deserve a speeding ticket?

If a central angle is very small, there is little difference in length between an arc and the inscribed chord. See the figure. Approximate each of the following lengths by finding the necessary arc length. (Note: When a central angle intercepts an arc, the arc is said to **subtend** *the angle.)*

Arc length ≈ length of inscribed chord

37. *Length of a Train* A railroad track in the desert is 3.5 km away. A train on the track subtends (horizontally) an angle of 3° 20′. Find the length of the train.

38. *Distance to a Boat* The mast of Brent Simon's boat is 32 ft high. If it subtends an angle of 2° 10′, how far away is it?

Concept Check Find the area of each sector.

39.

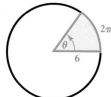

2π

θ

6

40.

4π

θ

8

41.

12π

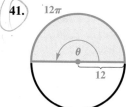

θ

12

42.

15π

θ

10

Concept Check Find the measure (in degrees) of each central angle. The number inside the sector is the area.

43.

6π sq units

6

44.

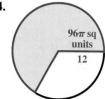

96π sq units

12

Concept Check *Find the measure (in radians) of each central angle. The number inside the sector is the area.*

45.
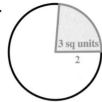

46.

Find the area of a sector of a circle having radius r and central angle θ. Express answers to the nearest tenth. See Example 5.

47. $r = 29.2$ m, $\theta = \dfrac{5\pi}{6}$ radians

48. $r = 59.8$ km, $\theta = \dfrac{2\pi}{3}$ radians

49. $r = 30.0$ ft, $\theta = \dfrac{\pi}{2}$ radians

50. $r = 90.0$ yd, $\theta = \dfrac{5\pi}{6}$ radians

51. $r = 12.7$ cm, $\theta = 81°$

52. $r = 18.3$ m, $\theta = 125°$

53. $r = 40.0$ mi, $\theta = 135°$

54. $r = 90.0$ km, $\theta = 270°$

Work each problem.

55. Find the measure (in radians) of a central angle of a sector of area 16 in.^{2} in a circle of radius 3.0 in.

56. Find the radius of a circle in which a central angle of $\frac{\pi}{6}$ radian determines a sector of area 64 m^2.

57. *Measures of a Structure* The figure shows Medicine Wheel, a Native American structure in northern Wyoming. This circular structure is perhaps 2500 yr old. There are 27 aboriginal spokes in the wheel, all equally spaced.

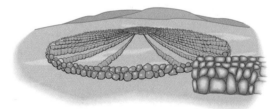

(a) Find the measure of each central angle in degrees and in radians.
(b) If the radius of the wheel is 76.0 ft, find the circumference.
(c) Find the length of each arc intercepted by consecutive pairs of spokes.
(d) Find the area of each sector formed by consecutive spokes.

58. *Area Cleaned by a Windshield Wiper* The Ford Model A, built from 1928 to 1931, had a single windshield wiper on the driver's side. The total arm and blade was 10 in. long and rotated back and forth through an angle of 95°. The shaded region in the figure is the portion of the windshield cleaned by the 7-in. wiper blade. What is the area of the region cleaned?

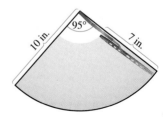

59. *Circular Railroad Curves* In the United States, circular railroad curves are designated by the **degree of curvature,** the central angle subtended by a chord of 100 ft. Suppose a portion of track has curvature 42.0°. (*Source:* Hay, W., *Railroad Engineering,* John Wiley and Sons, 1982.)

(a) What is the radius of the curve?
(b) What is the length of the arc determined by the 100-ft chord?

(c) What is the area of the portion of the circle bounded by the arc and the 100-ft chord?

60. *Land Required for a Solar-Power Plant*   A 300-megawatt solar-power plant requires approximately 950,000 m² of land area in order to collect the required amount of energy from sunlight. If this land area is circular, what is its radius? If this land area is a 35° sector of a circle, what is its radius?

61. *Area of a Lot*   A frequent problem in surveying city lots and rural lands adjacent to curves of highways and railways is that of finding the area when one or more of the boundary lines is the arc of a circle. Find the area of the lot (to two significant digits) shown in the figure. (*Source:* Anderson, J. and E. Michael, *Introduction to Surveying,* McGraw-Hill, 1985.)

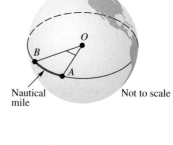

62. *Nautical Miles*   **Nautical miles** are used by ships and airplanes. They are different from **statute miles,** which equal 5280 ft. A nautical mile is defined to be the arc length along the equator intercepted by a central angle *AOB* of 1 min, as illustrated in the figure. If the equatorial radius of Earth is 3963 mi, use the arc length formula to approximate the number of statute miles in 1 nautical mile. Round your answer to two decimal places.

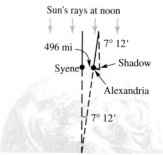

63. *Circumference of Earth*   The first accurate estimate of the distance around Earth was done by the Greek astronomer Eratosthenes (276–195 B.C.), who noted that the noontime position of the sun at the summer solstice differed by 7° 12′ from the city of Syene to the city of Alexandria. (See the figure.) The distance between these two cities is 496 mi. Use the arc length formula to estimate the radius of Earth. Then find the circumference of Earth. (*Source:* Zeilik, M., *Introductory Astronomy and Astrophysics,* Third Edition, Saunders College Publishers, 1992.)

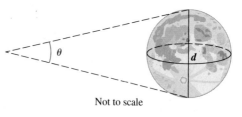

64. *Diameter of the Moon*   The distance to the moon is approximately 238,900 mi. Use the arc length formula to estimate the diameter *d* of the moon if angle θ in the figure is measured to be .5170°.

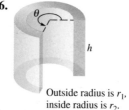

Not to scale

Volume of a Solid   *Multiply the area of the base by the height to find a formula for the volume V of each solid.*

65.

66.

Outside radius is r_1, inside radius is r_2.

RELATING CONCEPTS

For individual or collaborative investigation
(Exercises 67–70)

(Modeling) Measuring Paper Curl   Manufacturers of paper determine its quality by its curl. The curl of a sheet of paper is measured by holding it at the center of one edge and comparing the arc formed by the free end to arcs on a chart lying flat on a table. Each arc in the chart corresponds to a number d that gives the depth *of the arc. See the figure. (Source:* Tabakovic, H., J. Paullet, and R. Bertram, "Measuring the Curl of Paper," *The College Mathematics Journal,* Vol. 30 No. 4, September 1999.)

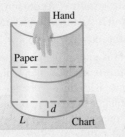
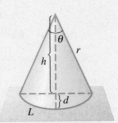

To produce the chart, it is necessary to find a function that relates d to the length of arc L. **Work Exercises 67–70 in order,** *to determine that function. Refer to the figure on the right.*

67. Express L in terms of r and θ, and then solve for r.

68. Use a right triangle to relate r, h, and θ. Solve for h.

69. Express d in terms of r and h, then substitute your answer from Exercise 68 for h. Factor out r.

70. Use your answer from Exercise 67 to substitute for r in the result from Exercise 69. This result is a formula that gives d for specific values of θ.

71. *Concept Check*   If the radius of a circle is doubled and the central angle of a sector is unchanged, how is the area of the sector changed?

72. *Concept Check*   Give the corresponding formula for the area of a sector when the angle is measured in degrees.

3.3 The Unit Circle and Circular Functions

Circular Functions ▪ Finding Values of Circular Functions ▪ Determining a Number with a Given Circular Function Value ▪ Applying Circular Functions

In **Section 1.3,** we defined the six trigonometric functions in such a way that the domain of each function was a set of *angles* in standard position. These angles can be measured in degrees or in radians. In advanced courses, such as calculus, it is necessary to modify the trigonometric functions so that their domains consist of *real numbers* rather than angles. We do this by using the relationship between an angle θ and an arc of length s on a circle.

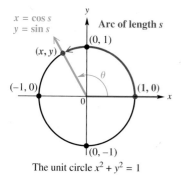

$x = \cos s$
$y = \sin s$

Arc of length s

$(0, 1)$

(x, y)

$(-1, 0)$

θ

$(1, 0)$

$(0, -1)$

The unit circle $x^2 + y^2 = 1$

Figure 11

Circular Functions

In Figure 11, we start at the point $(1, 0)$ and measure an arc of length s along the circle. If $s > 0$, then the arc is measured in a counterclockwise direction, and if $s < 0$, then the direction is clockwise. (If $s = 0$, then no arc is measured.) Let the endpoint of this arc be at the point (x, y). The circle in Figure 11 is the **unit circle**—it has center at the origin and radius 1 unit (hence the name *unit circle*). Recall from algebra that the equation of this circle is

$$x^2 + y^2 = 1. \quad \text{(Appendix B)}$$

Recall that the radian measure of θ is related to the arc length s. For θ measured in radians, we know that $s = r\theta$. Here, $r = 1$, so s, which is measured in linear units such as inches or centimeters, is equal to θ, measured in radians. Thus, the trigonometric functions of angle θ in radians found by choosing a point (x, y) on the unit circle can be rewritten as functions of the arc length s, a real number. When interpreted this way, they are called **circular functions.**

▼ LOOKING AHEAD TO CALCULUS

If you plan to study calculus, you must become very familiar with radian measure. In calculus, the trigonometric or circular functions are always understood to have real number domains.

CIRCULAR FUNCTIONS

For any real number s represented by a directed arc on the unit circle,

$$\sin s = y \qquad \cos s = x \qquad \tan s = \frac{y}{x} \quad (x \neq 0)$$

$$\csc s = \frac{1}{y} \quad (y \neq 0) \qquad \sec s = \frac{1}{x} \quad (x \neq 0) \qquad \cot s = \frac{x}{y} \quad (y \neq 0).$$

Since x represents the cosine of s and y represents the sine of s, and because of the discussion in **Section 3.1** on converting between degrees and radians, we can summarize a great deal of information in a concise manner, as seen in Figure 12.*

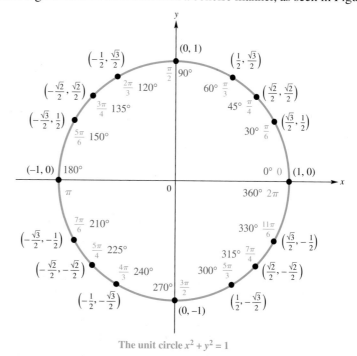

The unit circle $x^2 + y^2 = 1$

Figure 12

---

*The authors thank Professor Marvel Townsend of the University of Florida for her suggestion to include this figure.

The unit circle is symmetric with respect to the x-axis, the y-axis, and the origin. (See **Appendix D.**) Thus, if a point (a, b) lies on the unit circle, so does $(a, -b)$, $(-a, b)$, and $(-a, -b)$. Furthermore, each of these points has a *reference arc* of equal magnitude. For a point on the unit circle, its **reference arc** is the shortest arc from the point itself to the nearest point on the x-axis. (This concept is analogous to the reference angle concept introduced in **Chapter 2.**) Using the concept of symmetry makes determining sines and cosines of the real numbers identified in Figure 12 a relatively simple procedure if we know the coordinates of the points labeled in quadrant I.

For example, the quadrant I real number $\frac{\pi}{3}$ is associated with the point $\left(\frac{1}{2}, \frac{\sqrt{3}}{2}\right)$ on the unit circle. Therefore, we can use symmetry to identify the coordinates of the points associated with

$$\pi - \frac{\pi}{3} = \frac{2\pi}{3}, \qquad \pi + \frac{\pi}{3} = \frac{4\pi}{3}, \qquad \text{and} \qquad 2\pi - \frac{\pi}{3} = \frac{5\pi}{3}.$$

$$\uparrow \qquad\qquad\qquad \uparrow \qquad\qquad\qquad\qquad\qquad \uparrow$$

$$\text{Quadrant II} \qquad\qquad \text{Quadrant III} \qquad\qquad\qquad \text{Quadrant IV}$$

The following chart summarizes this information.

s	Quadrant of s	Symmetry Type and Corresponding Point	$\cos s$	$\sin s$
$\dfrac{\pi}{3}$	I	not applicable; $\left(\dfrac{1}{2}, \dfrac{\sqrt{3}}{2}\right)$	$\dfrac{1}{2}$	$\dfrac{\sqrt{3}}{2}$
$\pi - \dfrac{\pi}{3} = \dfrac{2\pi}{3}$	II	y-axis; $\left(-\dfrac{1}{2}, \dfrac{\sqrt{3}}{2}\right)$	$-\dfrac{1}{2}$	$\dfrac{\sqrt{3}}{2}$
$\pi + \dfrac{\pi}{3} = \dfrac{4\pi}{3}$	III	origin; $\left(-\dfrac{1}{2}, -\dfrac{\sqrt{3}}{2}\right)$	$-\dfrac{1}{2}$	$-\dfrac{\sqrt{3}}{2}$
$2\pi - \dfrac{\pi}{3} = \dfrac{5\pi}{3}$	IV	x-axis; $\left(\dfrac{1}{2}, -\dfrac{\sqrt{3}}{2}\right)$	$\dfrac{1}{2}$	$-\dfrac{\sqrt{3}}{2}$

▶ **Note** Since $\sin s = y$ and $\cos s = x$, we can replace x and y in the equation of the unit circle

$$x^2 + y^2 = 1$$

and obtain the Pythagorean identity

$$\cos^2 s + \sin^2 s = 1. \quad \text{(Section 1.4)}$$

The ordered pair (x, y) represents a point on the unit circle, and therefore

$$-1 \le x \le 1 \qquad \text{and} \qquad -1 \le y \le 1,$$

so $\qquad\qquad -1 \le \cos s \le 1 \qquad \text{and} \qquad -1 \le \sin s \le 1.$

For any value of s, both $\sin s$ and $\cos s$ exist, so the domain of these functions is the set of all real numbers.

For tan s, defined as $\frac{y}{x}$, x must not equal 0. The only way x can equal 0 is when the arc length s is $\frac{\pi}{2}$, $-\frac{\pi}{2}$, $\frac{3\pi}{2}$, $-\frac{3\pi}{2}$, and so on. To avoid a 0 denominator, the domain of the tangent function must be restricted to those values of s satisfying

$$s \neq (2n + 1)\frac{\pi}{2}, \qquad n \text{ any integer.}$$

The definition of secant also has x in the denominator, so the domain of secant is the same as the domain of tangent. Both cotangent and cosecant are defined with a denominator of y. To guarantee that $y \neq 0$, the domain of these functions must be the set of all values of s satisfying

$$s \neq n\pi, \qquad n \text{ any integer.}$$

DOMAINS OF THE CIRCULAR FUNCTIONS

The domains of the circular functions are as follows:

Sine and Cosine Functions: $(-\infty, \infty)$

Tangent and Secant Functions: $\left\{ s \mid s \neq (2n + 1)\dfrac{\pi}{2}, \text{ where } n \text{ is any integer} \right\}$

Cotangent and Cosecant Functions: $\{s \mid s \neq n\pi, \text{ where } n \text{ is any integer}\}$

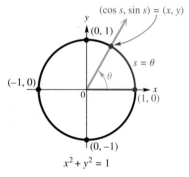

$(\cos s, \sin s) = (x, y)$

$s = \theta$

$x^2 + y^2 = 1$

Figure 13

Finding Values of Circular Functions

The circular functions of real numbers correspond to the trigonometric functions of angles measured in radians. Let us assume that angle θ is in standard position, superimposed on the unit circle. See Figure 13. Suppose that θ is the *radian* measure of this angle. Using the arc length formula $s = r\theta$ with $r = 1$, we have $s = \theta$. Thus, the length of the intercepted arc is the real number that corresponds to the radian measure of θ. Using the trigonometric function definitions from **Section 1.3,**

$$\sin \theta = \frac{y}{r} = \frac{y}{1} = y = \sin s, \qquad \text{and} \qquad \cos \theta = \frac{x}{r} = \frac{x}{1} = x = \cos s,$$

and so on. As shown here, the trigonometric functions and the circular functions lead to the same function values, provided we think of the angles as being in radian measure. This leads to the following important result.

EVALUATING A CIRCULAR FUNCTION

Circular function values of real numbers are obtained in the same manner as trigonometric function values of angles measured in radians. This applies both to methods of finding exact values (such as reference angle analysis) and to calculator approximations. *Calculators must be in radian mode when finding circular function values.*

▶ **EXAMPLE 1** FINDING EXACT CIRCULAR FUNCTION VALUES

Find the exact values of $\sin \frac{3\pi}{2}$, $\cos \frac{3\pi}{2}$, and $\tan \frac{3\pi}{2}$.

Solution Evaluating a circular function at the real number $\frac{3\pi}{2}$ is equivalent to evaluating it at $\frac{3\pi}{2}$ radians. An angle of $\frac{3\pi}{2}$ radians intersects the unit circle at the point $(0, -1)$, as shown in Figure 14. Since

$$\sin s = y, \quad \cos s = x, \quad \text{and} \quad \tan s = \frac{y}{x},$$

it follows that

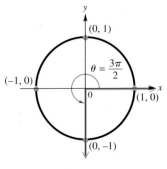

Figure 14

$$\sin \frac{3\pi}{2} = -1, \quad \cos \frac{3\pi}{2} = 0, \quad \text{and} \quad \tan \frac{3\pi}{2} \text{ is undefined.}$$

NOW TRY EXERCISE 1. ◀

▶ **EXAMPLE 2** FINDING EXACT CIRCULAR FUNCTION VALUES

(a) Use Figure 12 to find the exact values of $\cos \frac{7\pi}{4}$ and $\sin \frac{7\pi}{4}$.

(b) Use Figure 12 and the definition of tangent to find the exact value of $\tan\left(-\frac{5\pi}{3}\right)$.

(c) Use reference angles and degree/radian conversion to find the exact value of $\cos \frac{2\pi}{3}$.

Solution

(a) In Figure 12, we see that the real number $\frac{7\pi}{4}$ corresponds to the unit circle point $\left(\frac{\sqrt{2}}{2}, -\frac{\sqrt{2}}{2}\right)$. Thus,

$$\cos \frac{7\pi}{4} = \frac{\sqrt{2}}{2} \quad \text{and} \quad \sin \frac{7\pi}{4} = -\frac{\sqrt{2}}{2}.$$

(b) Moving around the unit circle $\frac{5\pi}{3}$ units in the *negative* direction yields the same ending point as moving around $\frac{\pi}{3}$ units in the positive direction. Thus, $-\frac{5\pi}{3}$ corresponds to $\left(\frac{1}{2}, \frac{\sqrt{3}}{2}\right)$, and

$$\tan\left(-\frac{5\pi}{3}\right) = \tan \frac{\pi}{3} = \frac{\frac{\sqrt{3}}{2}}{\frac{1}{2}} = \frac{\sqrt{3}}{2} \div \frac{1}{2} = \frac{\sqrt{3}}{2} \cdot \frac{2}{1} = \sqrt{3}.$$

$\boxed{\tan s = \frac{y}{x}}$

(c) An angle of $\frac{2\pi}{3}$ radians corresponds to an angle of 120°. In standard position, 120° lies in quadrant II with a reference angle of 60°, so

Cosine is negative in quadrant II.

$$\cos \frac{2\pi}{3} = \cos 120° = -\cos 60° = -\frac{1}{2}.$$

Reference angle **(Section 2.2)**

NOW TRY EXERCISES 7, 17, AND 21. ◀

▶ **EXAMPLE 3** APPROXIMATING CIRCULAR FUNCTION VALUES

Find a calculator approximation for each circular function value.

(a) cos 1.85   **(b)** cos .5149   **(c)** cot 1.3209   **(d)** sec(−2.9234)

Solution

cos(1.85)
               -.2756

Radian mode
This is how the TI-83/84 Plus calculator displays the result of Example 3(a), fixed to four decimal digits.

(a) With a calculator in radian mode, we find cos 1.85 ≈ −.2756.

(b) cos .5149 ≈ .8703   Use a calculator in radian mode.

(c) As before, to find cotangent, secant, and cosecant function values, we must use the appropriate reciprocal functions. To find cot 1.3209, first find tan 1.3209 and then find the reciprocal.

$$\cot 1.3209 = \frac{1}{\tan 1.3209} \approx .2552$$

(d) $\sec(-2.9234) = \dfrac{1}{\cos(-2.9234)} \approx -1.0243$

NOW TRY EXERCISES 23, 29, AND 33. ◀

▶ **Caution** A common error in trigonometry is using a calculator in degree mode when radian mode should be used. *Remember, if you are finding a circular function value of a real number, the calculator must be in radian mode.*

Determining a Number with a Given Circular Function Value

Recall from **Section 2.3** how we used a calculator to determine an angle measure, given a trigonometric function value of the angle. *Remember that the keys marked* **sin⁻¹, cos⁻¹,** *and* **tan⁻¹** *do not represent reciprocal functions.*

▶ **EXAMPLE 4**   FINDING A NUMBER GIVEN ITS CIRCULAR FUNCTION VALUE

(a) Approximate the value of s in the interval $\left[0, \frac{\pi}{2}\right]$, if cos s = .9685.

(b) Find the exact value of s in the interval $\left[\pi, \frac{3\pi}{2}\right]$, if tan s = 1.

Solution

(a) Since we are given a cosine value and want to determine the real number in $\left[0, \frac{\pi}{2}\right]$ having this cosine value, we use the *inverse cosine* function of a calculator. With the calculator in radian mode, we find

$$\cos^{-1}(.9685) \approx .2517. \quad \text{(Section 2.3)}$$

See Figure 15. (Refer to your owner's manual to determine how to evaluate the sin⁻¹, cos⁻¹, and tan⁻¹ functions with your calculator.)

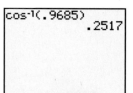

cos⁻¹(.9685)
                .2517

Figure 15

(b) Recall that $\tan \frac{\pi}{4} = 1$, and in quadrant III tan s is positive. Therefore,

$$\tan\left(\pi + \frac{\pi}{4}\right) = \tan \frac{5\pi}{4} = 1,$$

and $s = \frac{5\pi}{4}$. Figure 16 supports this result.

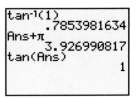

Figure 16

NOW TRY EXERCISES 55 AND 65. ◀

Applying Circular Functions

▶ **EXAMPLE 5**   MODELING THE ANGLE OF ELEVATION OF THE SUN

The angle of elevation θ of the sun in the sky at any latitude L is calculated with the formula

$$\sin \theta = \cos D \cos L \cos \omega + \sin D \sin L,$$

where $\theta = 0$ corresponds to sunrise and $\theta = \frac{\pi}{2}$ occurs if the sun is directly overhead. ω (the Greek letter *omega*) is the number of radians that Earth has rotated through since noon, when $\omega = 0$. D is the declination of the sun, which varies because Earth is tilted on its axis. (*Source:* Winter, C., R. Sizmann, and L. L. Vant-Hull, Editors, *Solar Power Plants*, Springer-Verlag, 1991.)

Sacramento, California, has latitude $L = 38.5°$ or .6720 radian. Find the angle of elevation θ of the sun at 3 P.M. on February 29, 2008, where at that time $D \approx -.1425$ and $\omega \approx .7854$.

Solution   Use the given formula for sin θ.

$$\sin \theta = \cos D \cos L \cos \omega + \sin D \sin L$$
$$= \cos(-.1425) \cos(.6720) \cos(.7854) + \sin(-.1425) \sin(.6720)$$
$$\approx .4593426188$$

Thus, $\theta \approx .4773$ radian or 27.3°.   Use inverse sine.

NOW TRY EXERCISE 77. ◀

CONNECTIONS   The diagram shown in Figure 17 illustrates a correspondence that ties together the right triangle ratio definitions of the trigonometric functions introduced in **Chapter 2** with the unit circle interpretation presented in this section. The arc SR is the first quadrant portion of the unit circle, and the standard-position angle POQ is designated θ. By definition, the coordinates of P are $(\cos \theta, \sin \theta)$. The six trigonometric functions of θ can be interpreted as lengths of line segments found in Figure 17.

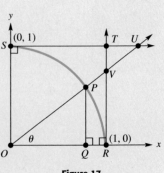

Figure 17

cos θ = OQ, sin θ = PQ

Use right triangle POQ and right triangle ratios:

$$\cos \theta = \frac{\text{side adjacent to } \theta}{\text{hypotenuse}} = \frac{OQ}{OP} = \frac{OQ}{1} = OQ,$$

$$\sin \theta = \frac{\text{side opposite } \theta}{\text{hypotenuse}} = \frac{PQ}{OP} = \frac{PQ}{1} = PQ.$$

tan θ = VR, sec θ = OV

Use right triangle VOR and right triangle ratios:

$$\tan \theta = \frac{\text{side opposite } \theta}{\text{side adjacent to } \theta} = \frac{VR}{OR} = \frac{VR}{1} = VR,$$

$$\sec \theta = \frac{\text{hypotenuse}}{\text{side adjacent to } \theta} = \frac{OV}{OR} = \frac{OV}{1} = OV.$$

cot θ = US, csc θ = OU

US and OR are parallel, and, thus, angle SUO is equal to θ as it is an alternate interior angle to angle $POQ = \theta$. Use right triangle USO and right triangle ratios:

$$\cot SUO = \cot \theta = \frac{\text{side adjacent to } \theta}{\text{side opposite } \theta} = \frac{US}{OS} = \frac{US}{1} = US,$$

$$\csc SUO = \csc \theta = \frac{\text{hypotenuse}}{\text{side opposite } \theta} = \frac{OU}{OS} = \frac{OU}{1} = OU.$$

Figure 18 uses color to illustrate the results found above.

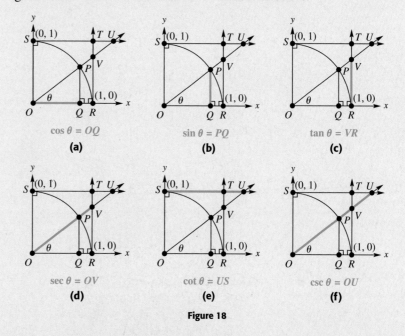

$\cos \theta = OQ$
(a)

$\sin \theta = PQ$
(b)

$\tan \theta = VR$
(c)

$\sec \theta = OV$
(d)

$\cot \theta = US$
(e)

$\csc \theta = OU$
(f)

Figure 18

FOR DISCUSSION OR WRITING

1. See if you can find other ways to justify the results given above.

2. How can this interpretation be extended to angles in other quadrants?

3.3 Exercises

For each value of the real number s, find **(a)** sin *s,* **(b)** cos *s, and* **(c)** tan *s. See Example 1.*

1. $s = \dfrac{\pi}{2}$

2. $s = \pi$

3. $s = 2\pi$

4. $s = 3\pi$

5. $s = -\pi$

6. $s = -\dfrac{3\pi}{2}$

Find the exact circular function value for each of the following. See Example 2.

7. $\sin \dfrac{7\pi}{6}$

8. $\cos \dfrac{5\pi}{3}$

9. $\tan \dfrac{3\pi}{4}$

10. $\sec \dfrac{2\pi}{3}$

11. $\csc \dfrac{11\pi}{6}$

12. $\cot \dfrac{5\pi}{6}$

13. $\cos\left(-\dfrac{4\pi}{3}\right)$

14. $\tan \dfrac{17\pi}{3}$

15. $\cos \dfrac{7\pi}{4}$

16. $\sec \dfrac{5\pi}{4}$

17. $\sin\left(-\dfrac{4\pi}{3}\right)$

18. $\sin\left(-\dfrac{5\pi}{6}\right)$

19. $\sec \dfrac{23\pi}{6}$

20. $\csc \dfrac{13\pi}{3}$

21. $\tan \dfrac{5\pi}{6}$

22. $\cos \dfrac{3\pi}{4}$

Find a calculator approximation for each circular function value. See Example 3.

23. sin .6109

24. sin .8203

25. cos(−1.1519)

26. cos(−5.2825)

27. tan 4.0203

28. tan 6.4752

29. csc(−9.4946)

30. csc 1.3875

31. sec 2.8440

32. sec(−8.3429)

33. cot 6.0301

34. cot 3.8426

Concept Check The figure displays a unit circle and an angle of 1 radian. The tick marks on the circle are spaced at every two-tenths radian. Use the figure to estimate each value.

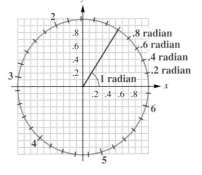

35. cos .8   **36.** cos .6   **37.** sin 2

38. sin 4   **39.** sin 3.8   **40.** cos 3.2

41. a positive angle whose cosine is −.65

42. a positive angle whose sine is −.95

43. a positive angle whose sine is .7

44. a positive angle whose cosine is .3

Concept Check Without using a calculator, decide whether each function value is positive or negative. (Hint: Consider the radian measures of the quadrantal angles.)

45. cos 2

46. sin(−1)

47. sin 5

48. cos 6

49. tan 6.29

50. tan(−6.29)

Concept Check Each figure in Exercises 51–54 shows an angle θ in standard position with its terminal side intersecting the unit circle. Evaluate the six circular function values of θ.

51.

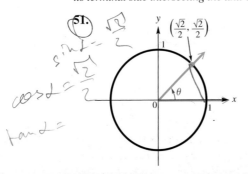

52.

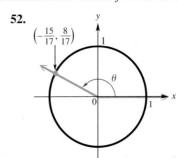

53.

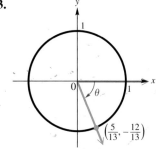

54.

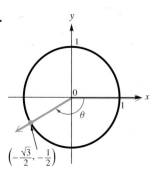

Find the value of s in the interval $\left[0, \frac{\pi}{2}\right]$ *that makes each statement true. See Example 4(a).*

55. $\tan s = .2126$ **56.** $\cos s = .7826$ **57.** $\sin s = .9918$

58. $\cot s = .2994$ **59.** $\sec s = 1.0806$ **60.** $\csc s = 1.0219$

Find the exact value of s in the given interval that has the given circular function value. Do not use a calculator. See Example 4(b).

61. $\left[\frac{\pi}{2}, \pi\right]$; $\sin s = \frac{1}{2}$

62. $\left[\frac{\pi}{2}, \pi\right]$; $\cos s = -\frac{1}{2}$

63. $\left[\pi, \frac{3\pi}{2}\right]$; $\tan s = \sqrt{3}$

64. $\left[\pi, \frac{3\pi}{2}\right]$; $\sin s = -\frac{1}{2}$

65. $\left[\frac{3\pi}{2}, 2\pi\right]$; $\tan s = -1$

66. $\left[\frac{3\pi}{2}, 2\pi\right]$; $\cos s = \frac{\sqrt{3}}{2}$

Suppose an arc of length s lies on the unit circle $x^2 + y^2 = 1$, *starting at the point* $(1, 0)$ *and terminating at the point* (x, y). *(See Figure 11.) Use a calculator to find the approximate coordinates for* (x, y). *(Hint:* $x = \cos s$ *and* $y = \sin s$.)

67. $s = 2.5$ **68.** $s = 3.4$ **69.** $s = -7.4$ **70.** $s = -3.9$

Concept Check *For each value of s, use a calculator to find* $\sin s$ *and* $\cos s$ *and then use the results to decide in which quadrant an angle of s radians lies.*

71. $s = 51$ **72.** $s = 49$ **73.** $s = 65$ **74.** $s = 79$

Concept Check *In Exercises 75 and 76, each graphing calculator screen shows a point on the unit circle. What is the length of the shortest arc of the circle from* $(1, 0)$ *to the point?*

75.

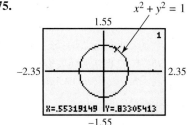

76.

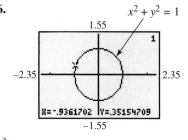

(Modeling) *Solve each problem. See Example 5.*

77. *Elevation of the Sun* Refer to Example 5.

    (a) Repeat the example for New Orleans, which has latitude $L = 30°$.

    (b) Compare your answers. Do they agree with your intuition?

78. *Length of a Day* The number of daylight hours H at any location can be calculated using the formula

$$\cos(.1309H) = -\tan D \tan L,$$

where D and L are defined in Example 5. Use this trigonometric equation to calculate the shortest and longest days in Minneapolis, Minnesota, if its latitude $L = 44.88°$, the shortest day occurs when $D = -23.44°$, and the longest day occurs when $D = 23.44°$. Remember to convert degrees to radians. (*Source:* Winter, C., R. Sizmann, and L. L. Vant-Hull, Editors, *Solar Power Plants,* Springer-Verlag, 1991.)

79. *Maximum Temperatures* Because the values of the circular functions repeat every 2π, they are used to describe things that repeat periodically. For example, the maximum afternoon temperature in a given city might be modeled by

$$t = 60 - 30 \cos \frac{x\pi}{6},$$

where t represents the maximum afternoon temperature in month x, with $x = 0$ representing January, $x = 1$ representing February, and so on. Find the maximum afternoon temperature for each of the following months.

(a) January   **(b)** April   **(c)** May

(d) June   **(e)** August   **(f)** October

80. *Temperature in Fairbanks* The temperature in Fairbanks is modeled by

$$T(x) = 37 \sin\left[\frac{2\pi}{365}(x - 101)\right] + 25,$$

where $T(x)$ is the temperature in degrees Fahrenheit on day x, with $x = 1$ corresponding to January 1 and $x = 365$ corresponding to December 31. Use a calculator to estimate the temperature on the following days. (*Source:* Lando, B. and C. Lando, "Is the Graph of Temperature Variation a Sine Curve?", *The Mathematics Teacher,* 70, September 1977.)

(a) March 1 (day 60)   **(b)** April 1 (day 91)   **(c)** Day 150

(d) June 15   **(e)** September 1   **(f)** October 31

CHAPTER 3 ▶ **Quiz** (Sections 3.1–3.3)

Convert each degree measure to radians.

1. $225°$ **2.** $-330°$

Convert each radian measure to degrees.

3. $\dfrac{5\pi}{3}$ **4.** $-\dfrac{7\pi}{6}$

A central angle of a circle with radius 300 in. intercepts an arc of 450 in. (These measures are accurate to the nearest inch.) Find each measure.

5. the radian measure of the angle   **6.** the area of the sector

Find each circular function value. Give exact values.

7. $\cos \dfrac{7\pi}{4}$ **8.** $\sin\left(-\dfrac{5\pi}{6}\right)$ **9.** $\tan 3\pi$

10. Find the exact value of s in the interval $\left[\frac{\pi}{2}, \pi\right]$ if $\sin s = \frac{\sqrt{3}}{2}$.

3.4 Linear and Angular Speed

Linear Speed ▪ Angular Speed

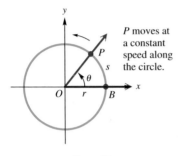

Figure 19

Linear Speed In many situations we need to know how fast a point on a circular disk is moving or how fast the central angle of such a disk is changing. Some examples occur with machinery involving gears or pulleys or the speed of a car around a curved portion of highway.

Suppose that point P moves at a constant speed along a circle of radius r and center O. See Figure 19. The measure of how fast the position of P is changing is called **linear speed.** If v represents linear speed, then

$$\text{speed} = \frac{\text{distance}}{\text{time}}, \quad \text{or} \quad v = \frac{s}{t},$$

where s is the length of the arc traced by point P at time t. (This formula is just a restatement of $d = rt$ with s as distance, v as rate (speed), and t as time.)

Angular Speed As point P in Figure 19 moves along the circle, ray OP rotates around the origin. Since ray OP is the terminal side of angle POB, the measure of the angle changes as P moves along the circle. The measure of how fast angle POB is changing is called **angular speed.** Angular speed, symbolized ω, is given as

$$\omega = \frac{\theta}{t}, \quad \theta \text{ in radians,}$$

where θ is the measure of angle POB at time t. As with earlier formulas in this chapter, θ must be measured in radians, with ω expressed as radians per unit of time. Angular speed is used in physics and engineering, among other applications.

In **Section 3.2,** the length s of the arc intercepted on a circle of radius r by a central angle of measure θ radians was found to be $s = r\theta$. Using this formula, the formula for linear speed, $v = \frac{s}{t}$, becomes

$$v = \frac{s}{t} = \frac{r\theta}{t} = r \cdot \frac{\theta}{t} = r\omega. \quad s = r\theta; \omega = \frac{\theta}{t}$$

The formula $v = r\omega$ relates linear and angular speeds.

As an example of linear and angular speeds, consider the following. The human joint that can be flexed the fastest is the wrist, which can rotate through $90°$, or $\frac{\pi}{2}$ radians, in .045 sec while holding a tennis racket. The angular speed of a human wrist swinging a tennis racket is

$$\omega = \frac{\theta}{t} = \frac{\frac{\pi}{2}}{.045} \approx 35 \text{ radians per sec.}$$

If the radius (distance) from the tip of the racket to the wrist joint is 2 ft, then the speed at the tip of the racket is

$$v = r\omega \approx 2(35) = 70 \text{ ft per sec,} \quad \text{or about 48 mph.}$$

In a tennis serve the arm rotates at the shoulder, so the final speed of the racket is considerably faster. (*Source:* Cooper, J. and R. Glassow, *Kinesiology,* Second Edition, C.V. Mosby, 1968.)

The formulas for angular and linear speed are summarized in the table.

Angular Speed	Linear Speed
$\omega = \dfrac{\theta}{t}$ (ω in radians per unit time, θ in radians)	$v = \dfrac{s}{t}$ $v = \dfrac{r\theta}{t}$ $v = r\omega$

▶ **EXAMPLE 1**   USING LINEAR AND ANGULAR SPEED FORMULAS

Suppose that point P is on a circle with radius 10 cm, and ray OP is rotating with angular speed $\frac{\pi}{18}$ radian per sec.

(a) Find the angle generated by P in 6 sec.

(b) Find the distance traveled by P along the circle in 6 sec.

(c) Find the linear speed of P in centimeters per second.

Solution

(a) The speed of ray OP is $\omega = \frac{\pi}{18}$ radian per sec. Since $\omega = \frac{\theta}{t}$, then in 6 sec

$$\frac{\pi}{18} = \frac{\theta}{6} \qquad \text{\footnotesize Let } \omega = \tfrac{\pi}{18} \text{ and } t = 6 \text{ in the}$$
$$\text{\footnotesize angular speed formula.}$$

$$\theta = \frac{6\pi}{18} = \frac{\pi}{3} \text{ radians.} \quad \text{\footnotesize Solve for } \theta.$$

(b) From part (a), P generates an angle of $\frac{\pi}{3}$ radians in 6 sec. The distance traveled by P along the circle is

$$s = r\theta = 10\left(\frac{\pi}{3}\right) = \frac{10\pi}{3} \text{ cm.} \quad \text{(Section 3.2)}$$

(c) From part (b), $s = \frac{10\pi}{3}$ for 6 sec, so for 1 sec we divide by 6.

$$v = \frac{s}{t} = \frac{\frac{10\pi}{3}}{6} = \frac{10\pi}{3} \div 6 = \frac{10\pi}{3} \cdot \frac{1}{6} = \frac{5\pi}{9} \text{ cm per sec}$$

> Be careful simplifying this complex fraction.

NOW TRY EXERCISE 3. ◀

▶ **EXAMPLE 2**   FINDING ANGULAR SPEED OF A PULLEY AND LINEAR SPEED OF A BELT

A belt runs a pulley of radius 6 cm at 80 revolutions per min.

(a) Find the angular speed of the pulley in radians per second.

(b) Find the linear speed of the belt in centimeters per second.

Solution

(a) In 1 min, the pulley makes 80 revolutions. Each revolution is 2π radians, so

$$80(2\pi) = 160\pi \text{ radians per min.}$$

Since there are 60 sec in 1 min, we find ω, the angular speed in radians per second, by dividing 160π by 60.

$$\omega = \frac{160\pi}{60} = \frac{8\pi}{3} \text{ radians per sec}$$

(b) The linear speed of the belt will be the same as that of a point on the circumference of the pulley. Thus,

$$v = r\omega = 6\left(\frac{8\pi}{3}\right) = 16\pi \approx 50 \text{ cm per sec.}$$

> NOW TRY EXERCISE 39. ◀

▶ **EXAMPLE 3** **FINDING LINEAR SPEED AND DISTANCE TRAVELED BY A SATELLITE**

A satellite traveling in a circular orbit 1600 km above the surface of Earth takes 2 hr to make an orbit. The radius of Earth is approximately 6400 km. See Figure 20.

(a) Approximate the linear speed of the satellite in kilometers per hour.

(b) Approximate the distance the satellite travels in 4.5 hr.

Solution

(a) The distance of the satellite from the center of Earth is approximately

$$r = 1600 + 6400 = 8000 \text{ km.}$$

For one orbit, $\theta = 2\pi$, and

$$s = r\theta = 8000(2\pi) \text{ km.}$$

Since it takes 2 hr to complete an orbit, the linear speed is approximately

$$v = \frac{s}{t} = \frac{8000(2\pi)}{2} = 8000\pi \approx 25{,}000 \text{ km per hr.}$$

(b) $s = vt = 8000\pi(4.5) = 36{,}000\pi \approx 110{,}000$ km

> NOW TRY EXERCISE 37. ◀

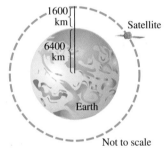

1600 km
6400 km
Satellite
Earth
Not to scale

Figure 20

3.4 Exercises

1. *Concept Check* If the point P moves around the circumference of the unit circle at an angular velocity of 1 radian per sec, how long will it take for P to move around the entire circle?

2. *Concept Check* If the point P moves around the circumference of the unit circle at a speed of 1 unit per sec, how long will it take for P to move around the entire circle?

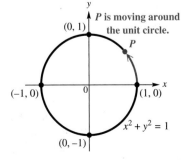

P is moving around the unit circle.

$(0, 1)$
$(-1, 0)$
$(1, 0)$
$(0, -1)$
$x^2 + y^2 = 1$

Suppose that point P is on a circle with radius r, and ray OP is rotating with angular speed ω. For the given values of r, ω, and t, find each of the following.

(a) *the angle generated by P in time t*
(b) *the distance traveled by P along the circle in time t*
(c) *the linear speed of P*

See Example 1.

3. $r = 20$ cm, $\omega = \dfrac{\pi}{12}$ radian per sec, $t = 6$ sec

4. $r = 30$ cm, $\omega = \dfrac{\pi}{10}$ radian per sec, $t = 4$ sec

Use the formula $\omega = \frac{\theta}{t}$ to find the value of the missing variable.

5. $\omega = \dfrac{2\pi}{3}$ radians per sec, $t = 3$ sec

6. $\omega = \dfrac{\pi}{4}$ radian per min, $t = 5$ min

7. $\theta = \dfrac{3\pi}{4}$ radians, $t = 8$ sec

8. $\theta = \dfrac{2\pi}{5}$ radians, $t = 10$ sec

9. $\theta = \dfrac{2\pi}{9}$ radian, $\omega = \dfrac{5\pi}{27}$ radian per min

10. $\theta = \dfrac{3\pi}{8}$ radians, $\omega = \dfrac{\pi}{24}$ radian per min

11. $\theta = 3.871142$ radians, $t = 21.4693$ sec

12. $\omega = .90674$ radian per min, $t = 11.876$ min

Use the formula $v = r\omega$ to find the value of the missing variable.

13. $r = 12$ m, $\omega = \dfrac{2\pi}{3}$ radians per sec

14. $r = 8$ cm, $\omega = \dfrac{9\pi}{5}$ radians per sec

15. $v = 9$ m per sec, $r = 5$ m

16. $v = 18$ ft per sec, $r = 3$ ft

17. $v = 107.692$ m per sec, $r = 58.7413$ m

18. $r = 24.93215$ cm, $\omega = .372914$ radian per sec

The formula $\omega = \frac{\theta}{t}$ can be rewritten as $\theta = \omega t$. Using ωt for θ changes $s = r\theta$ to $s = r\omega t$. Use the formula $s = r\omega t$ to find the value of the missing variable.

19. $r = 6$ cm, $\omega = \dfrac{\pi}{3}$ radians per sec, $t = 9$ sec

20. $r = 9$ yd, $\omega = \dfrac{2\pi}{5}$ radians per sec, $t = 12$ sec

21. $s = 6\pi$ cm, $r = 2$ cm, $\omega = \dfrac{\pi}{4}$ radian per sec

22. $s = \dfrac{12\pi}{5}$ m, $r = \dfrac{3}{2}$ m, $\omega = \dfrac{2\pi}{5}$ radians per sec

23. $s = \dfrac{3\pi}{4}$ km, $r = 2$ km, $t = 4$ sec   **24.** $s = \dfrac{8\pi}{9}$ m, $r = \dfrac{4}{3}$ m, $t = 12$ sec

Find ω for each of the following.

25. the hour hand of a clock

26. a line from the center to the edge of a CD revolving 300 times per min

27. the minute hand of a clock

28. the second hand of a clock

Find v for each of the following.

29. the tip of the minute hand of a clock, if the hand is 7 cm long

30. the tip of the second hand of a clock, if the hand is 28 mm long

31. a point on the edge of a flywheel of radius 2 m, rotating 42 times per min

32. a point on the tread of a tire of radius 18 cm, rotating 35 times per min

33. the tip of an airplane propeller 3 m long, rotating 500 times per min (*Hint: r = 1.5 m*)

34. a point on the edge of a gyroscope of radius 83 cm, rotating 680 times per min

Solve each problem. See Examples 1–3.

35. *Speed of a Bicycle*   The tires of a bicycle have radius 13 in. and are turning at the rate of 215 revolutions per min. See the figure. How fast is the bicycle traveling in miles per hour? (*Hint:* 5280 ft = 1 mi)

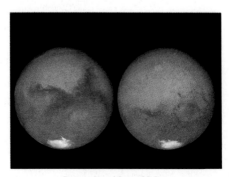

13 in.

36. *Hours in a Martian Day*   Mars rotates on its axis at the rate of about .2552 radian per hr. Approximately how many hours are in a Martian day (or *sol*)? (*Source:* Wright, J. W., General Editor, *The Universal Almanac,* Andrews and McMeel, 1997.)

Opposite sides of Mars

37. *Angular and Linear Speeds of Earth*   Earth travels about the sun in an orbit that is almost circular. Assume that the orbit is a circle with radius 93,000,000 mi. Its angular and linear speeds are used in designing solar-power facilities.

    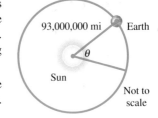

    (a) Assume that a year is 365 days, and find the angle formed by Earth's movement in one day.
    (b) Give the angular speed in radians per hour.
    (c) Find the linear speed of Earth in miles per hour.

38. *Angular and Linear Speeds of Earth*   Earth revolves on its axis once every 24 hr. Assuming that Earth's radius is 6400 km, find the following.

    (a) angular speed of Earth in radians per day and radians per hour
    (b) linear speed at the North Pole or South Pole
    (c) linear speed at Quito, Ecuador, a city on the equator
    (d) linear speed at Salem, Oregon (halfway from the equator to the North Pole)

39. *Speeds of a Pulley and a Belt*   The pulley shown has a radius of 12.96 cm. Suppose it takes 18 sec for 56 cm of belt to go around the pulley.

    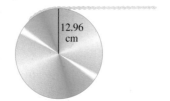

    (a) Find the linear speed of the belt in centimeters per second.

    (b) Find the angular speed of the pulley in radians per second.

40. *Angular Speeds of Pulleys*   The two pulleys in the figure have radii of 15 cm and 8 cm, respectively. The larger pulley rotates 25 times in 36 sec. Find the angular speed of each pulley in radians per second.

    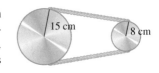

41. *Radius of a Spool of Thread*   A thread is being pulled off a spool at the rate of 59.4 cm per sec. Find the radius of the spool if it makes 152 revolutions per min.

42. *Time to Move Along a Railroad Track*   A railroad track is laid along the arc of a circle of radius 1800 ft. The circular part of the track subtends a central angle of 40°. How long (in seconds) will it take a point on the front of a train traveling 30.0 mph to go around this portion of the track?

43. *Angular Speed of a Motor Propeller*   A 90-horsepower outboard motor at full throttle will rotate its propeller at exactly 5000 revolutions per min. Find the angular speed of the propeller in radians per second.

44. *Linear Speed of a Golf Club*   The shoulder joint can rotate at 25.0 radians per sec. If a golfer's arm is straight and the distance from the shoulder to the club head is 5.00 ft, find the linear speed of the club head from shoulder rotation. (*Source:* Cooper, J. and R. Glassow, *Kinesiology,* Second Edition, C.V. Mosby, 1968.)

Chapter 3 Summary

QUICK REVIEW

CONCEPTS	EXAMPLES

3.1 Radian Measure

An angle with its vertex at the center of a circle that intercepts an arc on the circle equal in length to the radius of the circle has a measure of **1 radian.**

See Figure 1 on page 102.

Degree/Radian Relationship $180° = \pi$ **radians**

Converting Between Degrees and Radians
1. Multiply a degree measure by $\frac{\pi}{180}$ radian and simplify to convert to radians.

2. Multiply a radian measure by $\frac{180°}{\pi}$ and simplify to convert to degrees.

Convert 135° to radians.

$$135° = 135\left(\frac{\pi}{180} \text{ radian}\right) = \frac{3\pi}{4} \text{ radians}$$

Convert $-\frac{5\pi}{3}$ radians to degrees.

$$-\frac{5\pi}{3} \text{ radians} = -\frac{5\pi}{3}\left(\frac{180°}{\pi}\right) = -300°$$

3.2 Applications of Radian Measure

Arc Length

The length s of the arc intercepted on a circle of radius r by a central angle of measure θ radians is given by the product of the radius and the radian measure of the angle, or

$$s = r\theta, \qquad \theta \text{ in radians.}$$

In the figure, $s = r\theta$, so

$$\theta = \frac{s}{r} = \frac{3}{4} \text{ radian.}$$

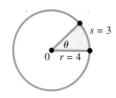

Area of a Sector

The area of a sector of a circle of radius r and central angle θ is given by

$$A = \frac{1}{2}r^2\theta, \qquad \theta \text{ in radians.}$$

The area of the sector in the figure above is

$$A = \frac{1}{2}(4)^2\left(\frac{3}{4}\right) = 6 \text{ sq units.}$$

(continued)

CONCEPTS	EXAMPLES

3.3 The Unit Circle and Circular Functions

Circular Functions

Start at the point $(1, 0)$ on the unit circle $x^2 + y^2 = 1$ and lay off an arc of length $|s|$ along the circle, going counterclockwise if s is positive and clockwise if s is negative. Let the endpoint of the arc be at the point (x, y). The six circular functions of s are defined as follows. (Assume that no denominators are 0.)

$$\sin s = y \qquad \cos s = x \qquad \tan s = \frac{y}{x}$$

$$\csc s = \frac{1}{y} \qquad \sec s = \frac{1}{x} \qquad \cot s = \frac{x}{y}$$

The Unit Circle

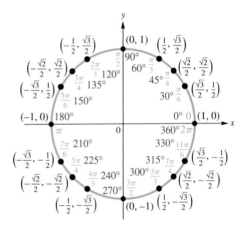

The unit circle $x^2 + y^2 = 1$

Use the unit circle to find each value.

$$\sin \frac{5\pi}{6} = \frac{1}{2}$$

$$\cos \frac{3\pi}{2} = 0$$

$$\tan \frac{\pi}{4} = \frac{\frac{\sqrt{2}}{2}}{\frac{\sqrt{2}}{2}} = 1$$

$$\csc \frac{7\pi}{4} = \frac{1}{-\frac{\sqrt{2}}{2}} = -\sqrt{2}$$

$$\sec \frac{7\pi}{6} = \frac{1}{-\frac{\sqrt{3}}{2}} = -\frac{2\sqrt{3}}{3}$$

$$\cot \frac{\pi}{3} = \frac{\frac{1}{2}}{\frac{\sqrt{3}}{2}} = \frac{\sqrt{3}}{3}$$

$$\sin 0 = 0$$

$$\cos \frac{\pi}{2} = 0$$

$$\tan \pi = \frac{0}{-1} = 0$$

$$\sin \left(-\frac{\pi}{2}\right) = \sin \frac{3\pi}{2} = -1$$

$$\cot \left(-\frac{3\pi}{2}\right) = \cot \frac{\pi}{2} = \frac{0}{1} = 0$$

$$\tan \frac{\pi}{2} \text{ is undefined.}$$

3.4 Linear and Angular Speed

Formulas for Angular and Linear Speed

Angular Speed	Linear Speed
$\omega = \dfrac{\theta}{t}$	$v = \dfrac{s}{t}$
	$v = \dfrac{r\theta}{t}$
(ω in radians per unit time, θ in radians)	$v = r\omega$

A belt runs a pulley of radius 8 in. at 60 revolutions per min. Find

(a) the angular speed ω in radians per minute, and

(b) the linear speed v of the belt in inches per minute.

(a) $\omega = 60(2\pi) = 120\pi$ radians per min

(b) $v = r\omega = 8(120\pi) = 960\pi$ in. per min

CHAPTER 3 ▶ Review Exercises

1. *Concept Check*   Which is larger—an angle of 1° or an angle of 1 radian?

2. *Concept Check*   Consider each angle in standard position having the given radian measure. In what quadrant does the terminal side lie?

   (a) 3   **(b)** 4   **(c)** −2   **(d)** 7

3. Find three angles coterminal with an angle of 1 radian.

4. Give an expression that generates all angles coterminal with an angle of $\frac{\pi}{6}$ radian. Let n represent any integer.

Convert each degree measure to radians. Leave answers as multiples of π.

5. 45°   6. 120°   7. 175°

8. 330°   9. 800°   10. 1020°

Convert each radian measure to degrees.

11. $\dfrac{5\pi}{4}$ 12. $\dfrac{9\pi}{10}$ 13. $\dfrac{8\pi}{3}$

14. $-\dfrac{6\pi}{5}$ 15. $-\dfrac{11\pi}{18}$ 16. $\dfrac{21\pi}{5}$

Suppose the tip of the minute hand of a clock is 2 in. from the center of the clock. For each duration, determine the distance traveled by the tip of the minute hand.

17. 15 min   18. 20 min

19. 3 hr   20. $10\frac{1}{2}$ hr

Solve each problem. Use a calculator as necessary.

21. *Arc Length*   The radius of a circle is 15.2 cm. Find the length of an arc of the circle intercepted by a central angle of $\frac{3\pi}{4}$ radians.

22. *Arc Length*   Find the length of an arc intercepted by a central angle of .769 radian on a circle with radius 11.4 cm.

23. *Arc Length*   A circle has radius 8.973 cm. Find the length of an arc on this circle intercepted by a central angle of 49.06°.

24. *Area of a Sector*   A central angle of $\frac{7\pi}{4}$ radians forms a sector of a circle. Find the area of the sector if the radius of the circle is 28.69 in.

25. *Area of a Sector*   Find the area of a sector of a circle having a central angle of 21° 40′ in a circle of radius 38.0 m.

26. *Height of a Tree*   A tree 2000 yd away subtends an angle of 1° 10′. Find the height of the tree to two significant digits.

Distance Between Cities Assume that the radius of Earth is 6400 km *in Exercises 27 and 28.*

27. Find the distance in kilometers between cities on a north-south line that are on latitudes 28° N and 12° S, respectively.

28. Two cities on the equator have longitudes of 72° E and 35° W, respectively. Find the distance between the cities.

Concept Check *In Exercises 29 and 30, find the measure of the central angle θ (in radians) and the area of the sector.*

29.
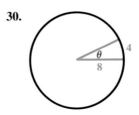

30.

31. *Concept Check* The hour hand of a wall clock measures 6 in. from its tip to the center of the clock.

   (a) Through what angle (in radians) does the hour hand pass between 1 o'clock and 3 o'clock?

   (b) What distance does the tip of the hour hand travel during the time period from 1 o'clock to 3 o'clock?

32. Describe what would happen to the central angle for a given arc length of a circle if the circle's radius were doubled. (Assume everything else is unchanged.)

Find each exact function value. Do not use a calculator.

33. $\tan \dfrac{\pi}{3}$

34. $\cos \dfrac{2\pi}{3}$

35. $\sin\left(-\dfrac{5\pi}{6}\right)$

36. $\tan\left(-\dfrac{7\pi}{3}\right)$

37. $\csc\left(-\dfrac{11\pi}{6}\right)$

38. $\cot\left(-\dfrac{17\pi}{3}\right)$

Without using a calculator, determine which of the following is greater.

39. tan 1 or tan 2

40. sin 1 or tan 1

41. cos 2 or sin 2

42. cos(sin 0) or sin(cos 0)

Use a calculator to find an approximation for each circular function value. Be sure your calculator is set in radian mode.

43. sin 1.0472

44. tan 1.2275

45. cos(−.2443)

46. cot 3.0543

47. sec 7.3159

48. csc 4.8386

Find the value of s in the interval $\left[0, \dfrac{\pi}{2}\right]$ that makes each statement true.

49. cos s = .9250

50. tan s = 4.0112

51. sin s = .4924

52. csc s = 1.2361

53. cot s = .5022

54. sec s = 4.5600

Find the exact value of s in the given interval that has the given circular function value. Do not use a calculator.

55. $\left[0, \dfrac{\pi}{2}\right]$; $\cos s = \dfrac{\sqrt{2}}{2}$

56. $\left[\dfrac{\pi}{2}, \pi\right]$; $\tan s = -\sqrt{3}$

57. $\left[\pi, \dfrac{3\pi}{2}\right]$; $\sec s = -\dfrac{2\sqrt{3}}{3}$ **58.** $\left[\dfrac{3\pi}{2}, 2\pi\right]$; $\sin s = -\dfrac{1}{2}$

*Solve each problem, where t, ω, θ, and s are as defined in **Section 3.4.***

59. Find t if $\theta = \dfrac{5\pi}{12}$ radians and $\omega = \dfrac{8\pi}{9}$ radians per sec.

60. Find θ if $t = 12$ sec and $\omega = 9$ radians per sec.

61. Find ω if $t = 8$ sec and $\theta = \dfrac{2\pi}{5}$ radians.

62. Find ω if $s = \dfrac{12\pi}{25}$ ft, $r = \dfrac{3}{5}$ ft, and $t = 15$ sec.

63. Find s if $r = 11.46$ cm, $\omega = 4.283$ radians per sec, and $t = 5.813$ sec.

64. *Linear Speed of a Flywheel* Find the linear speed of a point on the edge of a flywheel of radius 7 cm if the flywheel is rotating 90 times per sec.

65. *Angular Speed of a Ferris Wheel* A Ferris wheel has radius 25 ft. If it takes 30 sec for the wheel to turn $\dfrac{5\pi}{6}$ radians, what is the angular speed of the wheel?

(Modeling) *Solve each problem.*

66. *Phase Angle of the Moon* Because the moon orbits Earth, we observe different phases of the moon during the period of a month. In the figure, t is called the **phase angle.**

The **phase** F of the moon is modeled by

$$F(t) = \frac{1}{2}(1 - \cos t),$$

and gives the fraction of the moon's face that is illuminated by the sun. (*Source:* Duffet-Smith, P., *Practical Astronomy with Your Calculator,* Cambridge University Press, 1988.) Evaluate each expression and interpret the result.

(a) $F(0)$ **(b)** $F\left(\dfrac{\pi}{2}\right)$ **(c)** $F(\pi)$ **(d)** $F\left(\dfrac{3\pi}{2}\right)$

67. *Atmospheric Effect on Sunlight*   The shortest path for the sun's rays through Earth's atmosphere occurs when the sun is directly overhead. Disregarding the curvature of Earth, as the sun moves lower on the horizon, the distance that sunlight passes through the atmosphere increases by a factor of csc θ, where θ is the angle of elevation of the sun. This increased distance reduces both the intensity of the sun and the amount of ultraviolet light that reaches Earth's surface. See the figure below. (*Source:* Winter, C., R. Sizmann, and L. L. Vant-Hull, Editors, *Solar Power Plants*, Springer-Verlag, 1991.)

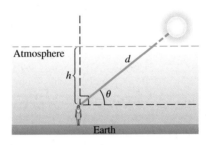

(a) Verify that $d = h \csc \theta$.
(b) Determine θ when $d = 2h$.
(c) The atmosphere filters out the ultraviolet light that causes skin to burn. Compare the difference between sunbathing when $\theta = \frac{\pi}{2}$ and when $\theta = \frac{\pi}{3}$. Which measure gives less ultraviolet light?

68. *Engine Specifications*   The figure shows a rotating wheel (crankshaft) with radius r and a connecting rod AP with length L. The piston slides back and forth along the x-axis as the wheel rotates counterclockwise at a rate of R revolutions per minute (RPM). The 1937 John Deere B engine, used in all models from 1935–1938, had the following specifications.

Maximum RPM: 1340 (no load) $\approx$ 505,168.1 radians per hr
Connecting rod: 10.500000 in.
Stroke: 5.250000 in.

(*Source:* Drost, J. P. and R. H. Kunferman, "Related Rates Challenge Problem: Calculate the Velocity of a Piston," *The AMATYC Review,* Vol. 21 No. 1, Fall 1999.)

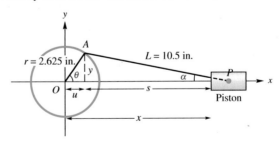

Find an expression for each measure.

(a) y **(b)** u **(c)** s **(d)** x
(e) The velocity v of the piston at the maximum RPM is given by

$$v = -2.625 \sin \theta \left(1 + \frac{\cos \theta}{\sqrt{15 + \cos^2 \theta}} \right) (505{,}168.1) \left(\frac{1}{12 \cdot 5280} \right),$$

where the last quantity changes inches to miles, so the velocity will be in miles per hour. Use the maximum-finding feature of a graphing calculator, with the window $[0, 6]$ by $[-25, 25]$, to find the value of θ that maximizes velocity. Give this maximum velocity.

CHAPTER 3 ▶ Test

Convert each degree measure to radians.

1. 120°

2. −45°

3. 5° (to the nearest hundredth)

Convert each radian measure to degrees.

4. $\dfrac{3\pi}{4}$

5. $-\dfrac{7\pi}{6}$

6. 4 (to the nearest hundredth)

7. A central angle of a circle with radius 150 cm intercepts an arc of 200 cm. Find each measure.

   (a) the radian measure of the angle
   (b) the area of a sector with that central angle

8. *Rotation of Gas Gauge Arrow*   The arrow on a car's gasoline gauge is $\frac{1}{2}$ in. long. See the figure. Through what angle does the arrow rotate when it moves 1 in. on the gauge?

Empty   Full

Find each circular function value.

9. $\sin \dfrac{3\pi}{4}$

10. $\cos\left(-\dfrac{7\pi}{6}\right)$

11. $\tan \dfrac{3\pi}{2}$

12. $\sec \dfrac{8\pi}{3}$

13. $\tan \pi$

14. $\cos \dfrac{3\pi}{2}$

15. Give the sine, cosine, and tangent of s.

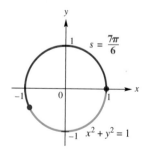

16. Give the domains of the six circular functions.

17. (a) Use a calculator to approximate s in the interval $\left[0, \frac{\pi}{2}\right]$, if $\sin s = .8258$.

   (b) Find the exact value of s in the interval $\left[0, \frac{\pi}{2}\right]$, if $\cos s = \frac{1}{2}$.

18. *Angular and Linear Speed of a Point*   Suppose that point P is on a circle with radius 60 cm, and ray OP is rotating with angular speed $\frac{\pi}{12}$ radian per sec.

   (a) Find the angle generated by P in 8 sec.
   (b) Find the distance traveled by P along the circle in 8 sec.
   (c) Find the linear speed of P.

19. *Orbital Speed of Jupiter* It takes Jupiter 11.64 yr to complete one orbit around the sun. See the figure. If Jupiter's average distance from the sun is 483,600,000 mi, find its orbital speed (speed along its orbital path) in miles per second. (*Source:* Wright, J. W., General Editor, *The Universal Almanac*, Andrews and McMeel, 1997.)

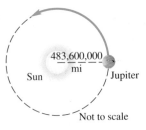

20. *Ferris Wheel* A Ferris wheel has radius 50.0 ft. A person takes a seat and then the wheel turns $\frac{2\pi}{3}$ radians.

(a) How far is the person above the ground?

(b) If it takes 30 sec for the wheel to turn $\frac{2\pi}{3}$ radians, what is the angular speed of the wheel?

CHAPTER 3 ▶ Quantitative Reasoning

Can you find the radius of an Indian artifact given an arc of a circle?

Suppose you find an interesting Indian pottery fragment. The archaeology professor at your college believes it is a piece of the edge of a ceremonial plate and shows you a formula that will give the radius of the original plate using measurements from your fragment, shown in Figure A. Measurements are in inches.

Figure A

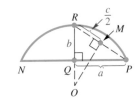

Figure B

In Figure B, a is $\frac{1}{2}$ the length of chord NP and b is the distance from the midpoint of chord NP to the circle. According to the professor's formula, the radius of the circle, OR, is given by

$$r = \frac{a^2 + b^2}{2b}.$$

Why does this formula work? See if you can use the trigonometry you've studied so far to explain it by answering the following.

1. Extend line segment RQ to a radius of the arc, and draw line segment RP and line segment MO, which is the perpendicular bisector of line segment RP. See Figure B. Find a pair of similar triangles in the figure. How do you know they are similar?

2. Find two pairs of corresponding sides, and set their ratios equal. Solve the equation for r.

3. How are a, b, and c related? Use this relationship to substitute for c in terms of a and b.

4. What is the radius of the original plate from which your fragment came?

4 Graphs of the Circular Functions

Phenomena that repeat in a regular pattern, such as rotation of a planet on its axis, high and low tides, and average monthly temperature, can be modeled by *periodic functions*. In Example 6 of Section 4.2, we model the average monthly temperature in New Orleans, a city famous for its Mardi Gras celebration, using a graph of the sine function. Since the 18th century, residents of New Orleans have hosted this annual carnival, featuring festive parades and pageants, elaborate costumes, and dancing in the streets, to mark the final days before the Christian season of Lent. (*Source: Microsoft Encarta Encyclopedia.*)

4.1 Graphs of the Sine and Cosine Functions

Periodic Functions ▪ Graph of the Sine Function ▪ Graph of the Cosine Function ▪
Graphing Techniques, Amplitude, and Period ▪ Using a Trigonometric Model

Periodic Functions Many things in daily life repeat with a predictable
pattern, such as weather, tides, and hours of daylight. Because the sine and cosine
functions repeat their values in a regular pattern, they are *periodic functions*.
Figure 1 shows a periodic graph that represents a normal heartbeat.

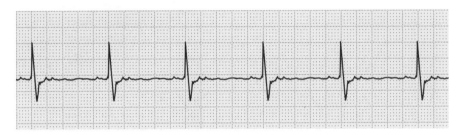

Figure 1

▼ LOOKING AHEAD TO CALCULUS
Periodic functions are used throughout
calculus, so you will need to know
their characteristics. One use of these
functions is to describe the location of
a point in the plane using **polar coordinates,** an alternative to rectangular
coordinates. (See **Chapter 8.**)

PERIODIC FUNCTION

A **periodic function** is a function f such that

$$f(x) = f(x + np),$$

for every real number x in the domain of f, every integer n, and some positive real number p. The least possible positive value of p is the **period** of the
function.

The circumference of the unit circle is 2π, so the least value of p for which
the sine and cosine functions repeat is 2π. ***Therefore, the sine and cosine functions are periodic functions with period 2π.***

Graph of the Sine Function In **Section 3.3** we saw that for a real number s, the point on the unit circle corresponding to s has coordinates $(\cos s, \sin s)$.
See Figure 2. Trace along the circle to verify the results shown in the table.

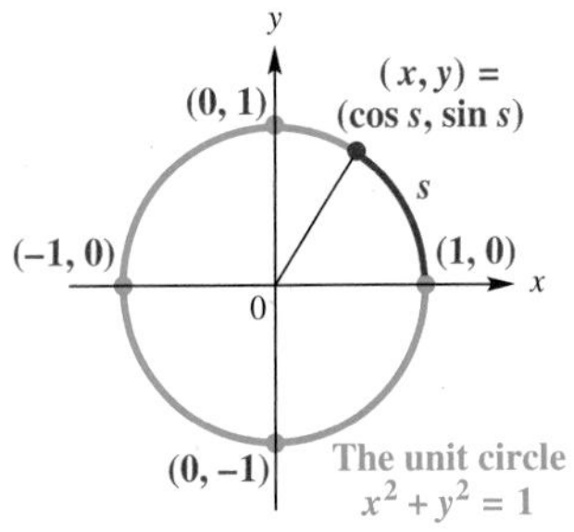

Figure 2

As s **Increases from**	$\sin s$	$\cos s$
0 to $\frac{\pi}{2}$	Increases from 0 to 1	Decreases from 1 to 0
$\frac{\pi}{2}$ to π	Decreases from 1 to 0	Decreases from 0 to -1
π to $\frac{3\pi}{2}$	Decreases from 0 to -1	Increases from -1 to 0
$\frac{3\pi}{2}$ to 2π	Increases from -1 to 0	Increases from 0 to 1

To avoid confusion when graphing the sine function, we use x rather than s; this corresponds to the letters in the xy-coordinate system. Selecting key values of x and finding the corresponding values of $\sin x$ leads to the table in Figure 3.

To obtain the traditional graph in Figure 3, we plot the points from the table, use symmetry, and join them with a smooth curve. Since $y = \sin x$ is periodic with period 2π and has domain $(-\infty, \infty)$, the graph continues in the same pattern in both directions. This graph is called a **sine wave,** or **sinusoid.**

SINE FUNCTION $f(x) = \sin x$

Domain: $(-\infty, \infty)$ Range: $[-1, 1]$

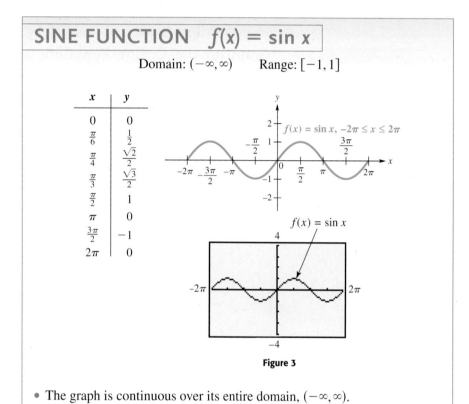

x	y
0	0
$\frac{\pi}{6}$	$\frac{1}{2}$
$\frac{\pi}{4}$	$\frac{\sqrt{2}}{2}$
$\frac{\pi}{3}$	$\frac{\sqrt{3}}{2}$
$\frac{\pi}{2}$	1
π	0
$\frac{3\pi}{2}$	-1
2π	0

Figure 3

- The graph is continuous over its entire domain, $(-\infty, \infty)$.
- Its x-intercepts are of the form $n\pi$, where n is an integer.
- Its period is 2π.
- The graph is symmetric with respect to the origin, so the function is an odd function. For all x in the domain, $\sin(-x) = -\sin x$.

▶ **Note** In algebra, we say that a function f is an **odd function** if for all x in the domain of f, $f(-x) = -f(x)$. The graph of an odd function is symmetric with respect to the origin (Appendix D); that is, if (x, y) belongs to the function, then $(-x, -y)$ also belongs to the function. For example, $\left(\frac{\pi}{2}, 1\right)$ and $\left(-\frac{\pi}{2}, -1\right)$ are points on the graph of $y = \sin x$, illustrating the property $\sin(-x) = -\sin x$.

Sine graphs occur in many practical applications. For example, look back at Figure 2 and assume that the line from the origin to some point (p, q) on the circle is part of the pedal of a bicycle, with a foot placed at (p, q). Since $q = \sin x$, the height of the pedal from the horizontal axis in Figure 2 is given by $\sin x$. By choosing various angles for the pedal and calculating q for each angle, the height of the pedal leads to the sine curve shown in Figure 4 on the next page.

Figure 4

Graph of the Cosine Function   The graph of $y = \cos x$ in Figure 5 has the same shape as the graph of $y = \sin x$. *The graph of the cosine function is, in fact, the graph of the sine function shifted, or translated, $\frac{\pi}{2}$ units to the left.*

COSINE FUNCTION $f(x) = \cos x$

Domain: $(-\infty, \infty)$ Range: $[-1, 1]$

x	y
0	1
$\frac{\pi}{6}$	$\frac{\sqrt{3}}{2}$
$\frac{\pi}{4}$	$\frac{\sqrt{2}}{2}$
$\frac{\pi}{3}$	$\frac{1}{2}$
$\frac{\pi}{2}$	0
π	-1
$\frac{3\pi}{2}$	0
2π	1

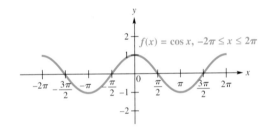

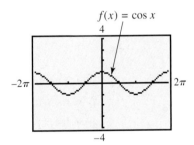

Figure 5

- The graph is continuous over its entire domain, $(-\infty, \infty)$.
- Its x-intercepts are of the form $(2n + 1)\frac{\pi}{2}$, where n is an integer.
- Its period is 2π.
- The graph is symmetric with respect to the y-axis, so the function is an even function. For all x in the domain, $\cos(-x) = \cos x$.

▶ **Note**   A function f is an **even function** if for all x in the domain of f, $f(-x) = f(x)$. The graph of an even function is symmetric with respect to the y-axis (Appendix D); that is, if (x, y) belongs to the function, then $(-x, y)$ also belongs to the function. For example, $\left(\frac{\pi}{2}, 0\right)$ and $\left(-\frac{\pi}{2}, 0\right)$ are points on the graph of $y = \cos x$, illustrating the property $\cos(-x) = \cos x$.

⊞ The calculator graphs of $f(x) = \sin x$ in Figure 3 and $f(x) = \cos x$ in Figure 5 are graphed in the window $[-2\pi, 2\pi]$ by $[-4, 4]$, with Xscl $= \frac{\pi}{2}$ and Yscl $= 1$. This is called the **trig viewing window.** (Your model may use a different "standard" trigonometric viewing window. Consult your owner's manual.) ■

Graphing Techniques, Amplitude, and Period

The examples that follow show graphs that are "stretched" or "compressed" (shrunk) either vertically, horizontally, or both when compared with the graphs of $y = \sin x$ or $y = \cos x$.

▶ EXAMPLE 1   GRAPHING $y = a \sin x$

Graph $y = 2 \sin x$, and compare to the graph of $y = \sin x$.

Solution   For a given value of x, the value of y is twice as large as it would be for $y = \sin x$, as shown in the table of values. The only change in the graph is the range, which becomes $[-2, 2]$. See Figure 6, which includes a graph of $y = \sin x$ for comparison.

x	0	$\frac{\pi}{2}$	π	$\frac{3\pi}{2}$	2π
$\sin x$	0	1	0	-1	0
$2 \sin x$	0	2	0	-2	0

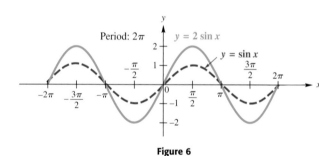

Figure 6

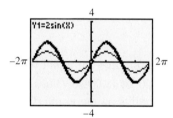

The thick graph style represents the function $y = 2 \sin x$ in Example 1.

The **amplitude** of a periodic function is half the difference between the maximum and minimum values. It describes the height of the graph both above and below a horizontal line passing through the "middle" of the graph. Thus, for both the basic sine and cosine functions, the amplitude is

$$\frac{1}{2}[1 - (-1)] = \frac{1}{2}(2) = 1.$$

Generalizing from Example 1 gives the following.

AMPLITUDE

The graph of $y = a \sin x$ or $y = a \cos x$, with $a \neq 0$, will have the same shape as the graph of $y = \sin x$ or $y = \cos x$, respectively, except with range $[-|a|, |a|]$. The amplitude is $|a|$.

NOW TRY EXERCISE 15. ◀

No matter what the value of the amplitude, the periods of $y = a \sin x$ and $y = a \cos x$ are still 2π. Consider $y = \sin 2x$. We can complete a table of values for the interval $[0, 2\pi]$.

x	0	$\frac{\pi}{4}$	$\frac{\pi}{2}$	$\frac{3\pi}{4}$	π	$\frac{5\pi}{4}$	$\frac{3\pi}{2}$	$\frac{7\pi}{4}$	2π
$\sin 2x$	0	1	0	-1	0	1	0	-1	0

Note that one complete cycle occurs in π units, not 2π units. Therefore, the period here is π, which equals $\frac{2\pi}{2}$. Now consider $y = \sin 4x$. Look at the next table.

x	0	$\frac{\pi}{8}$	$\frac{\pi}{4}$	$\frac{3\pi}{8}$	$\frac{\pi}{2}$	$\frac{5\pi}{8}$	$\frac{3\pi}{4}$	$\frac{7\pi}{8}$	π
$\sin 4x$	0	1	0	-1	0	1	0	-1	0

These values suggest that one complete cycle is achieved in $\frac{\pi}{2}$ or $\frac{2\pi}{4}$ units, which is reasonable since

$$\sin\left(4 \cdot \frac{\pi}{2}\right) = \sin 2\pi = 0.$$

In general, the graph of a function of the form $y = \sin bx$ or $y = \cos bx$, for $b > 0$, will have a period different from 2π when $b \neq 1$. To see why this is so, remember that the values of $\sin bx$ or $\cos bx$ will take on all possible values as bx ranges from 0 to 2π. Therefore, to find the period of either of these functions, we must solve the three-part inequality

$$0 \leq bx \leq 2\pi \quad \text{(Appendix A)}$$

$$0 \leq x \leq \frac{2\pi}{b}. \quad \begin{array}{l}\text{Divide each part by the}\\\text{positive number } b.\end{array}$$

Thus, the period is $\frac{2\pi}{b}$. By dividing the interval $\left[0, \frac{2\pi}{b}\right]$ into four equal parts, we obtain the values for which $\sin bx$ or $\cos bx$ is $-1, 0,$ or 1. These values will give minimum points, x-intercepts, and maximum points on the graph. Once these points are determined, we can sketch the graph by joining the points with a smooth sinusoidal curve. (If a function has $b < 0$, then the identities of the next chapter can be used to rewrite the function so that $b > 0$.)

> ▶ **Note** One method to divide an interval into four equal parts is as follows.
>
> *Step 1* Find the midpoint of the interval by adding the x-values of the endpoints and dividing by 2. (Appendix B)
>
> *Step 2* Find the midpoints of the two intervals found in Step 1, using the same procedure.

▶ **EXAMPLE 2** GRAPHING $y = \sin bx$

Graph $y = \sin 2x$, and compare to the graph of $y = \sin x$.

Solution In this function the coefficient of x is 2, so $b = 2$ and the period is $\frac{2\pi}{2} = \pi$. Therefore, the graph will complete one period over the interval $[0, \pi]$. The endpoints are 0 and π, and the three points between the endpoints are

$$\frac{1}{4}(0 + \pi), \quad \frac{1}{2}(0 + \pi), \quad \text{and} \quad \frac{3}{4}(0 + \pi),$$

which give the following x-values:

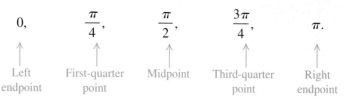

0,	$\frac{\pi}{4},$	$\frac{\pi}{2},$	$\frac{3\pi}{4},$	$\pi.$
Left endpoint	First-quarter point	Midpoint	Third-quarter point	Right endpoint

We plot the points from the table of values given on page 148, and join them with a smooth sinusoidal curve. More of the graph can be sketched by repeating this cycle, as shown in Figure 7. The amplitude is not changed.

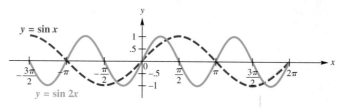

Figure 7

NOW TRY EXERCISE 27. ◄

We can think of the graph of $y = \sin bx$ as a horizontal stretching of the graph of $y = \sin x$ when $0 < b < 1$ and a horizontal shrinking when $b > 1$.

PERIOD

For $b > 0$, the graph of $y = \sin bx$ will resemble that of $y = \sin x$, but with period $\frac{2\pi}{b}$. Also, the graph of $y = \cos bx$ will resemble that of $y = \cos x$, but with period $\frac{2\pi}{b}$.

► **EXAMPLE 3** GRAPHING $y = \cos bx$

Graph $y = \cos \frac{2}{3}x$ over one period.

Solution The period is $\dfrac{2\pi}{\frac{2}{3}} = 2\pi \div \frac{2}{3} = 2\pi \cdot \frac{3}{2} = 3\pi$. We divide the interval

$[0, 3\pi]$ into four equal parts to get the x-values $0, \frac{3\pi}{4}, \frac{3\pi}{2}, \frac{9\pi}{4}$, and 3π that yield minimum points, maximum points, and x-intercepts. We use these values to obtain a table of key points for one period.

x	0	$\frac{3\pi}{4}$	$\frac{3\pi}{2}$	$\frac{9\pi}{4}$	3π
$\frac{2}{3}x$	0	$\frac{\pi}{2}$	π	$\frac{3\pi}{2}$	2π
$\cos \frac{2}{3}x$	1	0	-1	0	1

The amplitude is 1 because the maximum value is 1, the minimum value is -1, and $\frac{1}{2}(1 - (-1)) = \frac{1}{2}(2) = 1$. We plot these points and join them with a smooth curve. The graph is shown in Figure 8.

NOW TRY EXERCISE 25. ◄

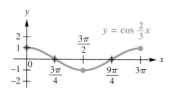

Figure 8

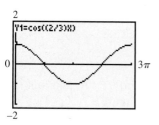

This screen shows a graph of the function in Example 3. By choosing Xscl $= \frac{3\pi}{4}$, the x-intercepts, maxima, and minima coincide with tick marks on the x-axis.

▶ **Note** Look back at the middle row of the table in Example 3. Dividing the interval $\left[0, \frac{2\pi}{b}\right]$ into four equal parts will always give the values $0, \frac{\pi}{2}, \pi, \frac{3\pi}{2}$, and 2π for this row, in this case resulting in values of $-1, 0$, or 1. These values lead to key points on the graph, which can then be easily sketched.

GUIDELINES FOR SKETCHING GRAPHS OF SINE AND COSINE FUNCTIONS

To graph $y = a \sin bx$ or $y = a \cos bx$, with $b > 0$, follow these steps.

Step 1 Find the period, $\frac{2\pi}{b}$. Start at 0 on the x-axis, and lay off a distance of $\frac{2\pi}{b}$.

Step 2 Divide the interval into four equal parts. (See the Note preceding Example 2 on page 148.)

Step 3 Evaluate the function for each of the five x-values resulting from Step 2. The points will be maximum points, minimum points, and x-intercepts.

Step 4 Plot the points found in Step 3, and join them with a sinusoidal curve having amplitude $|a|$.

Step 5 Draw the graph over additional periods as needed.

▶ **EXAMPLE 4** GRAPHING $y = a \sin bx$

Graph $y = -2 \sin 3x$ over one period using the preceding guidelines.

Solution

Step 1 For this function, $b = 3$, so the period is $\frac{2\pi}{3}$. The function will be graphed over the interval $\left[0, \frac{2\pi}{3}\right]$.

Step 2 Divide the interval $\left[0, \frac{2\pi}{3}\right]$ into four equal parts to get the x-values 0, $\frac{\pi}{6}, \frac{\pi}{3}, \frac{\pi}{2}$, and $\frac{2\pi}{3}$.

Step 3 Make a table of values determined by the x-values from Step 2.

x	0	$\frac{\pi}{6}$	$\frac{\pi}{3}$	$\frac{\pi}{2}$	$\frac{2\pi}{3}$
$3x$	0	$\frac{\pi}{2}$	π	$\frac{3\pi}{2}$	2π
$\sin 3x$	0	1	0	-1	0
$-2 \sin 3x$	0	-2	0	2	0

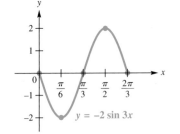

Figure 9

Step 4 Plot the points $(0,0)$, $\left(\frac{\pi}{6}, -2\right)$, $\left(\frac{\pi}{3}, 0\right)$, $\left(\frac{\pi}{2}, 2\right)$, and $\left(\frac{2\pi}{3}, 0\right)$, and join them with a sinusoidal curve with amplitude 2. See Figure 9.

Step 5 The graph can be extended by repeating the cycle.

Notice that when a is negative, the graph of $y = a \sin bx$ is the reflection across the x-axis of the graph of $y = |a| \sin bx$.

NOW TRY EXERCISE 29. ◀

▶ **EXAMPLE 5** GRAPHING $y = a \cos bx$ FOR b EQUAL TO A MULTIPLE OF π

Graph $y = -3 \cos \pi x$ over one period.

Solution

Step 1 Since $b = \pi$, the period is $\frac{2\pi}{\pi} = 2$, so we will graph the function over the interval $[0, 2]$.

Step 2 Dividing $[0, 2]$ into four equal parts yields the x-values $0, \frac{1}{2}, 1, \frac{3}{2}$, and 2.

Step 3 Make a table using these x-values.

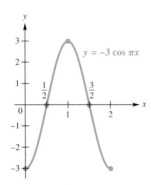

$y = -3 \cos \pi x$

Figure 10

x	0	$\frac{1}{2}$	1	$\frac{3}{2}$	2
πx	0	$\frac{\pi}{2}$	π	$\frac{3\pi}{2}$	2π
$\cos \pi x$	1	0	-1	0	1
$-3 \cos \pi x$	-3	0	3	0	-3

Step 4 Plot the points $(0, -3)$, $\left(\frac{1}{2}, 0\right)$, $(1, 3)$, $\left(\frac{3}{2}, 0\right)$, and $(2, -3)$, and join them with a sinusoidal curve having amplitude $|-3| = 3$. See Figure 10.

Step 5 The graph can be extended by repeating the cycle.

Notice that when b is an integer multiple of π, the x-intercepts of the graph are rational numbers.

NOW TRY EXERCISE 37. ◀

Using a Trigonometric Model

▶ **EXAMPLE 6** INTERPRETING A SINE FUNCTION MODEL

The average temperature (in °F) at Mould Bay, Canada, can be approximated by the function defined by

$$f(x) = 34 \sin\left[\frac{\pi}{6}(x - 4.3)\right],$$

where x is the month and $x = 1$ corresponds to January, $x = 2$ to February, and so on.

(a) To observe the graph over a two-year interval and to see the maximum and minimum points, graph f in the window $[0, 25]$ by $[-45, 45]$.

(b) What is the average temperature during the month of May?

(c) What would be an approximation for the average *yearly* temperature at Mould Bay?

Solution

(a) The graph of $f(x) = 34 \sin\left[\frac{\pi}{6}(x - 4.3)\right]$ is shown in Figure 11 on the next page. Its amplitude is 34, and the period is $\dfrac{2\pi}{\frac{\pi}{6}} = 2\pi \div \frac{\pi}{6} = 2\pi \cdot \frac{6}{\pi} = 12$.

The function f has a period of 12 months, or 1 year, which agrees with the changing of the seasons.

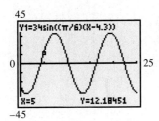

Figure 11

(b) May is the fifth month, so the average temperature during May is

$$f(5) = 34 \sin\left[\frac{\pi}{6}(5 - 4.3)\right] \approx 12°F. \quad \text{Let } x = 5. \text{ (Appendix C)}$$

See the display at the bottom of the screen in Figure 11.

(c) From the graph, it appears that the average yearly temperature is about 0°F since the graph is centered vertically about the line $y = 0$.

NOW TRY EXERCISE 55. ◀

CONNECTIONS Using a TI-83/84 Plus calculator, adjust the settings to correspond to the following screens.

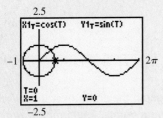 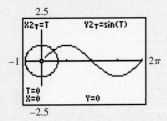

Tmax is 2π,
Tstep is $\frac{\pi}{40}$,
Xmax is 2π,
Xscl is $\frac{\pi}{2}$.

Graph the two equations (which are in **parametric form**), and watch as the unit circle and the sine function are graphed simultaneously. Press the TRACE key once to get the screen shown on the left below, and then press the up-arrow key to get the screen shown on the right below. The screen on the left gives a unit circle interpretation of $\cos 0 = 1$ and $\sin 0 = 0$. The screen on the right gives a rectangular coordinate graph interpretation of $\sin 0 = 0$.

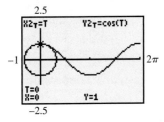

Now go back and redefine Y_{2T} as cos(T). Graph both equations again; the second screen will look like the one in the margin, after the TRACE and up-arrow keys are pressed. This screen indicates that $\cos 0 = 1$.

FOR DISCUSSION OR WRITING

1. On the unit circle, let T = 2. Find X and Y, and interpret their values.

2. On the sine graph, trace to the point where T = 1.9. What values of X and Y are displayed? Interpret these values with an equation in X and Y. Then repeat, but use the cosine graph.

4.1 Exercises

Concept Check   *In Exercises 1–8, match each function with its graph.*

1. $y = \sin x$ **2.** $y = \cos x$ **3.** $y = -\sin x$ **4.** $y = -\cos x$

5. $y = \sin 2x$ **6.** $y = \cos 2x$ **7.** $y = 2 \sin x$ **8.** $y = 2 \cos x$

A. 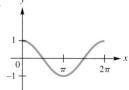   **B.** 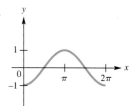   **C.**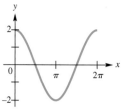

D. 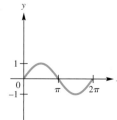   **E.** 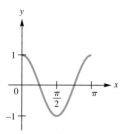   **F.**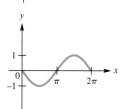

G. 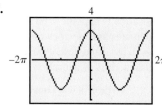   **H.**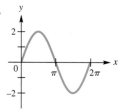

Concept Check   *In Exercises 9–12, match each function with its calculator graph.*

9. $y = \sin 3x$ **10.** $y = \cos 3x$ **11.** $y = 3 \cos x$ **12.** $y = 3 \sin x$

A. 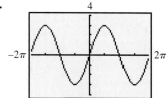   **B.**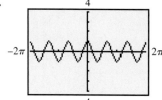

C.   **D.**

Graph each function over the interval $[-2\pi, 2\pi]$. *Give the amplitude. See Example 1.*

13. $y = 2 \cos x$ **14.** $y = 3 \sin x$ **15.** $y = \dfrac{2}{3} \sin x$

16. $y = \dfrac{3}{4} \cos x$ **17.** $y = -\cos x$ **18.** $y = -\sin x$

19. $y = -2 \sin x$ **20.** $y = -3 \cos x$ **21.** $y = \sin(-x)$

22. *Concept Check* In Exercise 21, why is the graph the same as that of $y = -\sin x$?

Graph each function over a two-period interval. Give the period and amplitude. See Examples 2–5.

23. $y = \sin \dfrac{1}{2}x$

24. $y = \sin \dfrac{2}{3}x$

25. $y = \cos \dfrac{3}{4}x$

26. $y = \cos \dfrac{1}{3}x$

27. $y = \sin 3x$

28. $y = \cos 2x$

29. $y = 2 \sin \dfrac{1}{4}x$

30. $y = 3 \sin 2x$

31. $y = -2 \cos 3x$

32. $y = -5 \cos 2x$

33. $y = \cos \pi x$

34. $y = -\sin \pi x$

35. $y = -2 \sin 2\pi x$

36. $y = 3 \cos 2\pi x$

37. $y = \dfrac{1}{2} \cos \dfrac{\pi}{2}x$

38. $y = -\dfrac{2}{3} \sin \dfrac{\pi}{4}x$

39. $y = \pi \sin \pi x$

40. $y = -\pi \cos \pi x$

Connecting Graphs with Equations *Each function graphed is of the form $y = a \sin bx$ or $y = a \cos bx$, where $b > 0$. Determine the equation of the graph.*

41.

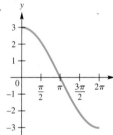

42.

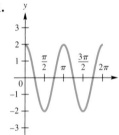

43.

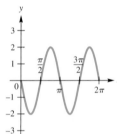

44.

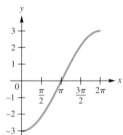

45.

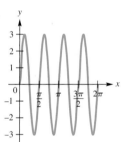

46.

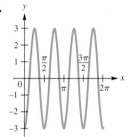

(Modeling) *Solve each problem.*

47. *Average Annual Temperature* Scientists believe that the average annual temperature in a given location is periodic. The average temperature at a given place during a given season fluctuates as time goes on, from colder to warmer, and back to colder. The graph shows an idealized description of the temperature (in °F) for approximately the last 150 thousand years of a location at the same latitude as Anchorage, Alaska.

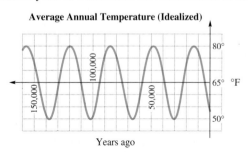

Average Annual Temperature (Idealized)

Years ago

(a) Find the highest and lowest temperatures recorded.
(b) Use these two numbers to find the amplitude.
(c) Find the period of the function.
(d) What is the trend of the temperature now?

48. *Blood Pressure Variation*   The graph gives the variation in blood pressure for a typical person. **Systolic** and **diastolic pressures** are the upper and lower limits of the periodic changes in pressure that produce the pulse. The length of time between peaks is called the period of the pulse.

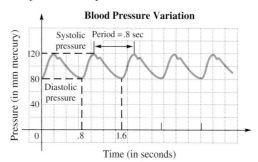

Blood Pressure Variation

(a) Find the amplitude of the graph.
(b) Find the pulse rate (the number of pulse beats in 1 min) for this person.

49. *Activity of a Nocturnal Animal*   Many of the activities of living organisms are periodic. For example, the graph at the right shows the time that a certain nocturnal animal begins its evening activity.

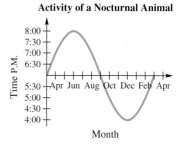

Activity of a Nocturnal Animal

(a) Find the amplitude of this graph.
(b) Find the period.

50. *Position of a Moving Arm*   The figure shows schematic diagrams of a rhythmically moving arm. The upper arm RO rotates back and forth about the point R; the position of the arm is measured by the angle y between the actual position and the downward vertical position. (*Source:* De Sapio, R., *Calculus for the Life Sciences.* Copyright © 1978 by W. H. Freeman and Company. Reprinted by permission.)

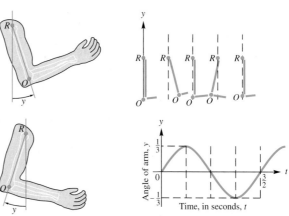

This graph shows the relationship between angle y and time t in seconds.

(a) Find an equation of the form $y = a \sin kt$ for the graph shown.
(b) How long does it take for a complete movement of the arm?

51. *Voltage of an Electrical Circuit* The voltage E in an electrical circuit is modeled by

$$E = 5 \cos 120\pi t,$$

where t is time measured in seconds.

(a) Find the amplitude and the period.

(b) How many cycles are completed in 1 sec? (The number of cycles (periods) completed in 1 sec is the **frequency** of the function.)

(c) Find E when $t = 0, .03, .06, .09, .12$.

(d) Graph E for $0 \le t \le \frac{1}{30}$.

52. *Voltage of an Electrical Circuit* For another electrical circuit, the voltage E is modeled by

$$E = 3.8 \cos 40\pi t,$$

where t is time measured in seconds.

(a) Find the amplitude and the period.

(b) Find the frequency. See Exercise 51(b).

(c) Find E when $t = .02, .04, .08, .12, .14$.

(d) Graph one period of E.

53. *Atmospheric Carbon Dioxide* At Mauna Loa, Hawaii, atmospheric carbon dioxide levels in parts per million (ppm) have been measured regularly since 1958. The function defined by

$$L(x) = .022x^2 + .55x + 316 + 3.5 \sin 2\pi x$$

can be used to model these levels, where x is in years and $x = 0$ corresponds to 1960. (*Source:* Nilsson, A., *Greenhouse Earth,* John Wiley and Sons, 1992.)

(a) Graph L in the window $[15, 35]$ by $[325, 365]$.

(b) When do the seasonal maximum and minimum carbon dioxide levels occur?

(c) L is the sum of a quadratic function and a sine function. What is the significance of each of these functions? Discuss what physical phenomena may be responsible for each function.

54. *Atmospheric Carbon Dioxide* Refer to Exercise 53. The carbon dioxide content in the atmosphere at Barrow, Alaska, in parts per million (ppm) can be modeled using the function defined by

$$C(x) = .04x^2 + .6x + 330 + 7.5 \sin 2\pi x,$$

where $x = 0$ corresponds to 1970. (*Source:* Zeilik, M. and S. Gregory, *Introductory Astronomy and Astrophysics,* Brooks/Cole, 1998.)

(a) Graph C in the window $[5, 25]$ by $[320, 380]$.

(b) Discuss possible reasons why the amplitude of the oscillations in the graph of C is larger than the amplitude of the oscillations in the graph of L in Exercise 53, which models Hawaii.

(c) Define a new function C that is valid if x represents the actual year, where $1970 \le x \le 1995$. (See horizontal translations in **Appendix D.**)

55. *Average Annual Temperature* The temperature in a certain city in Alaska is modeled by

$$T(x) = 25 + 37 \sin\left[\frac{2\pi}{365}(x - 101) \right],$$

where $T(x)$ is the temperature in degrees Fahrenheit on day x, with $x = 1$ corresponding to January 1 and $x = 365$ corresponding to December 31. Use a calculator to estimate the temperature on the following days. (*Source:* Lando, B. and C. Lando, "Is the Graph of Temperature Variation a Sine Curve?", *The Mathematics Teacher*, 70, September 1977.)

(a) March 15 (day 74)   (b) April 5 (day 95)   (c) Day 200
(d) June 25   (e) October 1   (f) December 31

56. *Fluctuation in the Solar Constant*   The **solar constant** S is the amount of energy per unit area that reaches Earth's atmosphere from the sun. It is equal to 1367 watts per sq m but varies slightly throughout the seasons. This fluctuation ΔS in S can be calculated using the formula

$$\Delta S = .034S \sin\left[\frac{2\pi(82.5 - N)}{365.25}\right].$$

In this formula, N is the day number covering a four-year period, where $N = 1$ corresponds to January 1 of a leap year and $N = 1461$ corresponds to December 31 of the fourth year. (*Source:* Winter, C., R. Sizmann, and L. L.Vant-Hull, Editors, *Solar Power Plants,* Springer-Verlag, 1991.)

(a) Calculate ΔS for $N = 80$, which is the spring equinox in the first year.
(b) Calculate ΔS for $N = 1268$, which is the summer solstice in the fourth year.
(c) What is the maximum value of ΔS?
(d) Find a value for N where ΔS is equal to 0.

Tides for Kahului Harbor   The chart shows the tides for Kahului Harbor (on the island of Maui, Hawaii). To identify high and low tides and times for other Maui areas, the following adjustments must be made.

Hana: High, +40 min, +.1 ft;   Makena: High, +1:21, −.5 ft;
          Low, +18 min, −.2 ft   Low, +1:09, −.2 ft
Maalaea: High, +1:52, −.1 ft;   Lahaina: High, +1:18, −.2 ft;
          Low, +1:19, −.2 ft   Low, +1:01, −.1 ft

JANUARY

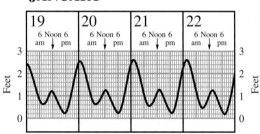

Source: Maui News. Original chart prepared by Edward K. Noda and Associates.

Use the graph to work Exercises 57–62.

57. The graph is an example of a periodic function. What is the period (in hours)?

58. What is the amplitude?

59. At what time on January 20 was low tide at Kahului? What was the height?

60. Repeat Exercise 59 for Maalaea.

61. At what time on January 22 was high tide at Kahului? What was the height?

62. Repeat Exercise 61 for Lahaina.

Musical Sound Waves Pure sounds produce single sine waves on an oscilloscope. Find the amplitude and period of each sine wave graph in Exercises 63 and 64. On the vertical scale, each square represents .5; on the horizontal scale, each square represents 30° or $\frac{\pi}{6}$.

63.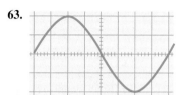

64.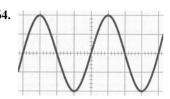

65. (a) Compare the graphs of $y = \sin 2x$ and $y = 2 \sin x$ over the interval $[0, 2\pi]$. Can we say that, in general, $\sin bx = b \sin x$? Explain.

(b) Compare the graphs of $y = \cos 3x$ and $y = 3 \cos x$ over the interval $[0, 2\pi]$. Can we say that, in general, $\cos bx = b \cos x$? Explain.

66. Refer to the graph of $y = \sin x$ in Figure 3. The graph completes one cycle between $x = 0$ and $x = 2\pi$. Consider the statement, "The function $y = \sin bx$ completes b cycles between 0 and 2π." Use your graphing calculator to confirm the statement for some positive integer values of b, such as 3, 4, and 5. Interpret and confirm the statement for $b = \frac{1}{2}$ and $b = \frac{3}{2}$.

4.2 Translations of the Graphs of the Sine and Cosine Functions

Horizontal Translations ▪ Vertical Translations ▪ Combinations of Translations ▪ Determining a Trigonometric Model Using Curve Fitting

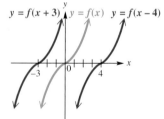

Horizontal translations of $y = f(x)$

(Appendix D)

Figure 12

Horizontal Translations The graph of the function defined by $y = f(x - d)$ is translated *horizontally* compared to the graph of $y = f(x)$. The translation is d units to the right if $d > 0$ and $|d|$ units to the left if $d < 0$. See Figure 12. With circular functions, a horizontal translation is called a **phase shift.** In the function $y = f(x - d)$, the expression $x - d$ is called the **argument.** In Examples 1–3, we give two methods that can be used to sketch the graph of a circular function involving a phase shift.

▶ **EXAMPLE 1** GRAPHING $y = \sin(x - d)$

Graph $y = \sin(x - \frac{\pi}{3})$ over one period.

Solution **Method 1** For the argument $x - \frac{\pi}{3}$ to result in all possible values throughout one period, it must take on all values between 0 and 2π, inclusive. Therefore, to find an interval of one period, we solve the three-part inequality

$$0 \le x - \frac{\pi}{3} \le 2\pi \quad \text{(Appendix A)}$$

$$\frac{\pi}{3} \le x \le \frac{7\pi}{3}. \qquad \text{Add } \frac{\pi}{3} \text{ to each part.}$$

Use the method described in the Note on page 148 to divide the interval $\left[\frac{\pi}{3}, \frac{7\pi}{3}\right]$ into four equal parts, obtaining the following x-values.

$$\frac{\pi}{3}, \quad \frac{5\pi}{6}, \quad \frac{4\pi}{3}, \quad \frac{11\pi}{6}, \quad \frac{7\pi}{3} \quad \boxed{\text{These are } key \text{ } x\text{-values.}}$$

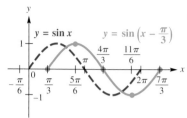

Figure 13

A table of values using these x-values follows.

x	$\frac{\pi}{3}$	$\frac{5\pi}{6}$	$\frac{4\pi}{3}$	$\frac{11\pi}{6}$	$\frac{7\pi}{3}$
$x - \frac{\pi}{3}$	0	$\frac{\pi}{2}$	π	$\frac{3\pi}{2}$	2π
$\sin\left(x - \frac{\pi}{3}\right)$	0	1	0	-1	0

We join the corresponding points with a smooth curve to get the solid blue graph shown in Figure 13. The period is 2π, and the amplitude is 1.

Method 2 We can also graph $y = \sin\left(x - \frac{\pi}{3}\right)$ by using a horizontal translation of the graph of $y = \sin x$. The argument $x - \frac{\pi}{3}$ indicates that the graph will be translated $\frac{\pi}{3}$ units to the *right* (the phase shift) as compared to the graph of $y = \sin x$. See Figure 13.

Therefore, to graph a function using this method, first graph the basic circular function, and then graph the desired function by using the appropriate translation.

NOW TRY EXERCISE 35. ◀

▶ **Note** The graph in Figure 13 of Example 1 can be extended through additional periods by repeating the given portion of the graph, as necessary.

▶ **EXAMPLE 2** GRAPHING $y = a \cos(x - d)$

Graph $y = 3 \cos\left(x + \frac{\pi}{4}\right)$ over one period.

Solution *Method 1* The graph can be sketched over one period by first solving the three-part inequality

$$0 \le x + \frac{\pi}{4} \le 2\pi$$

$$-\frac{\pi}{4} \le x \le \frac{7\pi}{4}. \qquad \text{Subtract } \tfrac{\pi}{4} \text{ from each part.}$$

Dividing this interval into four equal parts gives x-values of

$$-\frac{\pi}{4}, \quad \frac{\pi}{4}, \quad \frac{3\pi}{4}, \quad \frac{5\pi}{4}, \quad \frac{7\pi}{4}. \quad \text{Key } x\text{-values}$$

A table of points for these x-values leads to maximum points, minimum points, and x-intercepts.

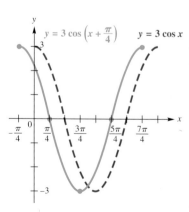

Figure 14

x	$-\frac{\pi}{4}$	$\frac{\pi}{4}$	$\frac{3\pi}{4}$	$\frac{5\pi}{4}$	$\frac{7\pi}{4}$
$x + \frac{\pi}{4}$	0	$\frac{\pi}{2}$	π	$\frac{3\pi}{2}$	2π
$\cos\left(x + \frac{\pi}{4}\right)$	1	0	-1	0	1
$3 \cos\left(x + \frac{\pi}{4}\right)$	3	0	-3	0	3

We join the corresponding points with a smooth curve to get the solid blue graph shown in Figure 14. The period is 2π, and the amplitude is 3.

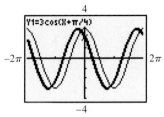

This screen shows the graph of

$$Y_1 = 3 \cos\left(X + \frac{\pi}{4}\right)$$

in Example 2 as a thick line. The graph of

$$Y_2 = 3 \cos X$$

is shown as a thin line for comparison.

Method 2 Write $3 \cos\left(x + \frac{\pi}{4}\right)$ in the form $a \cos(x - d)$.

$$y = 3 \cos\left(x + \frac{\pi}{4}\right) = 3 \cos\left[x - \left(-\frac{\pi}{4}\right)\right]$$

This result shows that $d = -\frac{\pi}{4}$. Since $-\frac{\pi}{4}$ is negative, the phase shift is $\left|-\frac{\pi}{4}\right| = \frac{\pi}{4}$ unit to the left. The graph is the same as that of $y = 3 \cos x$ (the dashed red graph in Figure 14 on the preceding page), except that it is translated $\frac{\pi}{4}$ unit to the left (the solid blue graph in Figure 14).

NOW TRY EXERCISE 37. ◀

▶ **EXAMPLE 3** GRAPHING $y = a \cos b(x - d)$

Graph $y = -2 \cos(3x + \pi)$ over two periods.

Solution ***Method 1*** The function can be sketched over one period by solving the three-part inequality

$$0 \leq 3x + \pi \leq 2\pi$$

to get the interval $\left[-\frac{\pi}{3}, \frac{\pi}{3}\right]$. Divide this interval into four equal parts to get the points $\left(-\frac{\pi}{3}, -2\right)$, $\left(-\frac{\pi}{6}, 0\right)$, $(0, 2)$, $\left(\frac{\pi}{6}, 0\right)$, and $\left(\frac{\pi}{3}, -2\right)$. Plot these points and join them with a smooth curve. By graphing an additional half period to the left and to the right, we obtain the graph shown in Figure 15.

Method 2 First write the expression in the form $a \cos b(x - d)$.

$$y = -2 \cos(3x + \pi) = -2 \cos 3\left(x + \frac{\pi}{3}\right) \quad \text{Rewrite } 3x + \pi \text{ as } 3\left(x + \frac{\pi}{3}\right).$$

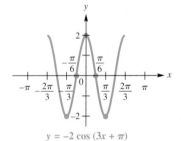

$$y = -2 \cos(3x + \pi)$$

Figure 15

Then $a = -2$, $b = 3$, and $d = -\frac{\pi}{3}$. The amplitude is $|-2| = 2$, and the period is $\frac{2\pi}{3}$ (since the value of b is 3). The phase shift is $\left|-\frac{\pi}{3}\right| = \frac{\pi}{3}$ units to the left as compared to the graph of $y = -2 \cos 3x$. Again, see Figure 15.

NOW TRY EXERCISE 43. ◀

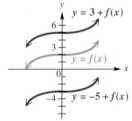

Vertical translations of $y = f(x)$
(Appendix D)

Figure 16

Vertical Translations The graph of a function of the form $y = c + f(x)$ is translated *vertically* as compared with the graph of $y = f(x)$. See Figure 16. The translation is c units up if $c > 0$ and $|c|$ units down if $c < 0$.

▶ **EXAMPLE 4** GRAPHING $y = c + a \cos bx$

Graph $y = 3 - 2 \cos 3x$ over two periods.

Solution The values of y will be 3 greater than the corresponding values of y in $y = -2 \cos 3x$. This means that the graph of $y = 3 - 2 \cos 3x$ is the same as the graph of $y = -2 \cos 3x$, vertically translated 3 units up. Since the period of $y = -2 \cos 3x$ is $\frac{2\pi}{3}$, the key points have x-values

$$0, \quad \frac{\pi}{6}, \quad \frac{\pi}{3}, \quad \frac{\pi}{2}, \quad \frac{2\pi}{3}. \quad \text{Key } x\text{-values}$$

Use these x-values to make a table of points.

x	0	$\frac{\pi}{6}$	$\frac{\pi}{3}$	$\frac{\pi}{2}$	$\frac{2\pi}{3}$
$\cos 3x$	1	0	-1	0	1
$2\cos 3x$	2	0	-2	0	2
$3 - 2\cos 3x$	1	3	5	3	1

The key points are shown on the graph in Figure 17, along with more of the graph, sketched using the fact that the function is periodic.

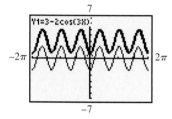

The function in Example 4 is shown using the thick graph style. Notice also the thin graph style for $y = -2\cos 3x$.

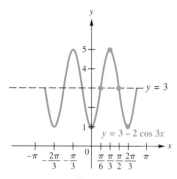

Figure 17

NOW TRY EXERCISE 47. ◀

Combinations of Translations   A function of the form

$$y = c + a \sin b(x - d) \quad \text{or} \quad y = c + a \cos b(x - d), \quad b > 0,$$

which involves stretching, shrinking, and translating, can be graphed according to the following guidelines.

FURTHER GUIDELINES FOR SKETCHING GRAPHS OF SINE AND COSINE FUNCTIONS

Method 1   Follow these steps.

Step 1   Find an interval whose length is one period $\frac{2\pi}{b}$ by solving the three-part inequality $0 \le b(x - d) \le 2\pi$.   (Appendix A)

Step 2   Divide the interval into four equal parts. (See the Note on page 148.)

Step 3   Evaluate the function for each of the five x-values resulting from Step 2. The points will be maximum points, minimum points, and points that intersect the line $y = c$ ("middle" points of the wave).

Step 4   Plot the points found in Step 3, and join them with a sinusoidal curve having amplitude $|a|$.

Step 5   Draw the graph over additional periods, as needed.

Method 2   First graph $y = a \sin bx$ or $y = a \cos bx$. The amplitude of the function is $|a|$, and the period is $\frac{2\pi}{b}$. Then use translations to graph the desired function. The vertical translation is c units up if $c > 0$ and $|c|$ units down if $c < 0$. The horizontal translation (phase shift) is d units to the right if $d > 0$ and $|d|$ units to the left if $d < 0$.

▶ **EXAMPLE 5**   GRAPHING $y = c + a \sin b(x - d)$

Graph $y = -1 + 2 \sin(4x + \pi)$ over two periods.

Solution   We use Method 1. First write the expression on the right side in the form $c + a \sin b(x - d)$.

$$y = -1 + 2 \sin(4x + \pi) = -1 + 2 \sin\left[4\left(x + \frac{\pi}{4}\right)\right]$$

Step 1 Find an interval whose length is one period.

$$0 \le 4\left(x + \frac{\pi}{4}\right) \le 2\pi$$

$$0 \le x + \frac{\pi}{4} \le \frac{\pi}{2} \qquad \text{Divide each part by 4.}$$

$$-\frac{\pi}{4} \le x \le \frac{\pi}{4} \qquad \text{Subtract } \tfrac{\pi}{4} \text{ from each part.}$$

Step 2 Divide the interval $\left[-\frac{\pi}{4}, \frac{\pi}{4}\right]$ into four equal parts to get the *x*-values

$$-\frac{\pi}{4}, \quad -\frac{\pi}{8}, \quad 0, \quad \frac{\pi}{8}, \quad \frac{\pi}{4}. \quad \text{Key } x\text{-values}$$

Step 3 Make a table of values.

x	$-\frac{\pi}{4}$	$-\frac{\pi}{8}$	0	$\frac{\pi}{8}$	$\frac{\pi}{4}$
$x + \frac{\pi}{4}$	0	$\frac{\pi}{8}$	$\frac{\pi}{4}$	$\frac{3\pi}{8}$	$\frac{\pi}{2}$
$4\left(x + \frac{\pi}{4}\right)$	0	$\frac{\pi}{2}$	π	$\frac{3\pi}{2}$	2π
$\sin 4\left(x + \frac{\pi}{4}\right)$	0	1	0	-1	0
$2 \sin 4\left(x + \frac{\pi}{4}\right)$	0	2	0	-2	0
$-1 + 2\sin(4x + \pi)$	-1	1	-1	-3	-1

Steps 4 and 5 Plot the points found in the table and join them with a sinusoidal curve. Figure 18 shows the graph, extended to the right and left to include two full periods.

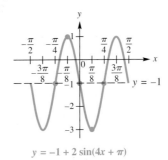

$y = -1 + 2 \sin(4x + \pi)$

Figure 18

NOW TRY EXERCISE 53. ◀

Determining a Trigonometric Model Using Curve Fitting

A sinusoidal function is often a good approximation of a set of real data points.

 ▶ **EXAMPLE 6**   MODELING TEMPERATURE WITH A SINE FUNCTION

The maximum average monthly temperature in New Orleans is 82°F and the minimum is 54°F. The table shows the average monthly temperatures. The scatter diagram for a two-year interval in Figure 19 strongly suggests that the temperatures can be modeled with a sine curve.

Month	°F	Month	°F
Jan	54	July	82
Feb	55	Aug	81
Mar	61	Sept	77
Apr	69	Oct	71
May	73	Nov	59
June	79	Dec	55

Source: Miller, A., J. Thompson, and R. Peterson, *Elements of Meteorology, Fourth Edition,* Charles E. Merrill Publishing Co., 1983.

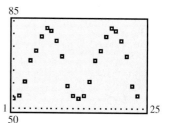

Figure 19

(a) Using only the maximum and minimum temperatures, determine a function of the form $f(x) = a \sin[b(x - d)] + c$, where a, b, c, and d are constants, that models the average monthly temperature in New Orleans. Let x represent the month, with January corresponding to $x = 1$.

(b) On the same coordinate axes, graph f for a two-year period together with the actual data values found in the table.

(c) Use the **sine regression** feature of a graphing calculator to determine a second model for these data.

Solution

(a) We use the maximum and minimum average monthly temperatures to find the amplitude a.

$$a = \frac{82 - 54}{2} = 14$$

The average of the maximum and minimum temperatures is a good choice for c. The average is

$$\frac{82 + 54}{2} = 68.$$

Since temperatures repeat every 12 months, b is $\frac{2\pi}{12} = \frac{\pi}{6}$. The coldest month is January, when $x = 1$, and the hottest month is July, when $x = 7$, so we should choose d to be about 4. The table shows that temperatures are actually a little warmer after July than before, so we experiment with values just greater than 4 to find d. Trial and error using a calculator leads to $d = 4.2$. Thus,

$$f(x) = a \sin[b(x - d)] + c = 14 \sin\left[\frac{\pi}{6}(x - 4.2)\right] + 68.$$

(b) Figure 20 shows the data points from the table, the graph of $y = 14 \sin\left[\frac{\pi}{6}(x - 4.2)\right] + 68$, and the graph of $y = 14 \sin \frac{\pi}{6}x + 68$ for comparison.

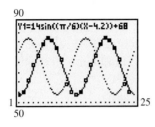

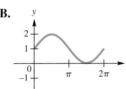

Values are rounded to the nearest hundredth.

(a)

(b)

Figure 20

Figure 21

(c) We used the given data for a two-year period to produce the model described in Figure 21(a). Figure 21(b) shows its graph along with the data points.

NOW TRY EXERCISE 57. ◀

4.2 Exercises

Concept Check In Exercises 1–8, match each function with its graph in A–H.

1. $y = \sin\left(x - \frac{\pi}{4}\right)$
2. $y = \sin\left(x + \frac{\pi}{4}\right)$
3. $y = \cos\left(x - \frac{\pi}{4}\right)$

4. $y = \cos\left(x + \frac{\pi}{4}\right)$
5. $y = 1 + \sin x$
6. $y = -1 + \sin x$

7. $y = 1 + \cos x$
8. $y = -1 + \cos x$

A.
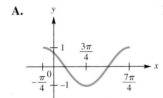

B.
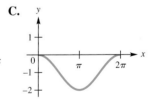

C.

D.
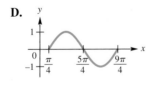

E.
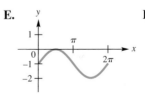

F.
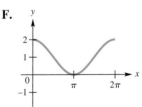

G.
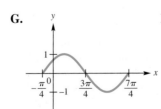

H.
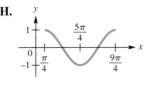

Concept Check   *In Exercises 9–12, match each function with its calculator graph in the standard trig window.*

9. $y = \cos\left(x - \dfrac{\pi}{4}\right)$

10. $y = \sin\left(x - \dfrac{\pi}{4}\right)$

11. $y = 1 + \sin x$

12. $y = -1 + \cos x$

A.

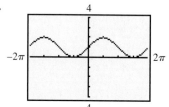

B.

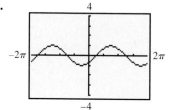

C.

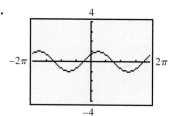

D.

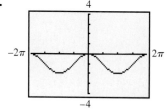

 13. The graphs of $y = \sin x + 1$ and $y = \sin(x + 1)$ are **NOT** the same. Explain why this is so.

14. *Concept Check*   Refer to Exercise 13. Which one of the two graphs is the same as that of $y = 1 + \sin x$?

Concept Check   *Match each function in Column I with the appropriate description in Column II.*

I	**II**
15. $y = 3\sin(2x - 4)$	**A.** amplitude $= 2$, period $= \dfrac{\pi}{2}$, phase shift $= \dfrac{3}{4}$
16. $y = 2\sin(3x - 4)$	**B.** amplitude $= 3$, period $= \pi$, phase shift $= 2$
17. $y = 4\sin(3x - 2)$	**C.** amplitude $= 4$, period $= \dfrac{2\pi}{3}$, phase shift $= \dfrac{2}{3}$
18. $y = 2\sin(4x - 3)$	**D.** amplitude $= 2$, period $= \dfrac{2\pi}{3}$, phase shift $= \dfrac{4}{3}$

Concept Check   *In Exercises 19 and 20, fill in the blanks with the word* right *or the word* left.

19. If the graph of $y = \cos x$ is translated $\frac{\pi}{2}$ units horizontally to the _____, it will coincide with the graph of $y = \sin x$.

20. If the graph of $y = \sin x$ is translated $\frac{\pi}{2}$ units horizontally to the _____, it will coincide with the graph of $y = \cos x$.

Connecting Graphs with Equations Each function graphed is of the form $y = c + \cos x$, $y = c + \sin x$, $y = \cos(x - d)$, or $y = \sin(x - d)$, where d is the least possible positive value. Determine the equation of the graph.

21.
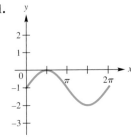

22.
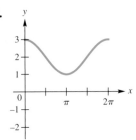

23.
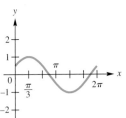

24.
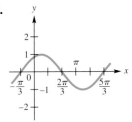

Find the amplitude, the period, any vertical translation, and any phase shift of the graph of each function. See Examples 1–5.

25. $y = 2 \sin(x - \pi)$

26. $y = \dfrac{2}{3} \sin\left(x + \dfrac{\pi}{2}\right)$

27. $y = 4 \cos\left(\dfrac{1}{2}x + \dfrac{\pi}{2}\right)$

28. $y = \dfrac{1}{2} \sin\left(\dfrac{1}{2}x + \pi\right)$

29. $y = 3 \cos \dfrac{\pi}{2}\left(x - \dfrac{1}{2}\right)$

30. $y = -\cos \pi\left(x - \dfrac{1}{3}\right)$

31. $y = 2 - \sin\left(3x - \dfrac{\pi}{5}\right)$

32. $y = -1 + \dfrac{1}{2} \cos(2x - 3\pi)$

Graph each function over a two-period interval. See Examples 1 and 2.

33. $y = \cos\left(x - \dfrac{\pi}{2}\right)$

34. $y = \sin\left(x - \dfrac{\pi}{4}\right)$

35. $y = \sin\left(x + \dfrac{\pi}{4}\right)$

36. $y = \cos\left(x - \dfrac{\pi}{3}\right)$

37. $y = 2 \cos\left(x - \dfrac{\pi}{3}\right)$

38. $y = 3 \sin\left(x - \dfrac{3\pi}{2}\right)$

Graph each function over a one-period interval. See Example 3.

39. $y = \dfrac{3}{2} \sin 2\left(x + \dfrac{\pi}{4}\right)$

40. $y = -\dfrac{1}{2} \cos 4\left(x + \dfrac{\pi}{2}\right)$

41. $y = -4 \sin(2x - \pi)$

42. $y = 3 \cos(4x + \pi)$

43. $y = \dfrac{1}{2} \cos\left(\dfrac{1}{2}x - \dfrac{\pi}{4}\right)$

44. $y = -\dfrac{1}{4} \sin\left(\dfrac{3}{4}x + \dfrac{\pi}{8}\right)$

Graph each function over a two-period interval. See Example 4.

45. $y = -3 + 2 \sin x$

46. $y = 2 - 3 \cos x$

47. $y = -1 - 2 \cos 5x$

48. $y = 1 - \dfrac{2}{3} \sin \dfrac{3}{4} x$

49. $y = 1 - 2 \cos \dfrac{1}{2} x$

50. $y = -3 + 3 \sin \dfrac{1}{2} x$

51. $y = -2 + \dfrac{1}{2} \sin 3x$

52. $y = 1 + \dfrac{2}{3} \cos \dfrac{1}{2} x$

Graph each function over a one-period interval. See Example 5.

53. $y = -3 + 2 \sin\left(x + \dfrac{\pi}{2}\right)$

54. $y = 4 - 3 \cos(x - \pi)$

55. $y = \dfrac{1}{2} + \sin 2\left(x + \dfrac{\pi}{4}\right)$

56. $y = -\dfrac{5}{2} + \cos 3\left(x - \dfrac{\pi}{6}\right)$

(Modeling) Solve each problem. See Example 6.

57. *Average Monthly Temperature* The average monthly temperature (in °F) in Vancouver, Canada, is shown in the table.

(a) Plot the average monthly temperature over a two-year period letting $x = 1$ correspond to the month of January during the first year. Do the data seem to indicate a translated sine graph?

(b) The highest average monthly temperature is 64°F in July, and the lowest average monthly temperature is 36°F in January. Their average is 50°F. Graph the data together with the line $y = 50$. What does this line represent with regard to temperature in Vancouver?

Month	°F	Month	°F
Jan	36	July	64
Feb	39	Aug	63
Mar	43	Sept	57
Apr	48	Oct	50
May	55	Nov	43
June	59	Dec	39

Source: Miller, A. and J. Thompson, *Elements of Meteorology, Fourth Edition,* Charles E. Merrill Publishing Co., 1983.

(c) Approximate the amplitude, period, and phase shift of the translated sine wave.

(d) Determine a function of the form $f(x) = a \sin b(x - d) + c$, where a, b, c, and d are constants, that models the data.

(e) Graph f together with the data on the same coordinate axes. How well does f model the given data?

(f) Use the sine regression capability of a graphing calculator to find the equation of a sine curve that fits these data.

58. *Average Monthly Temperature* The average monthly temperature (in °F) in Phoenix, Arizona, is shown in the table.

(a) Predict the average yearly temperature and compare it to the actual value of 70°F.

(b) Plot the average monthly temperature over a two-year period by letting $x = 1$ correspond to January of the first year.

(c) Determine a function of the form $f(x) = a \cos b(x - d) + c$, where a, b, c, and d are constants, that models the data.

(d) Graph f together with the data on the same coordinate axes. How well does f model the data?

(e) Use the sine regression capability of a graphing calculator to find the equation of a sine curve that fits these data (two years).

Month	°F	Month	°F
Jan	51	July	90
Feb	55	Aug	90
Mar	63	Sept	84
Apr	67	Oct	71
May	77	Nov	59
June	86	Dec	52

Source: Miller, A. and J. Thompson, *Elements of Meteorology, Fourth Edition,* Charles E. Merrill Publishing Co., 1983.

CHAPTER 4 ▶ Quiz (Sections 4.1–4.2)

Graph each function over a two-period interval. Give the period and amplitude.

1. $y = -4 \sin x$

2. $y = -\dfrac{1}{2} \cos 2x$

3. $y = 3 \sin \pi x$

4. $y = -2 \cos\left(x + \dfrac{\pi}{4}\right)$

5. $y = 2 + \sin(2x - \pi)$

Connecting Graphs with Equations *Each function graphed is of the form $y = a \cos bx$ or $y = a \sin bx$, where $b > 0$. Determine the equation of the graph.*

6.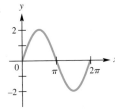

7.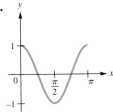

8.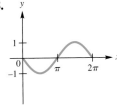

Average Monthly Temperature *The average temperature (in °F) at a certain location can be approximated by the function defined by*

$$f(x) = 12 \sin\left[\dfrac{\pi}{6}(x - 3.9)\right] + 72,$$

where $x = 1$ represents January, $x = 2$ represents February, and so on.

9. What is the average temperature in April?

10. What is the lowest average monthly temperature? What is the highest?

4.3 Graphs of the Tangent and Cotangent Functions

Graph of the Tangent Function ▪ Graph of the Cotangent Function ▪ Graphing Techniques

Graph of the Tangent Function Consider the table of selected points accompanying the graph of the tangent function in Figure 22 on the next page. These points include special values between $-\dfrac{\pi}{2}$ and $\dfrac{\pi}{2}$. The tangent function is undefined for odd multiples of $\dfrac{\pi}{2}$ and, thus, has *vertical asymptotes* for such values. A **vertical asymptote** is a vertical line that the graph approaches but does not intersect, while function values increase or decrease without bound as x-values get closer and closer to the line. Furthermore, since $\tan(-x) = -\tan x$ (see Exercise 43), the graph of the tangent function is symmetric with respect to the origin.

x	$y = \tan x$
$-\frac{\pi}{3}$	$-\sqrt{3} \approx -1.7$
$-\frac{\pi}{4}$	-1
$-\frac{\pi}{6}$	$-\frac{\sqrt{3}}{3} \approx -.6$
0	0
$\frac{\pi}{6}$	$\frac{\sqrt{3}}{3} \approx .6$
$\frac{\pi}{4}$	1
$\frac{\pi}{3}$	$\sqrt{3} \approx 1.7$

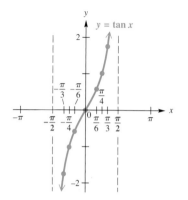

Figure 22

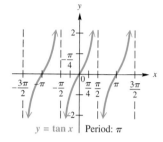

$y = \tan x$ | Period: π

Figure 23

The tangent function has period π. Because $\tan x = \frac{\sin x}{\cos x}$, tangent values are 0 when sine values are 0, and undefined when cosine values are 0. As x-values go from $-\frac{\pi}{2}$ to $\frac{\pi}{2}$, tangent values go from $-\infty$ to ∞ and increase throughout the interval. Those same values are repeated as x goes from $\frac{\pi}{2}$ to $\frac{3\pi}{2}$, $\frac{3\pi}{2}$ to $\frac{5\pi}{2}$, and so on. The graph of $y = \tan x$ from $-\frac{3\pi}{2}$ to $\frac{3\pi}{2}$ is shown in Figure 23.

TANGENT FUNCTION $\quad f(x) = \tan x$

Domain: $\left\{ x \mid x \neq (2n + 1)\frac{\pi}{2}, \text{ where } n \text{ is any integer} \right\}$ Range: $(-\infty, \infty)$

x	y
$-\frac{\pi}{2}$	undefined
$-\frac{\pi}{4}$	-1
0	0
$\frac{\pi}{4}$	1
$\frac{\pi}{2}$	undefined

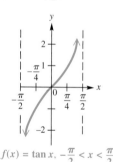

$f(x) = \tan x, \ -\frac{\pi}{2} < x < \frac{\pi}{2}$

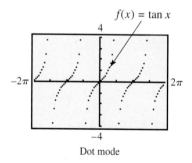

Dot mode

Figure 24

- The graph is discontinuous at values of x of the form $x = (2n + 1)\frac{\pi}{2}$ and has vertical asymptotes at these values.
- Its x-intercepts are of the form $x = n\pi$.
- Its period is π.
- Its graph has no amplitude, since there are no minimum or maximum values.
- The graph is symmetric with respect to the origin, so the function is an odd function. For all x in the domain, $\tan(-x) = -\tan x$.

Graph of the Cotangent Function
A similar analysis for selected points between 0 and π for the graph of the cotangent function yields the graph in Figure 25 on the next page. Here the vertical asymptotes are at x-values that are integer multiples of π. Because $\cot(-x) = -\cot x$ (see Exercise 44), this graph is also symmetric with respect to the origin. (This can be seen when more of the graph is plotted.)

x	$y = \cot x$
$\frac{\pi}{6}$	$\sqrt{3} \approx 1.7$
$\frac{\pi}{4}$	1
$\frac{\pi}{3}$	$\frac{\sqrt{3}}{3} \approx .6$
$\frac{\pi}{2}$	0
$\frac{2\pi}{3}$	$-\frac{\sqrt{3}}{3} \approx -.6$
$\frac{3\pi}{4}$	-1
$\frac{5\pi}{6}$	$-\sqrt{3} \approx -1.7$

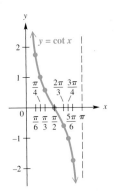

Figure 25

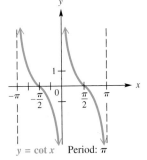

$y = \cot x$ Period: π

Figure 26

The cotangent function also has period π. Cotangent values are 0 when cosine values are 0, and undefined when sine values are 0. As x-values go from 0 to π, cotangent values go from ∞ to $-\infty$ and decrease throughout the interval. Those same values are repeated as x goes from π to 2π, 2π to 3π, and so on. The graph of $y = \cot x$ from $-\pi$ to π is shown in Figure 26. The graph continues in this pattern.

COTANGENT FUNCTION $f(x) = \cot x$

Domain: $\{x \mid x \neq n\pi, \text{ where } n \text{ is any integer}\}$ Range: $(-\infty, \infty)$

x	y
0	undefined
$\frac{\pi}{4}$	1
$\frac{\pi}{2}$	0
$\frac{3\pi}{4}$	-1
π	undefined

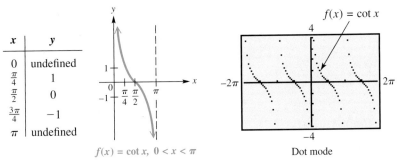

$f(x) = \cot x, \; 0 < x < \pi$

$f(x) = \cot x$

Dot mode

Figure 27

- The graph is discontinuous at values of x of the form $x = n\pi$ and has vertical asymptotes at these values.
- Its x-intercepts are of the form $x = (2n + 1)\frac{\pi}{2}$.
- Its period is π.
- Its graph has no amplitude, since there are no minimum or maximum values.
- The graph is symmetric with respect to the origin, so the function is an odd function. For all x in the domain, $\cot(-x) = -\cot x$.

The tangent function can be graphed directly with a graphing calculator, using the tangent key. To graph the cotangent function, however, we must use one of the identities $\cot x = \frac{1}{\tan x}$ or $\cot x = \frac{\cos x}{\sin x}$, since graphing calculators generally do not have cotangent keys. ■

Graphing Techniques

> ### GUIDELINES FOR SKETCHING GRAPHS OF TANGENT AND COTANGENT FUNCTIONS
>
> To graph $y = a \tan bx$ or $y = a \cot bx$, with $b > 0$, follow these steps.
>
> **Step 1** Determine the period, $\frac{\pi}{b}$. To locate two adjacent vertical asymptotes, solve the following equations for x:
>
> For $y = a \tan bx$: $\quad bx = -\dfrac{\pi}{2}$ and $bx = \dfrac{\pi}{2}$.
>
> For $y = a \cot bx$: $\quad bx = 0$ and $bx = \pi$.
>
> **Step 2** Sketch the two vertical asymptotes found in Step 1.
>
> **Step 3** Divide the interval formed by the vertical asymptotes into four equal parts. (See the Note on page 148.)
>
> **Step 4** Evaluate the function for the first-quarter point, midpoint, and third-quarter point, using the x-values found in Step 3.
>
> **Step 5** Join the points with a smooth curve, approaching the vertical asymptotes. Indicate additional asymptotes and periods of the graph as necessary.

> ▶ **EXAMPLE 1** GRAPHING $y = \tan bx$

Graph $y = \tan 2x$.

Solution

Step 1 The period of this function is $\frac{\pi}{2}$. To locate two adjacent vertical asymptotes, solve $2x = -\frac{\pi}{2}$ and $2x = \frac{\pi}{2}$ (since this is a tangent function). The two asymptotes have equations $x = -\frac{\pi}{4}$ and $x = \frac{\pi}{4}$.

Step 2 Sketch the two vertical asymptotes $x = \pm\frac{\pi}{4}$, as shown in Figure 28.

Step 3 Divide the interval $\left(-\frac{\pi}{4}, \frac{\pi}{4}\right)$ into four equal parts. This gives the following key x-values.

first-quarter value: $-\dfrac{\pi}{8}$, middle value: 0, third-quarter value: $\dfrac{\pi}{8}$ Key x-values

Step 4 Evaluate the function for the x-values found in Step 3.

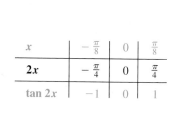

x	$-\frac{\pi}{8}$	0	$\frac{\pi}{8}$
$2x$	$-\frac{\pi}{4}$	0	$\frac{\pi}{4}$
$\tan 2x$	-1	0	1

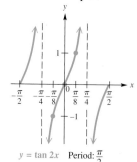

$y = \tan 2x$ Period: $\dfrac{\pi}{2}$

Figure 28

Step 5 Join these points with a smooth curve, approaching the vertical asymptotes. See Figure 28. Another period has been graphed, one half period to the left and one half period to the right.

NOW TRY EXERCISE 7. ◀

▶ **EXAMPLE 2** GRAPHING $y = a \tan bx$

Graph $y = -3 \tan \frac{1}{2}x$.

Solution The period is $\dfrac{\pi}{\frac{1}{2}} = \pi \div \frac{1}{2} = \pi \cdot \frac{2}{1} = 2\pi$. Adjacent asymptotes are at $x = -\pi$ and $x = \pi$. Dividing the interval $(-\pi, \pi)$ into four equal parts gives key x-values of $-\frac{\pi}{2}, 0$, and $\frac{\pi}{2}$. Evaluating the function at these x-values gives the following key points.

$$\left(-\frac{\pi}{2}, 3\right), \qquad (0,0), \qquad \left(\frac{\pi}{2}, -3\right) \quad \text{Key points}$$

By plotting these points and joining them with a smooth curve, we obtain the graph shown in Figure 29. Because the coefficient -3 is negative, the graph is reflected across the x-axis compared to the graph of $y = 3 \tan \frac{1}{2}x$.

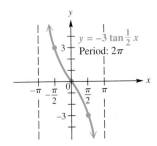

Figure 29

NOW TRY EXERCISE 15. ◀

▶ **Note** The function defined by $y = -3 \tan \frac{1}{2}x$ in Example 2, graphed in Figure 29, has a graph that compares to the graph of $y = \tan x$ as follows.

1. The period is larger because $b = \frac{1}{2}$, and $\frac{1}{2} < 1$.

2. The graph is "stretched" because $a = -3$, and $|-3| > 1$.

3. Each branch of the graph goes down from left to right (that is, the function decreases) between each pair of adjacent asymptotes because $a = -3$, and $-3 < 0$. When $a < 0$, the graph is reflected across the x-axis compared to the graph of $y = |a| \tan bx$.

▶ **EXAMPLE 3** GRAPHING $y = a \cot bx$

Graph $y = \frac{1}{2} \cot 2x$.

Solution Because this function involves the cotangent, we can locate two adjacent asymptotes by solving the equations $2x = 0$ and $2x = \pi$. The lines $x = 0$ (the y-axis) and $x = \frac{\pi}{2}$ are two such asymptotes. Divide the interval $\left(0, \frac{\pi}{2}\right)$ into four equal parts, getting key x-values of $\frac{\pi}{8}, \frac{\pi}{4}$, and $\frac{3\pi}{8}$. Evaluating the function at these x-values gives the key points $\left(\frac{\pi}{8}, \frac{1}{2}\right), \left(\frac{\pi}{4}, 0\right), \left(\frac{3\pi}{8}, -\frac{1}{2}\right)$. Joining these points with a smooth curve approaching the asymptotes gives the graph shown in Figure 30.

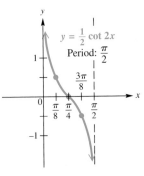

Figure 30

NOW TRY EXERCISE 17. ◀

Like the other circular functions, the graphs of the tangent and cotangent functions may be translated horizontally and vertically.

▶ EXAMPLE 4 GRAPHING A TANGENT FUNCTION WITH A VERTICAL TRANSLATION

Graph $y = 2 + \tan x$.

Analytic Solution

Every value of y for this function will be 2 units more than the corresponding value of y in $y = \tan x$, causing the graph of $y = 2 + \tan x$ to be translated 2 units up compared with the graph of $y = \tan x$. See Figure 31.

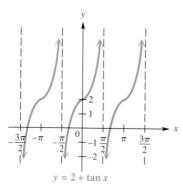

$y = 2 + \tan x$

Figure 31

Graphing Calculator Solution

To see the vertical translation, observe the coordinates displayed at the bottoms of the screens in Figures 32 and 33. For $X = \frac{\pi}{4} \approx .78539816$,

$$Y_1 = \tan X = 1,$$

while for the same X-value,

$$Y_2 = 2 + \tan X = 2 + 1 = 3.$$

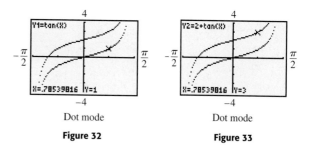

Dot mode | Dot mode

Figure 32 | **Figure 33**

NOW TRY EXERCISE 23. ◀

▶ EXAMPLE 5 GRAPHING A COTANGENT FUNCTION WITH VERTICAL AND HORIZONTAL TRANSLATIONS

Graph $y = -2 - \cot\left(x - \frac{\pi}{4}\right)$.

Solution Here $b = 1$, so the period is π. The graph will be translated down 2 units (because $c = -2$), reflected across the x-axis (because of the negative sign in front of the cotangent), and will have a phase shift (horizontal translation) $\frac{\pi}{4}$ unit to the right $\left(\text{because of the argument } \left(x - \frac{\pi}{4}\right)\right)$. To locate adjacent asymptotes, since this function involves the cotangent, we solve the following equations:

$$x - \frac{\pi}{4} = 0, \quad \text{so } x = \frac{\pi}{4} \quad \text{and} \quad x - \frac{\pi}{4} = \pi, \quad \text{so } x = \frac{5\pi}{4}.$$

Dividing the interval $\left(\frac{\pi}{4}, \frac{5\pi}{4}\right)$ into four equal parts and evaluating the function at the three key x-values within the interval gives these points.

$$\left(\frac{\pi}{2}, -3\right), \qquad \left(\frac{3\pi}{4}, -2\right), \qquad (\pi, -1) \quad \text{Key points}$$

Join these points with a smooth curve. This period of the graph, along with the one in the domain interval $\left(-\frac{3\pi}{4}, \frac{\pi}{4}\right)$, is shown in Figure 34.

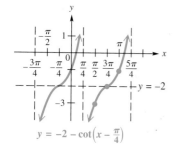

$y = -2 - \cot\left(x - \frac{\pi}{4}\right)$

Figure 34

NOW TRY EXERCISE 31. ◀

4.3 Exercises

Concept Check *In Exercises 1–6, match each function with its graph from choices A–F.*

1. $y = -\tan x$

2. $y = -\cot x$

3. $y = \tan\left(x - \dfrac{\pi}{4}\right)$

4. $y = \cot\left(x - \dfrac{\pi}{4}\right)$

5. $y = \cot\left(x + \dfrac{\pi}{4}\right)$

6. $y = \tan\left(x + \dfrac{\pi}{4}\right)$

A.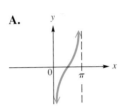

B.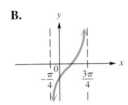

C.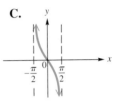

D.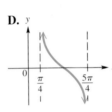

E.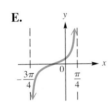

F.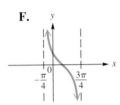

Graph each function over a one-period interval. See Examples 1–3.

7. $y = \tan 4x$

8. $y = \tan \dfrac{1}{2}x$

9. $y = 2 \tan x$

10. $y = 2 \cot x$

11. $y = 2 \tan \dfrac{1}{4}x$

12. $y = \dfrac{1}{2} \cot x$

13. $y = \cot 3x$

14. $y = -\cot \dfrac{1}{2}x$

15. $y = -2 \tan \dfrac{1}{4}x$

16. $y = 3 \tan \dfrac{1}{2}x$

17. $y = \dfrac{1}{2} \cot 4x$

18. $y = -\dfrac{1}{2} \cot 2x$

Graph each function over a two-period interval. See Examples 4 and 5.

19. $y = \tan(2x - \pi)$

20. $y = \tan\left(\dfrac{x}{2} + \pi\right)$

21. $y = \cot\left(3x + \dfrac{\pi}{4}\right)$

22. $y = \cot\left(2x - \dfrac{3\pi}{2}\right)$

23. $y = 1 + \tan x$

24. $y = 1 - \tan x$

25. $y = 1 - \cot x$

26. $y = -2 - \cot x$

27. $y = -1 + 2 \tan x$

28. $y = 3 + \dfrac{1}{2} \tan x$

29. $y = -1 + \dfrac{1}{2} \cot(2x - 3\pi)$

30. $y = -2 + 3 \tan(4x + \pi)$

31. $y = 1 - 2 \cot 2\left(x + \dfrac{\pi}{2}\right)$

32. $y = \dfrac{2}{3} \tan\left(\dfrac{3}{4}x - \pi\right) - 2$

Connecting Graphs with Equations   *Each function graphed is of the form* $y = a \tan bx$ *or* $y = a \cot bx$, *where* $b > 0$. *Determine the equation of the graph. (Half- and quarter-points are identified by dots.)*

33.

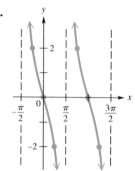

34.

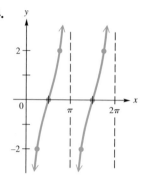

35.

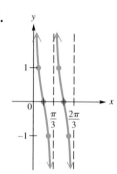

36.

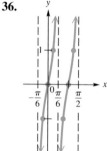

Concept Check   *In Exercises 37–40, tell whether each statement is* true *or* false. *If false,* tell why.

37. The least positive number k for which $x = k$ is an asymptote for the tangent function is $\frac{\pi}{2}$.

38. The least positive number k for which $x = k$ is an asymptote for the cotangent function is $\frac{\pi}{2}$.

39. The graph of $y = \tan x$ in Figure 23 suggests that $\tan(-x) = \tan x$ for all x in the domain of $\tan x$.

40. The graph of $y = \cot x$ in Figure 26 suggests that $\cot(-x) = -\cot x$ for all x in the domain of $\cot x$.

41. *Concept Check*   If c is any number, then how many solutions does the equation $c = \tan x$ have in the interval $(-2\pi, 2\pi]$?

42. Consider the function defined by $f(x) = -4 \tan(2x + \pi)$. What is the domain of f? What is its range?

43. Show that $\tan(-x) = -\tan x$ by writing $\tan(-x)$ as $\frac{\sin(-x)}{\cos(-x)}$ and then using the relationships for $\sin(-x)$ and $\cos(-x)$.

44. Show that $\cot(-x) = -\cot x$ by writing $\cot(-x)$ as $\frac{\cos(-x)}{\sin(-x)}$ and then using the relationships for $\cos(-x)$ and $\sin(-x)$.

45. *(Modeling) Distance of a Rotating Beacon*   A rotating beacon is located at point A next to a long wall. (See the figure on the next page.) The beacon is 4 m from the wall. The distance d is given by

$$d = 4 \tan 2\pi t,$$

where t is time measured in seconds since the beacon started rotating. (When $t = 0$, the beacon is aimed at point R. When the beacon is aimed to the right of R, the value of d is positive; d is negative if the beacon is aimed to the left of R.) Find d for each time.

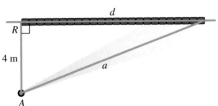

(a) $t = 0$
(b) $t = .4$
(c) $t = .8$
(d) $t = 1.2$
(e) Why is .25 a meaningless value for t?

 46. Simultaneously graph $y = \tan x$ and $y = x$ in the window $[-1, 1]$ by $[-1, 1]$ with a graphing calculator. Write a short description of the relationship between $\tan x$ and x for small x-values.

RELATING CONCEPTS

For individual or collaborative investigation
(Exercises 47–52)

Consider the function defined by

$$y = -2 - \cot\left(x - \frac{\pi}{4}\right)$$

from Example 5. **Work these exercises in order.**

47. What is the least positive number for which $y = \cot x$ is undefined?

48. Let k represent the number you found in Exercise 47. Set $x - \frac{\pi}{4}$ equal to k, and solve to find a positive number for which $\cot\left(x - \frac{\pi}{4}\right)$ is undefined.

49. Based on your answer in Exercise 48 and the fact that the cotangent function has period π, give the general form of the equations of the asymptotes of the graph of $y = -2 - \cot\left(x - \frac{\pi}{4}\right)$. Let n represent any integer.

 50. Use the capabilities of your calculator to find the least positive x-intercept of the graph of this function.

51. Use the fact that the period of this function is π to find the next positive x-intercept.

52. Give the solution set of the equation $-2 - \cot\left(x - \frac{\pi}{4}\right) = 0$ over all real numbers. Let n represent any integer.

4.4 Graphs of the Secant and Cosecant Functions

Graph of the Secant Function ▪ Graph of the Cosecant Function ▪ Graphing Techniques ▪ Addition of Ordinates ▪ Connecting Graphs with Equations

Graph of the Secant Function Consider the table of selected points accompanying the graph of the secant function in Figure 35 on the next page. These points include special values between $-\pi$ and π. The secant function is undefined for odd multiples of $\frac{\pi}{2}$ and, thus, like the tangent function has vertical asymptotes for such values. Furthermore, since $\sec(-x) = \sec x$ (see Exercise 31), the graph of the secant function is symmetric with respect to the y-axis.

x	$y = \sec x$	x	$y = \sec x$
$\pm\frac{\pi}{3}$	2	$\pm\frac{2\pi}{3}$	-2
$\pm\frac{\pi}{4}$	$\sqrt{2} \approx 1.4$	$\pm\frac{3\pi}{4}$	$-\sqrt{2} \approx -1.4$
$\pm\frac{\pi}{6}$	$\frac{2\sqrt{3}}{3} \approx 1.2$	$\pm\frac{5\pi}{6}$	$-\frac{2\sqrt{3}}{3} \approx -1.2$
0	1	$\pm\pi$	-1

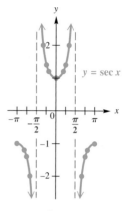

Figure 35

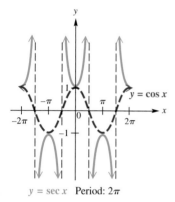

$y = \sec x$ Period: 2π

Figure 36

Because secant values are reciprocals of corresponding cosine values, the period of the secant function is 2π, the same as for $y = \cos x$. When $\cos x = 1$, the value of $\sec x$ is also 1; likewise, when $\cos x = -1$, $\sec x = -1$ as well. For all x, $-1 \le \cos x \le 1$, and thus, $|\sec x| \ge 1$ for all x in its domain. Figure 36 shows how the graphs of $y = \cos x$ and $y = \sec x$ are related.

SECANT FUNCTION $f(x) = \sec x$

Domain: $\left\{x \mid x \ne (2n + 1)\frac{\pi}{2},\right.$ Range: $(-\infty, -1] \cup [1, \infty)$
where n is any integer$\}$

x	y
$-\frac{\pi}{2}$	undefined
$-\frac{\pi}{4}$	$\sqrt{2}$
0	1
$\frac{\pi}{4}$	$\sqrt{2}$
$\frac{\pi}{2}$	undefined
$\frac{3\pi}{4}$	$-\sqrt{2}$
π	-1
$\frac{3\pi}{2}$	undefined

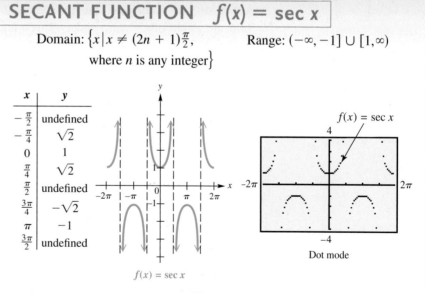

$f(x) = \sec x$

Dot mode

Figure 37

- The graph is discontinuous at values of x of the form $x = (2n + 1)\frac{\pi}{2}$ and has vertical asymptotes at these values.
- There are no x-intercepts.
- Its period is 2π.
- Its graph has no amplitude, since there are no maximum or minimum values.
- The graph is symmetric with respect to the y-axis, so the function is an even function. For all x in the domain, $\sec(-x) = \sec x$.

Graph of the Cosecant Function A similar analysis for selected points between $-\pi$ and π for the graph of the cosecant function yields the graph in Figure 38. The vertical asymptotes are at x-values that are integer multiples of π. Because $\csc(-x) = -\csc x$ (see Exercise 32), this graph is symmetric with respect to the origin.

x	$y = \csc x$		x	$y = \csc x$
$\frac{\pi}{6}$	2		$-\frac{\pi}{6}$	-2
$\frac{\pi}{4}$	$\sqrt{2} \approx 1.4$		$-\frac{\pi}{4}$	$-\sqrt{2} \approx -1.4$
$\frac{\pi}{3}$	$\frac{2\sqrt{3}}{3} \approx 1.2$		$-\frac{\pi}{3}$	$-\frac{2\sqrt{3}}{3} \approx -1.2$
$\frac{\pi}{2}$	1		$-\frac{\pi}{2}$	-1
$\frac{2\pi}{3}$	$\frac{2\sqrt{3}}{3} \approx 1.2$		$-\frac{2\pi}{3}$	$-\frac{2\sqrt{3}}{3} \approx -1.2$
$\frac{3\pi}{4}$	$\sqrt{2} \approx 1.4$		$-\frac{3\pi}{4}$	$-\sqrt{2} \approx -1.4$
$\frac{5\pi}{6}$	2		$-\frac{5\pi}{6}$	-2

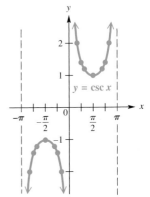

Figure 38

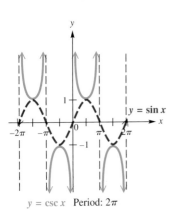

$y = \csc x$ Period: 2π

Figure 39

Because cosecant values are reciprocals of corresponding sine values, the period of the cosecant function is 2π, the same as for $y = \sin x$. When $\sin x = 1$, the value of $\csc x$ is also 1; likewise, when $\sin x = -1$, $\csc x = -1$. For all x, $-1 \le \sin x \le 1$, and thus $|\csc x| \ge 1$ for all x in its domain. Figure 39 shows how the graphs of $y = \sin x$ and $y = \csc x$ are related.

COSECANT FUNCTION $f(x) = \csc x$

Domain: $\{x \mid x \ne n\pi, \text{ where } n \text{ is any integer}\}$ Range: $(-\infty, -1] \cup [1, \infty)$

x	y
0	undefined
$\frac{\pi}{6}$	2
$\frac{\pi}{3}$	$\frac{2\sqrt{3}}{3}$
$\frac{\pi}{2}$	1
$\frac{2\pi}{3}$	$\frac{2\sqrt{3}}{3}$
π	undefined
$\frac{3\pi}{2}$	-1
2π	undefined

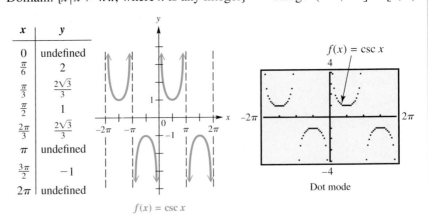

$f(x) = \csc x$

Dot mode

Figure 40

- The graph is discontinuous at values of x of the form $x = n\pi$ and has vertical asymptotes at these values.
- There are no x-intercepts.
- Its period is 2π.
- Its graph has no amplitude, since there are no maximum or minimum values.
- The graph is symmetric with respect to the origin, so the function is an odd function. For all x in the domain, $\csc(-x) = -\csc x$.

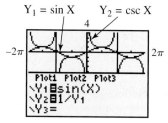

Trig window; connected mode

Figure 41

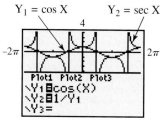

Trig window; connected mode

Figure 42

Typically, calculators do not have keys for the cosecant and secant functions. To graph $y = \csc x$ with a graphing calculator, use the fact that

$$\csc x = \frac{1}{\sin x}.$$

The graphs of $Y_1 = \sin X$ and $Y_2 = \csc X$ are shown in Figure 41. The calculator is in split screen and connected modes. Similarly, the secant function is graphed by using the identity

$$\sec x = \frac{1}{\cos x},$$

as shown in Figure 42.

Using dot mode for graphing will eliminate the vertical lines that appear in Figures 41 and 42. While they suggest asymptotes and are sometimes called **pseudo-asymptotes,** they are not actually parts of the graphs. ∎

Graphing Techniques In the previous section, we gave guidelines for sketching graphs of tangent and cotangent functions. We now present similar guidelines for graphing cosecant and secant functions.

GUIDELINES FOR SKETCHING GRAPHS OF COSECANT AND SECANT FUNCTIONS

To graph $y = a \csc bx$ or $y = a \sec bx,$ with $b > 0$, follow these steps.

Step 1 Graph the corresponding reciprocal function as a guide, using a dashed curve.

To Graph	Use as a Guide
$y = a \csc bx$	$y = a \sin bx$
$y = a \sec bx$	$y = a \cos bx$

Step 2 Sketch the vertical asymptotes. They will have equations of the form $x = k$, where k is an x-intercept of the graph of the guide function.

Step 3 Sketch the graph of the desired function by drawing the typical U-shaped branches between the adjacent asymptotes. The branches will be above the graph of the guide function when the guide function values are positive and below the graph of the guide function when the guide function values are negative. The graph will resemble those in Figures 37 and 40 in the function boxes on pages 177 and 178.

Like graphs of the sine and cosine functions, graphs of the secant and cosecant functions may be translated vertically and horizontally. The period of both basic functions is 2π.

▶ **EXAMPLE 1** GRAPHING $y = a \sec bx$

Graph $y = 2 \sec \frac{1}{2}x$.

Solution

Step 1 This function involves the secant, so the corresponding reciprocal function will involve the cosine. The guide function to graph is

$$y = 2 \cos \frac{1}{2}x.$$

Using the guidelines of **Section 4.1,** we find that this guide function has amplitude 2 and one period of the graph lies along the interval that satisfies the inequality

$$0 \le \frac{1}{2}x \le 2\pi, \quad \text{or} \quad [0, 4\pi]. \quad \text{(Appendix A)}$$

Dividing this interval into four equal parts gives the key points

$$(0, 2), \quad (\pi, 0), \quad (2\pi, -2), \quad (3\pi, 0), \quad (4\pi, 2), \quad \text{Key points}$$

which are joined with a dashed red curve to indicate that this graph is only a guide. An additional period is graphed as seen in Figure 43(a).

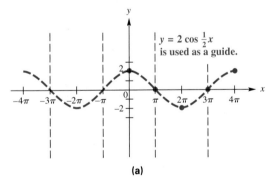
(a)

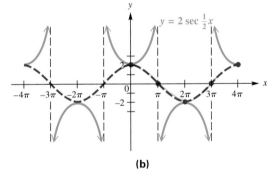
(b)

Figure 43

Step 2 Sketch the vertical asymptotes. These occur at x-values for which the guide function equals 0, such as

$$x = -3\pi, \quad x = -\pi, \quad x = \pi, \quad x = 3\pi.$$

See Figure 43(a).

Step 3 Sketch the graph of $y = 2 \sec \frac{1}{2}x$ by drawing the typical U-shaped branches, approaching the asymptotes. See the solid blue graph in Figure 43(b).

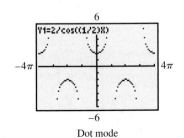
Dot mode

This is a calculator graph of the function in Example 1.

NOW TRY EXERCISE 5. ◀

▶ **EXAMPLE 2**   GRAPHING $y = a \csc(x - d)$

Graph $y = \frac{3}{2} \csc(x - \frac{\pi}{2})$.

Solution

Step 1 Use the guidelines of **Section 4.2** to graph the corresponding reciprocal function defined by

$$y = \frac{3}{2} \sin\left(x - \frac{\pi}{2}\right),$$

shown as a red dashed curve in Figure 44.

Step 2 Sketch the vertical asymptotes through the x-intercepts of the graph of $y = \frac{3}{2} \sin\left(x - \frac{\pi}{2}\right)$. These have the form $x = (2n + 1)\frac{\pi}{2}$, where n is any integer. See the black dashed lines in Figure 44.

Step 3 Sketch the graph of $y = \frac{3}{2} \csc\left(x - \frac{\pi}{2}\right)$ by drawing the typical U-shaped branches between adjacent asymptotes. See the solid blue graph in Figure 44.

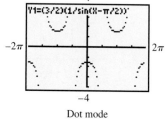

Dot mode

This is a calculator graph of the function in Example 2.

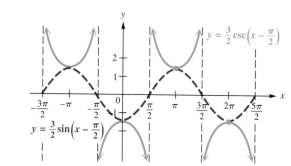

Figure 44

NOW TRY EXERCISE 7. ◀

Addition of Ordinates

New functions can be formed by adding or subtracting other functions. A function formed by combining two other functions, such as

$$y = \cos x + \sin x,$$

has historically been graphed using a method known as **addition of ordinates.** (The x-value of a point is sometimes called its **abscissa,** while its y-value is called its **ordinate.**) To apply this method to this function, we graph the functions $y = \cos x$ and $y = \sin x$. Then, for selected values of x, we add $\cos x$ and $\sin x$, and plot the points $(x, \cos x + \sin x)$. Joining the resulting points with a sinusoidal curve gives the graph of the desired function. While this method illustrates some valuable concepts involving the arithmetic of functions, it is time-consuming.

With graphing calculators, this technique is easily illustrated. Let $Y_1 = \cos X$, $Y_2 = \sin X$, and $Y_3 = Y_1 + Y_2$. Figure 45 on the next page shows the result when Y_1 and Y_2 are graphed in thin graph style, and $Y_3 = \cos X + \sin X$ is graphed in thick graph style. Notice that for $X = \frac{\pi}{6} \approx .52359878$, $Y_1 + Y_2 = Y_3$.

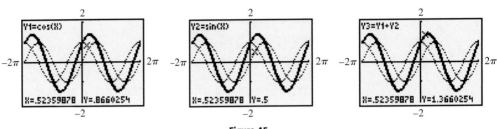

Figure 45

NOW TRY EXERCISE 35. ◄

Connecting Graphs with Equations   Now that the graphs of the six circular functions have been introduced, we can apply the concepts to give an equation of a given graph.

▶ EXAMPLE 3   DETERMINING AN EQUATION FOR A GRAPH

Determine an equation for each graph.

(a) 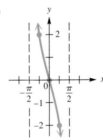   **(b)** 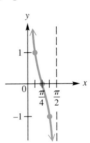   **(c)**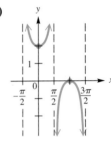

Solution

(a) This graph is that of $y = \tan x$ but reflected across the x-axis and stretched vertically by a factor of 2. Therefore, an equation for this graph is

$$y = -2 \tan x.$$

↑ x-axis reflection   ↖ Vertical stretch

(b) This graph is that of $y = \cot x$, but the period is $\frac{\pi}{2}$ rather than π. Therefore, if $y = \cot bx$, where $b > 0$, we must have $b = 2$. So an equation for this graph is

$$y = \cot 2x.$$

(c) This is the graph of $y = \sec x$, translated one unit upward. An equation is

$$y = 1 + \sec x.$$

NOW TRY EXERCISES 19, 21, AND 23. ◄

▶ **Note** Because the circular functions are periodic, there are actually infinitely many equations that correspond to each graph in Example 3. We have given only one such equation in each case. For instance, confirm that $y = -\tan\left(2x - \frac{\pi}{2}\right)$ is another equation for the graph in Example 3(b).

4.4 Exercises

Concept Check   In Exercises 1–4, match each function with its graph from choices A–D.

1. $y = -\csc x$ **2.** $y = -\sec x$ **3.** $y = \sec\left(x - \dfrac{\pi}{2}\right)$ **4.** $y = \csc\left(x + \dfrac{\pi}{2}\right)$

A.

B.

C.

D.

Graph each function over a one-period interval. See Examples 1 and 2.

5. $y = 3 \sec \dfrac{1}{4}x$ **6.** $y = -2 \sec \dfrac{1}{2}x$ **7.** $y = -\dfrac{1}{2} \csc\left(x + \dfrac{\pi}{2}\right)$

8. $y = \dfrac{1}{2} \csc\left(x - \dfrac{\pi}{2}\right)$ **9.** $y = \csc\left(x - \dfrac{\pi}{4}\right)$ **10.** $y = \sec\left(x + \dfrac{3\pi}{4}\right)$

11. $y = \sec\left(x + \dfrac{\pi}{4}\right)$ **12.** $y = \csc\left(x + \dfrac{\pi}{3}\right)$

13. $y = \sec\left(\dfrac{1}{2}x + \dfrac{\pi}{3}\right)$ **14.** $y = \csc\left(\dfrac{1}{2}x - \dfrac{\pi}{4}\right)$

15. $y = 2 + 3 \sec(2x - \pi)$ **16.** $y = 1 - 2 \csc\left(x + \dfrac{\pi}{2}\right)$

17. $y = 1 - \dfrac{1}{2} \csc\left(x - \dfrac{3\pi}{4}\right)$ **18.** $y = 2 + \dfrac{1}{4} \sec\left(\dfrac{1}{2}x - \pi\right)$

Connecting Graphs with Equations   Determine an equation for each graph. See Example 3.

19.

20.

21.

22.

23.

24.

Concept Check *In Exercises 25–28, tell whether each statement is* true *or* false. *If* false, *tell why.*

25. The tangent and secant functions are undefined for the same values.

26. The secant and cosecant functions are undefined for the same values.

27. The graph of $y = \sec x$ in Figure 37 suggests that $\sec(-x) = \sec x$ for all x in the domain of $\sec x$.

28. The graph of $y = \csc x$ in Figure 40 suggests that $\csc(-x) = -\csc x$ for all x in the domain of $\csc x$.

29. *Concept Check* If c is any number such that $-1 < c < 1$, then how many solutions does the equation $c = \sec x$ have over the entire domain of the secant function?

30. Consider the function defined by $g(x) = -2 \csc(4x + \pi)$. What is the domain of g? What is its range?

31. Show that $\sec(-x) = \sec x$ by writing $\sec(-x)$ as $\frac{1}{\cos(-x)}$ and then using the relationship between $\cos(-x)$ and $\cos x$.

32. Show that $\csc(-x) = -\csc x$ by writing $\csc(-x)$ as $\frac{1}{\sin(-x)}$ and then using the relationship between $\sin(-x)$ and $\sin x$.

33. *(Modeling) Distance of a Rotating Beacon* In the figure for Exercise 45 in **Section 4.3,** the distance a is given by

$$a = 4|\sec 2\pi t|.$$

Find a for each time.

(a) $t = 0$ (b) $t = .86$ (c) $t = 1.24$

34. Between each pair of successive asymptotes, a portion of the graph of $y = \sec x$ or $y = \csc x$ resembles a parabola. Can each of these portions actually be a parabola? Explain.

Use a graphing calculator to graph Y_1, Y_2, and $Y_1 + Y_2$ on the same screen. Evaluate each of the three functions at $X = \frac{\pi}{6}$, and verify that $Y_1(\frac{\pi}{6}) + Y_2(\frac{\pi}{6}) = (Y_1 + Y_2)(\frac{\pi}{6})$. See the discussion on addition of ordinates.

35. $Y_1 = \sin X, \quad Y_2 = \sin 2X$

36. $Y_1 = \cos X, \quad Y_2 = \sec X$

Summary Exercises on Graphing Circular Functions

These summary exercises provide practice with the various graphing techniques presented in this chapter. Graph each function over a one-period interval.

1. $y = 2 \sin \pi x$

2. $y = 4 \cos 1.5x$

3. $y = -2 + .5 \cos \frac{\pi}{4} x$

4. $y = 3 \sec \frac{\pi}{2} x$

5. $y = -4 \csc .5x$

6. $y = 3 \tan\left(\frac{\pi}{2} x + \pi\right)$

Graph each function over a two-period interval.

7. $y = -5 \sin \frac{x}{3}$

8. $y = 10 \cos\left(\frac{x}{4} + \frac{\pi}{2}\right)$

9. $y = 3 - 4 \sin(2.5x + \pi)$

10. $y = 2 - \sec[\pi(x - 3)]$

4.5 Harmonic Motion

Simple Harmonic Motion ▪ Damped Oscillatory Motion

Simple Harmonic Motion In part A of Figure 46, a spring with a weight attached to its free end is in equilibrium (or rest) position. If the weight is pulled down a units and released (part B of the figure), the spring's elasticity causes the weight to rise a units ($a > 0$) above the equilibrium position, as seen in part C, and then oscillate about the equilibrium position. If friction is neglected, this oscillatory motion is described mathematically by a sinusoid. Other applications of this type of motion include sound, electric current, and electromagnetic waves.

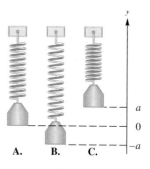

Figure 46

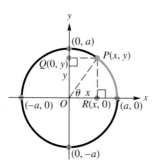

Figure 47

To develop a general equation for such motion, consider Figure 47. Suppose the point $P(x, y)$ moves around the circle counterclockwise at a uniform angular speed ω. Assume that at time $t = 0$, P is at $(a, 0)$. The angle swept out by ray OP at time t is given by $\theta = \omega t$. The coordinates of point P at time t are

$$x = a \cos \theta = a \cos \omega t \qquad \text{and} \qquad y = a \sin \theta = a \sin \omega t.$$

As P moves around the circle from the point $(a, 0)$, the point $Q(0, y)$ oscillates back and forth along the y-axis between the points $(0, a)$ and $(0, -a)$. Similarly, the point $R(x, 0)$ oscillates back and forth between $(a, 0)$ and $(-a, 0)$. This oscillatory motion is called **simple harmonic motion.**

The amplitude of the motion is $|a|$, and the period is $\frac{2\pi}{\omega}$. The moving points P and Q or P and R complete one oscillation or cycle per period. The number of cycles per unit of time, called the **frequency,** is the reciprocal of the period, $\frac{\omega}{2\pi}$, where $\omega > 0$.

SIMPLE HARMONIC MOTION

The position of a point oscillating about an equilibrium position at time t is modeled by either

$$s(t) = a \cos \omega t \qquad \text{or} \qquad s(t) = a \sin \omega t,$$

where a and ω are constants, with $\omega > 0$. The amplitude of the motion is $|a|$, the period is $\frac{2\pi}{\omega}$, and the frequency is $\frac{\omega}{2\pi}$ oscillations per time unit.

▶ **EXAMPLE 1**   MODELING THE MOTION OF A SPRING

Suppose that an object is attached to a coiled spring such as the one in Figure 46 on the preceding page. It is pulled down a distance of 5 in. from its equilibrium position, and then released. The time for one complete oscillation is 4 sec.

(a) Give an equation that models the position of the object at time t.

(b) Determine the position at $t = 1.5$ sec.

(c) Find the frequency.

Solution

(a) When the object is released at $t = 0$, the distance of the object from the equilibrium position is 5 in. below equilibrium. If $s(t)$ is to model the motion, then $s(0)$ must equal -5. We use

$$s(t) = a \cos \omega t,$$

with $a = -5$. We choose the cosine function because $\cos \omega(0) = \cos 0 = 1$, and $-5 \cdot 1 = -5$. (Had we chosen the sine function, a phase shift would have been required.) The period is 4, so

$$\frac{2\pi}{\omega} = 4, \qquad \text{or} \qquad \omega = \frac{\pi}{2}. \quad \text{Solve for } \omega. \text{ (Appendix A)}$$

Thus, the motion is modeled by

$$s(t) = -5 \cos \frac{\pi}{2} t.$$

(b) After 1.5 sec, the position is

$$s(1.5) = -5 \cos \left[\frac{\pi}{2} (1.5) \right] \approx 3.54 \text{ in.} \quad \text{(Appendix C)}$$

Since $3.54 > 0$, the object is above the equilibrium position.

(c) The frequency is the reciprocal of the period, or $\frac{1}{4}$ oscillation per sec.

NOW TRY EXERCISE 9. ◀

▶ **EXAMPLE 2**   ANALYZING HARMONIC MOTION

Suppose that an object oscillates according to the model

$$s(t) = 8 \sin 3t,$$

where t is in seconds and $s(t)$ is in feet. Analyze the motion.

Solution   The motion is harmonic because the model is of the form $s(t) = a \sin \omega t$. Because $a = 8$, the object oscillates 8 ft in either direction from its starting point. The period $\frac{2\pi}{3} \approx 2.1$ is the time, in seconds, it takes for one complete oscillation. The frequency is the reciprocal of the period, so the object completes $\frac{3}{2\pi} \approx .48$ oscillation per sec.

NOW TRY EXERCISE 15. ◀

Damped Oscillatory Motion   In the example of the stretched spring, we disregard the effect of friction. Friction causes the amplitude of the motion to diminish gradually until the weight comes to rest. In this situation, we say that the motion has been *damped* by the force of friction. Most oscillatory motions are damped, and the decrease in amplitude follows the pattern of exponential decay. A typical example of **damped oscillatory motion** is provided by the function defined by

$$s(t) = e^{-t} \sin t.$$

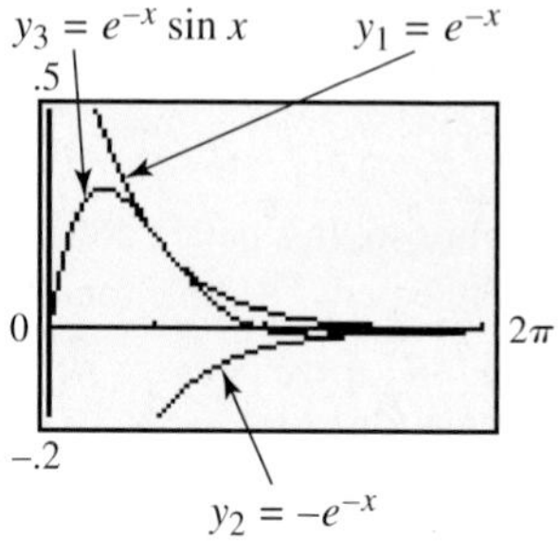

$y_3 = e^{-x} \sin x$ $y_1 = e^{-x}$

$y_2 = -e^{-x}$

Figure 48

(The number $e \approx 2.718$ is the base of the natural logarithm function, first studied in college algebra courses.) Figure 48 shows how the graph of $y_3 = e^{-x} \sin x$ is bounded above by the graph of $y_1 = e^{-x}$ and below by the graph of $y_2 = -e^{-x}$. The damped motion curve dips below the x-axis at $x = \pi$ but stays above the graph of y_2. Figure 49 shows a traditional graph of $s(t) = e^{-t} \sin t$, along with the graph of $y = \sin t$.

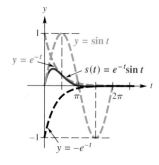

Figure 49

Shock absorbers are put on an automobile in order to damp oscillatory motion. Instead of oscillating up and down for a long while after hitting a bump or pothole, the oscillations of the car are quickly damped out for a smoother ride.

NOW TRY EXERCISE 21. ◀

4.5 Exercises

(Modeling) Springs   *Suppose that a weight on a spring has initial position $s(0)$ and period P.*

(a) *Find a function s given by $s(t) = a \cos \omega t$ that models the displacement of the weight.*

(b) *Evaluate $s(1)$. Is the weight moving upward, downward, or neither when $t = 1$? Support your results graphically or numerically.*

1. $s(0) = 2$ in.; $P = .5$ sec   **2.** $s(0) = 5$ in.; $P = 1.5$ sec

3. $s(0) = -3$ in.; $P = .8$ sec   **4.** $s(0) = -4$ in.; $P = 1.2$ sec

(Modeling) Music *A note on the piano has given frequency F. Suppose the maximum displacement at the center of the piano wire is given by s(0). Find constants a and ω so that the equation s(t) = a cos ωt models this displacement. Graph s in the viewing window* $[0, .05]$ *by* $[-.3, .3]$.

5. $F = 27.5; s(0) = .21$ **6.** $F = 110; s(0) = .11$

7. $F = 55; s(0) = .14$ **8.** $F = 220; s(0) = .06$

(Modeling) *Solve each problem. See Examples 1 and 2.*

9. *Spring* An object is attached to a coiled spring, as in Figure 46. It is pulled down a distance of 4 units from its equilibrium position, and then released. The time for one complete oscillation is 3 sec.

(a) Give an equation that models the position of the object at time t.
(b) Determine the position at $t = 1.25$ sec.
(c) Find the frequency.

10. *Spring* Repeat Exercise 9, but assume that the object is pulled down 6 units and the time for one complete oscillation is 4 sec.

11. *Particle Movement* Write the equation and then determine the amplitude, period, and frequency of the simple harmonic motion of a particle moving uniformly around a circle of radius 2 units, with angular speed

(a) 2 radians per sec   (b) 4 radians per sec.

12. *Pendulum* What are the period P and frequency T of oscillation of a pendulum of length $\frac{1}{2}$ ft? $\left(Hint: P = 2\pi\sqrt{\frac{L}{32}}, \text{where } L \text{ is the length of the pendulum in feet and } P \text{ is in seconds.}\right)$

13. *Pendulum* In Exercise 12, how long should the pendulum be to have period 1 sec?

14. *Spring* The formula for the up and down motion of a weight on a spring is given by

$$s(t) = a \sin\sqrt{\frac{k}{m}}\,t.$$

If the spring constant k is 4, what mass m must be used to produce a period of 1 sec?

15. *Spring* The height attained by a weight attached to a spring set in motion is

$$s(t) = -4 \cos 8\pi t$$

inches after t seconds.

(a) Find the maximum height that the weight rises above the equilibrium position of $s(t) = 0$.
(b) When does the weight first reach its maximum height, if $t \geq 0$?
(c) What are the frequency and period?

16. *Spring* (See Exercise 14.) A spring with spring constant $k = 2$ and a 1-unit mass m attached to it is stretched and then allowed to come to rest.

    (a) If the spring is stretched $\frac{1}{2}$ ft and released, what are the amplitude, period, and frequency of the resulting oscillatory motion?

    (b) What is the equation of the motion?

17. *Spring* The position of a weight attached to a spring is

$$s(t) = -5 \cos 4\pi t$$

inches after t seconds.

    (a) What is the maximum height that the weight rises above the equilibrium position?

    (b) What are the frequency and period?

    (c) When does the weight first reach its maximum height?

    (d) Calculate and interpret $s(1.3)$.

18. *Spring* The position of a weight attached to a spring is

$$s(t) = -4 \cos 10t$$

inches after t seconds.

    (a) What is the maximum height that the weight rises above the equilibrium position?

    (b) What are the frequency and period?

    (c) When does the weight first reach its maximum height?

    (d) Calculate and interpret $s(1.466)$.

19. *Spring* A weight attached to a spring is pulled down 3 in. below the equilibrium position.

    (a) Assuming that the frequency is $\frac{6}{\pi}$ cycles per sec, determine a model that gives the position of the weight at time t seconds.

    (b) What is the period?

20. *Spring* A weight attached to a spring is pulled down 2 in. below the equilibrium position.

    (a) Assuming that the period is $\frac{1}{3}$ sec, determine a model that gives the position of the weight at time t seconds.

    (b) What is the frequency?

Use a graphing calculator to graph $y_1 = e^{-t} \sin t$, $y_2 = e^{-t}$, and $y_3 = -e^{-t}$ in the viewing window $[0, \pi]$ by $[-.5, .5]$.

21. Find the t-intercepts of the graph of y_1. Explain the relationship of these intercepts with the x-intercepts of the graph of $y = \sin x$.

22. Find any points of intersection of y_1 and y_2 or y_1 and y_3. How are these points related to the graph of $y = \sin x$?

Chapter 4 Summary

KEY TERMS

4.1 periodic function period sine wave (sinusoid) amplitude	**4.2** phase shift argument **4.3** vertical asymptote **4.4** addition of ordinates	**4.5** simple harmonic motion frequency	damped oscillatory motion

QUICK REVIEW

CONCEPTS	EXAMPLES

4.1 Graphs of the Sine and Cosine Functions

4.2 Translations of the Graphs of the Sine and Cosine Functions

Sine and Cosine Functions

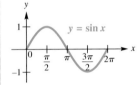

Domain: $(-\infty, \infty)$
Range: $[-1, 1]$
Amplitude: 1
Period: 2π

Domain: $(-\infty, \infty)$
Range: $[-1, 1]$
Amplitude: 1
Period: 2π

The graph of

$$y = c + a \sin b(x - d) \quad \text{or} \quad y = c + a \cos b(x - d),$$

$b > 0$, has

1. amplitude $|a|$,   2. period $\frac{2\pi}{b}$,
3. vertical translation c units up if $c > 0$ or $|c|$ units down if $c < 0$, and
4. phase shift d units to the right if $d > 0$ or $|d|$ units to the left if $d < 0$.

See pages 150 and 161 for a summary of graphing techniques.

Graph $y = \sin 3x$.

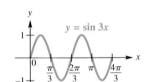

period: $\frac{2\pi}{3}$ amplitude: 1
domain: $(-\infty, \infty)$ range: $[-1, 1]$

Graph $y = -2 \cos x$.

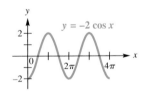

period: 2π amplitude: 2
domain: $(-\infty, \infty)$ range: $[-2, 2]$

CONCEPTS | EXAMPLES

4.3 Graphs of the Tangent and Cotangent Functions

Tangent and Cotangent Functions

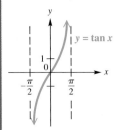 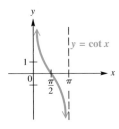

Domain: $\{x \mid x \neq (2n + 1)\frac{\pi}{2},$
where n is any integer $\}$
Range: $(-\infty, \infty)$
Period: π

Domain: $\{x \mid x \neq n\pi,$
where n is any integer$\}$
Range: $(-\infty, \infty)$
Period: π

See page 171 for a summary of graphing techniques.

Graph one period of $y = 2 \tan x$.

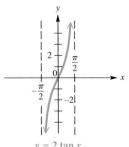

period: π
domain: $\{x \mid x \neq (2n + 1)\frac{\pi}{2},$
where n is any integer$\}$
range: $(-\infty, \infty)$

4.4 Graphs of the Secant and Cosecant Functions

Secant and Cosecant Functions

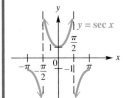 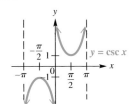

Domain: $\{x \mid x \neq (2n + 1)\frac{\pi}{2},$
where n is any integer$\}$
Range: $(-\infty, -1] \cup [1, \infty)$
Period: 2π

Domain: $\{x \mid x \neq n\pi,$
where n is any integer$\}$
Range: $(-\infty, -1] \cup [1, \infty)$
Period: 2π

See page 179 for a summary of graphing techniques.

Graph one period of $y = \sec\left(x + \frac{\pi}{4}\right)$.

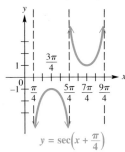

period: 2π
phase shift: $-\frac{\pi}{4}$
domain: $\{x \mid x \neq \frac{\pi}{4} + n\pi,$
where n is any integer$\}$
range: $(-\infty, -1] \cup [1, \infty)$

4.5 Harmonic Motion

Simple Harmonic Motion
The position of a point oscillating about an equilibrium position at time t is modeled by either

$$s(t) = a \cos \omega t \quad \text{or} \quad s(t) = a \sin \omega t,$$

where a and ω are constants, with $\omega > 0$. The amplitude of the motion is $|a|$, the period is $\frac{2\pi}{\omega}$, and the frequency is $\frac{\omega}{2\pi}$ oscillations per time unit.

A spring oscillates according to

$$s(t) = -5 \cos 6t,$$

where t is in seconds and $s(t)$ is in inches. Find the amplitude, period, and frequency.

$$\text{amplitude} = |-5| = 5 \text{ in.}; \quad \text{period} = \frac{2\pi}{6} = \frac{\pi}{3} \text{ sec};$$

$$\text{frequency} = \frac{3}{\pi} \text{ oscillation per sec}$$

CHAPTER 4	**Review Exercises**

1. *Concept Check* Which one of the following is true about the graph of $y = 4 \sin 2x$?
   - **A.** It has amplitude 2 and period $\frac{\pi}{2}$.   **B.** It has amplitude 4 and period π.
   - **C.** Its range is $[0, 4]$.   **D.** Its range is $[-4, 0]$.

2. *Concept Check* Which one of the following is false about the graph of $y = -3 \cos \frac{1}{2}x$?
   - **A.** Its range is $[-3, 3]$.
   - **B.** Its domain is $(-\infty, \infty)$.
   - **C.** Its amplitude is 3, and its period is 4π.
   - **D.** Its amplitude is 3, and its period is π.

3. *Concept Check* Which of the basic circular functions can have y-value $\frac{1}{2}$?

4. *Concept Check* Which of the basic circular functions can have y-value 2?

For each function, give the amplitude, period, vertical translation, and phase shift, as applicable.

5. $y = 2 \sin x$ 6. $y = \tan 3x$ 7. $y = -\dfrac{1}{2} \cos 3x$

8. $y = 2 \sin 5x$ 9. $y = 1 + 2 \sin \dfrac{1}{4}x$ 10. $y = 3 - \dfrac{1}{4} \cos \dfrac{2}{3}x$

11. $y = 3 \cos\left(x + \dfrac{\pi}{2}\right)$ 12. $y = -\sin\left(x - \dfrac{3\pi}{4}\right)$ 13. $y = \dfrac{1}{2} \csc\left(2x - \dfrac{\pi}{4}\right)$

14. $y = 2 \sec(\pi x - 2\pi)$ 15. $y = \dfrac{1}{3} \tan\left(3x - \dfrac{\pi}{3}\right)$ 16. $y = \cot\left(\dfrac{x}{2} + \dfrac{3\pi}{4}\right)$

Concept Check *Identify the circular function that satisfies each description.*

17. period is π, x-intercepts are of the form $n\pi$, where n is any integer

18. period is 2π, graph passes through the origin

19. period is 2π, graph passes through the point $\left(\frac{\pi}{2}, 0\right)$

20. period is 2π, domain is $\{x \mid x \neq n\pi$, where n is any integer$\}$

21. period is π, function is decreasing on the interval $(0, \pi)$

22. period is 2π, has vertical asymptotes of the form $x = (2n + 1)\frac{\pi}{2}$, where n is any integer

23. Suppose that f is a sine function with period 10 and $f(5) = 2$. Explain why $f(25) = 2$.

24. Suppose that f is a sine function with period π and $f\left(\frac{6\pi}{5}\right) = 1$. Explain why $f\left(-\frac{4\pi}{5}\right) = 1$.

Graph each function over a one-period interval.

25. $y = 3 \sin x$ 26. $y = \dfrac{1}{2} \sec x$ 27. $y = -\tan x$

28. $y = -2 \cos x$ 29. $y = 2 + \cot x$ 30. $y = -1 + \csc x$

31. $y = \sin 2x$ 32. $y = \tan 3x$ 33. $y = 3 \cos 2x$

34. $y = \dfrac{1}{2} \cot 3x$ **35.** $y = \cos\left(x - \dfrac{\pi}{4}\right)$ **36.** $y = \tan\left(x - \dfrac{\pi}{2}\right)$

37. $y = \sec\left(2x + \dfrac{\pi}{3}\right)$ **38.** $y = \sin\left(3x + \dfrac{\pi}{2}\right)$

39. $y = 1 + 2\cos 3x$ **40.** $y = -1 - 3\sin 2x$

41. $y = 2\sin \pi x$ **42.** $y = -\dfrac{1}{2}\cos(\pi x - \pi)$

43. Explain why a function of the form $f(x) = 2\sin(bx + c)$ has range $[-2, 2]$.

44. Explain why a function of the form $f(x) = 2\csc(bx + c)$ has range $(-\infty, -2] \cup [2, \infty)$.

Solve each problem.

45. *Viewing Angle to an Object* Let a person whose eyes are h_1 feet from the ground stand d feet from an object h_2 feet tall, where $h_2 > h_1$. Let θ be the angle of elevation to the top of the object. See the figure.

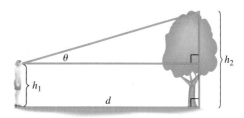

   (a) Show that $d = (h_2 - h_1)\cot \theta$.
   (b) Let $h_2 = 55$ and $h_1 = 5$. Graph d for the interval $0 < \theta \le \frac{\pi}{2}$.

46. *(Modeling) Tides* The figure shows a function f that models the tides in feet at Clearwater Beach, Florida, x hours after midnight starting on August 26, 2006. (*Source:* Pentcheff, D., *WWW Tide and Current Predictor.*)

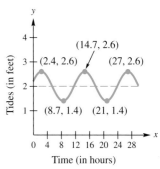

   (a) Find the time between high tides.
   (b) What is the difference in water levels between high tide and low tide?
   (c) The tides can be modeled by

$$f(x) = .6\cos[.511(x - 2.4)] + 2.$$

   Estimate the tides when $x = 10$.

47. *(Modeling) Maximum Temperatures* The maximum afternoon temperature in a given city might be modeled by

$$t = 60 - 30\cos\frac{x\pi}{6},$$

   where t represents the maximum afternoon temperature in month x, with $x = 0$ representing January, $x = 1$ representing February, and so on. Find the maximum afternoon temperature to the nearest degree for each month.

   (a) January   **(b)** April   **(c)** May
   (d) June   **(e)** August   **(f)** October

3. Give a short answer to each of the following.

(a) What is the domain of the cosine function?

(b) What is the range of the sine function?

(c) What is the least positive value for which the tangent function is undefined?

(d) What is the range of the secant function?

4. Consider the function defined by $y = 3 - 6 \sin\left(2x + \frac{\pi}{2}\right)$.

(a) What is its period? (b) What is the amplitude of its graph?

(c) What is its range? (d) What is the y-intercept of its graph?

(e) What is its phase shift?

Graph each function over a two-period interval. Identify asymptotes when applicable.

5. $y = \sin(2x + \pi)$ **6.** $y = -\cos 2x$

7. $y = 2 + \cos x$ **8.** $y = -1 + 2 \sin(x + \pi)$

9. $y = \tan\left(x - \frac{\pi}{2}\right)$ **10.** $y = -2 - \cot\left(x - \frac{\pi}{2}\right)$

11. $y = -\csc 2x$ **12.** $y = 3 \csc \pi x$

13. *(Modeling) Average Monthly Temperature* The average monthly temperature (in °F) in Austin, Texas, can be modeled using the circular function defined by

$$f(x) = 17.5 \sin\left[\frac{\pi}{6}(x - 4)\right] + 67.5,$$

where x is the month and $x = 1$ corresponds to January. (*Source:* Miller, A., J. Thompson, and R. Peterson, *Elements of Meteorology, Fourth Edition,* Charles E. Merrill Publishing Co., 1983.)

(a) Graph f in the window $[1, 25]$ by $[45, 90]$.

(b) Determine the amplitude, period, phase shift, and vertical translation of f.

(c) What is the average monthly temperature for the month of December?

(d) Determine the maximum and minimum average monthly temperatures and the months when they occur.

(e) What would be an approximation for the average *yearly* temperature in Austin? How is this related to the vertical translation of the sine function in the formula for f?

14. *(Modeling) Spring* The height of a weight attached to a spring is

$$s(t) = -4 \cos 8\pi t$$

inches after t seconds.

(a) Find the maximum height that the weight rises above the equilibrium position of $s(t) = 0$.

(b) When does the weight first reach its maximum height, if $t \geq 0$?

(c) What are the frequency and period?

15. Explain why the domains of the tangent and secant functions are the same, and then give a similar explanation for the cotangent and cosecant functions.

CHAPTER 4 ▶ Quantitative Reasoning

Does the fact that average monthly temperatures are periodic affect your utility bills?

In an article entitled "I Found Sinusoids in My Gas Bill" (*Mathematics Teacher,* January 2000), Cathy G. Schloemer presents the following graph that accompanied her gas bill.

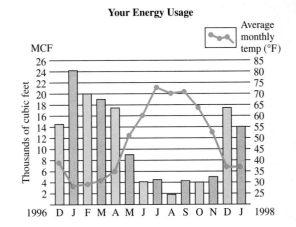

Notice that two sinusoids are suggested here: one for the behavior of the average monthly temperature and another for gas use in MCF (thousands of cubic feet).

1. If January 1997 is represented by $x = 1$, the data of estimated ordered pairs (month, temperature) are given in the list shown on the two graphing calculator screens below.

L1	L2	L3	1
1	28	------	
2	29		
3	31		
4	35		
5	51		
6	60		
7	73		

L1(1)=1

L1	L2	L3	1
7	73		
8	70		
9	71		
10	64		
11	53		
12	37	------	

L1(13) =

Use the sine regression feature of a graphing calculator to find a sine function that fits these data points. Then make a scatter diagram, and graph the function.

2. If January 1997 is again represented by $x = 1$, the data of estimated ordered pairs (month, gas use in MCF) are given in the list shown on the two graphing calculator screens below.

L1	L2	L3	1
1	24.2	------	
2	20		
3	18.8		
4	17.5		
5	9.2		
6	4.2		
7	4.8		

L1(1)=1

L1	L2	L3	1
7	4.8		
8	1.8		
9	4.8		
10	4		
11	5		
12	17.5	------	

L1(13) =

Use the sine regression feature of a graphing calculator to find a sine function that fits these data points. Then make a scatter diagram, and graph the function.

3. Answer the question posed at the top of the page, in the form of a short paragraph.

5 Trigonometric Identities

In 1831 Michael Faraday discovered that when a wire passes by a magnet, a small electric current is produced in the wire. Now we generate massive amounts of electricity by simultaneously rotating thousands of wires near large electromagnets. Because electric current alternates its direction on electrical wires, it is modeled accurately by either the sine or the cosine function.

We give many examples of applications of the trigonometric functions to electricity and other phenomena in the examples and exercises in this chapter, including a model of the wattage consumption of a toaster in Section 5.5, Example 6.

5.1 Fundamental Identities

Fundamental Identities ▪ **Using the Fundamental Identities**

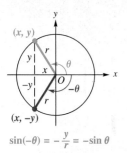

Figure 1

$$\sin(-\theta) = -\frac{y}{r} = -\sin\theta$$

Recall that an **identity** is an equation that is satisfied by *every* value in the domain of its variable. (Appendix A)

Fundamental Identities As suggested by the circle shown in Figure 1, an angle θ having the point (x, y) on its terminal side has a corresponding angle $-\theta$ with the point $(x, -y)$ on its terminal side. From the definition of sine,

$$\sin(-\theta) = \frac{-y}{r} \qquad \text{and} \qquad \sin\theta = \frac{y}{r}, \qquad \text{(Section 1.3)}$$

so $\sin(-\theta)$ and $\sin\theta$ are negatives of each other, or

$$\sin(-\theta) = -\sin\theta.$$

Figure 1 shows an angle θ in quadrant II, but the same result holds for θ in any quadrant. Also, by definition,

$$\cos(-\theta) = \frac{x}{r} \qquad \text{and} \qquad \cos\theta = \frac{x}{r}, \qquad \text{(Section 1.3)}$$

so

$$\cos(-\theta) = \cos\theta.$$

We use the identities for $\sin(-\theta)$ and $\cos(-\theta)$ to find $\tan(-\theta)$ in terms of $\tan\theta$:

$$\tan(-\theta) = \frac{\sin(-\theta)}{\cos(-\theta)} = \frac{-\sin\theta}{\cos\theta} = -\frac{\sin\theta}{\cos\theta}, \qquad \text{or} \qquad \tan(-\theta) = -\tan\theta.$$

Similar reasoning gives the remaining three **negative-angle** or **negative-number identities,** which, together with the reciprocal, quotient, and Pythagorean identities from **Chapter 1,** are called the **fundamental identities.**

FUNDAMENTAL IDENTITIES

Reciprocal Identities

$$\cot\theta = \frac{1}{\tan\theta} \qquad \sec\theta = \frac{1}{\cos\theta} \qquad \csc\theta = \frac{1}{\sin\theta}$$

Quotient Identities

$$\tan\theta = \frac{\sin\theta}{\cos\theta} \qquad \cot\theta = \frac{\cos\theta}{\sin\theta}$$

Pythagorean Identities

$$\sin^2\theta + \cos^2\theta = 1 \qquad \tan^2\theta + 1 = \sec^2\theta \qquad 1 + \cot^2\theta = \csc^2\theta$$

Negative-Angle Identities

$$\sin(-\theta) = -\sin\theta \qquad \cos(-\theta) = \cos\theta \qquad \tan(-\theta) = -\tan\theta$$
$$\csc(-\theta) = -\csc\theta \qquad \sec(-\theta) = \sec\theta \qquad \cot(-\theta) = -\cot\theta$$

> ▶ **Note** The most commonly recognized forms of the fundamental identities are given in the preceding box. Throughout this chapter you must also recognize alternative forms of these identities. *For example, two other forms of* $\sin^2\theta + \cos^2\theta = 1$ *are*
>
> $$\sin^2\theta = 1 - \cos^2\theta \quad and \quad \cos^2\theta = 1 - \sin^2\theta.$$

Using the Fundamental Identities   One way we use these identities is to find the values of other trigonometric functions from the value of a given trigonometric function. Although we could find such values using a right triangle, this is a good way to practice using the fundamental identities.

▶ **EXAMPLE 1**   FINDING TRIGONOMETRIC FUNCTION VALUES GIVEN ONE VALUE AND THE QUADRANT

If $\tan\theta = -\frac{5}{3}$ and θ is in quadrant II, find each function value.

(a) $\sec\theta$ **(b)** $\sin\theta$ **(c)** $\cot(-\theta)$

Solution

(a) Look for an identity that relates tangent and secant.

$$\tan^2\theta + 1 = \sec^2\theta \qquad \text{Pythagorean identity}$$

$$\left(-\frac{5}{3}\right)^2 + 1 = \sec^2\theta \qquad \tan\theta = -\frac{5}{3}$$

$$\frac{25}{9} + 1 = \sec^2\theta \qquad \left(-\frac{5}{3}\right)^2 = -\frac{5}{3}\left(-\frac{5}{3}\right) = \frac{25}{9}$$

$$\frac{34}{9} = \sec^2\theta \qquad \text{Combine terms.}$$

Choose the correct sign. $\longrightarrow$

$$-\sqrt{\frac{34}{9}} = \sec\theta \qquad \text{Take the negative square root. (Appendix A)}$$

$$-\frac{\sqrt{34}}{3} = \sec\theta \qquad \text{Simplify the radical; } -\sqrt{\frac{34}{9}} = -\frac{\sqrt{34}}{\sqrt{9}} = -\frac{\sqrt{34}}{3}.$$

We chose the negative square root since $\sec\theta$ is negative in quadrant II.

(b)

$$\tan\theta = \frac{\sin\theta}{\cos\theta} \qquad \text{Quotient identity}$$

$$\cos\theta\,\tan\theta = \sin\theta \qquad \text{Multiply each side by } \cos\theta.$$

$$\left(\frac{1}{\sec\theta}\right)\tan\theta = \sin\theta \qquad \text{Reciprocal identity}$$

$$\left(-\frac{3\sqrt{34}}{34}\right)\left(-\frac{5}{3}\right) = \sin\theta \qquad \begin{array}{l}\frac{1}{\sec\theta} = \frac{1}{-\frac{\sqrt{34}}{3}} = -\frac{3}{\sqrt{34}} = -\frac{3}{\sqrt{34}}\cdot\frac{\sqrt{34}}{\sqrt{34}} = -\frac{3\sqrt{34}}{34}; \\ \tan\theta = -\frac{5}{3}\end{array}$$

$$\sin\theta = \frac{5\sqrt{34}}{34} \qquad \text{Multiply; rewrite.}$$

(c) $\qquad \cot(-\theta) = \dfrac{1}{\tan(-\theta)}$ $\qquad\qquad$ Reciprocal identity

$\qquad\qquad \cot(-\theta) = \dfrac{1}{-\tan\,\theta}$ $\qquad\qquad$ Negative-angle identity

$\qquad\qquad \cot(-\theta) = \dfrac{1}{-\left(-\frac{5}{3}\right)} = \dfrac{3}{5}$ $\qquad$ $\tan\theta = -\frac{5}{3}$; simplify the complex fraction.

> NOW TRY EXERCISES 7, 11, AND 25. ◄

> ▶ **Caution** *To avoid a common error, when taking the square root, be sure to choose the sign based on the quadrant of θ and the function being evaluated.*

Any trigonometric function of a number or angle can be expressed in terms of any other function.

▶ **EXAMPLE 2** **EXPRESSING ONE FUNCTION IN TERMS OF ANOTHER**

Express $\cos x$ in terms of $\tan x$.

Solution Since $\sec x$ is related to both $\cos x$ and $\tan x$ by identities, start with $1 + \tan^2 x = \sec^2 x$.

$$\frac{1}{1 + \tan^2 x} = \frac{1}{\sec^2 x} \qquad\qquad \text{Take reciprocals.}$$

$$\frac{1}{1 + \tan^2 x} = \cos^2 x \qquad\qquad \text{Reciprocal identity}$$

> Remember both the positive and negative roots.

$$\pm\sqrt{\frac{1}{1 + \tan^2 x}} = \cos x \qquad\qquad \text{Take the square root of each side.}$$

$$\cos x = \frac{\pm 1}{\sqrt{1 + \tan^2 x}} \qquad \text{Quotient rule for radicals: } \sqrt[n]{\tfrac{a}{b}} = \tfrac{\sqrt[n]{a}}{\sqrt[n]{b}};\text{ rewrite.}$$

$$\cos x = \frac{\pm\sqrt{1 + \tan^2 x}}{1 + \tan^2 x} \qquad \text{Rationalize the denominator.}$$

Choose the $+$ sign or the $-$ sign, depending on the quadrant of x.

> NOW TRY EXERCISE 47. ◄

We can use a graphing calculator to decide whether two functions are identical. See Figure 2, which supports the identity $\sin^2 x + \cos^2 x = 1$. With an identity, you should see no difference in the two graphs.

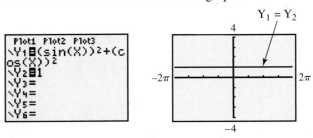

Figure 2

Each of the functions $\tan\theta$, $\cot\theta$, $\sec\theta$, and $\csc\theta$ can easily be expressed in terms of $\sin\theta$ and/or $\cos\theta$. We often make such substitutions in an expression to simplify it.

▶ **EXAMPLE 3**   REWRITING AN EXPRESSION IN TERMS OF SINE AND COSINE

Write $\tan\theta + \cot\theta$ in terms of $\sin\theta$ and $\cos\theta$, and then simplify the expression.

Solution

$$\tan\theta + \cot\theta = \frac{\sin\theta}{\cos\theta} + \frac{\cos\theta}{\sin\theta} \qquad \text{Quotient identities}$$

$$= \frac{\sin\theta}{\cos\theta}\cdot\frac{\sin\theta}{\sin\theta} + \frac{\cos\theta}{\sin\theta}\cdot\frac{\cos\theta}{\cos\theta} \qquad \text{Write each fraction with the least common denominator (LCD).}$$

$$= \frac{\sin^2\theta}{\cos\theta\sin\theta} + \frac{\cos^2\theta}{\cos\theta\sin\theta} \qquad \text{Multiply.}$$

$$= \frac{\sin^2\theta + \cos^2\theta}{\cos\theta\sin\theta} \qquad \text{Add fractions; } \frac{a}{c}+\frac{b}{c}=\frac{a+b}{c}.$$

$$\tan\theta + \cot\theta = \frac{1}{\cos\theta\sin\theta} \qquad \text{Pythagorean identity}$$

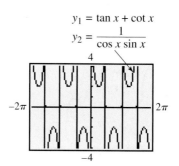

$y_1 = \tan x + \cot x$
$y_2 = \dfrac{1}{\cos x \sin x}$

The graph supports the result in Example 3. The graphs of y_1 and y_2 appear to be identical.

NOW TRY EXERCISE 59. ◀

▶ **Caution** *When working with trigonometric expressions and identities, be sure to write the argument of the function.* For example, we would *not* write $\sin^2 + \cos^2 = 1$; an argument such as θ is necessary in this identity.

5.1 Exercises

Concept Check   In Exercises 1–6, use identities to fill in the blanks.

1. If $\tan\theta = 2.6$, then $\tan(-\theta) = $ _____.
2. If $\cos\theta = -.65$, then $\cos(-\theta) = $ _____.
3. If $\tan\theta = 1.6$, then $\cot\theta = $ _____.
4. If $\cos\theta = .8$ and $\sin\theta = .6$, then $\tan(-\theta) = $ _____.
5. If $\sin\theta = \frac{2}{3}$, then $-\sin(-\theta) = $ _____.
6. If $\cos\theta = -\frac{1}{5}$, then $-\cos(-\theta) = $ _____.

Find $\sin\theta$. See Example 1.

7. $\cos\theta = \dfrac{3}{4}$, θ in quadrant I
8. $\cot\theta = -\dfrac{1}{3}$, θ in quadrant IV
9. $\cos(-\theta) = \dfrac{\sqrt5}{5}$, $\tan\theta < 0$
10. $\tan\theta = -\dfrac{\sqrt7}{2}$, $\sec\theta > 0$
11. $\sec\theta = \dfrac{11}{4}$, $\tan\theta < 0$
12. $\csc\theta = -\dfrac{8}{5}$

13. Why is it unnecessary to give the quadrant of θ in Exercise 12?

14. *Concept Check*   What is **WRONG** with the statement of this problem?

$$\text{Find } \cos(-\theta) \text{ if } \cos \theta = 3.$$

RELATING CONCEPTS

For individual or collaborative investigation
(Exercises 15–20)

*A function is called an **even function** if $f(-x) = f(x)$ for all x in the domain of f. A function is called an **odd function** if $f(-x) = -f(x)$ for all x in the domain of f.* **Work Exercises 15–20 in order,** *to see the connection between the negative-angle identities and even and odd functions.*

15. Complete the statement: $\sin(-x) = \underline{\hspace{1cm}}$.

16. Is the function defined by $f(x) = \sin x$ *even* or *odd*?

17. Complete the statement: $\cos(-x) = \underline{\hspace{1cm}}$.

18. Is the function defined by $f(x) = \cos x$ *even* or *odd*?

19. Complete the statement: $\tan(-x) = \underline{\hspace{1cm}}$.

20. Is the function defined by $f(x) = \tan x$ *even* or *odd*?

Concept Check   *For each graph of a circular function $y = f(x)$ in dot mode, determine whether $f(-x) = f(x)$ or $f(-x) = -f(x)$ is true.*

21.
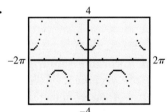

22.
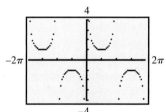

23.
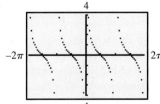

24.
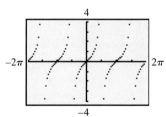

Find the remaining five trigonometric functions of θ. See Example 1.

25. $\sin \theta = \dfrac{2}{3}$, θ in quadrant II

26. $\cos \theta = \dfrac{1}{5}$, θ in quadrant I

27. $\tan \theta = -\dfrac{1}{4}$, θ in quadrant IV

28. $\csc \theta = -\dfrac{5}{2}$, θ in quadrant III

29. $\cot \theta = \dfrac{4}{3}$, $\sin \theta > 0$

30. $\sin \theta = -\dfrac{4}{5}$, $\cos \theta < 0$

31. $\sec \theta = \dfrac{4}{3}$, $\sin \theta < 0$

32. $\cos \theta = -\dfrac{1}{4}$, $\sin \theta > 0$

Concept Check For each expression in Column I, choose the expression from Column II that completes an identity.

I

II

33. $\dfrac{\cos x}{\sin x} =$ _____

A. $\sin^2 x + \cos^2 x$

34. $\tan x =$ _____

B. $\cot x$

35. $\cos(-x) =$ _____

C. $\sec^2 x$

36. $\tan^2 x + 1 =$ _____

D. $\dfrac{\sin x}{\cos x}$

37. $1 =$ _____

E. $\cos x$

Concept Check For each expression in Column I, choose the expression from Column II that completes an identity. You may have to rewrite one or both expressions.

I

II

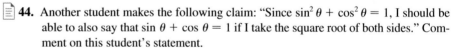

$\cos^2 x + \sin^2 x = 1$

38. $-\tan x \cos x =$ _____

A. $\dfrac{\sin^2 x}{\cos^2 x}$

39. $\sec^2 x - 1 =$ _____

B. $\dfrac{1}{\sec^2 x}$

40. $\dfrac{\sec x}{\csc x} =$ _____

C. $\sin(-x)$

41. $1 + \sin^2 x = \pm\cos^2 x$

D. $\csc^2 x - \cot^2 x + \sin^2 x$

42. $\cos^2 x =$ _____

E. $\tan x$

43. A student writes "$1 + \cot^2 = \csc^2$." Comment on this student's work.

44. Another student makes the following claim: "Since $\sin^2 \theta + \cos^2 \theta = 1$, I should be able to also say that $\sin \theta + \cos \theta = 1$ if I take the square root of both sides." Comment on this student's statement.

45. *Concept Check* Suppose that $\cos \theta = \frac{x}{x+1}$. Find an expression in x for $\sin \theta$.

46. *Concept Check* Suppose that $\sec \theta = \frac{p+4}{p}$. Find an expression in p for $\tan \theta$.

Write the first trigonometric function in terms of the second trigonometric function. See Example 2.

47. $\sin x;$ $\cos x$

48. $\cot x;$ $\sin x$

49. $\tan x;$ $\sec x$

50. $\cot x;$ $\csc x$

51. $\csc x;$ $\cos x$

52. $\sec x;$ $\sin x$

Write each expression in terms of sine and cosine, and simplify so that no quotients appear in the final expression. See Example 3.

53. $\cot \theta \sin \theta$

54. $\sec \theta \cot \theta \sin \theta$

55. $\cos \theta \csc \theta$

56. $\cot^2 \theta (1 + \tan^2 \theta)$

57. $\sin^2 \theta (\csc^2 \theta - 1)$

58. $(\sec \theta - 1)(\sec \theta + 1)$

59. $(1 - \cos \theta)(1 + \sec \theta)$

60. $\dfrac{\cos \theta + \sin \theta}{\sin \theta}$

61. $\dfrac{\cos^2 \theta - \sin^2 \theta}{\sin \theta \cos \theta}$

62. $\dfrac{1 - \sin^2 \theta}{1 + \cot^2 \theta}$

63. $\sec \theta - \cos \theta$

64. $(\sec \theta + \csc \theta)(\cos \theta - \sin \theta)$

65. $\sin \theta (\csc \theta - \sin \theta)$

66. $\dfrac{1 + \tan^2 \theta}{1 + \cot^2 \theta}$

67. $\sin^2 \theta + \tan^2 \theta + \cos^2 \theta$

68. $\dfrac{\tan(-\theta)}{\sec \theta}$

69. Let $\cos x = \frac{1}{5}$. Find all possible values of $\dfrac{\sec x - \tan x}{\sin x}$.

70. Let $\csc x = -3$. Find all possible values of $\dfrac{\sin x + \cos x}{\sec x}$.

RELATING CONCEPTS

For individual or collaborative investigation
(Exercises 71–76)

In **Chapter 4** *we graphed functions defined by*

$$y = c + a \cdot f[b(x - d)]$$

*with the assumption that $b > 0$. To see what happens when $b < 0$, **work Exercises 71–76 in order.***

71. Use a negative-angle identity to write $y = \sin(-2x)$ as a function of $2x$.

72. How does your answer to Exercise 71 relate to $y = \sin(2x)$?

73. Use a negative-angle identity to write $y = \cos(-4x)$ as a function of $4x$.

74. How does your answer to Exercise 73 relate to $y = \cos(4x)$?

75. Use your results from Exercises 71–74 to rewrite the following with a positive value of b.

(a) $y = \sin(-4x)$ **(b)** $y = \cos(-2x)$ **(c)** $y = -5 \sin(-3x)$

 76. Write a short response to this statement, often used by one of the authors of this text in trigonometry classes: *Students who tend to ignore negative signs should enjoy graphing functions involving the cosine and the secant.*

Use a graphing calculator to make a conjecture as to whether each equation is an identity. (Hint: In Exercises 81 and 82, graph as a function of x for a few different values of y (in radians).)

77. $\cos 2x = 1 - 2\sin^2 x$

78. $2 \sin s = \sin 2s$

79. $\sin x = \sqrt{1 - \cos^2 x}$

80. $\cos 2x = \cos^2 x - \sin^2 x$

81. $\cos(x - y) = \cos x - \cos y$

82. $\sin(x + y) = \sin x + \sin y$

5.2 Verifying Trigonometric Identities

Verifying Identities by Working with One Side ▪ Verifying Identities by Working with Both Sides

Recall that an identity is an equation that is satisfied for all meaningful replacements of the variable. One of the skills required for more advanced work in mathematics, especially in calculus, is the ability to use identities to write expressions in alternative forms. We develop this skill by using the fundamental identities to verify that a trigonometric equation is an identity (for those values of the variable for which it is defined). Here are some hints to help you get started.

▼ LOOKING AHEAD TO CALCULUS
Trigonometric identities are used in
calculus to simplify trigonometric expressions, determine derivatives of
trigonometric functions, and change
the form of some integrals.

HINTS FOR VERIFYING IDENTITIES

1. ***Learn the fundamental identities given in Section 5.1.*** Whenever you see either side of a fundamental identity, the other side should come to mind. ***Also, be aware of equivalent forms of the fundamental identities.*** For example, $\sin^2 \theta = 1 - \cos^2 \theta$ is an alternative form of the identity $\sin^2 \theta + \cos^2 \theta = 1$.

2. ***Try to rewrite the more complicated side*** of the equation so that it is identical to the simpler side.

3. ***It is sometimes helpful to express all trigonometric functions in the equation in terms of sine and cosine*** and then simplify the result.

4. ***Usually, any factoring or indicated algebraic operations should be performed.*** For example, the expression

$$\sin^2 x + 2 \sin x + 1 \quad \text{can be factored as} \quad (\sin x + 1)^2.$$

The sum or difference of two trigonometric expressions, such as $\frac{1}{\sin \theta} + \frac{1}{\cos \theta}$, can be added or subtracted in the same way as any other rational expression.

$$\frac{1}{\sin \theta} + \frac{1}{\cos \theta} = \frac{\cos \theta}{\sin \theta \cos \theta} + \frac{\sin \theta}{\sin \theta \cos \theta} \qquad \text{Write with the LCD.}$$

$$= \frac{\cos \theta + \sin \theta}{\sin \theta \cos \theta} \qquad \frac{a}{c} + \frac{b}{c} = \frac{a+b}{c}$$

5. ***As you select substitutions, keep in mind the side you are not changing, because it represents your goal.*** For example, to verify the identity

$$\tan^2 x + 1 = \frac{1}{\cos^2 x},$$

try to think of an identity that relates $\tan x$ to $\cos x$. In this case, since $\sec x = \frac{1}{\cos x}$ and $\sec^2 x = \tan^2 x + 1$, the secant function is the best link between the two sides.

6. If an expression contains $1 + \sin x$, ***multiplying both numerator and denominator*** by $1 - \sin x$ would give $1 - \sin^2 x$, which could be replaced with $\cos^2 x$. Similar results for $1 - \sin x$, $1 + \cos x$, and $1 - \cos x$ may be useful.

▶ **Caution** *Verifying identities is not the same as solving equations.* Techniques used in solving equations, such as adding the same terms to both sides, or multiplying both sides by the same term, should not be used when working with identities since you are starting with a statement (to be verified) that may not be true.

Verifying Identities by Working with One Side   To avoid the temptation to use algebraic properties of equations to verify identities, ***one strategy is to work with only one side and rewrite it to match the other side,*** as shown in Examples 1–4.

▶ **EXAMPLE 1**   **VERIFYING AN IDENTITY (WORKING WITH ONE SIDE)**

Verify that the following equation is an identity.

$$\cot \theta + 1 = \csc \theta (\cos \theta + \sin \theta)$$

Solution   We use the fundamental identities from **Section 5.1** to rewrite one side of the equation so that it is identical to the other side. Since the right side is more complicated, we work with it, using the third hint to change all functions to sine or cosine.

Steps   **Reasons**

Right side of given equation

$$\overbrace{\csc \theta (\cos \theta + \sin \theta)}^{} = \frac{1}{\sin \theta}(\cos \theta + \sin \theta) \qquad \csc \theta = \frac{1}{\sin \theta}$$

$$= \frac{\cos \theta}{\sin \theta} + \frac{\sin \theta}{\sin \theta} \qquad \text{Distributive property;}$$
$$a(b + c) = ab + ac$$

$$= \underbrace{\cot \theta + 1}_{} \qquad \frac{\cos \theta}{\sin \theta} = \cot \theta; \frac{\sin \theta}{\sin \theta} = 1$$

Left side of given equation

The given equation is an identity. The right side is identical to the left side.

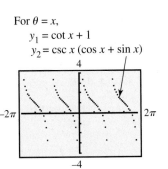

For $\theta = x$,
$$y_1 = \cot x + 1$$
$$y_2 = \csc x \,(\cos x + \sin x)$$

The graphs coincide, supporting the conclusion in Example 1.

NOW TRY EXERCISE 35. ◀

▶ **EXAMPLE 2**   **VERIFYING AN IDENTITY (WORKING WITH ONE SIDE)**

Verify that the following equation is an identity.

$$\tan^2 x(1 + \cot^2 x) = \frac{1}{1 - \sin^2 x}$$

Solution   We work with the more complicated left side, as suggested in the second hint. Again, we use the fundamental identities from **Section 5.1.**

$$\tan^2 x(1 + \cot^2 x) = \tan^2 x + \tan^2 x \cot^2 x \qquad \text{Distributive property}$$

$$= \tan^2 x + \tan^2 x \cdot \frac{1}{\tan^2 x} \qquad \cot^2 x = \frac{1}{\tan^2 x}$$

$$= \tan^2 x + 1 \qquad \tan^2 x \cdot \frac{1}{\tan^2 x} = 1$$

$$= \sec^2 x \qquad \tan^2 x + 1 = \sec^2 x$$

$$= \frac{1}{\cos^2 x} \qquad \sec^2 x = \frac{1}{\cos^2 x}$$

$$= \frac{1}{1 - \sin^2 x} \qquad \cos^2 x = 1 - \sin^2 x$$

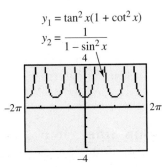

$$y_1 = \tan^2 x(1 + \cot^2 x)$$
$$y_2 = \frac{1}{1 - \sin^2 x}$$

The screen supports the conclusion in Example 2.

Since the left side is identical to the right side, the given equation is an identity.

NOW TRY EXERCISE 39. ◀

▶ **EXAMPLE 3** VERIFYING AN IDENTITY (WORKING WITH ONE SIDE)

Verify that the following equation is an identity.

$$\frac{\tan t - \cot t}{\sin t \cos t} = \sec^2 t - \csc^2 t$$

Solution We transform the more complicated left side to match the right side.

$$\frac{\tan t - \cot t}{\sin t \cos t} = \frac{\tan t}{\sin t \cos t} - \frac{\cot t}{\sin t \cos t} \qquad \frac{a - b}{c} = \frac{a}{c} - \frac{b}{c}$$

$$= \tan t \cdot \frac{1}{\sin t \cos t} - \cot t \cdot \frac{1}{\sin t \cos t} \qquad \frac{a}{b} = a \cdot \frac{1}{b}$$

$$= \frac{\sin t}{\cos t} \cdot \frac{1}{\sin t \cos t} - \frac{\cos t}{\sin t} \cdot \frac{1}{\sin t \cos t} \qquad \tan t = \frac{\sin t}{\cos t}; \cot t = \frac{\cos t}{\sin t}$$

$$= \frac{1}{\cos^2 t} - \frac{1}{\sin^2 t} \qquad \text{Multiply.}$$

$$= \sec^2 t - \csc^2 t \qquad \frac{1}{\cos^2 t} = \sec^2 t; \frac{1}{\sin^2 t} = \csc^2 t$$

The third hint about writing all trigonometric functions in terms of sine and cosine was used in the third line of the solution.

NOW TRY EXERCISE 43. ◀

▶ **EXAMPLE 4** VERIFYING AN IDENTITY (WORKING WITH ONE SIDE)

Verify that the following equation is an identity.

$$\frac{\cos x}{1 - \sin x} = \frac{1 + \sin x}{\cos x}$$

Solution We work on the right side, using the last hint in the list given earlier to multiply numerator and denominator on the right by $1 - \sin x$.

$$\frac{1 + \sin x}{\cos x} = \frac{(1 + \sin x)(1 - \sin x)}{\cos x(1 - \sin x)} \qquad \text{Multiply by 1 in the form } \frac{1 - \sin x}{1 - \sin x}.$$

$$= \frac{1 - \sin^2 x}{\cos x(1 - \sin x)} \qquad (x + y)(x - y) = x^2 - y^2$$

$$= \frac{\cos^2 x}{\cos x(1 - \sin x)} \qquad 1 - \sin^2 x = \cos^2 x$$

$$= \frac{\cos x}{1 - \sin x} \qquad \text{Lowest terms}$$

NOW TRY EXERCISE 49. ◀

Verifying Identities by Working with Both Sides
If both sides of an identity appear to be equally complex, the identity can be verified by working independently on the left side and on the right side, until each side is changed into some common third result. *Each step, on each side, must be reversible.* With all

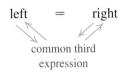

left = right

common third
expression

steps reversible, the procedure is as shown in the margin. The left side leads to a common third expression, which leads back to the right side. This procedure is just a shortcut for the procedure used in Examples 1–4: one side is changed into the other side, but by going through an intermediate step.

▶ EXAMPLE 5   VERIFYING AN IDENTITY (WORKING WITH BOTH SIDES)

Verify that the following equation is an identity.

$$\frac{\sec\alpha + \tan\alpha}{\sec\alpha - \tan\alpha} = \frac{1 + 2\sin\alpha + \sin^2\alpha}{\cos^2\alpha}$$

Solution   Both sides appear equally complex, so we verify the identity by changing each side into a common third expression. We work first on the left, multiplying numerator and denominator by $\cos\alpha$.

$$\frac{\sec\alpha + \tan\alpha}{\sec\alpha - \tan\alpha} = \frac{(\sec\alpha + \tan\alpha)\cos\alpha}{(\sec\alpha - \tan\alpha)\cos\alpha} \qquad \text{Multiply by 1 in the form } \tfrac{\cos\alpha}{\cos\alpha}.$$

$$= \frac{\sec\alpha\cos\alpha + \tan\alpha\cos\alpha}{\sec\alpha\cos\alpha - \tan\alpha\cos\alpha} \qquad \text{Distributive property}$$

$$= \frac{1 + \tan\alpha\cos\alpha}{1 - \tan\alpha\cos\alpha} \qquad \sec\alpha\cos\alpha = 1$$

$$= \frac{1 + \dfrac{\sin\alpha}{\cos\alpha}\cdot\cos\alpha}{1 - \dfrac{\sin\alpha}{\cos\alpha}\cdot\cos\alpha} \qquad \tan\alpha = \tfrac{\sin\alpha}{\cos\alpha}$$

$$= \frac{1 + \sin\alpha}{1 - \sin\alpha} \qquad \text{Simplify.}$$

On the right side of the original equation, begin by factoring.

$$\frac{1 + 2\sin\alpha + \sin^2\alpha}{\cos^2\alpha} = \frac{(1 + \sin\alpha)^2}{\cos^2\alpha} \qquad x^2 + 2xy + y^2 = (x + y)^2$$

$$= \frac{(1 + \sin\alpha)^2}{1 - \sin^2\alpha} \qquad \cos^2\alpha = 1 - \sin^2\alpha$$

$$= \frac{(1 + \sin\alpha)^2}{(1 + \sin\alpha)(1 - \sin\alpha)} \qquad \begin{array}{l}\text{Factor the denominator;}\\ x^2 - y^2 = (x + y)(x - y).\end{array}$$

$$= \frac{1 + \sin\alpha}{1 - \sin\alpha} \qquad \text{Lowest terms}$$

We have shown that

Left side of   Common third   Right side of
given equation   expression   given equation

$$\underbrace{\frac{\sec\alpha + \tan\alpha}{\sec\alpha - \tan\alpha}} = \underbrace{\frac{1 + \sin\alpha}{1 - \sin\alpha}} = \underbrace{\frac{1 + 2\sin\alpha + \sin^2\alpha}{\cos^2\alpha}},$$

verifying that the given equation is an identity.

NOW TRY EXERCISE 65. ◀

> ▶ **Caution** Use the method of Example 5 *only* if the steps are reversible.

There are usually several ways to verify a given identity. For instance, another way to begin verifying the identity in Example 5 is to work on the left as follows.

$$\frac{\sec \alpha + \tan \alpha}{\sec \alpha - \tan \alpha} = \frac{\dfrac{1}{\cos \alpha} + \dfrac{\sin \alpha}{\cos \alpha}}{\dfrac{1}{\cos \alpha} - \dfrac{\sin \alpha}{\cos \alpha}} \quad \text{Fundamental identities (Section 5.1)}$$

$$= \frac{\dfrac{1 + \sin \alpha}{\cos \alpha}}{\dfrac{1 - \sin \alpha}{\cos \alpha}} \quad \text{Add and subtract fractions.}$$

Multiply by $\frac{\cos \alpha}{\cos \alpha}$ here.

$$= \frac{1 + \sin \alpha}{1 - \sin \alpha} \quad \text{Simplify the complex fraction.}$$

Compare this with the result shown in Example 5 for the right side to see that the two sides indeed agree.

▶ **EXAMPLE 6** APPLYING A PYTHAGOREAN IDENTITY TO RADIOS

Tuners in radios select a radio station by adjusting the frequency. A tuner may contain an inductor L and a capacitor C, as illustrated in Figure 3. The energy stored in the inductor at time t is given by

$$L(t) = k \sin^2(2\pi Ft)$$

and the energy stored in the capacitor is given by

$$C(t) = k \cos^2(2\pi Ft),$$

where F is the frequency of the radio station and k is a constant. The total energy E in the circuit is given by

$$E(t) = L(t) + C(t).$$

Show that E is a constant function. (*Source:* Weidner, R. and R. Sells, *Elementary Classical Physics,* Vol. 2, Allyn & Bacon, 1973.)

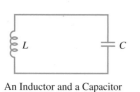

An Inductor and a Capacitor

Figure 3

Solution

$$
\begin{aligned}
E(t) &= L(t) + C(t) && \text{Given equation} \\
&= k \sin^2(2\pi Ft) + k \cos^2(2\pi Ft) && \text{Substitute.} \\
&= k[\sin^2(2\pi Ft) + \cos^2(2\pi Ft)] && \text{Factor out } k. \\
&= k(1) && \sin^2 \theta + \cos^2 \theta = 1 \text{ (Here } \theta = 2\pi Ft.) \\
&= k
\end{aligned}
$$

Since k is a constant, $E(t)$ is a constant function.

NOW TRY EXERCISE 95. ◀

5.2 Exercises

Perform each indicated operation and simplify the result.

1. $\cot \theta + \dfrac{1}{\cot \theta}$

2. $\dfrac{\sec x}{\csc x} + \dfrac{\csc x}{\sec x}$

3. $\tan s(\cot s + \csc s)$

4. $\cos \beta(\sec \beta + \csc \beta)$

5. $\dfrac{1}{\csc^2 \theta} + \dfrac{1}{\sec^2 \theta}$

6. $\dfrac{1}{\sin \alpha - 1} - \dfrac{1}{\sin \alpha + 1}$

7. $\dfrac{\cos x}{\sec x} + \dfrac{\sin x}{\csc x}$

8. $\dfrac{\cos \theta}{\sin \theta} + \dfrac{\sin \theta}{1 + \cos \theta}$

9. $(1 + \sin t)^2 + \cos^2 t$

10. $(1 + \tan s)^2 - 2 \tan s$

11. $\dfrac{1}{1 + \cos x} - \dfrac{1}{1 - \cos x}$

12. $(\sin \alpha - \cos \alpha)^2$

Factor each trigonometric expression.

13. $\sin^2 \theta - 1$

14. $\sec^2 \theta - 1$

15. $(\sin x + 1)^2 - (\sin x - 1)^2$

16. $(\tan x + \cot x)^2 - (\tan x - \cot x)^2$

17. $2 \sin^2 x + 3 \sin x + 1$

18. $4 \tan^2 \beta + \tan \beta - 3$

19. $\cos^4 x + 2 \cos^2 x + 1$

20. $\cot^4 x + 3 \cot^2 x + 2$

21. $\sin^3 x - \cos^3 x$

22. $\sin^3 \alpha + \cos^3 \alpha$

Each expression simplifies to a constant, a single function, or a power of a function. Use fundamental identities to simplify each expression.

23. $\tan \theta \cos \theta$

24. $\cot \alpha \sin \alpha$

25. $\sec r \cos r$

26. $\cot t \tan t$

27. $\dfrac{\sin \beta \tan \beta}{\cos \beta}$

28. $\dfrac{\csc \theta \sec \theta}{\cot \theta}$

29. $\sec^2 x - 1$

30. $\csc^2 t - 1$

31. $\dfrac{\sin^2 x}{\cos^2 x} + \sin x \csc x$

32. $\dfrac{1}{\tan^2 \alpha} + \cot \alpha \tan \alpha$

33. $1 - \dfrac{1}{\csc^2 x}$

34. $1 - \dfrac{1}{\sec^2 x}$

In Exercises 35–78, verify that each trigonometric equation is an identity. See Examples 1–5.

35. $\dfrac{\cot \theta}{\csc \theta} = \cos \theta$

36. $\dfrac{\tan \alpha}{\sec \alpha} = \sin \alpha$

37. $\dfrac{1 - \sin^2 \beta}{\cos \beta} = \cos \beta$

38. $\dfrac{\tan^2 \alpha + 1}{\sec \alpha} = \sec \alpha$

39. $\cos^2 \theta(\tan^2 \theta + 1) = 1$

40. $\sin^2 \beta(1 + \cot^2 \beta) = 1$

41. $\cot s + \tan s = \sec s \csc s$

42. $\sin^2 \alpha + \tan^2 \alpha + \cos^2 \alpha = \sec^2 \alpha$

43. $\dfrac{\cos \alpha}{\sec \alpha} + \dfrac{\sin \alpha}{\csc \alpha} = \sec^2 \alpha - \tan^2 \alpha$

44. $\dfrac{\sin^2 \theta}{\cos \theta} = \sec \theta - \cos \theta$

45. $\sin^4 \theta - \cos^4 \theta = 2 \sin^2 \theta - 1$

46. $\dfrac{\cos \theta}{\sin \theta \cot \theta} = 1$

47. $\dfrac{1 - \cos x}{1 + \cos x} = (\cot x - \csc x)^2$

48. $\sin^2 \theta(1 + \cot^2 \theta) - 1 = 0$

49. $\dfrac{\cos \theta + 1}{\tan^2 \theta} = \dfrac{\cos \theta}{\sec \theta - 1}$

50. $\dfrac{(\sec \theta - \tan \theta)^2 + 1}{\sec \theta \csc \theta - \tan \theta \csc \theta} = 2 \tan \theta$

51. $\dfrac{1}{1 - \sin \theta} + \dfrac{1}{1 + \sin \theta} = 2 \sec^2 \theta$

52. $\dfrac{1}{\sec \alpha - \tan \alpha} = \sec \alpha + \tan \alpha$

53. $\dfrac{\cot \alpha + 1}{\cot \alpha - 1} = \dfrac{1 + \tan \alpha}{1 - \tan \alpha}$

54. $\dfrac{\csc \theta + \cot \theta}{\tan \theta + \sin \theta} = \cot \theta \csc \theta$

55. $\sec^4 x - \sec^2 x = \tan^4 x + \tan^2 x$

56. $(\sec \alpha - \tan \alpha)^2 = \dfrac{1 - \sin \alpha}{1 + \sin \alpha}$

57. $\dfrac{\sec^4 s - \tan^4 s}{\sec^2 s + \tan^2 s} = \sec^2 s - \tan^2 s$

58. $\dfrac{\cot^2 t - 1}{1 + \cot^2 t} = 1 - 2 \sin^2 t$

59. $\dfrac{\tan^2 t - 1}{\sec^2 t} = \dfrac{\tan t - \cot t}{\tan t + \cot t}$

60. $\dfrac{\sin^4 \alpha - \cos^4 \alpha}{\sin^2 \alpha - \cos^2 \alpha} = 1$

61. $(1 - \cos^2 \alpha)(1 + \cos^2 \alpha) = 2 \sin^2 \alpha - \sin^4 \alpha$

62. $\tan^2 \alpha \sin^2 \alpha = \tan^2 \alpha + \cos^2 \alpha - 1$

63. $\sin^2 \alpha \sec^2 \alpha + \sin^2 \alpha \csc^2 \alpha = \sec^2 \alpha$

64. $\dfrac{-1}{\tan \alpha - \sec \alpha} + \dfrac{-1}{\tan \alpha + \sec \alpha} = 2 \tan \alpha$

65. $\dfrac{\tan s}{1 + \cos s} + \dfrac{\sin s}{1 - \cos s} = \cot s + \sec s \csc s$

66. $\dfrac{1 - \cos x}{1 + \cos x} = \csc^2 x - 2 \csc x \cot x + \cot^2 x$

67. $\dfrac{1 - \sin \theta}{1 + \sin \theta} = \sec^2 \theta - 2 \sec \theta \tan \theta + \tan^2 \theta$

68. $\sin \theta + \cos \theta = \dfrac{\sin \theta}{1 - \dfrac{\cos \theta}{\sin \theta}} + \dfrac{\cos \theta}{1 - \dfrac{\sin \theta}{\cos \theta}}$

69. $\dfrac{\sin \theta}{1 - \cos \theta} - \dfrac{\sin \theta \cos \theta}{1 + \cos \theta} = \csc \theta (1 + \cos^2 \theta)$

70. $(1 + \sin x + \cos x)^2 = 2(1 + \sin x)(1 + \cos x)$

71. $\dfrac{1 + \cos x}{1 - \cos x} - \dfrac{1 - \cos x}{1 + \cos x} = 4 \cot x \csc x$

72. $(\sec \alpha + \csc \alpha)(\cos \alpha - \sin \alpha) = \cot \alpha - \tan \alpha$

73. $\dfrac{1 + \sin \theta}{1 - \sin \theta} - \dfrac{1 - \sin \theta}{1 + \sin \theta} = 4 \tan \theta \sec \theta$

74. $\dfrac{1 - \cos \theta}{1 + \cos \theta} = 2 \csc^2 \theta - 2 \csc \theta \cot \theta - 1$

75. $(2 \sin x + \cos x)^2 + (2 \cos x - \sin x)^2 = 5$

76. $\sin^2 x(1 + \cot x) + \cos^2 x(1 - \tan x) + \cot^2 x = \csc^2 x$

77. $\sec x - \cos x + \csc x - \sin x - \sin x \tan x = \cos x \cot x$

78. $\sin^3 \theta + \cos^3 \theta = (\cos \theta + \sin \theta)(1 - \cos \theta \sin \theta)$

Graph each expression and make a conjecture, predicting what might be an identity. Then verify your conjecture algebraically.

79. $(\sec \theta + \tan \theta)(1 - \sin \theta)$

80. $(\csc \theta + \cot \theta)(\sec \theta - 1)$

81. $\dfrac{\cos \theta + 1}{\sin \theta + \tan \theta}$

82. $\tan \theta \sin \theta + \cos \theta$

Graph the expressions on each side of the equals symbol to determine whether the equation might be an identity. (Note: Use a domain whose length is at least 2π.) If the equation looks like an identity, verify it algebraically. See Example 1.

83. $\dfrac{2 + 5 \cos x}{\sin x} = 2 \csc x + 5 \cot x$

84. $1 + \cot^2 x = \dfrac{\sec^2 x}{\sec^2 x - 1}$

85. $\dfrac{\tan x - \cot x}{\tan x + \cot x} = 2 \sin^2 x$

86. $\dfrac{1}{1 + \sin x} + \dfrac{1}{1 - \sin x} = \sec^2 x$

By substituting a number for s or t, show that the equation is not an identity.

87. $\sin(\csc s) = 1$

88. $\sqrt{\cos^2 s} = \cos s$

89. $\csc t = \sqrt{1 + \cot^2 t}$

90. $\cos t = \sqrt{1 - \sin^2 t}$

91. *Concept Check* When is $\sin x = \sqrt{1 - \cos^2 x}$ a true statement?

92. *Concept Check* When is $\cos x = \sqrt{1 - \sin^2 x}$ a true statement?

(Modeling) Work each problem.

93. *Intensity of a Lamp* According to Lambert's law, the intensity of light from a single source on a flat surface at point P is given by

$$I = k \cos^2 \theta,$$

where k is a constant. (*Source:* Winter, C., *Solar Power Plants*, Springer-Verlag, 1991.)

(a) Write I in terms of the sine function.

(b) Why does the maximum value of I occur when $\theta = 0$?

94. *Oscillating Spring* The distance or displacement y of a weight attached to an oscillating spring from its natural position is modeled by

$$y = 4 \cos(2\pi t),$$

where t is time in seconds. Potential energy is the energy of position and is given by

$$P = ky^2,$$

where k is a constant. The weight has the greatest potential energy when the spring is stretched the most. (*Source:* Weidner, R. and R. Sells, *Elementary Classical Physics*, Vol. 1, Allyn & Bacon, 1973.)

(a) Write an expression for P that involves the cosine function.

(b) Use a fundamental identity to write P in terms of $\sin(2\pi t)$.

95. *Radio Tuners* Refer to Example 6. Let the energy stored in the inductor be given by

$$L(t) = 3 \cos^2(6,000,000t)$$

and the energy in the capacitor be given by

$$C(t) = 3 \sin^2(6,000,000t),$$

where t is time in seconds. The total energy E in the circuit is given by $E(t) = L(t) + C(t)$.

(a) Graph L, C, and E in the window $[0, 10^{-6}]$ by $[-1, 4]$, with $\text{Xscl} = 10^{-7}$ and $\text{Yscl} = 1$. Interpret the graph.

(b) Make a table of values for L, C, and E starting at $t = 0$, incrementing by 10^{-7}. Interpret your results.

(c) Use a fundamental identity to derive a simplified expression for $E(t)$.

5.3 Sum and Difference Identities for Cosine

Difference Identity for Cosine ▪ **Sum Identity for Cosine** ▪ **Cofunction Identities** ▪ **Applying the Sum and Difference Identities**

Difference Identity for Cosine Several examples presented earlier should have convinced you by now that $\cos(A - B)$ *does not equal* $\cos A - \cos B$. For example, if $A = \frac{\pi}{2}$ and $B = 0$, then

$$\cos(A - B) = \cos\left(\frac{\pi}{2} - 0\right) = \cos\frac{\pi}{2} = 0,$$

while $\qquad \cos A - \cos B = \cos\frac{\pi}{2} - \cos 0 = 0 - 1 = -1.$

We can now derive a formula for $\cos(A - B)$. We start by locating angles A and B in standard position on a unit circle, with $B < A$. Let S and Q be the points where the terminal sides of angles A and B, respectively, intersect the circle. Let P be the point $(1, 0)$, and locate point R on the unit circle so that angle POR equals the difference $A - B$. See Figure 4.

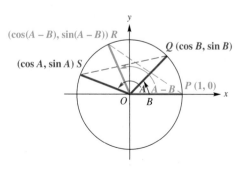

Figure 4

Point Q is on the unit circle, thus the x-coordinate of Q is the cosine of angle B, while the y-coordinate of Q is the sine of angle B.

$$Q \text{ has coordinates } (\cos B, \sin B).$$

In the same way,

$$S \text{ has coordinates } (\cos A, \sin A),$$

and $\qquad\qquad R \text{ has coordinates } (\cos(A - B), \sin(A - B)).$

Angle SOQ also equals $A - B$. Since the central angles SOQ and POR are equal, chords PR and SQ are equal. By the distance formula, since $PR = SQ$,

$$\sqrt{[\cos(A - B) - 1]^2 + [\sin(A - B) - 0]^2}$$
$$= \sqrt{(\cos A - \cos B)^2 + (\sin A - \sin B)^2}. \quad \text{(Appendix B)}$$

Squaring both sides and clearing parentheses gives

$$\cos^2(A - B) - 2\cos(A - B) + 1 + \sin^2(A - B)$$
$$= \cos^2 A - 2\cos A \cos B + \cos^2 B + \sin^2 A - 2\sin A \sin B + \sin^2 B.$$

Since $\sin^2 x + \cos^2 x = 1$ for any value of x, we can rewrite the equation as

$$2 - 2\cos(A - B) = 2 - 2\cos A \cos B - 2\sin A \sin B$$
$$\mathbf{\cos(A - B) = \cos A \cos B + \sin A \sin B.} \qquad \text{Subtract 2; divide by } -2.$$

This is the identity for $\cos(A - B)$. Although Figure 4 shows angles A and B in the second and first quadrants, respectively, this result is the same for any values of these angles.

Sum Identity for Cosine
To find a similar expression for $\cos(A + B)$, rewrite $A + B$ as $A - (-B)$ and use the identity for $\cos(A - B)$.

$$\cos(A + B) = \cos[A - (-B)]$$
$$= \cos A \cos(-B) + \sin A \sin(-B) \qquad \text{Cosine difference identity}$$
$$= \cos A \cos B + \sin A(-\sin B) \qquad \text{Negative-angle identities} \\ \text{(Section 5.1)}$$

$$\mathbf{\cos(A + B) = \cos A \cos B - \sin A \sin B}$$

COSINE OF A SUM OR DIFFERENCE

$$\mathbf{\cos(A + B) = \cos A \cos B - \sin A \sin B}$$
$$\mathbf{\cos(A - B) = \cos A \cos B + \sin A \sin B}$$

These identities are important in calculus and useful in certain applications. For example, the method shown in Example 1 can be applied to get an exact value for $\cos 15°$, as well as to practice using the sum and difference identities.

▶ EXAMPLE 1   FINDING EXACT COSINE FUNCTION VALUES

Find the *exact* value of each expression.

(a) $\cos 15°$ **(b)** $\cos \dfrac{5\pi}{12}$ **(c)** $\cos 87° \cos 93° - \sin 87° \sin 93°$

Solution

(a) To find $\cos 15°$, we write $15°$ as the sum or difference of two angles with known function values, such as $45°$ and $30°$, since $15° = 45° - 30°$. (We could also use $60° - 45°$.) Then we use the cosine difference identity.

$$\begin{aligned}
\cos 15° &= \cos(45° - 30°) && \text{\footnotesize $15° = 45° - 30°$} \\
&= \cos 45° \cos 30° + \sin 45° \sin 30° && \text{\footnotesize Cosine difference identity} \\
&= \frac{\sqrt{2}}{2} \cdot \frac{\sqrt{3}}{2} + \frac{\sqrt{2}}{2} \cdot \frac{1}{2} && \text{\footnotesize Substitute known values. (Section 2.1)} \\
&= \frac{\sqrt{6} + \sqrt{2}}{4} && \text{\footnotesize Multiply; add fractions.}
\end{aligned}$$

(b)
$$\begin{aligned}
\cos \frac{5\pi}{12} &= \cos\left(\frac{\pi}{6} + \frac{\pi}{4}\right) && \text{\footnotesize $\frac{\pi}{6} = \frac{2\pi}{12}, \frac{\pi}{4} = \frac{3\pi}{12}$} \\
&= \cos \frac{\pi}{6} \cos \frac{\pi}{4} - \sin \frac{\pi}{6} \sin \frac{\pi}{4} && \text{\footnotesize Cosine sum identity} \\
&= \frac{\sqrt{3}}{2} \cdot \frac{\sqrt{2}}{2} - \frac{1}{2} \cdot \frac{\sqrt{2}}{2} && \text{\footnotesize Substitute known values. (Section 3.1)} \\
&= \frac{\sqrt{6} - \sqrt{2}}{4} && \text{\footnotesize Multiply; subtract fractions.}
\end{aligned}$$

```
cos(5π/12)
      .2588190451
(√(6)-√(2))/4
      .2588190451
```

The screen supports the solution in Example 1(b) by showing that

$\cos \dfrac{5\pi}{12} = \dfrac{\sqrt{6} - \sqrt{2}}{4}$.

(c)
$$\begin{aligned}
\cos 87° \cos 93° - \sin 87° \sin 93° &= \cos(87° + 93°) && \text{\footnotesize Cosine sum identity} \\
&= \cos 180° && \text{\footnotesize Add.} \\
&= -1 && \text{\footnotesize (Section 1.3)}
\end{aligned}$$

NOW TRY EXERCISES 7, 9, AND 11. ◀

Cofunction Identities

We can use the identity for the cosine of the difference of two angles and the fundamental identities to derive the *cofunction identities,* presented in **Section 2.1** for values of θ in the interval $[0°, 90°]$.

COFUNCTION IDENTITIES

$$\cos(90° - \theta) = \sin \theta \qquad \cot(90° - \theta) = \tan \theta$$
$$\sin(90° - \theta) = \cos \theta \qquad \sec(90° - \theta) = \csc \theta$$
$$\tan(90° - \theta) = \cot \theta \qquad \csc(90° - \theta) = \sec \theta$$

Similar identities can be obtained for a real number domain by replacing $90°$ with $\frac{\pi}{2}$.

Substituting 90° for A and θ for B in the identity for $\cos(A - B)$ gives

$$\cos(90° - \theta) = \cos 90° \cos \theta + \sin 90° \sin \theta$$
$$= 0 \cdot \cos \theta + 1 \cdot \sin \theta$$
$$= \sin \theta.$$

This result is true for *any* value of θ since the identity for $\cos(A - B)$ is true for any values of A and B.

▶ **EXAMPLE 2** **USING COFUNCTION IDENTITIES TO FIND θ**

Find an angle θ that satisfies each of the following.

(a) $\cot \theta = \tan 25°$ **(b)** $\sin \theta = \cos(-30°)$ **(c)** $\csc \dfrac{3\pi}{4} = \sec \theta$

Solution

(a) Since tangent and cotangent are cofunctions, $\tan(90° - \theta) = \cot \theta$.

$$\cot \theta = \tan 25°$$
$$\tan(90° - \theta) = \tan 25° \quad \text{Cofunction identity}$$
$$90° - \theta = 25° \quad \text{Set angle measures equal.}$$
$$\theta = 65° \quad \text{Solve for } \theta.$$

(b)
$$\sin \theta = \cos(-30°)$$
$$\cos(90° - \theta) = \cos(-30°) \quad \text{Cofunction identity}$$
$$90° - \theta = -30°$$
$$\theta = 120°$$

(c)
$$\csc \frac{3\pi}{4} = \sec \theta$$
$$\sec\left(\frac{\pi}{2} - \frac{3\pi}{4}\right) = \sec \theta \quad \text{Cofunction identity}$$
$$\sec\left(-\frac{\pi}{4}\right) = \sec \theta \quad \text{Combine terms.}$$
$$-\frac{\pi}{4} = \theta$$

NOW TRY EXERCISES 33 AND 37. ◀

▶ **Note** Because trigonometric (circular) functions are periodic, the solutions in Example 2 are not unique. We give only one of infinitely many possibilities.

Applying the Sum and Difference Identities

If one of the angles A or B in the identities for $\cos(A + B)$ and $\cos(A - B)$ is a quadrantal angle, then the identity allows us to write the expression in terms of a single function of A or B.

▶ **EXAMPLE 3** **REDUCING cos(A − B) TO A FUNCTION OF A SINGLE VARIABLE**

Write $\cos(180° − \theta)$ as a trigonometric function of θ alone.

Solution $\cos(180° − \theta) = \cos 180° \cos \theta + \sin 180° \sin \theta$

Cosine difference identity

$$= (−1) \cos \theta + (0) \sin \theta \quad \text{(Section 1.3)}$$

$$= −\cos \theta$$

NOW TRY EXERCISE 39. ◀

CONNECTIONS   (This discussion applies to functions of both angles and real numbers.)

The result of Example 3 can be written as an identity:

$$\cos(180° − \theta) = −\cos \theta.$$

This is an example of a **reduction formula,** which is an identity that *reduces* a function of a quadrantal angle plus or minus θ to a function of θ alone. Another example of a reduction formula is $\cos(270° + \theta) = \sin \theta$. (Verify this using the formula for the cosine of the sum.)

Here is an interesting method for quickly determining a reduction formula for a trigonometric function f of the form $f(Q \pm \theta)$, where Q is a quadrantal angle. *There are two cases to consider, and in each case, think of θ as a small positive angle* so that you can determine the quadrant in which $Q \pm \theta$ will lie.

Case 1: **Suppose that Q is a quadrantal angle whose terminal side lies along the *x*-axis.** Determine the quadrant in which $Q \pm \theta$ will lie for a small positive angle θ. If the given function f is positive in that quadrant, use a + sign on the reduced form; if f is negative in that quadrant, use a − sign. The reduced form will have that sign, f as the function, and θ as the argument. For example:

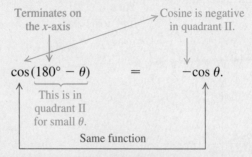

Case 2: **Suppose that Q is a quadrantal angle whose terminal side lies along the *y*-axis.** Determine the quadrant in which $Q \pm \theta$ will lie for a small positive angle θ. If the given function f is positive in that quadrant, use a + sign on the reduced form; if f is negative in that quadrant, use a − sign. The reduced form will have that sign, the *cofunction of f* as the function, and θ as the argument.

(continued)

For example:

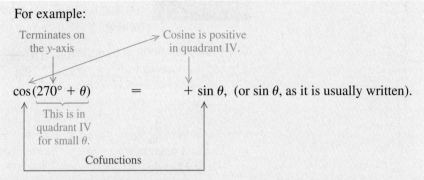

Terminates on
the *y*-axis

Cosine is positive
in quadrant IV.

$$\cos(270° + \theta) \qquad = \qquad + \sin\theta, \text{ (or } \sin\theta, \text{ as it is usually written).}$$

This is in
quadrant IV
for small θ.

Cofunctions

FOR DISCUSSION OR WRITING

Use these ideas to write reduction formulas for the following. (Those involving
sine and tangent can be verified after studying the identities in **Section 5.4.**)

1. $\cos(90° + \theta)$ **2.** $\cos(270° - \theta)$ **3.** $\cos(180° + \theta)$

4. $\cos(180° - \theta)$ **5.** $\sin(180° + \theta)$ **6.** $\tan(270° - \theta)$

▶ **EXAMPLE 4**   FINDING cos(*s* + *t*) GIVEN INFORMATION ABOUT *s*
AND *t*

Suppose that $\sin s = \frac{3}{5}$, $\cos t = -\frac{12}{13}$, and both s and t are in quadrant II. Find
$\cos(s + t)$.

Solution   By the cosine sum identity, $\cos(s + t) = \cos s \cos t - \sin s \sin t$.
The values of $\sin s$ and $\cos t$ are given, so we can find $\cos(s + t)$ if we know the
values of $\cos s$ and $\sin t$. To find $\cos s$ and $\sin t$, we sketch two angles in the second quadrant, one with $\sin s = \frac{3}{5}$ and the other with $\cos t = -\frac{12}{13}$. See Figure 5.

In Figure 5(a), since $\sin s = \frac{3}{5} = \frac{y}{r}$, we let $y = 3$ and $r = 5$. Substituting in
the Pythagorean theorem, we get $x^2 + 3^2 = 5^2$ and solve to find $x = -4$. Thus,
$\cos s = -\frac{4}{5}$. In Figure 5(b), $\cos t = -\frac{12}{13} = \frac{x}{r}$, so we let $x = -12$ and $r = 13$.
Then $(-12)^2 + y^2 = 13^2$; we solve to get $y = 5$. Thus, $\sin t = \frac{5}{13}$. Now we can
find $\cos(s + t)$.

$$\cos(s + t) = \cos s \cos t - \sin s \sin t \qquad \text{Cosine sum identity}$$

$$= -\frac{4}{5}\left(-\frac{12}{13}\right) - \frac{3}{5} \cdot \frac{5}{13} \qquad \text{Substitute.}$$

$$= \frac{48}{65} - \frac{15}{65} = \frac{33}{65}$$

NOW TRY EXERCISE 47. ◀

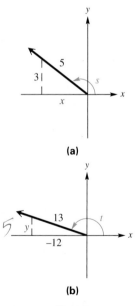

(a)

(b)

Figure 5

▶ **Note** In Example 4, the values of $\cos s$ and $\sin t$ could also be found by
using the Pythagorean identities. The problem could then be solved using the
identity for $\cos(s + t)$ as shown in the example.

▶ **EXAMPLE 5** APPLYING THE COSINE DIFFERENCE IDENTITY TO VOLTAGE

Common household electric current is called **alternating current** because the current alternates direction within the wires. The voltage V in a typical 115-volt outlet can be expressed by the function

$$V(t) = 163 \sin \omega t,$$

where ω is the angular speed (in radians per second) of the rotating generator at the electrical plant and t is time measured in seconds. (*Source:* Bell, D., *Fundamentals of Electric Circuits,* Fourth Edition, Prentice-Hall, 1988.)

(a) It is essential for electric generators to rotate at precisely 60 cycles per sec so household appliances and computers will function properly. Determine ω for these electric generators.

(b) Graph V in the window $[0, .05]$ by $[-200, 200]$.

(c) Determine a value of ϕ so that the graph of $V(t) = 163 \cos(\omega t - \phi)$ is the same as the graph of $V(t) = 163 \sin \omega t$.

Solution

(a) Each cycle is 2π radians at 60 cycles per sec, so the angular speed is $\omega = 60(2\pi) = 120\pi$ radians per sec.

(b) $V(t) = 163 \sin \omega t = 163 \sin 120\pi t$. Because the amplitude of the function is 163 (from **Section 4.1**), $[-200, 200]$ is an appropriate interval for the range, as shown in Figure 6.

(c) Using the negative-angle identity for cosine and a cofunction identity,

$$\cos\left(x - \frac{\pi}{2}\right) = \cos\left[-\left(\frac{\pi}{2} - x\right)\right] = \cos\left(\frac{\pi}{2} - x\right) = \sin x.$$

Therefore, if $\phi = \frac{\pi}{2}$, then

$$V(t) = 163 \cos(\omega t - \phi) = 163 \cos\left(\omega t - \frac{\pi}{2}\right) = 163 \sin \omega t.$$

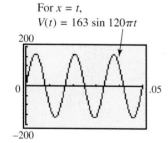

For $x = t$,
$V(t) = 163 \sin 120\pi t$

Figure 6

NOW TRY EXERCISE 71. ◀

5.3 Exercises

Concept Check *Match each expression in Column I with the correct expression in Column II to form an identity.*

I	II
1. $\cos(x + y) =$ _____	**A.** $\cos x \cos y + \sin x \sin y$
2. $\cos(x - y) =$ _____	**B.** $\cos x$
3. $\cos(90° - x) =$ _____	**C.** $\cos x + \sin x$
4. $\sin(90° - x) =$ _____	**D.** $\cos x - \sin x$
	E. $\sin x$
	F. $\cos x \cos y - \sin x \sin y$

Use identities to find each exact value. (Do not use a calculator.) See Example 1.

5. $\cos 75°$

6. $\cos(-15°)$

7. $\cos 105°$
   (*Hint:* $105° = 60° + 45°$)

8. $\cos(-105°)$
   (*Hint:* $-105° = -60° + (-45°)$)

9. $\cos \dfrac{7\pi}{12}$

10. $\cos\left(-\dfrac{\pi}{12}\right)$

11. $\cos 40° \cos 50° - \sin 40° \sin 50°$

12. $\cos \dfrac{7\pi}{9} \cos \dfrac{2\pi}{9} - \sin \dfrac{7\pi}{9} \sin \dfrac{2\pi}{9}$

Use a graphing or scientific calculator to support your answer for each of the following. See Example 1.

13. Exercise 11

14. Exercise 12

Write each function value in terms of the cofunction of a complementary angle. See Example 2.

15. $\tan 87°$

16. $\sin 15°$

17. $\cos \dfrac{\pi}{12}$

18. $\sin \dfrac{2\pi}{5}$

19. $\csc(-14° \, 24')$

20. $\sin 142° \, 14'$

21. $\sin \dfrac{5\pi}{8}$

22. $\cot \dfrac{9\pi}{10}$

23. $\sec 146° \, 42'$

24. $\tan 174° \, 3'$

25. $\cot 176.9814°$

26. $\sin 98.0142°$

Use identities to fill in each blank with the appropriate trigonometric function name. See Example 2.

27. $\cot \dfrac{\pi}{3} = $ _____ $\dfrac{\pi}{6}$

28. $\sin \dfrac{2\pi}{3} = $ _____ $\left(-\dfrac{\pi}{6}\right)$

29. _____ $33° = \sin 57°$

30. _____ $72° = \cot 18°$

31. $\cos 70° = \dfrac{1}{\text{_____} \, 20°}$

32. $\tan 24° = \dfrac{1}{\text{_____} \, 66°}$

Find an angle θ that makes each statement true. See Example 2.

33. $\tan \theta = \cot(45° + 2\theta)$

34. $\sin \theta = \cos(2\theta - 10°)$

35. $\sec \theta = \csc\left(\dfrac{\theta}{2} + 20°\right)$

36. $\cos \theta = \sin(3\theta + 10°)$

37. $\sin(3\theta - 15°) = \cos(\theta + 25°)$

38. $\cot(\theta - 10°) = \tan(2\theta + 20°)$

Use identities to write each expression as a function of θ. See Example 3.

39. $\cos(0° - \theta)$

40. $\cos(90° - \theta)$

41. $\cos(180° - \theta)$

42. $\cos(270° - \theta)$

43. $\cos(0° + \theta)$

44. $\cos(90° + \theta)$

45. $\cos(180° + \theta)$

46. $\cos(270° + \theta)$

Find $\cos(s + t)$ and $\cos(s - t)$. See Example 4.

47. $\cos s = -\dfrac{1}{5}$ and $\sin t = \dfrac{3}{5}$, s and t in quadrant II

48. $\sin s = \dfrac{2}{3}$ and $\sin t = -\dfrac{1}{3}$, s in quadrant II and t in quadrant IV

49. $\sin s = \dfrac{3}{5}$ and $\sin t = -\dfrac{12}{13}$, s in quadrant I and t in quadrant III

50. $\cos s = -\dfrac{8}{17}$ and $\cos t = -\dfrac{3}{5}$, s and t in quadrant III

51. $\sin s = \dfrac{\sqrt{5}}{7}$ and $\sin t = \dfrac{\sqrt{6}}{8}$, s and t in quadrant I

52. $\cos s = \dfrac{\sqrt{2}}{4}$ and $\sin t = -\dfrac{\sqrt{5}}{6}$, s and t in quadrant IV

Concept Check   *Tell whether each statement is* true *or* false.

53. $\cos 42° = \cos(30° + 12°)$

54. $\cos(-24°) = \cos 16° - \cos 40°$

55. $\cos 74° = \cos 60° \cos 14° + \sin 60° \sin 14°$

56. $\cos 140° = \cos 60° \cos 80° - \sin 60° \sin 80°$

57. $\cos \dfrac{\pi}{3} = \cos \dfrac{\pi}{12} \cos \dfrac{\pi}{4} - \sin \dfrac{\pi}{12} \sin \dfrac{\pi}{4}$

58. $\cos \dfrac{2\pi}{3} = \cos \dfrac{11\pi}{12} \cos \dfrac{\pi}{4} + \sin \dfrac{11\pi}{12} \sin \dfrac{\pi}{4}$

59. $\cos 70° \cos 20° - \sin 70° \sin 20° = 0$

60. $\cos 85° \cos 40° + \sin 85° \sin 40° = \dfrac{\sqrt{2}}{2}$

61. $\tan\left(\theta - \dfrac{\pi}{2}\right) = \cot \theta$ **62.** $\sin\left(\theta - \dfrac{\pi}{2}\right) = \cos \theta$

Verify that each equation is an identity.

63. $\cos\left(\dfrac{\pi}{2} + x\right) = -\sin x$ **64.** $\sec(\pi - x) = -\sec x$

65. $\cos 2x = \cos^2 x - \sin^2 x$ (*Hint:* $\cos 2x = \cos(x + x)$.)

66. $1 + \cos 2x - \cos^2 x = \cos^2 x$ (*Hint:* Use the result from Exercise 65.)

RELATING CONCEPTS

For individual or collaborative investigation
(Exercises 67–70)

The identities for $\cos(A + B)$ *and* $\cos(A - B)$ *can be used to find exact values of expressions like* $\cos 195°$ *and* $\cos 255°$, *where the angle is not in the first quadrant.* **Work Exercises 67–70 in order,** *to see how this is done.*

67. By writing 195° as 180° + 15°, use the identity for $\cos(A + B)$ to express $\cos 195°$ as $-\cos 15°$.

68. Use the identity for $\cos(A - B)$ to find $-\cos 15°$.

69. By the results of Exercises 67 and 68, $\cos 195° =$ _____.

70. Find each exact value using the method shown in Exercises 67–69.

(a) $\cos 255°$ (b) $\cos \dfrac{11\pi}{12}$

(Modeling) *Solve each problem.*

71. *Electric Current* Refer to Example 5.

(a) How many times does the current oscillate in .05 sec?

(b) What are the maximum and minimum voltages in this outlet? Is the voltage always equal to 115 volts?

72. *Sound Waves* Sound is a result of waves applying pressure to a person's eardrum. For a pure sound wave radiating outward in a spherical shape, the trigonometric function defined by

$$P = \frac{a}{r} \cos\left(\frac{2\pi r}{\lambda} - ct\right)$$

can be used to model the sound pressure at a radius of r feet from the source, where t is time in seconds, λ is length of the sound wave in feet, c is speed of sound in feet per second, and a is maximum sound pressure at the source measured in pounds per square foot. (*Source:* Beranek, L., *Noise and Vibration Control,* Institute of Noise Control Engineering, Washington, D.C., 1988.) Let $\lambda = 4.9$ ft and $c = 1026$ ft per sec.

(a) Let $a = .4$ lb per ft^2. Graph the sound pressure at distance $r = 10$ ft from its source in the window $[0, .05]$ by $[-.05, .05]$. Describe P at this distance.

(b) Now let $a = 3$ and $t = 10$. Graph the sound pressure in the window $[0, 20]$ by $[-2, 2]$. What happens to pressure P as radius r increases?

(c) Suppose a person stands at a radius r so that $r = n\lambda$, where n is a positive integer. Use the difference identity for cosine to simplify P in this situation.

5.4 Sum and Difference Identities for Sine and Tangent

Sum and Difference Identities for Sine ▪ **Sum and Difference Identities for Tangent** ▪ **Applying the Sum and Difference Identities**

Sum and Difference Identities for Sine
We can use the cosine sum and difference identities to derive similar identities for sine and tangent. Since $\sin \theta = \cos(90° - \theta)$, we replace θ with $A + B$ to get

$\sin(A + B) = \cos[90° - (A + B)]$ Cofunction identity **(Section 5.3)**

$\qquad\quad = \cos[(90° - A) - B]$

$\qquad\quad = \cos(90° - A) \cos B + \sin(90° - A) \sin B$

Cosine difference identity **(Section 5.3)**

$\sin(A + B) = \sin A \cos B + \cos A \sin B.$ Cofunction identities

Now we write $\sin(A - B)$ as $\sin[A + (-B)]$ and use the identity for $\sin(A + B)$.

$\sin(A - B) = \sin[A + (-B)]$

$\qquad\quad = \sin A \cos(-B) + \cos A \sin(-B)$ Sine sum identity

$\sin(A - B) = \sin A \cos B - \cos A \sin B$ Negative-angle identities
(Section 5.1)

SINE OF A SUM OR DIFFERENCE

$$\sin(A + B) = \sin A \cos B + \cos A \sin B$$
$$\sin(A - B) = \sin A \cos B - \cos A \sin B$$

Sum and Difference Identities for Tangent

To derive the identity for $\tan(A + B)$, we start with

$$\tan(A + B) = \frac{\sin(A + B)}{\cos(A + B)}$$ Fundamental identity (Section 5.1)

$$= \frac{\sin A \cos B + \cos A \sin B}{\cos A \cos B - \sin A \sin B}.$$ Sum identities

We express this result in terms of the tangent function by multiplying both numerator and denominator by $\frac{1}{\cos A \cos B}$.

$$\tan(A + B) = \frac{\dfrac{\sin A \cos B + \cos A \sin B}{1}}{\dfrac{\cos A \cos B - \sin A \sin B}{1}} \cdot \frac{\dfrac{1}{\cos A \cos B}}{\dfrac{1}{\cos A \cos B}}$$ Simplify the complex fraction.

$$= \frac{\dfrac{\sin A \cos B}{\cos A \cos B} + \dfrac{\cos A \sin B}{\cos A \cos B}}{\dfrac{\cos A \cos B}{\cos A \cos B} - \dfrac{\sin A \sin B}{\cos A \cos B}}$$ Multiply numerators; multiply denominators.

$$= \frac{\dfrac{\sin A}{\cos A} + \dfrac{\sin B}{\cos B}}{1 - \dfrac{\sin A}{\cos A} \cdot \dfrac{\sin B}{\cos B}}$$ Simplify.

Since $\frac{\sin \theta}{\cos \theta} = \tan \theta$, we have

$$\tan(A + B) = \frac{\tan A + \tan B}{1 - \tan A \tan B}.$$

Replacing B with $-B$ and using the fact that $\tan(-B) = -\tan B$ gives the identity for the tangent of the difference of two angles.

TANGENT OF A SUM OR DIFFERENCE

$$\tan(A + B) = \frac{\tan A + \tan B}{1 - \tan A \tan B} \qquad \tan(A - B) = \frac{\tan A - \tan B}{1 + \tan A \tan B}$$

Applying the Sum and Difference Identities

▶ **EXAMPLE 1** FINDING EXACT SINE AND TANGENT FUNCTION VALUES

Find the *exact* value of each expression.

(a) $\sin 75°$ **(b)** $\tan \dfrac{7\pi}{12}$ **(c)** $\sin 40° \cos 160° - \cos 40° \sin 160°$

Solution

(a) $\sin 75° = \sin(45° + 30°)$ $75° = 45° + 30°$

$\qquad\qquad = \sin 45° \cos 30° + \cos 45° \sin 30°$ Sine sum identity

$\qquad\qquad = \dfrac{\sqrt{2}}{2} \cdot \dfrac{\sqrt{3}}{2} + \dfrac{\sqrt{2}}{2} \cdot \dfrac{1}{2}$ Substitute known values.
(Section 2.1)

$\qquad\qquad = \dfrac{\sqrt{6} + \sqrt{2}}{4}$ Multiply; add fractions.

(b) $\tan \dfrac{7\pi}{12} = \tan\left(\dfrac{\pi}{3} + \dfrac{\pi}{4}\right)$ $\frac{\pi}{3} = \frac{4\pi}{12}; \frac{\pi}{4} = \frac{3\pi}{12}$

$\qquad\qquad = \dfrac{\tan \frac{\pi}{3} + \tan \frac{\pi}{4}}{1 - \tan \frac{\pi}{3} \tan \frac{\pi}{4}}$ Tangent sum identity

$\qquad\qquad = \dfrac{\sqrt{3} + 1}{1 - \sqrt{3} \cdot 1}$ Substitute known values.
(Section 3.1)

$\qquad\qquad = \dfrac{\sqrt{3} + 1}{1 - \sqrt{3}} \cdot \dfrac{1 + \sqrt{3}}{1 + \sqrt{3}}$ Rationalize the denominator.

$\qquad\qquad = \dfrac{\sqrt{3} + 3 + 1 + \sqrt{3}}{1 - 3}$ Multiply.

$\qquad\qquad = \dfrac{4 + 2\sqrt{3}}{-2}$ Combine terms.

> Factor first. Then divide out the common factor.

$\qquad\qquad = \dfrac{2(2 + \sqrt{3})}{2(-1)}$ Factor out 2.

$\qquad\qquad = -2 - \sqrt{3}$ Lowest terms

(c) $\sin 40° \cos 160° - \cos 40° \sin 160° = \sin(40° - 160°)$

$\qquad\qquad\qquad\qquad\qquad\qquad\qquad\qquad$ Sine difference identity

$\qquad\qquad\qquad\qquad\qquad\qquad = \sin(-120°)$ Subtract.

$\qquad\qquad\qquad\qquad\qquad\qquad = -\sin 120°$ Negative-angle identity

$\qquad\qquad\qquad\qquad\qquad\qquad = -\dfrac{\sqrt{3}}{2}$ (Section 2.2)

NOW TRY EXERCISES 9, 11, AND 15. ◀

▶ **EXAMPLE 2** WRITING FUNCTIONS AS EXPRESSIONS INVOLVING FUNCTIONS OF θ

Write each function as an expression involving functions of θ.

(a) $\sin(30° + \theta)$ **(b)** $\tan(45° - \theta)$ **(c)** $\sin(180° + \theta)$

Solution

(a) Using the identity for $\sin(A + B)$,

$$\sin(30° + \theta) = \sin 30° \cos \theta + \cos 30° \sin \theta$$

$$= \frac{1}{2} \cos \theta + \frac{\sqrt{3}}{2} \sin \theta.$$

(b) $\tan(45° - \theta) = \dfrac{\tan 45° - \tan \theta}{1 + \tan 45° \tan \theta}$ Tangent difference identity

$$= \frac{1 - \tan \theta}{1 + \tan \theta}$$

(c) $\sin(180° + \theta) = \sin 180° \cos \theta + \cos 180° \sin \theta$ Sine sum identity

$$= 0 \cdot \cos \theta + (-1) \sin \theta \qquad \text{(Section 1.3)}$$

$$= -\sin \theta$$

NOW TRY EXERCISES 29 AND 33. ◀

▶ **EXAMPLE 3** FINDING FUNCTION VALUES AND THE QUADRANT OF $A + B$

Suppose that A and B are angles in standard position, with $\sin A = \frac{4}{5}$, $\frac{\pi}{2} < A < \pi$, and $\cos B = -\frac{5}{13}$, $\pi < B < \frac{3\pi}{2}$. Find each of the following.

(a) $\sin(A + B)$ **(b)** $\tan(A + B)$ **(c)** the quadrant of $A + B$

Solution

(a) The identity for $\sin(A + B)$ requires $\sin A$, $\cos A$, $\sin B$, and $\cos B$. We are given values of $\sin A$ and $\cos B$. We must find values of $\cos A$ and $\sin B$.

$$\sin^2 A + \cos^2 A = 1 \qquad \text{Fundamental identity (Section 5.1)}$$

$$\frac{16}{25} + \cos^2 A = 1 \qquad \sin A = \frac{4}{5}$$

$$\cos^2 A = \frac{9}{25} \qquad \text{Subtract } \frac{16}{25}.$$

$$\cos A = -\frac{3}{5} \qquad \begin{array}{l}\text{Take square roots (Appendix A); since}\\ A \text{ is in quadrant II, } \cos A < 0.\end{array}$$

In the same way, $\sin B = -\frac{12}{13}$. Now use the formula for $\sin(A + B)$.

$$\sin(A + B) = \frac{4}{5}\left(-\frac{5}{13}\right) + \left(-\frac{3}{5}\right)\left(-\frac{12}{13}\right)$$

$$= -\frac{20}{65} + \frac{36}{65} = \frac{16}{65}$$

(b) To find $\tan(A + B)$, first use the values of sine and cosine from part (a), $\sin A = \frac{4}{5}$, $\cos A = -\frac{3}{5}$, $\sin B = -\frac{12}{13}$, and $\cos B = -\frac{5}{13}$, to get $\tan A = -\frac{4}{3}$ and $\tan B = \frac{12}{5}$.

$$\tan(A + B) = \frac{-\frac{4}{3} + \frac{12}{5}}{1 - \left(-\frac{4}{3}\right)\left(\frac{12}{5}\right)} = \frac{\frac{16}{15}}{1 + \frac{48}{15}} = \frac{\frac{16}{15}}{\frac{63}{15}} = \frac{16}{15} \div \frac{63}{15} = \frac{16}{15} \cdot \frac{15}{63} = \frac{16}{63}$$

(c) From parts (a) and (b),

$$\sin(A + B) = \frac{16}{65} \quad \text{and} \quad \tan(A + B) = \frac{16}{63},$$

both positive. Therefore, $A + B$ must be in quadrant I, since it is the only quadrant in which both sine and tangent are positive.

NOW TRY EXERCISE 41. ◀

▶ **EXAMPLE 4** **VERIFYING AN IDENTITY USING SUM AND DIFFERENCE IDENTITIES**

Verify that the equation is an identity.

$$\sin\left(\frac{\pi}{6} + \theta\right) + \cos\left(\frac{\pi}{3} + \theta\right) = \cos\theta$$

Solution Work on the left side, using the sum identities for $\sin(A + B)$ and $\cos(A + B)$.

$$\sin\left(\frac{\pi}{6} + \theta\right) + \cos\left(\frac{\pi}{3} + \theta\right)$$

$$= \left(\sin\frac{\pi}{6}\cos\theta + \cos\frac{\pi}{6}\sin\theta\right) \qquad \text{Sine sum identity}$$

$$+ \left(\cos\frac{\pi}{3}\cos\theta - \sin\frac{\pi}{3}\sin\theta\right) \qquad \text{Cosine sum identity (Section 5.3)}$$

$$= \left(\frac{1}{2}\cos\theta + \frac{\sqrt{3}}{2}\sin\theta\right) \qquad \sin\frac{\pi}{6} = \frac{1}{2}; \cos\frac{\pi}{6} = \frac{\sqrt{3}}{2}$$

$$+ \left(\frac{1}{2}\cos\theta - \frac{\sqrt{3}}{2}\sin\theta\right) \qquad \cos\frac{\pi}{3} = \frac{1}{2}; \sin\frac{\pi}{3} = \frac{\sqrt{3}}{2}$$

$$= \frac{1}{2}\cos\theta + \frac{1}{2}\cos\theta \qquad \text{Simplify.}$$

$$= \cos\theta$$

NOW TRY EXERCISE 59. ◀

5.4 Exercises

Concept Check Match each expression in Column I with its value in Column II. See Example 1.

I

1. sin 15°

2. sin 105°

3. tan 15°

4. tan 105°

5. sin(−105°)

6. tan(−105°)

II

A. $\dfrac{\sqrt{6} + \sqrt{2}}{4}$

B. $\dfrac{-\sqrt{6} - \sqrt{2}}{4}$

C. $\dfrac{\sqrt{6} - \sqrt{2}}{4}$

D. $2 + \sqrt{3}$

E. $2 - \sqrt{3}$

F. $-2 - \sqrt{3}$

7. Compare the formulas for sin(A − B) and sin(A + B). How do they differ? How are they alike?

8. Compare the formulas for tan(A − B) and tan(A + B). How do they differ? How are they alike?

Use identities to find each exact value. See Example 1.

9. $\sin \dfrac{5\pi}{12}$

10. $\tan \dfrac{5\pi}{12}$

11. $\tan \dfrac{\pi}{12}$

12. $\sin \dfrac{\pi}{12}$

13. $\sin\left(-\dfrac{7\pi}{12}\right)$

14. $\tan\left(-\dfrac{7\pi}{12}\right)$

15. $\sin 76° \cos 31° - \cos 76° \sin 31°$

16. $\sin 40° \cos 50° + \cos 40° \sin 50°$

17. $\dfrac{\tan 80° + \tan 55°}{1 - \tan 80° \tan 55°}$

18. $\dfrac{\tan 80° - \tan(-55°)}{1 + \tan 80° \tan(-55°)}$

19. $\dfrac{\tan 100° + \tan 80°}{1 - \tan 100° \tan 80°}$

20. $\sin 100° \cos 10° - \cos 100° \sin 10°$

21. $\sin \dfrac{\pi}{5} \cos \dfrac{3\pi}{10} + \cos \dfrac{\pi}{5} \sin \dfrac{3\pi}{10}$

22. $\dfrac{\tan \frac{5\pi}{12} + \tan \frac{\pi}{4}}{1 - \tan \frac{5\pi}{12} \tan \frac{\pi}{4}}$

Use identities to write each expression as a single function of x or θ. See Example 2.

23. $\cos(30° + \theta)$

24. $\cos(45° - \theta)$

25. $\cos(60° + \theta)$

26. $\cos(\theta - 30°)$

27. $\cos\left(\dfrac{3\pi}{4} - x\right)$

28. $\sin(45° + \theta)$

29. $\tan(\theta + 30°)$

30. $\tan\left(\dfrac{\pi}{4} + x\right)$

31. $\sin\left(\dfrac{\pi}{4} + x\right)$

32. $\sin(180° - \theta)$

33. $\sin(270° - \theta)$

34. $\tan(180° + \theta)$

35. $\tan(360° - \theta)$

36. $\sin(\pi + \theta)$

37. $\tan(\pi - \theta)$

38. Why is it not possible to use the method of Example 2 to find a formula for tan(270° − θ)?

39. Why is it that standard trigonometry texts usually do not develop formulas for the cotangent, secant, and cosecant of the sum and difference of two numbers or angles?

40. Show that if A, B, and C are the angles of a triangle, then $\sin(A + B + C) = 0$.

Use the given information to find **(a)** $\sin(s + t)$, **(b)** $\tan(s + t)$, *and* **(c)** *the quadrant of* $s + t$. *See Example 3.*

41. $\cos s = \dfrac{3}{5}$ and $\sin t = \dfrac{5}{13}$, s and t in quadrant I

42. $\cos s = -\dfrac{1}{5}$ and $\sin t = \dfrac{3}{5}$, s and t in quadrant II

43. $\sin s = \dfrac{2}{3}$ and $\sin t = -\dfrac{1}{3}$, s in quadrant II and t in quadrant IV

44. $\sin s = \dfrac{3}{5}$ and $\sin t = -\dfrac{12}{13}$, s in quadrant I and t in quadrant III

45. $\cos s = -\dfrac{8}{17}$ and $\cos t = -\dfrac{3}{5}$, s and t in quadrant III

46. $\cos s = -\dfrac{15}{17}$ and $\sin t = \dfrac{4}{5}$, s in quadrant II and t in quadrant I

Find each exact value. Use the technique developed in **Section 5.3,** *Exercises 67–70.*

47. $\sin 165°$ **48.** $\tan 165°$ **49.** $\sin 255°$

50. $\tan 285°$ **51.** $\tan \dfrac{11\pi}{12}$ **52.** $\sin\left(-\dfrac{13\pi}{12}\right)$

Graph each expression and use the graph to make a conjecture, predicting what might be an identity. Then verify your conjecture algebraically.

53. $\sin\left(\dfrac{\pi}{2} + \theta\right)$ **54.** $\sin\left(\dfrac{3\pi}{2} + \theta\right)$ **55.** $\tan\left(\dfrac{\pi}{2} + \theta\right)$ **56.** $\dfrac{1 + \tan\theta}{1 - \tan\theta}$

Verify that each equation is an identity. See Example 4.

57. $\sin 2x = 2\sin x \cos x$ (*Hint:* $\sin 2x = \sin(x + x)$)

58. $\sin(x + y) + \sin(x - y) = 2\sin x \cos y$

59. $\sin(210° + x) - \cos(120° + x) = 0$

60. $\tan(x - y) - \tan(y - x) = \dfrac{2(\tan x - \tan y)}{1 + \tan x \tan y}$

61. $\dfrac{\cos(\alpha - \beta)}{\cos\alpha \sin\beta} = \tan\alpha + \cot\beta$ **62.** $\dfrac{\sin(s + t)}{\cos s \cos t} = \tan s + \tan t$

63. $\dfrac{\sin(x - y)}{\sin(x + y)} = \dfrac{\tan x - \tan y}{\tan x + \tan y}$ **64.** $\dfrac{\sin(x + y)}{\cos(x - y)} = \dfrac{\cot x + \cot y}{1 + \cot x \cot y}$

65. $\dfrac{\sin(s - t)}{\sin t} + \dfrac{\cos(s - t)}{\cos t} = \dfrac{\sin s}{\sin t \cos t}$ **66.** $\dfrac{\tan(\alpha + \beta) - \tan\beta}{1 + \tan(\alpha + \beta)\tan\beta} = \tan\alpha$

RELATING CONCEPTS

For individual or collaborative investigation
(Exercises 67–72)

Refer to the figure on the left below. By the definition of tan θ, *m* = tan θ, *where m is the slope and* θ *is the angle of inclination of the line. The following exercises, which depend on properties of triangles, refer to triangle ABC in the figure on the right below.* **Work Exercises 67–72 in order.** *Assume that all angles are measured in degrees.*

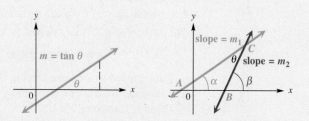

67. In terms of β, what is the measure of angle *ABC*?

68. Use the fact that the sum of the angles in a triangle is 180° to express θ in terms of α and β.

69. Apply the formula for tan(*A* − *B*) to obtain an expression for tan θ in terms of tan α and tan β.

70. Replace tan α with m_1 and tan β with m_2 to obtain $\tan \theta = \dfrac{m_2 - m_1}{1 + m_1 m_2}$.

 In Exercises 71 and 72, use the result from Exercise 70 to find the acute angle between each pair of lines. (Note that the tangent of the angle will be positive.) Use a calculator and round to the nearest tenth of a degree.

71. $x + y = 9$, $2x + y = -1$

72. $5x - 2y + 4 = 0$, $3x + 5y = 6$

(Modeling)   *Solve each problem.*

73. *Back Stress* If a person bends at the waist with a straight back making an angle of θ degrees with the horizontal, then the force *F* exerted on the back muscles can be modeled by the equation

$$F = \frac{.6W \sin(\theta + 90°)}{\sin 12°},$$

where *W* is the weight of the person. (*Source:* Metcalf, H., *Topics in Classical Biophysics,* Prentice-Hall, 1980.)

(a) Calculate *F* when *W* = 170 lb and θ = 30°.

(b) Use an identity to show that *F* is approximately equal to 2.9*W* cos θ.

(c) For what value of θ is *F* maximum?

74. *Back Stress*   Refer to Exercise 73.

   (a) Suppose a 200-lb person bends at the waist so that $\theta = 45°$. Estimate the force exerted on the person's back muscles.

   (b) Approximate graphically the value of θ that results in the back muscles exerting a force of 400 lb.

75. *Voltage*   A coil of wire rotating in a magnetic field induces a voltage

$$E = 20 \sin\left(\frac{\pi t}{4} - \frac{\pi}{2}\right).$$

Use an identity from this section to express this in terms of $\cos \frac{\pi t}{4}$.

76. *Voltage of a Circuit*   When the two voltages

$$V_1 = 30 \sin 120\pi t \qquad \text{and} \qquad V_2 = 40 \cos 120\pi t$$

are applied to the same circuit, the resulting voltage V will be equal to their sum. (*Source:* Bell, D., *Fundamentals of Electric Circuits,* Second Edition, Reston Publishing Company, 1981.)

   (a) Graph the sum in the window $[0, .05]$ by $[-60, 60]$.

   (b) Use the graph to estimate values for a and ϕ so that $V = a \sin(120\pi t + \phi)$.

   (c) Use identities to verify that your expression for V is valid.

CHAPTER 5 ▶   Quiz (Sections 5.1–5.4)

1. If $\sin \theta = -\frac{7}{25}$ and θ is in quadrant IV, find the remaining five trigonometric function values of θ.

2. Express $\cot^2 x + \csc^2 x$ in terms of $\sin x$ and $\cos x$, and simplify.

3. Find the exact value of $\sin\left(-\frac{7\pi}{12}\right)$.

4. Express $\cos(180° - \theta)$ as a function of θ alone.

5. If $\cos A = \frac{3}{5}$, $\sin B = -\frac{5}{13}$, $0 < A < \frac{\pi}{2}$, and $\pi < B < \frac{3\pi}{2}$, find each of the following.

   (a) $\cos(A + B)$ **(b)** $\sin(A + B)$ **(c)** the quadrant of $A + B$

6. Express $\tan\left(\frac{3\pi}{4} + x\right)$ as a function of x alone.

Verify each identity.

7. $\dfrac{1 + \sin \theta}{\cot^2 \theta} = \dfrac{\sin \theta}{\csc \theta - 1}$

8. $\dfrac{\sin^2 \theta - \cos^2 \theta}{\sin^4 \theta - \cos^4 \theta} = 1$

9. $\sin\left(\dfrac{\pi}{3} + \theta\right) - \sin\left(\dfrac{\pi}{3} - \theta\right) = \sin \theta$

10. $\dfrac{\cos(x + y) + \cos(x - y)}{\sin(x - y) + \sin(x + y)} = \cot x$

5.5 Double-Angle Identities

Double-Angle Identities ▪ An Application ▪ Product-to-Sum and Sum-to-Product Identities

Double-Angle Identities   When $A = B$ in the identities for the sum of two angles, these identities are called the **double-angle identities.** For example, to derive an expression for $\cos 2A$, we let $B = A$ in the identity $\cos(A + B) = \cos A \cos B - \sin A \sin B$.

$$\cos 2A = \cos(A + A)$$
$$= \cos A \cos A - \sin A \sin A \quad \text{Cosine sum identity (Section 5.3)}$$
$$\mathbf{\cos 2A = \cos^2 A - \sin^2 A}$$

Two other useful forms of this identity can be obtained by substituting either $\cos^2 A = 1 - \sin^2 A$ or $\sin^2 A = 1 - \cos^2 A$. Replace $\cos^2 A$ with the expression $1 - \sin^2 A$ to get

$$\cos 2A = \cos^2 A - \sin^2 A$$
$$= (1 - \sin^2 A) - \sin^2 A \quad \text{Fundamental identity (Section 5.1)}$$
$$\mathbf{\cos 2A = 1 - 2 \sin^2 A,}$$

or replace $\sin^2 A$ with $1 - \cos^2 A$ to get

$$\cos 2A = \cos^2 A - \sin^2 A$$
$$= \cos^2 A - (1 - \cos^2 A) \quad \text{Fundamental identity}$$
$$= \cos^2 A - 1 + \cos^2 A$$
$$\mathbf{\cos 2A = 2 \cos^2 A - 1.}$$

We find $\sin 2A$ with the identity $\sin(A + B) = \sin A \cos B + \cos A \sin B$, letting $B = A$.

$$\sin 2A = \sin(A + A)$$
$$= \sin A \cos A + \cos A \sin A \quad \text{Sine sum identity (Section 5.4)}$$
$$\mathbf{\sin 2A = 2 \sin A \cos A}$$

Using the identity for $\tan(A + B)$, we find $\tan 2A$.

$$\tan 2A = \tan(A + A)$$
$$= \frac{\tan A + \tan A}{1 - \tan A \tan A} \quad \text{Tangent sum identity (Section 5.4)}$$
$$\mathbf{\tan 2A = \frac{2 \tan A}{1 - \tan^2 A}}$$

▼ **LOOKING AHEAD TO CALCULUS**

The identities

$$\cos 2A = 1 - 2 \sin^2 A$$

and $\cos 2A = 2 \cos^2 A - 1$

can be rewritten as

$$\sin^2 A = \frac{1}{2} (1 - \cos 2A)$$

and $\cos^2 A = \frac{1}{2} (1 + \cos 2A).$

These identities are used to integrate the functions $f(A) = \sin^2 A$ and $g(A) = \cos^2 A.$

DOUBLE-ANGLE IDENTITIES

$$\cos 2A = \cos^2 A - \sin^2 A \qquad \cos 2A = 1 - 2 \sin^2 A$$
$$\cos 2A = 2 \cos^2 A - 1 \qquad \sin 2A = 2 \sin A \cos A$$
$$\tan 2A = \frac{2 \tan A}{1 - \tan^2 A}$$

▶ **EXAMPLE 1** FINDING FUNCTION VALUES OF 2θ GIVEN INFORMATION ABOUT θ

Given $\cos \theta = \frac{3}{5}$ and $\sin \theta < 0$, find $\sin 2\theta$, $\cos 2\theta$, and $\tan 2\theta$.

Solution To find $\sin 2\theta$, we must first find the value of $\sin \theta$.

$$\sin^2 \theta + \left(\frac{3}{5}\right)^2 = 1 \qquad \sin^2 \theta + \cos^2 \theta = 1; \cos \theta = \frac{3}{5}$$

$$\sin^2 \theta = \frac{16}{25} \qquad \text{Simplify.}$$

Pay attention to signs here. ⟶ $\sin \theta = -\frac{4}{5}$ Take square roots (**Appendix A**); choose the negative square root since $\sin \theta < 0$.

Using the double-angle identity for sine,

$$\sin 2\theta = 2 \sin \theta \cos \theta = 2\left(-\frac{4}{5}\right)\left(\frac{3}{5}\right) = -\frac{24}{25}. \quad \sin \theta = -\frac{4}{5}; \cos \theta = \frac{3}{5}$$

Now we find $\cos 2\theta$, using the first of the double-angle identities for cosine.

Any of the three forms may be used. ⟶ $\cos 2\theta = \cos^2 \theta - \sin^2 \theta = \frac{9}{25} - \frac{16}{25} = -\frac{7}{25}$

The value of $\tan 2\theta$ can be found in either of two ways. We can use the double-angle identity and the fact that $\tan \theta = \frac{\sin \theta}{\cos \theta} = \frac{-\frac{4}{5}}{\frac{3}{5}} = -\frac{4}{5} \div \frac{3}{5} = -\frac{4}{5} \cdot \frac{5}{3} = -\frac{4}{3}$.

$$\tan 2\theta = \frac{2 \tan \theta}{1 - \tan^2 \theta} = \frac{2\left(-\frac{4}{3}\right)}{1 - \left(-\frac{4}{3}\right)^2} = \frac{-\frac{8}{3}}{-\frac{7}{9}} = \frac{24}{7}$$

Alternatively, we can find $\tan 2\theta$ by finding the quotient of $\sin 2\theta$ and $\cos 2\theta$.

$$\tan 2\theta = \frac{\sin 2\theta}{\cos 2\theta} = \frac{-\frac{24}{25}}{-\frac{7}{25}} = \frac{24}{7}$$

NOW TRY EXERCISE 15. ◀

▶ **EXAMPLE 2** FINDING FUNCTION VALUES OF θ GIVEN INFORMATION ABOUT 2θ

Find the values of the six trigonometric functions of θ if $\cos 2\theta = \frac{4}{5}$ and $90° < \theta < 180°$.

Solution We must obtain a trigonometric function value of θ alone.

$$\cos 2\theta = 1 - 2 \sin^2 \theta \qquad \text{Double-angle identity}$$

$$\frac{4}{5} = 1 - 2 \sin^2 \theta \qquad \cos 2\theta = \frac{4}{5}$$

$$-\frac{1}{5} = -2 \sin^2 \theta \qquad \text{Subtract 1.}$$

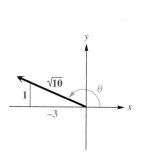

Figure 7

$$\frac{1}{10} = \sin^2 \theta \qquad \text{Multiply by } -\tfrac{1}{2}.$$

$$\sin \theta = \sqrt{\frac{1}{10}} \qquad \begin{array}{l}\text{Take square roots; choose the positive square} \\ \text{root since } \theta \text{ terminates in quadrant II.}\end{array}$$

$$\sin \theta = \frac{\sqrt{1}}{\sqrt{10}} \cdot \frac{\sqrt{10}}{\sqrt{10}} = \frac{\sqrt{10}}{10} \qquad \text{Quotient rule; rationalize the denominator.}$$

Now find values of $\cos \theta$ and $\tan \theta$ by sketching and labeling a right triangle in quadrant II. Since $\sin \theta = \frac{1}{\sqrt{10}}$, the triangle in Figure 7 is labeled accordingly. The Pythagorean theorem is used to find the remaining leg. Now,

$$\cos \theta = \frac{-3}{\sqrt{10}} = -\frac{3\sqrt{10}}{10}, \qquad \text{and} \qquad \tan \theta = \frac{1}{-3} = -\frac{1}{3}. \quad \text{(Section 1.3)}$$

Find the other three functions using reciprocals.

$$\csc \theta = \frac{1}{\sin \theta} = \sqrt{10}, \qquad \sec \theta = \frac{1}{\cos \theta} = -\frac{\sqrt{10}}{3}, \qquad \cot \theta = \frac{1}{\tan \theta} = -3$$

NOW TRY EXERCISE 9. ◀

▶ **EXAMPLE 3** VERIFYING A DOUBLE-ANGLE IDENTITY

Verify that the following equation is an identity.

$$\cot x \sin 2x = 1 + \cos 2x$$

Solution We start by working on the left side, using the hint from **Section 5.2** about writing all functions in terms of sine and cosine.

$$\cot x \sin 2x = \frac{\cos x}{\sin x} \cdot \sin 2x \qquad \text{Quotient identity}$$

$$= \frac{\cos x}{\sin x} (2 \sin x \cos x) \qquad \text{Double-angle identity}$$

> Be able to recognize alternative forms of identities.

$$= 2 \cos^2 x$$

$$= 1 + \cos 2x \qquad \begin{array}{l}\cos 2x = 2\cos^2 x - 1, \text{ so} \\ 2\cos^2 x = 1 + \cos 2x\end{array}$$

NOW TRY EXERCISE 37. ◀

▶ **EXAMPLE 4** SIMPLIFYING EXPRESSIONS USING DOUBLE-ANGLE IDENTITIES

Simplify each expression.

(a) $\cos^2 7x - \sin^2 7x$ **(b)** $\sin 15° \cos 15°$

Solution

(a) This expression suggests one of the double-angle identities for cosine: $\cos 2A = \cos^2 A - \sin^2 A$. Substituting $7x$ for A gives

$$\cos^2 7x - \sin^2 7x = \cos 2(7x) = \cos 14x.$$

(b) If the expression sin 15° cos 15° were 2 sin 15° cos 15°, we could apply the identity for sin 2A directly since

$$\sin 2A = 2 \sin A \cos A.$$

We can still apply the identity with $A = 15°$ by writing the multiplicative identity element 1 as $\frac{1}{2}(2)$.

$$\sin 15° \cos 15° = \frac{1}{2}(2) \sin 15° \cos 15° \qquad \text{Multiply by 1 in the form } \tfrac{1}{2}(2).$$

> This is not an obvious way to begin, but it is indeed valid.

$$= \frac{1}{2}(2 \sin 15° \cos 15°) \qquad \text{Associative property}$$

$$= \frac{1}{2} \sin(2 \cdot 15°) \qquad 2 \sin A \cos A = \sin 2A, \text{ with } A = 15°$$

$$= \frac{1}{2} \sin 30° \qquad \text{Multiply.}$$

$$= \frac{1}{2} \cdot \frac{1}{2} \qquad \sin 30° = \tfrac{1}{2} \text{ (Section 2.1)}$$

$$= \frac{1}{4} \qquad \text{Multiply.}$$

NOW TRY EXERCISES 17 AND 19. ◀

Identities involving larger multiples of the variable can be derived by repeated use of the double-angle identities and other identities.

▶ **EXAMPLE 5** DERIVING A MULTIPLE-ANGLE IDENTITY

Write sin 3x in terms of sin x.

Solution

$$\sin 3x = \sin(2x + x) \qquad \boxed{\text{Use the simple fact that } 3 = 2 + 1 \text{ here.}}$$

$$= \sin 2x \cos x + \cos 2x \sin x \qquad \text{Sine sum identity (Section 5.4)}$$

$$= (2 \sin x \cos x)\cos x + (\cos^2 x - \sin^2 x)\sin x \qquad \text{Double-angle identities}$$

$$= 2 \sin x \cos^2 x + \cos^2 x \sin x - \sin^3 x \qquad \text{Multiply.}$$

$$= 2 \sin x(1 - \sin^2 x) + (1 - \sin^2 x)\sin x - \sin^3 x \qquad \cos^2 x = 1 - \sin^2 x$$

$$= 2 \sin x - 2 \sin^3 x + \sin x - \sin^3 x - \sin^3 x \qquad \text{Distributive property}$$

$$= 3 \sin x - 4 \sin^3 x \qquad \text{Combine terms.}$$

NOW TRY EXERCISE 29. ◀

An Application

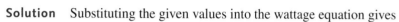

▶ **EXAMPLE 6** DETERMINING WATTAGE CONSUMPTION

If a toaster is plugged into a common household outlet, the wattage consumed is not constant. Instead, it varies at a high frequency according to the model

$$W = \frac{V^2}{R},$$

where V is the voltage and R is a constant that measures the resistance of the toaster in ohms. (*Source:* Bell, D., *Fundamentals of Electric Circuits,* Fourth Edition, Prentice-Hall, 1998.) Graph the wattage W consumed by a typical toaster with $R = 15$ and $V = 163 \sin 120\pi t$ in the window $[0, .05]$ by $[-500, 2000]$. How many oscillations are there?

Solution Substituting the given values into the wattage equation gives

$$W = \frac{V^2}{R} = \frac{(163 \sin 120\pi t)^2}{15}.$$

To determine the range of W, we note that $\sin 120\pi t$ has maximum value 1, so the expression for W has maximum value $\frac{163^2}{15} \approx 1771$. The minimum value is 0. The graph in Figure 8 shows that there are six oscillations.

For $x = t$,

$$W(t) = \frac{(163 \sin 120\pi t)^2}{15}$$

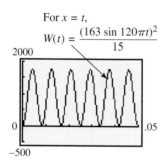

Figure 8

NOW TRY EXERCISE 67. ◀

Product-to-Sum and Sum-to-Product Identities

The identities for $\cos(A + B)$ and $\cos(A - B)$ can be added to derive an identity useful in calculus.

$$\cos(A + B) = \cos A \cos B - \sin A \sin B$$
$$\underline{\cos(A - B) = \cos A \cos B + \sin A \sin B}$$
$$\cos(A + B) + \cos(A - B) = 2 \cos A \cos B \qquad \text{Add.}$$

or

$$\cos A \cos B = \frac{1}{2}[\cos(A + B) + \cos(A - B)]$$

Similarly, subtracting $\cos(A + B)$ from $\cos(A - B)$ gives

$$\sin A \sin B = \frac{1}{2}[\cos(A - B) - \cos(A + B)].$$

Using the identities for $\sin(A + B)$ and $\sin(A - B)$ in the same way, we get two more identities. Those and the previous ones are now summarized.

▼ **LOOKING AHEAD TO CALCULUS**
The product-to-sum identities are used in calculus to find **integrals** of functions that are products of trigonometric functions. One classic calculus text includes the following example:

Evaluate $\int \cos 5x \cos 3x \, dx$.

The first solution line reads:
"We may write

$$\cos 5x \cos 3x = \frac{1}{2}[\cos 8x + \cos 2x]."$$

PRODUCT-TO-SUM IDENTITIES

$$\cos A \cos B = \frac{1}{2}[\cos(A + B) + \cos(A - B)]$$

$$\sin A \sin B = \frac{1}{2}[\cos(A - B) - \cos(A + B)]$$

$$\sin A \cos B = \frac{1}{2}[\sin(A + B) + \sin(A - B)]$$

$$\cos A \sin B = \frac{1}{2}[\sin(A + B) - \sin(A - B)]$$

▶ **EXAMPLE 7** **USING A PRODUCT-TO-SUM IDENTITY**

Write $4 \cos 75° \sin 25°$ as the sum or difference of two functions.

Solution Use the identity for $\cos A \sin B$, with $A = 75°$ and $B = 25°$.

$$4 \cos 75° \sin 25° = 4 \left[\frac{1}{2} \left(\sin (75° + 25°) - \sin (75° - 25°) \right) \right]$$

$$= 2 \sin 100° - 2 \sin 50°$$

NOW TRY EXERCISE 55. ◀

Another group of identities allows us to write sums as products.

SUM-TO-PRODUCT IDENTITIES

$$\sin A + \sin B = 2 \sin\left(\frac{A + B}{2}\right) \cos\left(\frac{A - B}{2}\right)$$

$$\sin A - \sin B = 2 \cos\left(\frac{A + B}{2}\right) \sin\left(\frac{A - B}{2}\right)$$

$$\cos A + \cos B = 2 \cos\left(\frac{A + B}{2}\right) \cos\left(\frac{A - B}{2}\right)$$

$$\cos A - \cos B = -2 \sin\left(\frac{A + B}{2}\right) \sin\left(\frac{A - B}{2}\right)$$

▶ **EXAMPLE 8** **USING A SUM-TO-PRODUCT IDENTITY**

Write $\sin 2\theta - \sin 4\theta$ as a product of two functions.

Solution Use the identity for $\sin A - \sin B$, with $2\theta = A$ and $4\theta = B$.

$$\sin 2\theta - \sin 4\theta = 2 \cos\left(\frac{2\theta + 4\theta}{2}\right) \sin\left(\frac{2\theta - 4\theta}{2}\right)$$

$$= 2 \cos \frac{6\theta}{2} \sin\left(\frac{-2\theta}{2}\right)$$

$$= 2 \cos 3\theta \sin(-\theta)$$

$$= -2 \cos 3\theta \sin \theta \quad \sin(-\theta) = -\sin \theta \text{ (Section 5.1)}$$

NOW TRY EXERCISE 61. ◀

5.5 Exercises

Concept Check *Match each expression in Column I with its value in Column II.*

I   **II**

1. $2 \cos^2 15° - 1$ **A.** $\dfrac{1}{2}$

2. $\dfrac{2 \tan 15°}{1 - \tan^2 15°}$ **B.** $\dfrac{\sqrt{2}}{2}$

3. $2 \sin 22.5° \cos 22.5°$ **C.** $\dfrac{\sqrt{3}}{2}$

4. $\cos^2 \dfrac{\pi}{6} - \sin^2 \dfrac{\pi}{6}$ **D.** $-\sqrt{3}$

5. $2 \sin \dfrac{\pi}{3} \cos \dfrac{\pi}{3}$ **E.** $\dfrac{\sqrt{3}}{3}$

6. $\dfrac{2 \tan \frac{\pi}{3}}{1 - \tan^2 \frac{\pi}{3}}$

Use identities to find values of the sine and cosine functions for each angle measure. See Examples 1 and 2.

7. θ, given $\cos 2\theta = \dfrac{3}{5}$ and θ terminates in quadrant I

8. θ, given $\cos 2\theta = \dfrac{3}{4}$ and θ terminates in quadrant III

9. θ, given $\cos 2\theta = -\dfrac{5}{12}$ and $\dfrac{\pi}{2} < \theta < \pi$

10. x, given $\cos 2x = \dfrac{2}{3}$ and $\dfrac{\pi}{2} < x < \pi$

11. 2θ, given $\sin \theta = \dfrac{2}{5}$ and $\cos \theta < 0$

12. 2θ, given $\cos \theta = -\dfrac{12}{13}$ and $\sin \theta > 0$

13. $2x$, given $\tan x = 2$ and $\cos x > 0$

14. $2x$, given $\tan x = \dfrac{5}{3}$ and $\sin x < 0$

15. 2θ, given $\sin \theta = -\dfrac{\sqrt{5}}{7}$ and $\cos \theta > 0$

16. 2θ, given $\cos \theta = \dfrac{\sqrt{3}}{5}$ and $\sin \theta > 0$

Use an identity to write each expression as a single trigonometric function value or as a single number. See Example 4.

17. $\cos^2 15° - \sin^2 15°$ **18.** $\dfrac{2 \tan 15°}{1 - \tan^2 15°}$ **19.** $1 - 2 \sin^2 15°$

20. $1 - 2 \sin^2 22\dfrac{1}{2}°$ **21.** $2 \cos^2 67\dfrac{1}{2}° - 1$ **22.** $\cos^2 \dfrac{\pi}{8} - \dfrac{1}{2}$

23. $\dfrac{\tan 51°}{1 - \tan^2 51°}$ **24.** $\dfrac{\tan 34°}{2(1 - \tan^2 34°)}$ **25.** $\dfrac{1}{4} - \dfrac{1}{2} \sin^2 47.1°$

26. $\dfrac{1}{8} \sin 29.5° \cos 29.5°$ **27.** $\sin^2 \dfrac{2\pi}{5} - \cos^2 \dfrac{2\pi}{5}$ **28.** $\cos^2 2x - \sin^2 2x$

Express each function as a trigonometric function of x. See Example 5.

29. $\sin 4x$ **30.** $\cos 3x$ **31.** $\tan 3x$ **32.** $\cos 4x$

Graph each expression and use the graph to make a conjecture as to what might be an identity. Then verify your conjecture algebraically.

33. $\cos^4 x - \sin^4 x$ **34.** $\dfrac{4 \tan x \cos^2 x - 2 \tan x}{1 - \tan^2 x}$

35. $\dfrac{2 \tan x}{2 - \sec^2 x}$ **36.** $\dfrac{\cot^2 x - 1}{2 \cot x}$

Verify that each equation is an identity. See Example 3.

37. $(\sin x + \cos x)^2 = \sin 2x + 1$ **38.** $\sec 2x = \dfrac{\sec^2 x + \sec^4 x}{2 + \sec^2 x - \sec^4 x}$

39. $\tan 8\theta - \tan 8\theta \tan^2 4\theta = 2 \tan 4\theta$ **40.** $\sin 2x = \dfrac{2 \tan x}{1 + \tan^2 x}$

41. $\cos 2\theta = \dfrac{2 - \sec^2 \theta}{\sec^2 \theta}$ **42.** $-\tan 2\theta = \dfrac{2 \tan \theta}{\sec^2 \theta - 2}$

43. $\sin 4x = 4 \sin x \cos x \cos 2x$ **44.** $\dfrac{1 + \cos 2x}{\sin 2x} = \cot x$

45. $\dfrac{2 \cos 2\theta}{\sin 2\theta} = \cot \theta - \tan \theta$ **46.** $\cot 4\theta = \dfrac{1 - \tan^2 2\theta}{2 \tan 2\theta}$

47. $\tan x + \cot x = 2 \csc 2x$ **48.** $\cos 2x = \dfrac{1 - \tan^2 x}{1 + \tan^2 x}$

49. $1 + \tan x \tan 2x = \sec 2x$ **50.** $\dfrac{\cot A - \tan A}{\cot A + \tan A} = \cos 2A$

51. $\sin 2A \cos 2A = \sin 2A - 4 \sin^3 A \cos A$

52. $\sin 4x = 4 \sin x \cos x - 8 \sin^3 x \cos x$

53. $\tan(\theta - 45°) + \tan(\theta + 45°) = 2 \tan 2\theta$

54. $\cot \theta \tan(\theta + \pi) - \sin(\pi - \theta) \cos\left(\dfrac{\pi}{2} - \theta\right) = \cos^2 \theta$

Write each expression as a sum or difference of trigonometric functions. See Example 7.

55. $2 \sin 58° \cos 102°$ **56.** $2 \cos 85° \sin 140°$ **57.** $2 \sin \dfrac{\pi}{6} \cos \dfrac{\pi}{3}$

58. $5 \cos 3x \cos 2x$ **59.** $6 \sin 4x \sin 5x$ **60.** $8 \sin 7x \sin 9x$

Write each expression as a product of trigonometric functions. See Example 8.

61. $\cos 4x - \cos 2x$ **62.** $\cos 5x + \cos 8x$ **63.** $\sin 25° + \sin(-48°)$

64. $\sin 102° - \sin 95°$ **65.** $\cos 4x + \cos 8x$ **66.** $\sin 9x - \sin 3x$

⊞ (*Modeling*) *Solve each problem.*

67. *Wattage Consumption* Refer to Example 6. Use an identity to determine values of *a, c,* and *ω* so that $W = a\cos(\omega t) + c$. Check your answer by graphing both expressions for *W* on the same coordinate axes.

68. *Amperage, Wattage, and Voltage* Amperage is a measure of the amount of electricity that is moving through a circuit, whereas voltage is a measure of the force pushing the electricity. The wattage *W* consumed by an electrical device can be determined by calculating the product of the amperage *I* and voltage *V*. (*Source:* Wilcox, G. and C. Hesselberth, *Electricity for Engineering Technology,* Allyn & Bacon, 1970.)

(a) A household circuit has voltage

$$V = 163\sin 120\pi t$$

when an incandescent light bulb is turned on with amperage

$$I = 1.23\sin 120\pi t.$$

Graph the wattage $W = VI$ consumed by the light bulb in the window $[0, .05]$ by $[-50, 300]$.

(b) Determine the maximum and minimum wattages used by the light bulb.

(c) Use identities to determine values for *a, c,* and *ω* so that $W = a\cos(\omega t) + c$.

(d) Check your answer by graphing both expressions for *W* on the same coordinate axes.

(e) Use the graph to estimate the average wattage used by the light. For how many watts (to the nearest integer) do you think this incandescent light bulb is rated?

5.6 Half-Angle Identities

Half-Angle Identities ▪ **Applying the Half-Angle Identities**

Half-Angle Identities From the alternative forms of the identity for cos 2*A*, we derive three additional identities for $\sin\frac{A}{2}$, $\cos\frac{A}{2}$, and $\tan\frac{A}{2}$. These are known as **half-angle identities.**

To derive the identity for $\sin\frac{A}{2}$, start with the following double-angle identity for cosine and solve for sin *x.*

$$\cos 2x = 1 - 2\sin^2 x$$

$$2\sin^2 x = 1 - \cos 2x \qquad \text{Add } 2\sin^2 x; \text{ subtract } \cos 2x.$$

> Remember both the positive and negative square roots.

$$\sin x = \pm\sqrt{\frac{1 - \cos 2x}{2}} \qquad \text{Divide by 2; take square roots. (\textbf{Appendix A})}$$

$$\sin\frac{A}{2} = \pm\sqrt{\frac{1 - \cos A}{2}} \qquad \text{Let } 2x = A, \text{ so } x = \tfrac{A}{2}; \text{ substitute.}$$

The ± sign in this identity indicates that the appropriate sign is chosen depending on the quadrant of $\frac{A}{2}$. For example, if $\frac{A}{2}$ is a quadrant III angle, we choose the negative sign since the sine function is negative in quadrant III.

We derive the identity for $\cos \frac{A}{2}$ using the double-angle identity $\cos 2x = 2 \cos^2 x - 1$.

$$1 + \cos 2x = 2 \cos^2 x \qquad \text{Add 1.}$$

$$\cos^2 x = \frac{1 + \cos 2x}{2} \qquad \text{Rewrite; divide by 2.}$$

$$\cos x = \pm \sqrt{\frac{1 + \cos 2x}{2}} \qquad \text{Take square roots.}$$

$$\cos \frac{A}{2} = \pm \sqrt{\frac{1 + \cos A}{2}} \qquad \text{Replace } x \text{ with } \frac{A}{2}.$$

An identity for $\tan \frac{A}{2}$ comes from the identities for $\sin \frac{A}{2}$ and $\cos \frac{A}{2}$.

$$\tan \frac{A}{2} = \frac{\sin \frac{A}{2}}{\cos \frac{A}{2}} = \frac{\pm \sqrt{\dfrac{1 - \cos A}{2}}}{\pm \sqrt{\dfrac{1 + \cos A}{2}}} = \pm \sqrt{\frac{1 - \cos A}{1 + \cos A}}$$

We derive an alternative identity for $\tan \frac{A}{2}$ using double-angle identities.

$$\tan \frac{A}{2} = \frac{\sin \frac{A}{2}}{\cos \frac{A}{2}} = \frac{2 \sin \frac{A}{2} \cos \frac{A}{2}}{2 \cos^2 \frac{A}{2}} \qquad \begin{array}{l}\text{Multiply by } 2 \cos \frac{A}{2} \text{ in numerator}\\ \text{and denominator.}\end{array}$$

$$= \frac{\sin 2\left(\frac{A}{2}\right)}{1 + \cos 2\left(\frac{A}{2}\right)} \qquad \begin{array}{l}\text{Double-angle identities}\\ \text{(Section 5.5)}\end{array}$$

$$\tan \frac{A}{2} = \frac{\sin A}{1 + \cos A} \qquad \text{Simplify.}$$

From the identity $\tan \dfrac{A}{2} = \dfrac{\sin A}{1 + \cos A}$, we can also derive $\tan \dfrac{A}{2} = \dfrac{1 - \cos A}{\sin A}$.

HALF-ANGLE IDENTITIES

$$\cos \frac{A}{2} = \pm \sqrt{\frac{1 + \cos A}{2}} \qquad \sin \frac{A}{2} = \pm \sqrt{\frac{1 - \cos A}{2}}$$

$$\tan \frac{A}{2} = \pm \sqrt{\frac{1 - \cos A}{1 + \cos A}} \qquad \tan \frac{A}{2} = \frac{\sin A}{1 + \cos A} \qquad \tan \frac{A}{2} = \frac{1 - \cos A}{\sin A}$$

▶ **Note** The final two identities for $\tan \frac{A}{2}$ do not require a sign choice. When using the other half-angle identities, select the plus or minus sign according to the quadrant in which $\frac{A}{2}$ terminates. For example, if an angle $A = 324°$, then $\frac{A}{2} = 162°$, which lies in quadrant II. So when $A = 324°$, $\cos \frac{A}{2}$ and $\tan \frac{A}{2}$ are negative, while $\sin \frac{A}{2}$ is positive.

Applying the Half-Angle Identities

▶ **EXAMPLE 1** USING A HALF-ANGLE IDENTITY TO FIND AN EXACT VALUE

Find the exact value of cos 15° using the half-angle identity for cosine.

Solution

$$\cos 15° = \cos \frac{1}{2}(30°) = \sqrt{\frac{1 + \cos 30°}{2}}$$

Choose the positive square root.

$$= \sqrt{\frac{1 + \frac{\sqrt{3}}{2}}{2}} = \sqrt{\frac{\left(1 + \frac{\sqrt{3}}{2}\right) \cdot 2}{2 \cdot 2}} = \frac{\sqrt{2 + \sqrt{3}}}{2}$$

Simplify the radicals.

NOW TRY EXERCISE 11. ◀

▶ **EXAMPLE 2** USING A HALF-ANGLE IDENTITY TO FIND AN EXACT VALUE

Find the exact value of tan 22.5° using the identity $\tan \frac{A}{2} = \frac{\sin A}{1 + \cos A}$.

Solution Since $22.5° = \frac{1}{2}(45°)$, replace A with 45°.

$$\tan 22.5° = \tan \frac{45°}{2} = \frac{\sin 45°}{1 + \cos 45°} = \frac{\frac{\sqrt{2}}{2}}{1 + \frac{\sqrt{2}}{2}} = \frac{\frac{\sqrt{2}}{2}}{1 + \frac{\sqrt{2}}{2}} \cdot \frac{2}{2}$$

$$= \frac{\sqrt{2}}{2 + \sqrt{2}} = \frac{\sqrt{2}}{2 + \sqrt{2}} \cdot \frac{2 - \sqrt{2}}{2 - \sqrt{2}} = \frac{2\sqrt{2} - 2}{2}$$

Rationalize the denominator.

$$= \frac{2(\sqrt{2} - 1)}{2} = \sqrt{2} - 1$$

Factor out 2.

Factor first; then divide out the common factor.

NOW TRY EXERCISE 13. ◀

▶ **EXAMPLE 3** FINDING FUNCTION VALUES OF $\frac{s}{2}$ GIVEN INFORMATION ABOUT s

Given $\cos s = \frac{2}{3}$, with $\frac{3\pi}{2} < s < 2\pi$, find $\cos \frac{s}{2}$, $\sin \frac{s}{2}$, and $\tan \frac{s}{2}$.

Solution The angle associated with $\frac{s}{2}$ terminates in quadrant II, since

$$\frac{3\pi}{2} < s < 2\pi$$

and $\qquad \frac{3\pi}{4} < \frac{s}{2} < \pi.$ Divide by 2. **(Appendix A)**

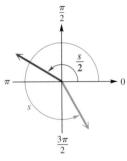

Figure 9

See Figure 9. In quadrant II, the values of $\cos \frac{s}{2}$ and $\tan \frac{s}{2}$ are negative and the value of $\sin \frac{s}{2}$ is positive. Now use the appropriate half-angle identities and simplify the radicals.

$$\sin \frac{s}{2} = \sqrt{\frac{1 - \frac{2}{3}}{2}} = \sqrt{\frac{1}{6}} = \frac{\sqrt{1}}{\sqrt{6}} \cdot \frac{\sqrt{6}}{\sqrt{6}} = \frac{\sqrt{6}}{6}$$

$$\cos \frac{s}{2} = -\sqrt{\frac{1 + \frac{2}{3}}{2}} = -\sqrt{\frac{5}{6}} = -\frac{\sqrt{5}}{\sqrt{6}} \cdot \frac{\sqrt{6}}{\sqrt{6}} = -\frac{\sqrt{30}}{6}$$

$$\tan \frac{s}{2} = \frac{\sin \frac{s}{2}}{\cos \frac{s}{2}} = \frac{\frac{\sqrt{6}}{6}}{-\frac{\sqrt{30}}{6}} = \frac{\sqrt{6}}{-\sqrt{30}} = -\frac{\sqrt{6}}{\sqrt{30}} \cdot \frac{\sqrt{30}}{\sqrt{30}} = -\frac{\sqrt{180}}{30} = -\frac{6\sqrt{5}}{6 \cdot 5} = -\frac{\sqrt{5}}{5}$$

Notice that it is not necessary to use a half-angle identity for $\tan \frac{s}{2}$ once we find $\sin \frac{s}{2}$ and $\cos \frac{s}{2}$. However, using this identity would provide an excellent check.

NOW TRY EXERCISE 19. ◀

▶ **EXAMPLE 4**   SIMPLIFYING EXPRESSIONS USING THE HALF-ANGLE IDENTITIES

Simplify each expression.

(a) $\pm\sqrt{\dfrac{1 + \cos 12x}{2}}$

(b) $\dfrac{1 - \cos 5\alpha}{\sin 5\alpha}$

Solution

(a) This matches part of the identity for $\cos \frac{A}{2}$. Replace A with $12x$ to get

$$\cos \frac{A}{2} = \pm\sqrt{\frac{1 + \cos A}{2}} = \pm\sqrt{\frac{1 + \cos 12x}{2}} = \cos \frac{12x}{2} = \cos 6x.$$

(b) Use the third identity for $\tan \frac{A}{2}$ given earlier with $A = 5\alpha$ to get

$$\frac{1 - \cos 5\alpha}{\sin 5\alpha} = \tan \frac{5\alpha}{2}.$$

NOW TRY EXERCISES 37 AND 39. ◀

▶ **EXAMPLE 5**   VERIFYING AN IDENTITY

Verify that the following equation is an identity.

$$\left(\sin \frac{x}{2} + \cos \frac{x}{2}\right)^2 = 1 + \sin x$$

Solution We work on the more complicated left side.

$$\left(\sin\frac{x}{2} + \cos\frac{x}{2}\right)^2$$

> Remember the term 2*ab* when squaring a binomial.

$$= \sin^2\frac{x}{2} + 2\sin\frac{x}{2}\cos\frac{x}{2} + \cos^2\frac{x}{2} \qquad (a+b)^2 = a^2 + 2ab + b^2$$

$$= 1 + 2\sin\frac{x}{2}\cos\frac{x}{2} \qquad \sin^2\frac{x}{2} + \cos^2\frac{x}{2} = 1$$

$$= 1 + \sin 2\left(\frac{x}{2}\right) \qquad 2\sin\frac{x}{2}\cos\frac{x}{2} = \sin 2\left(\frac{x}{2}\right)$$

$$= 1 + \sin x \qquad \text{Multiply.}$$

NOW TRY EXERCISE 51. ◀

5.6 Exercises

Concept Check Determine whether the positive or negative square root should be selected.

1. $\sin 195° = \pm\sqrt{\dfrac{1 - \cos 390°}{2}}$

2. $\cos 58° = \pm\sqrt{\dfrac{1 + \cos 116°}{2}}$

3. $\tan 225° = \pm\sqrt{\dfrac{1 - \cos 450°}{1 + \cos 450°}}$

4. $\sin(-10°) = \pm\sqrt{\dfrac{1 - \cos(-20°)}{2}}$

Match each expression in Column I with its value in Column II. See Examples 1 and 2.

I

5. $\sin 15°$ **6.** $\tan 15°$

7. $\cos\dfrac{\pi}{8}$ **8.** $\tan\left(-\dfrac{\pi}{8}\right)$

9. $\tan 67.5°$ **10.** $\cos 67.5°$

II

A. $2 - \sqrt{3}$ **B.** $\dfrac{\sqrt{2 - \sqrt{2}}}{2}$

C. $\dfrac{\sqrt{2 - \sqrt{3}}}{2}$ **D.** $\dfrac{\sqrt{2 + \sqrt{2}}}{2}$

E. $1 - \sqrt{2}$ **F.** $1 + \sqrt{2}$

Use a half-angle identity to find each exact value. See Examples 1 and 2.

11. $\sin 67.5°$ **12.** $\sin 195°$ **13.** $\cos 195°$

14. $\tan 195°$ **15.** $\cos 165°$ **16.** $\sin 165°$

17. Explain how you could use an identity of this section to find the exact value of $\sin 7.5°$. (*Hint:* $7.5 = \frac{1}{2}(\frac{1}{2})(30)$.)

18. The identity $\tan\frac{A}{2} = \pm\sqrt{\frac{1 - \cos A}{1 + \cos A}}$ can be used to find $\tan 22.5° = \sqrt{3 - 2\sqrt{2}}$, and the identity $\tan\frac{A}{2} = \frac{\sin A}{1 + \cos A}$ can be used to find $\tan 22.5° = \sqrt{2} - 1$. Show that these answers are the same, without using a calculator. (*Hint:* If $a > 0$ and $b > 0$ and $a^2 = b^2$, then $a = b$.)

Find each of the following. See Example 3.

19. $\cos\dfrac{x}{2}$, given $\cos x = \dfrac{1}{4}$, with $0 < x < \dfrac{\pi}{2}$

20. $\sin \dfrac{x}{2}$, given $\cos x = -\dfrac{5}{8}$, with $\dfrac{\pi}{2} < x < \pi$

21. $\tan \dfrac{\theta}{2}$, given $\sin \theta = \dfrac{3}{5}$, with $90° < \theta < 180°$

22. $\cos \dfrac{\theta}{2}$, given $\sin \theta = -\dfrac{4}{5}$, with $180° < \theta < 270°$

23. $\sin \dfrac{x}{2}$, given $\tan x = 2$, with $0 < x < \dfrac{\pi}{2}$

24. $\cos \dfrac{x}{2}$, given $\cot x = -3$, with $\dfrac{\pi}{2} < x < \pi$

25. $\tan \dfrac{\theta}{2}$, given $\tan \theta = \dfrac{\sqrt{7}}{3}$, with $180° < \theta < 270°$

26. $\cot \dfrac{\theta}{2}$, given $\tan \theta = -\dfrac{\sqrt{5}}{2}$, with $90° < \theta < 180°$

27. $\sin \theta$, given $\cos 2\theta = \dfrac{3}{5}$ and θ terminates in quadrant I

28. $\cos \theta$, given $\cos 2\theta = \dfrac{1}{2}$ and θ terminates in quadrant II

29. $\cos x$, given $\cos 2x = -\dfrac{5}{12}$, with $\dfrac{\pi}{2} < x < \pi$

30. $\sin x$, given $\cos 2x = \dfrac{2}{3}$, with $\pi < x < \dfrac{3\pi}{2}$

31. *Concept Check*   If $\cos x \approx .9682$ and $\sin x = .25$, then $\tan \frac{x}{2} \approx$ _____.

32. *Concept Check*   If $\cos x = -.75$ and $\sin x \approx .6614$, then $\tan \frac{x}{2} \approx$ _____.

Use an identity to write each expression as a single trigonometric function. See Example 4.

33. $\sqrt{\dfrac{1 - \cos 40°}{2}}$

34. $\sqrt{\dfrac{1 + \cos 76°}{2}}$

35. $\sqrt{\dfrac{1 - \cos 147°}{1 + \cos 147°}}$

36. $\sqrt{\dfrac{1 + \cos 165°}{1 - \cos 165°}}$

37. $\dfrac{1 - \cos 59.74°}{\sin 59.74°}$

38. $\dfrac{\sin 158.2°}{1 + \cos 158.2°}$

39. $\pm\sqrt{\dfrac{1 + \cos 18x}{2}}$

40. $\pm\sqrt{\dfrac{1 + \cos 20\alpha}{2}}$

41. $\pm\sqrt{\dfrac{1 - \cos 8\theta}{1 + \cos 8\theta}}$

42. $\pm\sqrt{\dfrac{1 - \cos 5A}{1 + \cos 5A}}$

43. $\pm\sqrt{\dfrac{1 + \cos \frac{x}{4}}{2}}$

44. $\pm\sqrt{\dfrac{1 - \cos \frac{3\theta}{5}}{2}}$

Graph each expression and use the graph to make a conjecture as to what might be an identity. Then verify your conjecture algebraically.

45. $\dfrac{\sin x}{1 + \cos x}$

46. $\dfrac{1 - \cos x}{\sin x}$

47. $\dfrac{\tan \frac{x}{2} + \cot \frac{x}{2}}{\cot \frac{x}{2} - \tan \frac{x}{2}}$

48. $1 - 8 \sin^2 \dfrac{x}{2} \cos^2 \dfrac{x}{2}$

Verify that each equation is an identity. See Example 5.

49. $\sec^2 \dfrac{x}{2} = \dfrac{2}{1 + \cos x}$

50. $\cot^2 \dfrac{x}{2} = \dfrac{(1 + \cos x)^2}{\sin^2 x}$

51. $\sin^2 \dfrac{x}{2} = \dfrac{\tan x - \sin x}{2 \tan x}$

52. $\dfrac{\sin 2x}{2 \sin x} = \cos^2 \dfrac{x}{2} - \sin^2 \dfrac{x}{2}$

53. $\dfrac{2}{1 + \cos x} - \tan^2 \dfrac{x}{2} = 1$

54. $\tan \dfrac{\theta}{2} = \csc \theta - \cot \theta$

55. $1 - \tan^2 \dfrac{\theta}{2} = \dfrac{2 \cos \theta}{1 + \cos \theta}$

56. $\cos x = \dfrac{1 - \tan^2 \frac{x}{2}}{1 + \tan^2 \frac{x}{2}}$

57. Use the identity $\tan \frac{A}{2} = \frac{\sin A}{1 + \cos A}$ to derive the equivalent identity $\tan \frac{A}{2} = \frac{1 - \cos A}{\sin A}$ by multiplying both the numerator and denominator by $1 - \cos A$.

(Modeling) Mach Number An airplane flying faster than sound sends out sound waves that form a cone, as shown in the figure. The cone intersects the ground to form a **hyperbola.** As this hyperbola passes over a particular point on the ground, a sonic boom is heard at that point. If θ is the angle at the vertex of the cone, then

$$\sin \frac{\theta}{2} = \frac{1}{m},$$

where m is the Mach number for the speed of the plane. (We assume $m > 1$.) The Mach number is the ratio of the speed of the plane and the speed of sound. Thus, a speed of Mach 1.4 means that the plane is flying at 1.4 times the speed of sound. In Exercises 58–61, one of the values θ or m is given. Find the other value.

58. $m = \dfrac{3}{2}$ **59.** $m = \dfrac{5}{4}$ **60.** $\theta = 30°$ **61.** $\theta = 60°$

62. *(Modeling) Railroad Curves* In the United States, circular railroad curves are designated by the **degree of curvature,** the central angle subtended by a chord of 100 ft. See the figure. (*Source:* Hay, W. W., *Railroad Engineering,* John Wiley and Sons, 1982.)

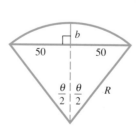

(a) Use the figure to write an expression for $\cos \frac{\theta}{2}$.

(b) Use the result of part (a) and the third half-angle identity for tangent to write an expression for $\tan \frac{\theta}{4}$.

(c) If $b = 12$, what is the measure of angle θ to the nearest degree?

RELATING CONCEPTS

For individual or collaborative investigation
(Exercises 63–70)

These exercises use results from plane geometry, instead of the half-angle formulas, to obtain exact values of the trigonometric functions of 15°. **Work Exercises 63–70 in order.**

   Start with a right triangle ACB having a 60° angle at A and a 30° angle at B. Let the hypotenuse of this triangle have length 2. Extend side BC and draw a semicircle with diameter along BC extended, center at B, and radius AB. Draw segment AE. (See the figure.) Since any angle inscribed in a semicircle is a right angle, triangle EAD is a right triangle.

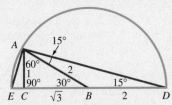

63. Why does $AB = BD$? Conclude that triangle ABD is isosceles.

64. Why does angle ABD have measure 150°?

65. Why do angles DAB and ADB both have measures of 15°?

66. What is the length DC?

67. Use the Pythagorean theorem to show that the length AD is $\sqrt{6} + \sqrt{2}$. (*Note:* $(\sqrt{6} + \sqrt{2})^2 = 8 + 4\sqrt{3}$.)

68. Use angle ADB of triangle EAD to find cos 15°.

69. Show that AE has length $\sqrt{6} - \sqrt{2}$ and find sin 15°.

70. Use triangle ACD to find tan 15°.

Summary Exercises on Verifying Trigonometric Identities

These summary exercises provide practice with the various types of trigonometric identities presented in this chapter. Verify that each equation is an identity.

1. $\tan \theta + \cot \theta = \sec \theta \csc \theta$

2. $\csc \theta \cos^2 \theta + \sin \theta = \csc \theta$

3. $\tan \dfrac{x}{2} = \csc x - \cot x$

4. $\sec(\pi - x) = -\sec x$

5. $\dfrac{\sin t}{1 + \cos t} = \dfrac{1 - \cos t}{\sin t}$

6. $\dfrac{1 - \sin t}{\cos t} = \dfrac{1}{\sec t + \tan t}$

7. $\sin 2\theta = \dfrac{2 \tan \theta}{1 + \tan^2 \theta}$

8. $\dfrac{2}{1 + \cos x} - \tan^2 \dfrac{x}{2} = 1$

9. $\cot \theta - \tan \theta = \dfrac{2 \cos^2 \theta - 1}{\sin \theta \cos \theta}$

10. $\dfrac{1}{\sec t - 1} + \dfrac{1}{\sec t + 1} = 2 \cot t \csc t$

11. $\dfrac{\sin(x + y)}{\cos(x - y)} = \dfrac{\cot x + \cot y}{1 + \cot x \cot y}$

12. $1 - \tan^2 \dfrac{\theta}{2} = \dfrac{2 \cos \theta}{1 + \cos \theta}$

13. $\dfrac{\sin\theta + \tan\theta}{1 + \cos\theta} = \tan\theta$

14. $\csc^4 x - \cot^4 x = \dfrac{1 + \cos^2 x}{1 - \cos^2 x}$

15. $\cos x = \dfrac{1 - \tan^2\frac{x}{2}}{1 + \tan^2\frac{x}{2}}$

16. $\cos 2x = \dfrac{2 - \sec^2 x}{\sec^2 x}$

17. $\dfrac{\tan^2 t + 1}{\tan t \csc^2 t} = \tan t$

18. $\dfrac{\sin s}{1 + \cos s} + \dfrac{1 + \cos s}{\sin s} = 2\csc s$

19. $\tan 4\theta = \dfrac{2\tan 2\theta}{2 - \sec^2 2\theta}$

20. $\tan\left(\dfrac{x}{2} + \dfrac{\pi}{4}\right) = \sec x + \tan x$

21. $\dfrac{\cot s - \tan s}{\cos s + \sin s} = \dfrac{\cos s - \sin s}{\sin s \cos s}$

22. $\dfrac{\tan\theta - \cot\theta}{\tan\theta + \cot\theta} = 1 - 2\cos^2\theta$

23. $\dfrac{\tan(x + y) - \tan y}{1 + \tan(x + y)\tan y} = \tan x$

24. $2\cos^2\dfrac{x}{2}\tan x = \tan x + \sin x$

25. $\dfrac{\cos^4 x - \sin^4 x}{\cos^2 x} = 1 - \tan^2 x$

26. $\dfrac{\csc t + 1}{\csc t - 1} = (\sec t + \tan t)^2$

27. $\dfrac{2(\sin x - \sin^3 x)}{\cos x} = \sin 2x$

28. $\dfrac{1}{2}\cot\dfrac{x}{2} - \dfrac{1}{2}\tan\dfrac{x}{2} = \cot x$

29. $\sin(60° + x) + \sin(60° - x) = \sqrt{3}\cos x$

30. $\sin(60° - x) - \sin(60° + x) = -\sin x$

31. $\dfrac{\cos(x + y) + \cos(y - x)}{\sin(x + y) - \sin(y - x)} = \cot x$

32. $\sin x + \sin 3x + \sin 5x + \sin 7x = 4\cos x \cos 2x \sin 4x$

33. $\sin^3\theta + \cos^3\theta + \sin\theta\cos^2\theta + \sin^2\theta\cos\theta = \sin\theta + \cos\theta$

34. $\dfrac{\cos x + \sin x}{\cos x - \sin x} - \dfrac{\cos x - \sin x}{\cos x + \sin x} = 2\tan 2x$

Chapter 5 Summary

QUICK REVIEW

CONCEPTS	EXAMPLES

5.1 Fundamental Identities

Reciprocal Identities

$$\cot\theta = \frac{1}{\tan\theta} \qquad \sec\theta = \frac{1}{\cos\theta} \qquad \csc\theta = \frac{1}{\sin\theta}$$

Quotient Identities

$$\tan\theta = \frac{\sin\theta}{\cos\theta} \qquad \cot\theta = \frac{\cos\theta}{\sin\theta}$$

Pythagorean Identities

$$\sin^2\theta + \cos^2\theta = 1 \qquad \tan^2\theta + 1 = \sec^2\theta$$
$$1 + \cot^2\theta = \csc^2\theta$$

Negative-Angle Identities

$$\sin(-\theta) = -\sin\theta \quad \cos(-\theta) = \cos\theta \quad \tan(-\theta) = -\tan\theta$$
$$\csc(-\theta) = -\csc\theta \quad \sec(-\theta) = \sec\theta \quad \cot(-\theta) = -\cot\theta$$

If θ is in quadrant IV and $\sin\theta = -\frac{3}{5}$, find $\csc\theta$, $\cos\theta$, and $\sin(-\theta)$.

$$\csc\theta = \frac{1}{\sin\theta} = \frac{1}{-\frac{3}{5}} = -\frac{5}{3}$$

$$\sin^2\theta + \cos^2\theta = 1$$

$$\left(-\frac{3}{5}\right)^2 + \cos^2\theta = 1$$

$$\cos^2\theta = 1 - \frac{9}{25} = \frac{16}{25}$$

$$\cos\theta = +\sqrt{\frac{16}{25}} = \frac{4}{5} \qquad \text{cos } \theta \text{ is positive in quadrant IV.}$$

$$\sin(-\theta) = -\sin\theta = \frac{3}{5}$$

5.2 Verifying Trigonometric Identities

See the box titled Hints for Verifying Identities on page 207.

5.3 Sum and Difference Identities for Cosine

5.4 Sum and Difference Identities for Sine and Tangent

Cofunction Identities

$$\cos(90° - \theta) = \sin\theta \qquad \cot(90° - \theta) = \tan\theta$$
$$\sin(90° - \theta) = \cos\theta \qquad \sec(90° - \theta) = \csc\theta$$
$$\tan(90° - \theta) = \cot\theta \qquad \csc(90° - \theta) = \sec\theta$$

Sum and Difference Identities

$$\cos(A - B) = \cos A\cos B + \sin A\sin B$$
$$\cos(A + B) = \cos A\cos B - \sin A\sin B$$
$$\sin(A + B) = \sin A\cos B + \cos A\sin B$$
$$\sin(A - B) = \sin A\cos B - \cos A\sin B$$
$$\tan(A + B) = \frac{\tan A + \tan B}{1 - \tan A\tan B}$$
$$\tan(A - B) = \frac{\tan A - \tan B}{1 + \tan A\tan B}$$

Find a value of θ such that $\tan\theta = \cot 78°$.

$$\tan\theta = \cot 78°$$
$$\cot(90° - \theta) = \cot 78°$$
$$90° - \theta = 78°$$
$$\theta = 12°$$

Find the exact value of $\cos(-15°)$.

$$\cos(-15°) = \cos(30° - 45°)$$
$$= \cos 30°\cos 45° + \sin 30°\sin 45°$$
$$= \frac{\sqrt{3}}{2}\cdot\frac{\sqrt{2}}{2} + \frac{1}{2}\cdot\frac{\sqrt{2}}{2}$$
$$= \frac{\sqrt{6}}{4} + \frac{\sqrt{2}}{4} = \frac{\sqrt{6} + \sqrt{2}}{4}$$

Write $\tan\left(\frac{\pi}{4} + \theta\right)$ in terms of $\tan\theta$.

$$\tan\left(\frac{\pi}{4} + \theta\right) = \frac{\tan\frac{\pi}{4} + \tan\theta}{1 - \tan\frac{\pi}{4}\tan\theta} = \frac{1 + \tan\theta}{1 - \tan\theta} \qquad \tan\frac{\pi}{4} = 1$$

CONCEPTS	EXAMPLES

5.5 Double-Angle Identities

Double-Angle Identities

$$\cos 2A = \cos^2 A - \sin^2 A \qquad \cos 2A = 1 - 2\sin^2 A$$

$$\cos 2A = 2\cos^2 A - 1 \qquad \sin 2A = 2\sin A \cos A$$

$$\tan 2A = \frac{2\tan A}{1 - \tan^2 A}$$

Given $\cos\theta = -\frac{5}{13}$ and $\sin\theta > 0$, find $\sin 2\theta$.

Sketch a triangle in quadrant II since $\cos\theta < 0$ and $\sin\theta > 0$. Use it to find that $\sin\theta = \frac{12}{13}$.

$$\sin 2\theta = 2\sin\theta\cos\theta$$

$$= 2\left(\frac{12}{13}\right)\left(-\frac{5}{13}\right) = -\frac{120}{169}$$

Product-to-Sum Identities

$$\cos A \cos B = \frac{1}{2}[\cos(A+B) + \cos(A-B)]$$

$$\sin A \sin B = \frac{1}{2}[\cos(A-B) - \cos(A+B)]$$

$$\sin A \cos B = \frac{1}{2}[\sin(A+B) + \sin(A-B)]$$

$$\cos A \sin B = \frac{1}{2}[\sin(A+B) - \sin(A-B)]$$

Write $\sin(-\theta)\sin 2\theta$ as the difference of two functions.

$$\sin(-\theta)\sin 2\theta = \frac{1}{2}[\cos(-\theta - 2\theta) - \cos(-\theta + 2\theta)]$$

$$= \frac{1}{2}[\cos(-3\theta) - \cos\theta]$$

$$= \frac{1}{2}\cos(-3\theta) - \frac{1}{2}\cos\theta$$

$$= \frac{1}{2}\cos 3\theta - \frac{1}{2}\cos\theta$$

Sum-to-Product Identities

$$\sin A + \sin B = 2\sin\left(\frac{A+B}{2}\right)\cos\left(\frac{A-B}{2}\right)$$

$$\sin A - \sin B = 2\cos\left(\frac{A+B}{2}\right)\sin\left(\frac{A-B}{2}\right)$$

$$\cos A + \cos B = 2\cos\left(\frac{A+B}{2}\right)\cos\left(\frac{A-B}{2}\right)$$

$$\cos A - \cos B = -2\sin\left(\frac{A+B}{2}\right)\sin\left(\frac{A-B}{2}\right)$$

Write $\cos\theta + \cos 3\theta$ as a product of two functions.

$$\cos\theta + \cos 3\theta = 2\cos\left(\frac{\theta + 3\theta}{2}\right)\cos\left(\frac{\theta - 3\theta}{2}\right)$$

$$= 2\cos\left(\frac{4\theta}{2}\right)\cos\left(\frac{-2\theta}{2}\right)$$

$$= 2\cos 2\theta \cos(-\theta)$$

$$= 2\cos 2\theta \cos\theta$$

5.6 Half-Angle Identities

Half-Angle Identities

$$\cos\frac{A}{2} = \pm\sqrt{\frac{1 + \cos A}{2}} \qquad \sin\frac{A}{2} = \pm\sqrt{\frac{1 - \cos A}{2}}$$

$$\tan\frac{A}{2} = \pm\sqrt{\frac{1 - \cos A}{1 + \cos A}} \qquad \tan\frac{A}{2} = \frac{\sin A}{1 + \cos A}$$

$$\tan\frac{A}{2} = \frac{1 - \cos A}{\sin A}$$

$\left(\text{In the first three identities, the sign is chosen based on the quadrant of } \frac{A}{2}.\right)$

Find the exact value of $\tan 67.5°$.

We choose the last form with $A = 135°$.

$$\tan 67.5° = \tan\frac{135°}{2} = \frac{1 - \cos 135°}{\sin 135°} = \frac{1 - \left(-\frac{\sqrt{2}}{2}\right)}{\frac{\sqrt{2}}{2}}$$

$$= \frac{1 + \frac{\sqrt{2}}{2}}{\frac{\sqrt{2}}{2}} \cdot \frac{2}{2} = \frac{2 + \sqrt{2}}{\sqrt{2}} \quad \text{or} \quad \sqrt{2} + 1$$

Rationalize the denominator; simplify.

CHAPTER 5 ▶ Review Exercises

Concept Check For each expression in Column I, choose the expression from Column II that completes an identity.

I				II

1. $\sec x = $ _____ **2.** $\csc x = $ _____   **A.** $\dfrac{1}{\sin x}$ **B.** $\dfrac{1}{\cos x}$

3. $\tan x = $ _____ **4.** $\cot x = $ _____   **C.** $\dfrac{\sin x}{\cos x}$ **D.** $\dfrac{1}{\cot^2 x}$

5. $\tan^2 x = $ _____ **6.** $\sec^2 x = $ _____   **E.** $\dfrac{1}{\cos^2 x}$ **F.** $\dfrac{\cos x}{\sin x}$

Use identities to write each expression in terms of $\sin \theta$ *and* $\cos \theta$*, and simplify.*

7. $\sec^2 \theta - \tan^2 \theta$ **8.** $\dfrac{\cot \theta}{\sec \theta}$ **9.** $\tan^2 \theta(1 + \cot^2 \theta)$

10. $\csc \theta + \cot \theta$ **11.** $\tan \theta - \sec \theta \csc \theta$ **12.** $\csc^2 \theta + \sec^2 \theta$

13. Use the trigonometric identities to find $\sin x$, $\tan x$, and $\cot(-x)$, given $\cos x = \frac{3}{5}$ and x in quadrant IV.

14. Given $\tan x = -\frac{5}{4}$, where $\frac{\pi}{2} < x < \pi$, use the trigonometric identities to find $\cot x$, $\csc x$, and $\sec x$.

15. Find the exact values of the six trigonometric functions of $165°$.

16. Find the exact values of $\sin x$, $\cos x$, and $\tan x$, for $x = \frac{\pi}{12}$, using
   (a) difference identities   **(b)** half-angle identities.

Concept Check For each expression in Column I, use an identity to choose an expression from Column II with the same value.

I			II

17. $\cos 210°$ **18.** $\sin 35°$ **A.** $\sin(-35°)$ **B.** $\cos 55°$

19. $\tan(-35°)$ **20.** $-\sin 35°$ **C.** $\sqrt{\dfrac{1 + \cos 150°}{2}}$ **D.** $2 \sin 150° \cos 150°$

21. $\cos 35°$ **22.** $\cos 75°$ **E.** $\cos 150° \cos 60° - \sin 150° \sin 60°$

23. $\sin 75°$ **24.** $\sin 300°$ **F.** $\cot(-35°)$

25. $\cos 300°$ **26.** $\cos(-55°)$ **G.** $\cos^2 150° - \sin^2 150°$

   H. $\sin 15° \cos 60° + \cos 15° \sin 60°$

   I. $\cos(-35°)$ **J.** $\cot 125°$

For each of the following, find $\sin(x + y)$, $\cos(x - y)$, $\tan(x + y)$, *and the quadrant of* $x + y$.

27. $\sin x = -\dfrac{3}{5}$, $\cos y = -\dfrac{7}{25}$, x and y in quadrant III

28. $\sin x = \dfrac{3}{5}$, $\cos y = \dfrac{24}{25}$, x in quadrant I, y in quadrant IV

29. $\sin x = -\dfrac{1}{2}$, $\cos y = -\dfrac{2}{5}$, x and y in quadrant III

30. $\sin y = -\dfrac{2}{3}$, $\cos x = -\dfrac{1}{5}$, x in quadrant II, y in quadrant III

31. $\sin x = \dfrac{1}{10}$, $\cos y = \dfrac{4}{5}$, x in quadrant I, y in quadrant IV

32. $\cos x = \dfrac{2}{9}$, $\sin y = -\dfrac{1}{2}$, x in quadrant IV, y in quadrant III

Find sine and cosine of each of the following.

33. θ, given $\cos 2\theta = -\dfrac{3}{4}$, $90° < 2\theta < 180°$

34. B, given $\cos 2B = \dfrac{1}{8}$, B in quadrant IV

35. $2x$, given $\tan x = 3$, $\sin x < 0$ **36.** $2y$, given $\sec y = -\dfrac{5}{3}$, $\sin y > 0$

Find each of the following.

37. $\cos\dfrac{\theta}{2}$, given $\cos\theta = -\dfrac{1}{2}$, with $90° < \theta < 180°$

38. $\sin\dfrac{A}{2}$, given $\cos A = -\dfrac{3}{4}$, with $90° < A < 180°$

39. $\tan x$, given $\tan 2x = 2$, with $\pi < x < \dfrac{3\pi}{2}$

40. $\sin y$, given $\cos 2y = -\dfrac{1}{3}$, with $\dfrac{\pi}{2} < y < \pi$

41. $\tan\dfrac{x}{2}$, given $\sin x = .8$, with $0 < x < \dfrac{\pi}{2}$

42. $\sin 2x$, given $\sin x = .6$, with $\dfrac{\pi}{2} < x < \pi$

Graph each expression and use the graph to make a conjecture as to what might be an identity. Then verify your conjecture algebraically.

43. $-\dfrac{\sin 2x + \sin x}{\cos 2x - \cos x}$ **44.** $\dfrac{1 - \cos 2x}{\sin 2x}$ **45.** $\dfrac{\sin x}{1 - \cos x}$

46. $\dfrac{\cos x \sin 2x}{1 + \cos 2x}$ **47.** $\dfrac{2(\sin x - \sin^3 x)}{\cos x}$ **48.** $\csc x - \cot x$

Verify that each equation is an identity.

49. $\sin^2 x - \sin^2 y = \cos^2 y - \cos^2 x$ **50.** $2\cos^3 x - \cos x = \dfrac{\cos^2 x - \sin^2 x}{\sec x}$

51. $\dfrac{\sin^2 x}{2 - 2\cos x} = \cos^2\dfrac{x}{2}$ **52.** $\dfrac{\sin 2x}{\sin x} = \dfrac{2}{\sec x}$

53. $2\cos A - \sec A = \cos A - \dfrac{\tan A}{\csc A}$ **54.** $\dfrac{2\tan B}{\sin 2B} = \sec^2 B$

55. $1 + \tan^2\alpha = 2\tan\alpha\csc 2\alpha$ **56.** $\dfrac{2\cot x}{\tan 2x} = \csc^2 x - 2$

57. $\tan\theta\sin 2\theta = 2 - 2\cos^2\theta$ **58.** $\csc A \sin 2A - \sec A = \cos 2A \sec A$

59. $2\tan x \csc 2x - \tan^2 x = 1$ **60.** $2\cos^2\theta - 1 = \dfrac{1 - \tan^2\theta}{1 + \tan^2\theta}$

61. $\tan\theta\cos^2\theta = \dfrac{2\tan\theta\cos^2\theta - \tan\theta}{1 - \tan^2\theta}$ **62.** $\sec^2\alpha - 1 = \dfrac{\sec 2\alpha - 1}{\sec 2\alpha + 1}$

63. $\dfrac{\sin^2 x - \cos^2 x}{\csc x} = 2\sin^3 x - \sin x$

64. $\sin^3 \theta = \sin \theta - \cos^2 \theta \sin \theta$

65. $\tan 4\theta = \dfrac{2\tan 2\theta}{2 - \sec^2 2\theta}$

66. $2\cos^2 \dfrac{x}{2}\tan x = \tan x + \sin x$

67. $\tan\left(\dfrac{x}{2} + \dfrac{\pi}{4}\right) = \sec x + \tan x$

68. $\dfrac{1}{2}\cot \dfrac{x}{2} - \dfrac{1}{2}\tan \dfrac{x}{2} = \cot x$

69. $-\cot \dfrac{x}{2} = \dfrac{\sin 2x + \sin x}{\cos 2x - \cos x}$

70. $\dfrac{\sin 3t + \sin 2t}{\sin 3t - \sin 2t} = \dfrac{\tan \frac{5t}{2}}{\tan \frac{t}{2}}$

(Modeling) Solve each problem.

71. *Distance Traveled by a Stone* The distance D of an object thrown (or projected) from height h (feet) at angle θ with initial velocity v is modeled by the formula

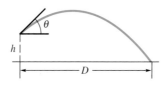

$$D = \frac{v^2 \sin \theta \cos \theta + v \cos \theta \sqrt{(v \sin \theta)^2 + 64h}}{32}.$$

See the figure. (*Source:* Kreighbaum, E. and K. Barthels, *Biomechanics,* Allyn & Bacon, 1996.) Also see the **Chapter 2** Quantitative Reasoning.

(a) Find D when $h = 0$; that is, when the object is projected from the ground.

(b) Suppose a car driving over loose gravel kicks up a small stone at a velocity of 36 ft per sec (about 25 mph) and an angle $\theta = 30°$. How far will the stone travel?

72. *Amperage, Wattage, and Voltage* Suppose that for an electric heater, voltage is given by $V = a \sin 2\pi\omega t$ and amperage by $I = b \sin 2\pi\omega t$, where t is time in seconds.

(a) Find the period of the graph for the voltage.

(b) Show that the graph of the wattage $W = VI$ will have half the period of the voltage. Interpret this result.

CHAPTER 5 ▶   Test

1. If $\cos \theta = \frac{24}{25}$ and θ is in quadrant IV, find the five remaining trigonometric function values of θ.

2. Express $\sec \theta - \sin \theta \tan \theta$ as a single function of θ.

3. Express $\tan^2 x - \sec^2 x$ in terms of $\sin x$ and $\cos x$, and simplify.

4. Find the exact value of $\cos \frac{5\pi}{12}$.

5. Express as a function of x alone.

(a) $\cos(270° - x)$ (b) $\tan(\pi + x)$

6. Use a half-angle identity to find the exact value of $\sin(-22.5°)$.

7. Graph $y = \cot \frac{1}{2}x - \cot x$, and make a conjecture as to what might be an identity. Then verify your conjecture algebraically.

8. Given that $\sin A = \frac{5}{13}$, $\cos B = -\frac{3}{5}$, A is a quadrant I angle, and B is a quadrant II angle, find each of the following.

(a) $\sin(A + B)$ (b) $\cos(A + B)$ (c) $\tan(A - B)$

(d) the quadrant of $A + B$

9. Given that $\cos \theta = -\frac{3}{5}$ and $\frac{\pi}{2} < \theta < \pi$, find each of the following.

(a) $\cos 2\theta$ (b) $\sin 2\theta$ (c) $\tan 2\theta$ (d) $\cos \dfrac{1}{2}\theta$ (e) $\tan \dfrac{1}{2}\theta$

Verify each identity.

10. $\sec^2 B = \dfrac{1}{1 - \sin^2 B}$

11. $\tan^2 x - \sin^2 x = (\tan x \sin x)^2$

12. $\dfrac{\tan x - \cot x}{\tan x + \cot x} = 2 \sin^2 x - 1$

13. $\cos 2A = \dfrac{\cot A - \tan A}{\csc A \sec A}$

14. $\dfrac{\sin 2x}{\cos 2x + 1} = \tan x$

15. *(Modeling) Voltage* The voltage in common household current is expressed as $V = 163 \sin \omega t$, where ω is the angular speed (in radians per second) of the generator at the electrical plant and t is time (in seconds).

(a) Use an identity to express V in terms of cosine.

(b) If $\omega = 120\pi$, what is the maximum voltage? Give the least positive value of t when the maximum voltage occurs.

CHAPTER 5 ▶ # Quantitative Reasoning

How does NASA calculate rotation and roll of a spacecraft?

With the spectacular success of the Mars explorers *Spirit* and *Opportunity*, interest in space travel has been revived. The drawing on the left below shows the three quantities that determine the motion of a spacecraft. A conventional three-dimensional spacecraft coordinate system is shown on the right below.

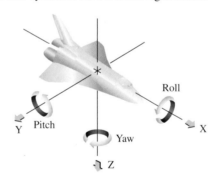

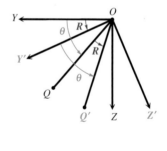

Angle $YOQ = \theta$ and $OQ = r$. The coordinates of Q are (x, y, z), where

$$y = r \cos \theta \qquad \text{and} \qquad z = r \sin \theta.$$

When the spacecraft performs a rotation, it is necessary to find the coordinates in the spacecraft system after the rotation takes place. For example, suppose the spacecraft undergoes roll through angle R. The coordinates (x, y, z) of point Q become (x', y', z'), the coordinates of the corresponding point Q'. In the new reference system, $OQ' = r$ and, since the roll is around the x-axis and angle $Y'OQ' = YOQ = \theta$,

$$x' = x, \qquad y' = r \cos(\theta + R), \qquad \text{and} \qquad z' = r \sin(\theta + R).$$

Use the cosine and sine sum identities to write expressions for y' and z' in terms of y, z, and R. (*Source:* Kastner, B., *Space Mathematics,* NASA.)

6 Inverse Circular Functions and Trigonometric Equations

On the night of October 20, 1955, a new sound exploded from a stage at Brooklyn High School in Cleveland, Ohio. The concert showcased rising stars in the music industry, including Elvis Presley and Bill Haley, and ushered in a new era of popular music—*rock and roll*. The distinctive melodic phrases, or *riffs,* generated by the musicians' electric guitars defined the new genre. (*Source:* Rock and Roll Hall of Fame.)

Sound waves, such as those initiated by musical instruments, travel in sinusoidal patterns that can be graphed as sine or cosine functions and described by trigonometric equations. When sound waves are combined and organized to have rhythm, melody, harmony, and dynamics, the brain interprets them as music.

In Section 6.2, Example 6, and Section 6.3, Example 5, various aspects of music are analyzed using trigonometric equations and graphs.

6.1 Inverse Circular Functions

Inverse Functions ▪ Inverse Sine Function ▪ Inverse Cosine Function ▪ Inverse Tangent Function ▪ Remaining Inverse Circular Functions ▪ Inverse Function Values

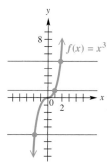

$f(x) = x^3$ is a one-to-one function. It satisfies the conditions of the horizontal line test.

$g(x) = x^2$ is not one-to-one. It does not satisfy the conditions of the horizontal line test.

Figure 1

Inverse Functions *Recall that for a function f, every element x in the domain corresponds to one and only one element y, or f(x), in the range.* (Appendix C) This means the following:

1. If point (a, b) lies on the graph of f, then there is no other point on the graph that has a as first coordinate.

2. Other points may have b as second coordinate, however, since the definition of function allows range elements to be used more than once.

If a function is defined so that *each range element is used only once,* then it is called a **one-to-one function.** For example, the function

$$f(x) = x^3 \text{ is a one-to-one function}$$

because every real number has exactly one real cube root. On the other hand,

$$g(x) = x^2 \text{ is not a one-to-one function}$$

because, for example, $g(2) = 4$ and $g(-2) = 4$. There are two domain elements, 2 and -2, that correspond to the range element 4.

The **horizontal line test** helps determine graphically whether a function is one-to-one.

HORIZONTAL LINE TEST

Any horizontal line will intersect the graph of a one-to-one function in at most one point.

This test is applied to the graphs of $f(x) = x^3$ and $g(x) = x^2$ in Figure 1.

By interchanging the components of the ordered pairs of a one-to-one function f, we obtain a new set of ordered pairs that satisfies the definition of function. This new function, called the *inverse function,* or *inverse,* of f, is symbolized f^{-1}.

INVERSE FUNCTION

The **inverse function** of the one-to-one function f is defined as

$$f^{-1} = \{(y, x) \,|\, (x, y) \text{ belongs to } f\}.$$

▶ **Caution** *The -1 in $f^{-1}(x)$ is not an exponent.* That is,

$$f^{-1}(x) \neq \frac{1}{f(x)}.$$

The following statements summarize our discussion of inverse functions.

SUMMARY OF INVERSE FUNCTIONS

1. In a one-to-one function, each x-value corresponds to only one y-value and each y-value corresponds to only one x-value.

2. If a function f is one-to-one, then f has an inverse function f^{-1}.

3. The domain of f is the range of f^{-1}, and the range of f is the domain of f^{-1}.

4. The graphs of f and f^{-1} are reflections of each other across the line $y = x$.

5. To find $f^{-1}(x)$ from $f(x)$, follow these steps.

Step 1 Replace $f(x)$ with y and interchange x and y.

Step 2 Solve for y.

Step 3 Replace y with $f^{-1}(x)$.

Figure 2 illustrates some of these concepts.

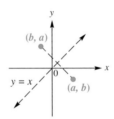

(b, a) is the reflection of (a, b) across the line $y = x$.

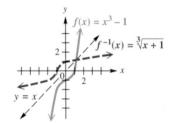

The graph of f^{-1} is the reflection of the graph of f across the line $y = x$.

Figure 2

We often restrict the domain of a function that is not one-to-one to make it one-to-one, without changing the range. For example, we saw in Figure 1 that $g(x) = x^2$, with its natural domain $(-\infty, \infty)$, is not one-to-one. However, if we restrict its domain to the set of nonnegative numbers $[0, \infty)$, we obtain a new function f that is one-to-one and has the same range as g, $[0, \infty)$. See Figure 3.

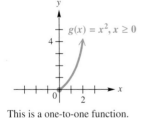

This is a one-to-one function.

Figure 3

▶ **Note** We could have chosen to restrict the domain of $g(x) = x^2$ to $(-\infty, 0]$ to obtain a different one-to-one function. For the trigonometric functions, such choices are made based on general agreement by mathematicians.

▼ LOOKING AHEAD TO CALCULUS

The **inverse circular functions** are used in calculus to solve certain types of related-rates problems and to integrate certain rational functions.

Inverse Sine Function Refer to the graph of the sine function in Figure 4 on the next page. Applying the horizontal line test, we see that $y = \sin x$ does not define a one-to-one function. If we restrict the domain to the interval $\left[-\frac{\pi}{2}, \frac{\pi}{2}\right]$, which is the part of the graph in Figure 4 shown in color, this restricted function is one-to-one and has an inverse function. The range of $y = \sin x$ is $[-1, 1]$, so the domain of the inverse function will be $[-1, 1]$, and its range will be $\left[-\frac{\pi}{2}, \frac{\pi}{2}\right]$.

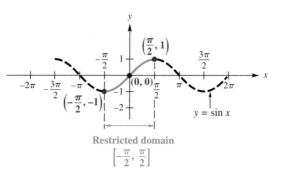

Figure 4

Reflecting the graph of $y = \sin x$ on the restricted domain, shown in Figure 5(a), across the line $y = x$ gives the graph of the inverse function, shown in Figure 5(b). Some key points are labeled on the graph. The equation of the inverse of $y = \sin x$ is found by interchanging x and y to get

$$x = \sin y.$$

This equation is solved for y by writing

$$y = \sin^{-1} x \quad \text{(read "inverse sine of } x\text{").}$$

As Figure 5(b) shows, the domain of $y = \sin^{-1} x$ is $[-1, 1]$, while the restricted domain of $y = \sin x$, $\left[-\frac{\pi}{2}, \frac{\pi}{2}\right]$, is the range of $y = \sin^{-1} x$. An alternative notation for $\sin^{-1} x$ is arcsin x.

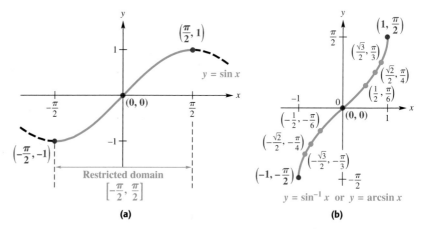

Figure 5

INVERSE SINE FUNCTION

$y = \sin^{-1} x$ or $y = \arcsin x$ means that $x = \sin y$, for $-\frac{\pi}{2} \le y \le \frac{\pi}{2}$.

We can think of $y = \sin^{-1} x$ or $y = \arcsin x$ as

"y is the number in the interval $\left[-\frac{\pi}{2}, \frac{\pi}{2}\right]$ whose sine is x."

Thus, we can write $y = \sin^{-1} x$ as $\sin y = x$ to evaluate it. We must pay close attention to the domain and range intervals.

▶ **EXAMPLE 1**   FINDING INVERSE SINE VALUES

Find y in each equation.

(a) $y = \arcsin \dfrac{1}{2}$ **(b)** $y = \sin^{-1}(-1)$ **(c)** $y = \sin^{-1}(-2)$

Algebraic Solution

(a) The graph of the function defined by $y = \arcsin x$ (Figure 5(b)) includes the point $\left(\frac{1}{2}, \frac{\pi}{6}\right)$. Therefore, $\arcsin \frac{1}{2} = \frac{\pi}{6}$.

Alternatively, we can think of $y = \arcsin \frac{1}{2}$ as "y is the number in $\left[-\frac{\pi}{2}, \frac{\pi}{2}\right]$ whose sine is $\frac{1}{2}$." Then we can write the given equation as $\sin y = \frac{1}{2}$. Since $\sin \frac{\pi}{6} = \frac{1}{2}$ and $\frac{\pi}{6}$ is in the range of the arcsine function, $y = \frac{\pi}{6}$.

(b) Writing the equation $y = \sin^{-1}(-1)$ in the form $\sin y = -1$ shows that $y = -\frac{\pi}{2}$. Notice that the point $\left(-1, -\frac{\pi}{2}\right)$ is on the graph of $y = \sin^{-1} x$.

(c) Because -2 is not in the domain of the inverse sine function, $\sin^{-1}(-2)$ does not exist.

Graphing Calculator Solution

We graph $Y_1 = \sin^{-1} X$ and find the points with X-values $\frac{1}{2} = .5$ and -1. For these X-values, Figure 6 shows that $Y = \frac{\pi}{6} \approx .52359878$ and $Y = -\frac{\pi}{2} \approx -1.570796$.

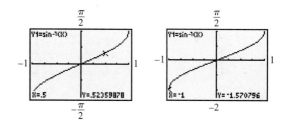

Figure 6

Since $\sin^{-1}(-2)$ does not exist, a calculator will give an error message for this input.

NOW TRY EXERCISES 13 AND 23. ◀

▶ **Caution** In Example 1(b), it is tempting to give the value of $\sin^{-1}(-1)$ as $\frac{3\pi}{2}$, since $\sin \frac{3\pi}{2} = -1$. Notice, however, that $\frac{3\pi}{2}$ is not in the range of the inverse sine function. *Be certain that the number given for an inverse function value is in the range of the particular inverse function being considered.*

INVERSE SINE FUNCTION
$y = \sin^{-1} x$ OR $y = \arcsin x$

Domain: $[-1, 1]$ Range: $\left[-\frac{\pi}{2}, \frac{\pi}{2}\right]$

x	y
-1	$-\frac{\pi}{2}$
$-\frac{\sqrt{2}}{2}$	$-\frac{\pi}{4}$
0	0
$\frac{\sqrt{2}}{2}$	$\frac{\pi}{4}$
1	$\frac{\pi}{2}$

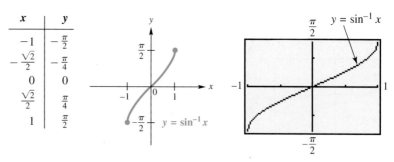

Figure 7

• The inverse sine function is increasing and continuous on its domain $[-1, 1]$.

• Its x-intercept is 0, and its y-intercept is 0.

• Its graph is symmetric with respect to the origin; it is an odd function.

Inverse Cosine Function The function $y = \cos^{-1} x$ (or $y = \arccos x$) is defined by restricting the domain of the function $y = \cos x$ to the interval $[0, \pi]$ as in Figure 8. This restricted function, which is the part of the graph in Figure 8 shown in color, is one-to-one and has an inverse function. The inverse function, $y = \cos^{-1} x$, is found by interchanging the roles of x and y. Reflecting the graph of $y = \cos x$ across the line $y = x$ gives the graph of the inverse function shown in Figure 9. Again, some key points are shown on the graph.

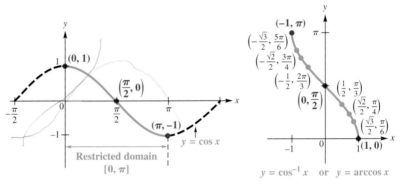

Figure 8   **Figure 9**

INVERSE COSINE FUNCTION

$y = \cos^{-1} x$ **or** $y = \arccos x$ **means that** $x = \cos y$, **for** $0 \leq y \leq \pi$.

We can think of $y = \cos^{-1} x$ *or* $y = \arccos x$ *as*

"y is the number in the interval $[0, \pi]$ *whose cosine is x."*

▶ **EXAMPLE 2**   FINDING INVERSE COSINE VALUES

Find y in each equation.

(a) $y = \arccos 1$

(b) $y = \cos^{-1}\left(-\dfrac{\sqrt{2}}{2}\right)$

Solution

(a) Since the point $(1, 0)$ lies on the graph of $y = \arccos x$ in Figure 9, the value of y is 0. Alternatively, we can think of $y = \arccos 1$ as

   "y is the number in $[0, \pi]$ whose cosine is 1,"   or $\cos y = 1$.

   Thus, $y = 0$, since $\cos 0 = 1$ and 0 is in the range of the arccosine function.

(b) We must find the value of y that satisfies $\cos y = -\dfrac{\sqrt{2}}{2}$, where y is in the interval $[0, \pi]$, the range of the function $y = \cos^{-1} x$. The only value for y that satisfies these conditions is $\frac{3\pi}{4}$. Again, this can be verified from the graph in Figure 9.

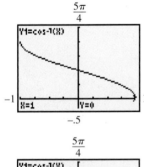

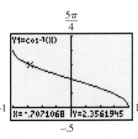

These screens support the results of Example 2, since

$-\dfrac{\sqrt{2}}{2} \approx -.7071068$ and

$\dfrac{3\pi}{4} \approx 2.3561945.$

NOW TRY EXERCISES 15 AND 25. ◀

Our observations about the inverse cosine function lead to the following generalizations.

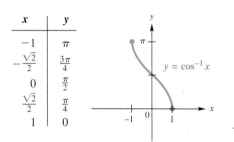 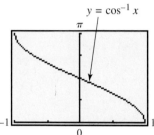

Figure 10

- The inverse cosine function is decreasing and continuous on its domain $[-1, 1]$.
- Its x-intercept is 1, and its y-intercept is $\frac{\pi}{2}$.
- Its graph is neither symmetric with respect to the y-axis nor the origin.

Inverse Tangent Function Restricting the domain of the function $y = \tan x$ to the open interval $\left(-\frac{\pi}{2}, \frac{\pi}{2}\right)$ yields a one-to-one function. By interchanging the roles of x and y, we obtain the inverse tangent function given by $y = \tan^{-1} x$ or $y = \arctan x$. Figure 11 shows the graph of the restricted tangent function. Figure 12 gives the graph of $y = \tan^{-1} x$.

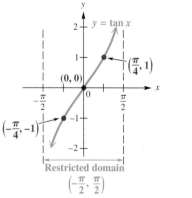

Figure 11

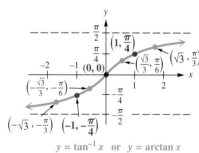

Figure 12

INVERSE TANGENT FUNCTION

$y = \tan^{-1} x$ or $y = \arctan x$ means that $x = \tan y$, for $-\frac{\pi}{2} < y < \frac{\pi}{2}$.

INVERSE TANGENT FUNCTION
$y = \tan^{-1}x$ OR $y = \arctan x$

Domain: $(-\infty, \infty)$ Range: $\left(-\frac{\pi}{2}, \frac{\pi}{2}\right)$

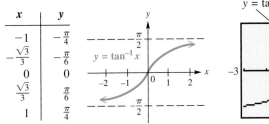

x	y
-1	$-\frac{\pi}{4}$
$-\frac{\sqrt{3}}{3}$	$-\frac{\pi}{6}$
0	0
$\frac{\sqrt{3}}{3}$	$\frac{\pi}{6}$
1	$\frac{\pi}{4}$

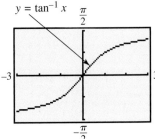

Figure 13

- The inverse tangent function is increasing and continuous on its domain $(-\infty, \infty)$.
- Its x-intercept is 0, and its y-intercept is 0.
- Its graph is symmetric with respect to the origin; it is an odd function.
- The lines $y = \frac{\pi}{2}$ and $y = -\frac{\pi}{2}$ are horizontal asymptotes.

Remaining Inverse Circular Functions The remaining three inverse trigonometric functions are defined similarly; their graphs are shown in Figure 14.

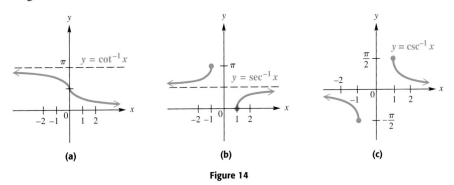

(a)   (b)   (c)

Figure 14

INVERSE COTANGENT, SECANT, AND COSECANT FUNCTIONS*

$y = \cot^{-1}x$ or $y = \operatorname{arccot} x$ means that $x = \cot y$, for $0 < y < \pi$.

$y = \sec^{-1}x$ or $y = \operatorname{arcsec} x$ means that $x = \sec y$, for $0 \le y \le \pi, y \ne \frac{\pi}{2}$.

$y = \csc^{-1}x$ or $y = \operatorname{arccsc} x$ means that $x = \csc y$, for $-\frac{\pi}{2} \le y \le \frac{\pi}{2}, y \ne 0$.

---

*The inverse secant and inverse cosecant functions are sometimes defined with different ranges. We use intervals that match their reciprocal functions (except for one missing point).

The table gives all six inverse trigonometric functions with their domains and ranges.

Inverse Function	Domain	Range	
		Interval	Quadrants of the Unit Circle
$y = \sin^{-1} x$	$[-1, 1]$	$\left[-\frac{\pi}{2}, \frac{\pi}{2}\right]$	I and IV
$y = \cos^{-1} x$	$[-1, 1]$	$[0, \pi]$	I and II
$y = \tan^{-1} x$	$(-\infty, \infty)$	$\left(-\frac{\pi}{2}, \frac{\pi}{2}\right)$	I and IV
$y = \cot^{-1} x$	$(-\infty, \infty)$	$(0, \pi)$	I and II
$y = \sec^{-1} x$	$(-\infty, -1] \cup [1, \infty)$	$[0, \pi], y \neq \frac{\pi}{2}$	I and II
$y = \csc^{-1} x$	$(-\infty, -1] \cup [1, \infty)$	$\left[-\frac{\pi}{2}, \frac{\pi}{2}\right], y \neq 0$	I and IV

Inverse Function Values   The inverse circular functions are formally defined with real number ranges. However, there are times when it may be convenient to find degree-measured angles equivalent to these real number values. It is also often convenient to think in terms of the unit circle and choose the inverse function values based on the quadrants given in the preceding table.

▶ **EXAMPLE 3**   **FINDING INVERSE FUNCTION VALUES (DEGREE-MEASURED ANGLES)**

Find the *degree measure* of θ in the following.

(a) $\theta = \arctan 1$

(b) $\theta = \sec^{-1} 2$

Solution

(a) Here θ must be in $(-90°, 90°)$, but since 1 is positive, θ must be in quadrant I. The alternative statement, $\tan \theta = 1$, leads to $\theta = 45°$.

(b) Write the equation as $\sec \theta = 2$. For $\sec^{-1} x$, θ is in quadrant I or II. Because 2 is positive, θ is in quadrant I and $\theta = 60°$, since $\sec 60° = 2$. Note that $60°$ $\left(\text{the degree equivalent of } \frac{\pi}{3}\right)$ is in the range of the inverse secant function.

NOW TRY EXERCISES 35 AND 41. ◀

The inverse trigonometric function keys on a calculator give results in the proper quadrant for the inverse sine, inverse cosine, and inverse tangent functions, according to the definitions of these functions. For example, on a calculator, in degrees, $\sin^{-1} .5 = 30°$, $\sin^{-1}(-.5) = -30°$, $\tan^{-1}(-1) = -45°$, and $\cos^{-1}(-.5) = 120°$.

Finding $\cot^{-1} x$, $\sec^{-1} x$, and $\csc^{-1} x$ with a calculator is not as straightforward, because these functions must be expressed in terms of $\tan^{-1} x$, $\cos^{-1} x$, and $\sin^{-1} x$, respectively. If $y = \sec^{-1} x$, for example, then $\sec y = x$, which must be written as a cosine function as follows:

If **sec** $y = x$, then $\dfrac{1}{\cos y} = x$ or $\cos y = \dfrac{1}{x}$,   and $y = \cos^{-1} \dfrac{1}{x}$.

In summary, to find $\sec^{-1} x$, we find $\cos^{-1} \frac{1}{x}$. Similar statements apply to $\csc^{-1} x$ and $\cot^{-1} x$. There is one additional consideration with $\cot^{-1} x$. Since we take the inverse tangent of the reciprocal to find inverse cotangent, the calculator gives values of inverse cotangent with the same range as inverse tangent, $\left(-\frac{\pi}{2}, \frac{\pi}{2}\right)$, which is not the correct range for inverse cotangent. For inverse cotangent, the proper range must be considered and the results adjusted accordingly.

▶ **EXAMPLE 4** FINDING INVERSE FUNCTION VALUES WITH A CALCULATOR

(a) Find y in radians if $y = \csc^{-1}(-3)$.

(b) Find θ in degrees if $\theta = \text{arccot}(-.3541)$.

Solution

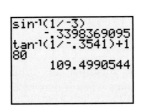

Figure 15

(a) With the calculator in radian mode, enter $\csc^{-1}(-3)$ as $\sin^{-1}\left(-\frac{1}{3}\right)$ to get $y \approx -.3398369095$. See Figure 15.

(b) Now set the calculator to degree mode. A calculator gives the inverse tangent value of a negative number as a quadrant IV angle. The restriction on the range of arccotangent implies that θ must be in quadrant II, so enter

$$\text{arccot}(-.3541) \quad \text{as} \quad \tan^{-1}\left(\frac{1}{-.3541}\right) + 180°.$$

As shown in Figure 15, $\theta \approx 109.4990544°$.

NOW TRY EXERCISES 51 AND 59. ◀

> ▶ **Caution** *Be careful when using your calculator to evaluate the inverse cotangent of a negative quantity.* To do this, we must enter the inverse tangent of the *reciprocal* of the negative quantity, which returns an angle in quadrant IV. Since inverse cotangent is negative in quadrant II, adjust your calculator result by adding 180° or π accordingly.

▶ **EXAMPLE 5** FINDING FUNCTION VALUES USING DEFINITIONS OF THE TRIGONOMETRIC FUNCTIONS

Evaluate each expression without using a calculator.

(a) $\sin\left(\tan^{-1} \dfrac{3}{2}\right)$

(b) $\tan\left(\cos^{-1}\left(-\dfrac{5}{13}\right)\right)$

Solution

(a) Let $\theta = \tan^{-1} \frac{3}{2}$, so $\tan \theta = \frac{3}{2}$. The inverse tangent function yields values only in quadrants I and IV, and since $\frac{3}{2}$ is positive, θ is in quadrant I. Sketch θ in quadrant I, and label a triangle, as shown in Figure 16 on the next page. By the Pythagorean theorem, the hypotenuse is $\sqrt{13}$. The value of sine is the quotient of the side opposite and the hypotenuse, so

$$\sin\left(\tan^{-1} \frac{3}{2}\right) = \sin \theta = \frac{3}{\sqrt{13}} = \frac{3}{\sqrt{13}} \cdot \frac{\sqrt{13}}{\sqrt{13}} = \frac{3\sqrt{13}}{13}. \quad \text{(Section 2.1)}$$

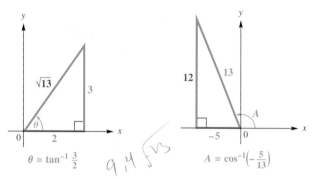

$\theta = \tan^{-1} \frac{3}{2}$

Figure 16

$A = \cos^{-1}\left(-\frac{5}{13}\right)$

Figure 17

(b) Let $A = \cos^{-1}\left(-\frac{5}{13}\right)$. Then, $\cos A = -\frac{5}{13}$. Since $\cos^{-1} x$ for a negative value of x is in quadrant II, sketch A in quadrant II, as shown in Figure 17.

$$\tan\left(\cos^{-1}\left(-\frac{5}{13}\right)\right) = \tan A = -\frac{12}{5}$$

NOW TRY EXERCISES 77 AND 79. ◀

▶ EXAMPLE 6   FINDING FUNCTION VALUES USING IDENTITIES

Evaluate each expression without using a calculator.

(a) $\cos\left(\arctan \sqrt{3} + \arcsin \frac{1}{3}\right)$ **(b)** $\tan\left(2 \arcsin \frac{2}{5}\right)$

Solution

(a) Let $A = \arctan \sqrt{3}$ and $B = \arcsin \frac{1}{3}$, so $\tan A = \sqrt{3}$ and $\sin B = \frac{1}{3}$. Sketch both A and B in quadrant I, as shown in Figure 18. Now, use the cosine sum identity.

$$\cos(A + B) = \cos A \cos B - \sin A \sin B \quad \text{(Section 5.3)}$$

$$\cos\left(\arctan \sqrt{3} + \arcsin \frac{1}{3}\right) = \cos(\arctan \sqrt{3}) \cos\left(\arcsin \frac{1}{3}\right)$$

$$- \sin(\arctan \sqrt{3}) \sin\left(\arcsin \frac{1}{3}\right) \quad (1)$$

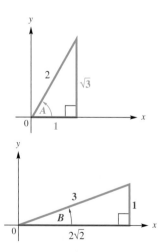

From Figure 18,

$$\cos(\arctan \sqrt{3}) = \cos A = \frac{1}{2}, \quad \cos\left(\arcsin \frac{1}{3}\right) = \cos B = \frac{2\sqrt{2}}{3},$$

$$\sin(\arctan \sqrt{3}) = \sin A = \frac{\sqrt{3}}{2}, \quad \sin\left(\arcsin \frac{1}{3}\right) = \sin B = \frac{1}{3}.$$

Substitute these values into equation (1) to get

$$\cos\left(\arctan \sqrt{3} + \arcsin \frac{1}{3}\right) = \frac{1}{2} \cdot \frac{2\sqrt{2}}{3} - \frac{\sqrt{3}}{2} \cdot \frac{1}{3} = \frac{2\sqrt{2} - \sqrt{3}}{6}.$$

Figure 18

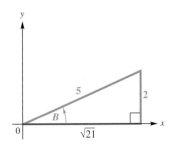

Figure 19

(b) Let $\arcsin \frac{2}{5} = B$. Then, from the double-angle tangent identity,

$$\tan\left(2 \arcsin \frac{2}{5}\right) = \tan 2B = \frac{2 \tan B}{1 - \tan^2 B}. \quad \text{(Section 5.5)}$$

Since $\arcsin \frac{2}{5} = B$, $\sin B = \frac{2}{5}$. Sketch a triangle in quadrant I, find the length of the third side, and then find $\tan B$. From the triangle in Figure 19, $\tan B = \frac{2}{\sqrt{21}}$, and

$$\tan\left(2 \arcsin \frac{2}{5}\right) = \frac{2\left(\frac{2}{\sqrt{21}}\right)}{1 - \left(\frac{2}{\sqrt{21}}\right)^2} = \frac{\frac{4}{\sqrt{21}}}{1 - \frac{4}{21}} = \frac{\frac{4}{\sqrt{21}} \cdot \frac{\sqrt{21}}{\sqrt{21}}}{\frac{17}{21}} = \frac{\frac{4\sqrt{21}}{21}}{\frac{17}{21}} = \frac{4\sqrt{21}}{17}.$$

> Be careful simplifying the complex fraction.

> NOW TRY EXERCISES 81 AND 89. ◀

While the work shown in Examples 5 and 6 does not rely on a calculator, we can support our algebraic work with one. By entering $\cos\left(\arctan \sqrt{3} + \arcsin \frac{1}{3}\right)$ from Example 6(a) into a calculator, we get the approximation .1827293862, the same approximation as when we enter $\frac{2\sqrt{2} - \sqrt{3}}{6}$ (the exact value we obtained algebraically). Similarly, we obtain the same approximation when we evaluate $\tan\left(2 \arcsin \frac{2}{5}\right)$ and $\frac{4\sqrt{21}}{17}$, supporting our answer in Example 6(b).

▶ **EXAMPLE 7** WRITING FUNCTION VALUES IN TERMS OF *u*

Write each trigonometric expression as an algebraic expression in *u*.

(a) $\sin(\tan^{-1} u)$ **(b)** $\cos(2 \sin^{-1} u)$

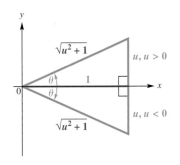

Figure 20

Solution

(a) Let $\theta = \tan^{-1} u$, so $\tan \theta = u$. Here, *u* may be positive or negative. Since $-\frac{\pi}{2} < \tan^{-1} u < \frac{\pi}{2}$, sketch θ in quadrants I and IV and label two triangles, as shown in Figure 20. Since sine is given by the quotient of the side opposite and the hypotenuse,

$$\sin(\tan^{-1} u) = \sin \theta = \frac{u}{\sqrt{u^2 + 1}} = \frac{u}{\sqrt{u^2 + 1}} \cdot \frac{\sqrt{u^2 + 1}}{\sqrt{u^2 + 1}} = \frac{u\sqrt{u^2 + 1}}{u^2 + 1}.$$

> Rationalize the denominator.

The result is positive when *u* is positive and negative when *u* is negative.

(b) Let $\theta = \sin^{-1} u$, so $\sin \theta = u$. To find $\cos 2\theta$, use the double-angle identity $\cos 2\theta = 1 - 2 \sin^2 \theta$.

$$\cos(2 \sin^{-1} u) = \cos 2\theta = 1 - 2 \sin^2 \theta = 1 - 2u^2 \quad \text{(Section 5.5)}$$

> NOW TRY EXERCISES 97 AND 101. ◀

▶ **EXAMPLE 8**   FINDING THE OPTIMAL ANGLE OF ELEVATION OF A SHOT PUT

The optimal angle of elevation θ a shot-putter should aim for to throw the greatest distance depends on the velocity v of the throw and the initial height h of the shot. See Figure 21. One model for θ that achieves this greatest distance is

$$\theta = \arcsin\left(\sqrt{\frac{v^2}{2v^2 + 64h}}\right).$$

(*Source:* Townend, M. S., *Mathematics in Sport,* Chichester, Ellis Horwood Limited, 1984.)

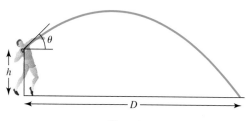

Figure 21

Suppose a shot-putter can consistently throw the steel ball with $h = 6.6$ ft and $v = 42$ ft per sec. At what angle should he throw the ball to maximize distance?

Solution   To find this angle, substitute and use a calculator in degree mode.

$$\theta = \arcsin\left(\sqrt{\frac{42^2}{2(42^2) + 64(6.6)}}\right) \approx 42° \quad h = 6.6, v = 42$$

NOW TRY EXERCISE 107. ◀

6.1   Exercises

Concept Check   *Complete each statement.*

1. For a function to have an inverse, it must be _____ .

2. The domain of $y = \arcsin x$ equals the _____ of $y = \sin x$.

3. The range of $y = \cos^{-1} x$ equals the _____ of $y = \cos x$.

4. The point $\left(\frac{\pi}{4}, 1\right)$ lies on the graph of $y = \tan x$. Therefore, the point _____ lies on the graph of _____ .

5. If a function f has an inverse and $f(\pi) = -1$, then $f^{-1}(-1) = $ _____ .

6. How can the graph of f^{-1} be sketched if the graph of f is known?

Concept Check   *In Exercises 7–10, write short answers.*

7. Consider the inverse sine function, defined by $y = \sin^{-1} x$ or $y = \arcsin x$.
   (a) What is its domain?   (b) What is its range?
   (c) Is this function increasing or decreasing?
   (d) Why is $\arcsin(-2)$ not defined?

8. Consider the inverse cosine function, defined by $y = \cos^{-1} x$ or $y = \arccos x$.

    (a) What is its domain?   **(b)** What is its range?

    (c) Is this function increasing or decreasing?

    (d) $\text{Arccos}\left(-\frac{1}{2}\right) = \frac{2\pi}{3}$. Why is $\arccos\left(-\frac{1}{2}\right)$ not equal to $-\frac{4\pi}{3}$?

9. Consider the inverse tangent function, defined by $y = \tan^{-1} x$ or $y = \arctan x$.

    (a) What is its domain?   **(b)** What is its range?

    (c) Is this function increasing or decreasing?

    (d) Is there any real number x for which $\arctan x$ is not defined? If so, what is it (or what are they)?

10. Give the domain and range of the three other inverse trigonometric functions, as defined in this section.

    (a) inverse cosecant function   **(b)** inverse secant function

    (c) inverse cotangent function

11. *Concept Check* Is $\sec^{-1} a$ calculated as $\cos^{-1} \frac{1}{a}$ or as $\frac{1}{\cos^{-1} a}$?

12. *Concept Check* For positive values of a, $\cot^{-1} a$ is calculated as $\tan^{-1} \frac{1}{a}$. How is $\cot^{-1} a$ calculated for negative values of a?

Find the exact value of each real number y. Do not use a calculator. See Examples 1 and 2.

13. $y = \sin^{-1} 0$ **14.** $y = \tan^{-1} 1$ **15.** $y = \cos^{-1}(-1)$

16. $y = \arctan(-1)$ **17.** $y = \sin^{-1}(-1)$ **18.** $y = \cos^{-1}\frac{1}{2}$

19. $y = \arctan 0$ **20.** $y = \arcsin\left(-\frac{\sqrt{3}}{2}\right)$ **21.** $y = \arccos 0$

22. $y = \tan^{-1}(-1)$ **23.** $y = \sin^{-1}\frac{\sqrt{2}}{2}$ **24.** $y = \cos^{-1}\left(-\frac{1}{2}\right)$

25. $y = \arccos\left(-\frac{\sqrt{3}}{2}\right)$ **26.** $y = \arcsin\left(-\frac{\sqrt{2}}{2}\right)$ **27.** $y = \cot^{-1}(-1)$

28. $y = \sec^{-1}(-\sqrt{2})$ **29.** $y = \csc^{-1}(-2)$ **30.** $y = \text{arccot}(-\sqrt{3})$

31. $y = \text{arcsec}\frac{2\sqrt{3}}{3}$ **32.** $y = \csc^{-1}\sqrt{2}$ **33.** $y = \sec^{-1} 1$

34. *Concept Check* Is there a value for y such that $y = \sec^{-1} 0$?

Give the degree measure of θ. Do not use a calculator. See Example 3.

35. $\theta = \arctan(-1)$ **36.** $\theta = \arccos\left(-\frac{1}{2}\right)$ **37.** $\theta = \arcsin\left(-\frac{\sqrt{3}}{2}\right)$

38. $\theta = \arcsin\left(-\frac{\sqrt{2}}{2}\right)$ **39.** $\theta = \cot^{-1}\left(-\frac{\sqrt{3}}{3}\right)$ **40.** $\theta = \csc^{-1}(-2)$

41. $\theta = \sec^{-1}(-2)$ **42.** $\theta = \csc^{-1}(-1)$ **43.** $\theta = \tan^{-1}\sqrt{3}$

44. $\theta = \cot^{-1}\frac{\sqrt{3}}{3}$ **45.** $\theta = \sin^{-1} 2$ **46.** $\theta = \cos^{-1}(-2)$

Use a calculator to give each value in decimal degrees. See Example 4.

47. $\theta = \sin^{-1}(-.13349122)$ **48.** $\theta = \cos^{-1}(-.13348816)$

49. $\theta = \arccos(-.39876459)$ **50.** $\theta = \arcsin .77900016$

51. $\theta = \csc^{-1} 1.9422833$

52. $\theta = \cot^{-1} 1.7670492$

53. $\theta = \cot^{-1}(-.60724226)$

54. $\theta = \cot^{-1}(-2.7733744)$

55. $\theta = \tan^{-1}(-7.7828641)$

56. $\theta = \sec^{-1}(-5.1180378)$

Use a calculator to give each real number value. (Be sure the calculator is in radian mode.) See Example 4.

57. $y = \arctan 1.1111111$

58. $y = \arcsin .81926439$

59. $y = \cot^{-1}(-.92170128)$

60. $y = \sec^{-1}(-1.2871684)$

61. $y = \arcsin .92837781$

62. $y = \arccos .44624593$

63. $y = \cos^{-1}(-.32647891)$

64. $y = \sec^{-1} 4.7963825$

65. $y = \cot^{-1}(-36.874610)$

66. $y = \cot^{-1}(1.0036571)$

The screen here shows how to define the inverse secant, cosecant, and cotangent functions in order to graph them using a TI-83/84 Plus graphing calculator.

Use this information to graph each inverse circular function and compare your graphs to those in Figure 14.

67. $y = \sec^{-1} x$

68. $y = \csc^{-1} x$

69. $y = \cot^{-1} x$

Graph each inverse circular function by hand.

70. $y = \text{arccsc } 2x$

71. $y = \text{arcsec } \dfrac{1}{2}x$

72. $y = 2 \cot^{-1} x$

73. *Concept Check*   Explain why attempting to find $\sin^{-1} 1.003$ on your calculator will result in an error message.

RELATING CONCEPTS

For individual or collaborative investigation
(Exercises 74–76)*

74. Consider the function defined by $f(x) = 3x - 2$ and its inverse $f^{-1}(x) = \frac{1}{3}x + \frac{2}{3}$. Simplify $f(f^{-1}(x))$ and $f^{-1}(f(x))$. What do you notice in each case? What would the graph look like in each case?

75. Use a graphing calculator to graph $y = \tan(\tan^{-1} x)$ in the standard viewing window, using radian mode. How does this compare to the graph you described in Exercise 74?

76. Use a graphing calculator to graph $y = \tan^{-1}(\tan x)$ in the standard viewing window, using radian and dot modes. Why does this graph not agree with the graph you found in Exercise 75?

---

*The authors wish to thank Carol Walker of Hinds Community College for making a suggestion on which these exercises are based.

Give the exact value of each expression without using a calculator. See Examples 5 and 6.

77. $\tan\left(\arccos \dfrac{3}{4}\right)$ **78.** $\sin\left(\arccos \dfrac{1}{4}\right)$ **79.** $\cos(\tan^{-1}(-2))$

80. $\sec\left(\sin^{-1}\left(-\dfrac{1}{5}\right)\right)$ **81.** $\sin\left(2\tan^{-1} \dfrac{12}{5}\right)$ **82.** $\cos\left(2\sin^{-1} \dfrac{1}{4}\right)$

83. $\cos\left(2\arctan \dfrac{4}{3}\right)$ **84.** $\tan\left(2\cos^{-1} \dfrac{1}{4}\right)$ **85.** $\sin\left(2\cos^{-1} \dfrac{1}{5}\right)$

86. $\cos(2\tan^{-1}(-2))$ **87.** $\sec(\sec^{-1} 2)$ **88.** $\csc\left(\csc^{-1} \sqrt{2}\right)$

89. $\cos\left(\tan^{-1} \dfrac{5}{12} - \tan^{-1} \dfrac{3}{4}\right)$ **90.** $\cos\left(\sin^{-1} \dfrac{3}{5} + \cos^{-1} \dfrac{5}{13}\right)$

91. $\sin\left(\sin^{-1} \dfrac{1}{2} + \tan^{-1}(-3)\right)$ **92.** $\tan\left(\cos^{-1} \dfrac{\sqrt{3}}{2} - \sin^{-1}\left(-\dfrac{3}{5}\right)\right)$

Use a calculator to find each value. Give answers as real numbers.

93. $\cos(\tan^{-1} .5)$ **94.** $\sin(\cos^{-1} .25)$

95. $\tan(\arcsin .12251014)$ **96.** $\cot(\arccos .58236841)$

Write each expression as an algebraic (nontrigonometric) expression in u, u > 0. See Example 7.

97. $\sin(\arccos u)$ **98.** $\tan(\arccos u)$ **99.** $\cos(\arcsin u)$

100. $\cot(\arcsin u)$ **101.** $\sin\left(2\sec^{-1} \dfrac{u}{2}\right)$ **102.** $\cos\left(2\tan^{-1} \dfrac{3}{u}\right)$

103. $\tan\left(\sin^{-1} \dfrac{u}{\sqrt{u^2 + 2}}\right)$ **104.** $\sec\left(\cos^{-1} \dfrac{u}{\sqrt{u^2 + 5}}\right)$

105. $\sec\left(\text{arccot} \dfrac{\sqrt{4 - u^2}}{u}\right)$ **106.** $\csc\left(\arctan \dfrac{\sqrt{9 - u^2}}{u}\right)$

(Modeling) *Solve each problem.*

107. *Angle of Elevation of a Shot Put*   Refer to Example 8.
   (a) What is the optimal angle when $h = 0$?
   (b) Fix h at 6 ft and regard θ as a function of v. As v gets larger and larger, the graph approaches an asymptote. Find the equation of that asymptote.

108. *Landscaping Formula*   A shrub is planted in a 100-ft-wide space between buildings measuring 75 ft and 150 ft tall. The location of the shrub determines how much sun it receives each day. Show that if θ is the angle in the figure and x is the distance of the shrub from the taller building, then the value of θ (in radians) is given by

$$\theta = \pi - \arctan\left(\dfrac{75}{100 - x}\right) - \arctan\left(\dfrac{150}{x}\right).$$

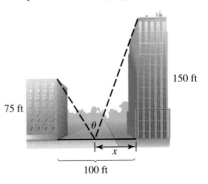

150 ft

75 ft

θ

x

100 ft

109. *Observation of a Painting*   A painting 1 m high and 3 m from the floor will cut off an angle θ to an observer, where

$$\theta = \tan^{-1}\left(\frac{x}{x^2 + 2}\right).$$

Assume that the observer is x meters from the wall where the painting is displayed and that the eyes of the observer are 2 m above the ground. (See the figure.) Find the value of θ for the following values of x. Round to the nearest degree.

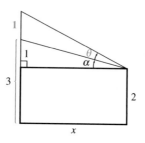

(a) 1   **(b)** 2   **(c)** 3

(d) Derive the formula given above. (*Hint:* Use the identity for $\tan(\theta + \alpha)$. Use right triangles.)

(e) Graph the function for θ with a graphing calculator, and determine the distance that maximizes the angle.

(f) The idea in part (e) was first investigated in 1471 by the astronomer Regiomontanus. (*Source:* Maor, E., *Trigonometric Delights,* Princeton University Press, 1998.) If the bottom of the picture is a meters above eye level and the top of the picture is b meters above eye level, then the optimum value of x is $\sqrt{ab}$ meters. Use this result to find the exact answer to part (e).

110. *Communications Satellite Coverage*   The figure shows a stationary communications satellite positioned 20,000 mi above the equator. What percent, to the nearest tenth, of the equator can be seen from the satellite? The diameter of Earth is 7927 mi at the equator.

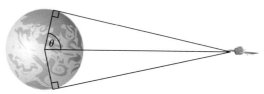

6.2 Trigonometric Equations I

Solving by Linear Methods ▪ **Solving by Factoring** ▪ **Solving by Quadratic Methods** ▪ **Solving by Using Trigonometric Identities**

▼ LOOKING AHEAD TO CALCULUS

There are many instances in calculus where it is necessary to solve trigonometric equations. Examples include solving related-rates problems and optimization problems.

In **Chapter 5,** we studied trigonometric equations that were identities. We now consider trigonometric equations that are *conditional;* that is, equations that are satisfied by some values but not others. (Appendix A)

Solving by Linear Methods   Conditional equations with trigonometric (or circular) functions can usually be solved using algebraic methods and trigonometric identities.

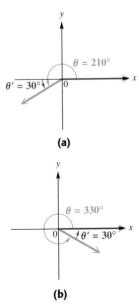

(a)

(b)

Figure 22

▶ **EXAMPLE 1** SOLVING A TRIGONOMETRIC EQUATION BY LINEAR METHODS

Solve $2 \sin \theta + 1 = 0$ over the interval $[0°, 360°)$.

Solution Because $\sin \theta$ is the first power of a trigonometric function, we use the same method as we would to solve the linear equation $2x + 1 = 0$.

$$2 \sin \theta + 1 = 0$$
$$2 \sin \theta = -1 \qquad \text{Subtract 1. (Appendix A)}$$
$$\sin \theta = -\frac{1}{2} \qquad \text{Divide by 2.}$$

To find values of θ that satisfy $\sin \theta = -\frac{1}{2}$, we observe that θ must be in either quadrant III or IV since the sine function is negative only in these two quadrants. Furthermore, the reference angle must be 30° since $\sin 30° = \frac{1}{2}$. The graphs in Figure 22 show the two possible values of θ, 210° and 330°. The solution set is $\{210°, 330°\}$.

Alternatively, we could determine the solutions by referring to Figure 12 in **Section 3.3** on page 119.

NOW TRY EXERCISE 11. ◀

Solving by Factoring

▶ **EXAMPLE 2** SOLVING A TRIGONOMETRIC EQUATION BY FACTORING

Solve $\sin \theta \tan \theta = \sin \theta$ over the interval $[0°, 360°)$.

Solution
$$\sin \theta \tan \theta = \sin \theta$$
$$\sin \theta \tan \theta - \sin \theta = 0 \qquad \text{Subtract } \sin \theta.$$
$$\sin \theta (\tan \theta - 1) = 0 \qquad \text{Factor out } \sin \theta.$$
$$\sin \theta = 0 \quad \text{or} \quad \tan \theta - 1 = 0 \quad \text{Zero-factor property (Appendix A)}$$
$$\tan \theta = 1$$
$$\theta = 0° \quad \text{or} \quad \theta = 180° \qquad \theta = 45° \quad \text{or} \quad \theta = 225°$$

The solution set is $\{0°, 45°, 180°, 225°\}$.

NOW TRY EXERCISE 31. ◀

▶ **Caution** There are four solutions in Example 2. Trying to solve the equation by dividing each side by $\sin \theta$ would lead to just $\tan \theta = 1$, which would give $\theta = 45°$ or $\theta = 225°$. The other two solutions would not appear. The missing solutions are the ones that make the divisor, $\sin \theta$, equal 0. *For this reason, we avoid dividing by a variable expression.*

Solving by Quadratic Methods An equation in the form $au^2 + bu + c = 0$, where u is an algebraic expression, is solved by quadratic methods. (Appendix A) The expression u may be a trigonometric function, as in the equation $\tan^2 x + \tan x - 2 = 0$ which we solve in the next example.

▶ **EXAMPLE 3**   SOLVING A TRIGONOMETRIC EQUATION BY FACTORING

Solve $\tan^2 x + \tan x - 2 = 0$ over the interval $[0, 2\pi)$.

Solution   This equation is quadratic in form and can be solved by factoring.

$$\tan^2 x + \tan x - 2 = 0$$

$$(\tan x - 1)(\tan x + 2) = 0 \qquad \text{Factor.}$$

$$\tan x - 1 = 0 \quad \text{or} \quad \tan x + 2 = 0 \qquad \text{Zero-factor property}$$

$$\tan x = 1 \quad \text{or} \quad \tan x = -2 \qquad \text{Solve each equation.}$$

The solutions for $\tan x = 1$ over the interval $[0, 2\pi)$ are $x = \frac{\pi}{4}$ and $x = \frac{5\pi}{4}$.

To solve $\tan x = -2$ over that interval, we use a scientific calculator set in *radian* mode. We find that $\tan^{-1}(-2) \approx -1.1071487$. This is a quadrant IV number, based on the range of the inverse tangent function. (Refer to Figure 12 in **Section 3.3** on page 119.) However, since we want solutions over the interval $[0, 2\pi)$, we must first add π to -1.1071487, and then add 2π.

$$x \approx -1.1071487 + \pi \approx 2.0344439$$

$$x \approx -1.1071487 + 2\pi \approx 5.1760366$$

The solutions over the required interval form the solution set

$$\left\{ \frac{\pi}{4}, \quad \frac{5\pi}{4}, \quad 2.0344, \quad 5.1760 \right\}.$$

Exact values

Approximate values to four decimal places

NOW TRY EXERCISE 21. ◀

▶ **EXAMPLE 4**   SOLVING A TRIGONOMETRIC EQUATION USING THE QUADRATIC FORMULA

Find all solutions of $\cot x(\cot x + 3) = 1$. Write the solution set.

Solution   We multiply the factors on the left and subtract 1 to get the equation in standard quadratic form.

$$\cot^2 x + 3\cot x - 1 = 0 \quad \text{(Appendix A)}$$

Since this equation cannot be solved by factoring, we use the quadratic formula, with $a = 1$, $b = 3$, $c = -1$, and $\cot x$ as the variable.

$$\cot x = \frac{-b \pm \sqrt{b^2 - 4ac}}{2a} \qquad \text{Quadratic formula (Appendix A)}$$

$$= \frac{-3 \pm \sqrt{3^2 - 4(1)(-1)}}{2(1)} \qquad a = 1, b = 3, c = -1$$

Be careful with signs.

$$= \frac{-3 \pm \sqrt{9 + 4}}{2} \qquad \text{Simplify.}$$

$$= \frac{-3 \pm \sqrt{13}}{2}$$

$$\cot x \approx -3.302775638 \quad \text{or} \quad \cot x \approx .3027756377 \qquad \text{Use a calculator.}$$

We cannot find inverse cotangent values directly on a calculator, so we use the fact that $\cot x = \frac{1}{\tan x}$, and take reciprocals to get

$$\tan x \approx \frac{1}{-3.302775638} \quad \text{or} \quad \tan x \approx \frac{1}{.3027756377}$$

$$\tan x \approx -.3027756377 \quad \text{or} \quad \tan x \approx 3.302775638$$

$$x \approx -.2940013018 \quad \text{or} \quad x \approx 1.276795025.$$

To find *all* solutions, we add integer multiples of the period of the tangent function, which is π, to each solution found previously. Thus, the solution set of the equation is written as

$$\{-.2940 + n\pi, \ 1.2768 + n\pi, \text{ where } n \text{ is any integer}\}.$$

NOW TRY EXERCISE 43. ◀

Solving by Using Trigonometric Identities
Recall that squaring both sides of an equation, such as $\sqrt{x + 4} = x + 2$, will yield all solutions but may also give extraneous values. (In this equation, 0 is a solution, while -3 is extraneous. Verify this.)

▶ **EXAMPLE 5** SOLVING A TRIGONOMETRIC EQUATION BY SQUARING

Solve $\tan x + \sqrt{3} = \sec x$ over the interval $[0, 2\pi)$.

Solution Since the tangent and secant functions are related by the identity $1 + \tan^2 x = \sec^2 x$, square both sides and express $\sec^2 x$ in terms of $\tan^2 x$.

$$(\tan x + \sqrt{3})^2 = (\sec x)^2$$

> **Don't forget the middle term.**

$$\tan^2 x + 2\sqrt{3}\tan x + 3 = \sec^2 x \qquad (x + y)^2 = x^2 + 2xy + y^2$$

$$\tan^2 x + 2\sqrt{3}\tan x + 3 = 1 + \tan^2 x \qquad \begin{array}{l}\text{Pythagorean identity}\\ \text{(Section 5.1)}\end{array}$$

$$2\sqrt{3}\tan x = -2 \qquad \text{Subtract } 3 + \tan^2 x.$$

$$\tan x = -\frac{1}{\sqrt{3}} = -\frac{\sqrt{3}}{3} \qquad \begin{array}{l}\text{Divide by } 2\sqrt{3}; \text{ rationalize}\\ \text{the denominator.}\end{array}$$

The possible solutions are $\frac{5\pi}{6}$ and $\frac{11\pi}{6}$. Now check them. Try $\frac{5\pi}{6}$ first.

Left side: $\quad \tan x + \sqrt{3} = \tan \dfrac{5\pi}{6} + \sqrt{3} = -\dfrac{\sqrt{3}}{3} + \sqrt{3} = \dfrac{2\sqrt{3}}{3}$

Right side: $\quad \sec x = \sec \dfrac{5\pi}{6} = -\dfrac{2\sqrt{3}}{3}$ ⟵———— Not equal

The check shows that $\frac{5\pi}{6}$ is not a solution. Now check $\frac{11\pi}{6}$.

Left side: $\quad \tan \dfrac{11\pi}{6} + \sqrt{3} = -\dfrac{\sqrt{3}}{3} + \sqrt{3} = \dfrac{2\sqrt{3}}{3}$

Right side: $\quad \sec \dfrac{11\pi}{6} = \dfrac{2\sqrt{3}}{3}$ ⟵———— Equal

This solution satisfies the equation, so $\left\{\frac{11\pi}{6}\right\}$ is the solution set.

NOW TRY EXERCISE 41. ◀

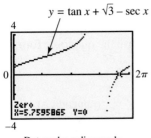

$y = \tan x + \sqrt{3} - \sec x$

Dot mode; radian mode

The graph shows that on the interval $[0, 2\pi)$, the only x-intercept of the graph of $y = \tan x + \sqrt{3} - \sec x$ is 5.7595865, which is an approximation for $\frac{11\pi}{6}$, the solution found in Example 5.

Methods for solving trigonometric equations can be summarized as follows.

SOLVING A TRIGONOMETRIC EQUATION

1. Decide whether the equation is linear or quadratic in form, so you can determine the solution method.

2. If only one trigonometric function is present, solve the equation for that function.

3. If more than one trigonometric function is present, rearrange the equation so that one side equals 0. Then try to factor and set each factor equal to 0 to solve.

4. If the equation is quadratic in form, but not factorable, use the quadratic formula. Check that solutions are in the desired interval.

5. Try using identities to change the form of the equation. It may be helpful to square both sides of the equation first. If this is done, check for extraneous solutions.

▶ **EXAMPLE 6**   **DESCRIBING A MUSICAL TONE FROM A GRAPH**

A basic component of music is a pure tone. The graph in Figure 23 models the sinusoidal pressure $y = P$ in pounds per square foot from a pure tone at time $x = t$ in seconds.

(a) The frequency of a pure tone is often measured in hertz. One hertz is equal to one cycle per second and is abbreviated Hz. What is the frequency f in hertz of the pure tone shown in the graph?

(b) The time for the tone to produce one complete cycle is called the **period.** Approximate the period T in seconds of the pure tone.

(c) An equation for the graph is $y = .004 \sin 300\pi x$. Use a calculator to estimate all solutions to the equation that make $y = .004$ over the interval $[0, .02]$.

Solution

(a) From the graph in Figure 23, we see that there are 6 cycles in .04 sec. This is equivalent to $\frac{6}{.04} = 150$ cycles per sec. The pure tone has a frequency of $f = 150$ Hz.

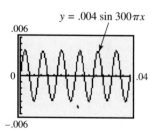

Figure 23

(b) Six periods cover a time of .04 sec. One period would be equal to $T = \frac{.04}{6} = \frac{1}{150}$, or $.00\overline{6}$ sec.

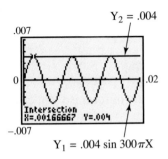

Figure 24

(c) If we reproduce the graph in Figure 23 on a calculator as Y_1 and also graph a second function as $Y_2 = .004$, we can determine that the approximate values of x at the points of intersection of the graphs over the interval $[0, .02]$ are

$$.0017, \quad .0083, \quad \text{and} \quad .015.$$

The first value is shown in Figure 24. These values represent time in seconds.

NOW TRY EXERCISE 53. ◀

6.2 Exercises

Concept Check *Refer to the summary box on solving a trigonometric equation on page 277. Decide on the appropriate technique to begin the solution of each equation. Do not solve the equation.*

1. $2 \cot x + 1 = -1$
2. $\sin x + 2 = 3$
3. $5 \sec^2 x = 6 \sec x$
4. $2 \cos^2 x - \cos x = 1$
5. $9 \sin^2 x - 5 \sin x = 1$
6. $\tan^2 x - 4 \tan x + 2 = 0$
7. $\tan x - \cot x = 0$
8. $\cos^2 x = \sin^2 x + 1$

9. Suppose in solving an equation over the interval $[0°, 360°)$, you reach the step $\sin \theta = -\frac{1}{2}$. Why is $-30°$ not a correct answer?

10. Lindsay solved the equation $\sin x = 1 - \cos x$ by squaring both sides to get $\sin^2 x = 1 - 2 \cos x + \cos^2 x$. Several steps later, using correct algebra, she determined that the solution set for solutions over the interval $[0, 2\pi)$ is $\left\{0, \frac{\pi}{2}, \frac{3\pi}{2}\right\}$. Explain why this is not the correct solution set.

Solve each equation for exact solutions over the interval $[0, 2\pi)$. See Examples 1–3.

11. $2 \cot x + 1 = -1$
12. $\sin x + 2 = 3$
13. $2 \sin x + 3 = 4$
14. $2 \sec x + 1 = \sec x + 3$
15. $\tan^2 x + 3 = 0$
16. $\sec^2 x + 2 = -1$
17. $(\cot x - 1)\left(\sqrt{3} \cot x + 1\right) = 0$
18. $(\csc x + 2)\left(\csc x - \sqrt{2}\right) = 0$
19. $\cos^2 x + 2 \cos x + 1 = 0$
20. $2 \cos^2 x - \sqrt{3} \cos x = 0$
21. $-2 \sin^2 x = 3 \sin x + 1$
22. $2 \cos^2 x - \cos x = 1$

Solve each equation for solutions over the interval $[0°, 360°)$. Give solutions to the nearest tenth as appropriate. See Examples 2–5.

23. $\left(\cot \theta - \sqrt{3}\right)\left(2 \sin \theta + \sqrt{3}\right) = 0$
24. $(\tan \theta - 1)(\cos \theta - 1) = 0$
25. $2 \sin \theta - 1 = \csc \theta$
26. $\tan \theta + 1 = \sqrt{3} + \sqrt{3} \cot \theta$
27. $\tan \theta - \cot \theta = 0$
28. $\cos^2 \theta = \sin^2 \theta + 1$
29. $\csc^2 \theta - 2 \cot \theta = 0$
30. $\sin^2 \theta \cos \theta = \cos \theta$
31. $2 \tan^2 \theta \sin \theta - \tan^2 \theta = 0$
32. $\sin^2 \theta \cos^2 \theta = 0$
33. $\sec^2 \theta \tan \theta = 2 \tan \theta$
34. $\cos^2 \theta - \sin^2 \theta = 0$
35. $9 \sin^2 \theta - 6 \sin \theta = 1$
36. $4 \cos^2 \theta + 4 \cos \theta = 1$
37. $\tan^2 \theta + 4 \tan \theta + 2 = 0$
38. $3 \cot^2 \theta - 3 \cot \theta - 1 = 0$
39. $\sin^2 \theta - 2 \sin \theta + 3 = 0$
40. $2 \cos^2 \theta + 2 \cos \theta - 1 = 0$
41. $\cot \theta + 2 \csc \theta = 3$
42. $2 \sin \theta = 1 - 2 \cos \theta$

Determine the solution set of each equation in radians (for x) to four decimal places or degrees (for θ) to the nearest tenth as appropriate. See Example 4.

43. $3 \sin^2 x - \sin x - 1 = 0$

44. $2 \cos^2 x + \cos x = 1$

45. $4 \cos^2 x - 1 = 0$

46. $2 \cos^2 x + 5 \cos x + 2 = 0$

47. $5 \sec^2 \theta = 6 \sec \theta$

48. $3 \sin^2 \theta - \sin \theta = 2$

49. $\dfrac{2 \tan \theta}{3 - \tan^2 \theta} = 1$

50. $\sec^2 \theta = 2 \tan \theta + 4$

The following equations cannot be solved by algebraic methods. Use a graphing calculator to find all solutions over the interval $[0, 2\pi)$. Express solutions to four decimal places.

51. $x^2 + \sin x - x^3 - \cos x = 0$

52. $x^3 - \cos^2 x = \dfrac{1}{2}x - 1$

(Modeling) *Solve each problem.*

53. *Pressure on the Eardrum*   See Example 6. No musical instrument can generate a true pure tone. A pure tone has a unique, constant frequency and amplitude that sounds rather dull and uninteresting. The pressures caused by pure tones on the eardrum are sinusoidal. The change in pressure P in pounds per square foot on a person's eardrum from a pure tone at time t in seconds can be modeled using the equation

$$P = A \sin(2\pi f t + \phi),$$

where f is the frequency in cycles per second, and ϕ is the phase angle. When P is positive, there is an increase in pressure and the eardrum is pushed inward; when P is negative, there is a decrease in pressure and the eardrum is pushed outward. (*Source:* Roederer, J., *Introduction to the Physics and Psychophysics of Music*, Second Edition, Springer-Verlag, 1975.) A graph of the tone middle C is shown in the figure.

(a) Determine algebraically the values of t for which $P = 0$ over $[0, .005]$.

(b) From the graph and your answer in part (a), determine the interval for which $P \le 0$ over $[0, .005]$.

(c) Would an eardrum hearing this tone be vibrating outward or inward when $P < 0$?

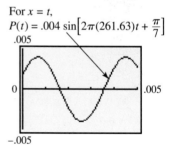

For $x = t$,
$P(t) = .004 \sin\left[2\pi(261.63)t + \dfrac{\pi}{7}\right]$

54. *Accident Reconstruction*   The model

$$.342D \cos \theta + h \cos^2 \theta = \dfrac{16D^2}{V_0^2}$$

is used to reconstruct accidents in which a vehicle vaults into the air after hitting an obstruction. V_0 is velocity in feet per second of the vehicle when it hits, D is distance (in feet) from the obstruction to the landing point, and h is the difference in height (in feet) between landing point and takeoff point. Angle θ is the takeoff angle, the angle between the horizontal and the path of the vehicle. Find θ to the nearest degree if $V_0 = 60$, $D = 80$, and $h = 2$.

55. *Electromotive Force*   In an electric circuit, let

$$V = \cos 2\pi t$$

model the electromotive force in volts at t seconds. Find the least value of t where $0 \le t \le \frac{1}{2}$ for each value of V.

(a) $V = 0$

(b) $V = .5$

(c) $V = .25$

56. *Voltage Induced by a Coil of Wire* A coil of wire rotating in a magnetic field induces a voltage modeled by

$$E = 20 \sin\left(\frac{\pi t}{4} - \frac{\pi}{2}\right),$$

where t is time in seconds. Find the least positive time to produce each voltage.

(a) 0 (b) $10\sqrt{3}$

57. *Movement of a Particle* A particle moves along a straight line. The distance of the particle from the origin at time t is modeled by

$$s(t) = \sin t + 2 \cos t.$$

Find a value of t that satisfies each equation.

(a) $s(t) = \dfrac{2 + \sqrt{3}}{2}$ (b) $s(t) = \dfrac{3\sqrt{2}}{2}$

58. Explain what is **WRONG** with the following solution for all x over the interval $[0, 2\pi)$ of the equation $\sin^2 x - \sin x = 0$.

$$\sin^2 x - \sin x = 0$$
$$\sin x - 1 = 0 \quad \text{Divide by } \sin x.$$
$$\sin x = 1 \quad \text{Add 1.}$$
$$x = \frac{\pi}{2}$$

The solution set is $\left\{\frac{\pi}{2}\right\}$.

6.3 Trigonometric Equations II

Equations with Half-Angles ▪ Equations with Multiple Angles

In this section, we discuss trigonometric equations that involve functions of half-angles and multiple angles. Solving these equations often requires adjusting solution intervals to fit given domains.

Equations with Half-Angles

▶ **EXAMPLE 1** SOLVING AN EQUATION USING A HALF-ANGLE IDENTITY

Solve $2 \sin \dfrac{x}{2} = 1$

(a) over the interval $[0, 2\pi)$, and (b) give all solutions.

Solution

(a) Write the interval $[0, 2\pi)$ as the inequality

$$0 \le x < 2\pi.$$

The corresponding interval for $\frac{x}{2}$ is

$$0 \le \frac{x}{2} < \pi. \quad \text{Divide by 2. (Appendix A)}$$

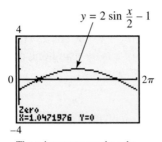

$y = 2 \sin \frac{x}{2} - 1$

Zero
X=1.0471976 Y=0

The x-intercepts are the solutions found in Example 1.
Using Xscl $= \frac{\pi}{3}$ makes it possible to support the exact solutions by counting the tick marks from 0 on the graph.

To find all values of $\frac{x}{2}$ over the interval $[0, \pi)$ that satisfy the given equation, first solve for $\sin \frac{x}{2}$.

$$2 \sin \frac{x}{2} = 1$$

$$\sin \frac{x}{2} = \frac{1}{2} \qquad \text{Divide by 2.}$$

The two numbers over the interval $[0, \pi)$ with sine value $\frac{1}{2}$ are $\frac{\pi}{6}$ and $\frac{5\pi}{6}$, so

$$\frac{x}{2} = \frac{\pi}{6} \qquad \text{or} \qquad \frac{x}{2} = \frac{5\pi}{6}$$

$$x = \frac{\pi}{3} \qquad \text{or} \qquad x = \frac{5\pi}{3}. \qquad \text{Multiply by 2.}$$

The solution set over the given interval is $\left\{\frac{\pi}{3}, \frac{5\pi}{3}\right\}$.

(b) Since this is a sine function with period 4π, the solution set is

$$\left\{\frac{\pi}{3} + 4n\pi, \frac{5\pi}{3} + 4n\pi, \text{ where } n \text{ is any integer}\right\}.$$

NOW TRY EXERCISE 15. ◀

Equations with Multiple Angles

▶ **EXAMPLE 2**   SOLVING AN EQUATION WITH A DOUBLE ANGLE

Solve $\cos 2x = \cos x$ over the interval $[0, 2\pi)$.

Solution   First change $\cos 2x$ to a trigonometric function of x. Use the identity $\cos 2x = 2 \cos^2 x - 1$ so the equation involves only $\cos x$. Then factor.

$$\cos 2x = \cos x$$

$$2 \cos^2 x - 1 = \cos x \qquad \text{Substitute; double-angle identity (Section 5.5)}$$

$$2 \cos^2 x - \cos x - 1 = 0 \qquad \text{Subtract } \cos x.$$

$$(2 \cos x + 1)(\cos x - 1) = 0 \qquad \text{Factor.}$$

$$2 \cos x + 1 = 0 \qquad \text{or} \qquad \cos x - 1 = 0 \qquad \text{Zero-factor property (Appendix A)}$$

$$\cos x = -\frac{1}{2} \qquad \text{or} \qquad \cos x = 1 \qquad \text{Solve each equation.}$$

Cosine is $-\frac{1}{2}$ in quadrants II and III with reference arc $\frac{\pi}{3}$, and has a value of 1 at 0 radians; thus, solutions over the required interval are

$$x = \frac{2\pi}{3} \qquad \text{or} \qquad x = \frac{4\pi}{3} \qquad \text{or} \qquad x = 0.$$

The solution set is $\left\{0, \frac{2\pi}{3}, \frac{4\pi}{3}\right\}$.

NOW TRY EXERCISE 17. ◀

▶ **Caution** In the solution of Example 2, cos 2x cannot be changed to cos x by dividing by 2 since 2 is not a factor of cos 2x, that is, $\frac{\cos 2x}{2} \neq \cos x$. The only way to change cos 2x to a trigonometric function of x is by using one of the identities for cos 2x.

▶ **EXAMPLE 3** SOLVING AN EQUATION USING A MULTIPLE-ANGLE IDENTITY

Solve $4 \sin \theta \cos \theta = \sqrt{3}$ over the interval $[0°, 360°)$.

Solution   The identity $2 \sin \theta \cos \theta = \sin 2\theta$ is useful here.

$$4 \sin \theta \cos \theta = \sqrt{3}$$
$$2(2 \sin \theta \cos \theta) = \sqrt{3} \qquad 4 = 2 \cdot 2$$
$$2 \sin 2\theta = \sqrt{3} \qquad 2 \sin \theta \cos \theta = \sin 2\theta \text{ (Section 5.5)}$$
$$\sin 2\theta = \frac{\sqrt{3}}{2} \qquad \text{Divide by 2.}$$

From the given interval $0° \leq \theta < 360°$, the interval for 2θ is $0° \leq 2\theta < 720°$. Since the sine is positive in quadrants I and II, solutions over this interval are

$$2\theta = 60°, 120°, 420°, 480°,$$
or $$\theta = 30°, 60°, 210°, 240°. \qquad \text{Divide by 2.}$$

The final two solutions for 2θ were found by adding 360° to 60° and 120°, respectively, giving the solution set $\{30°, 60°, 210°, 240°\}$.

NOW TRY EXERCISE 37. ◀

▶ **EXAMPLE 4** SOLVING AN EQUATION WITH A MULTIPLE ANGLE

Solve $\tan 3x + \sec 3x = 2$ over the interval $[0, 2\pi)$.

Solution   Since the tangent and secant functions are related by the identity $1 + \tan^2 \theta = \sec^2 \theta$, one way to begin is to express everything in terms of secant.

$$\tan 3x + \sec 3x = 2$$

> Don't forget the middle term.

$$\tan 3x = 2 - \sec 3x \qquad \text{Subtract sec } 3x.$$
$$\tan^2 3x = 4 - 4 \sec 3x + \sec^2 3x \qquad \begin{array}{l} \text{Square both sides;} \\ (x - y)^2 = x^2 - 2xy + y^2. \end{array}$$
$$\sec^2 3x - 1 = 4 - 4 \sec 3x + \sec^2 3x \qquad \begin{array}{l} \text{Replace } \tan^2 3x \text{ with } \sec^2 3x - 1. \\ \text{(Section 5.1)} \end{array}$$
$$4 \sec 3x = 5 \qquad \text{Simplify.}$$
$$\sec 3x = \frac{5}{4} \qquad \text{Divide by 4.}$$
$$\frac{1}{\cos 3x} = \frac{5}{4} \qquad \sec \theta = \frac{1}{\cos \theta} \text{ (Section 5.1)}$$
$$\cos 3x = \frac{4}{5} \qquad \text{Use reciprocals.}$$

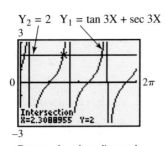

$Y_2 = 2$ $Y_1 = \tan 3X + \sec 3X$

Connected mode; radian mode

The screen shows that one solution is approximately 2.3089. An advantage of using a graphing calculator is that extraneous values do not appear.

Multiply each term of the inequality $0 \le x < 2\pi$ by 3 to find the interval for $3x$: $[0, 6\pi)$. Using a calculator and the fact that cosine is positive in quadrants I and IV,

$$3x \approx .6435, 5.6397, 6.9267, 11.9229, 13.2099, 18.2061$$
$$x \approx .2145, 1.8799, 2.3089, 3.9743, 4.4033, 6.0687. \qquad \text{Divide by 3.}$$

Since both sides of the equation were squared, each proposed solution must be checked. Verify by substitution in the given equation that the solution set is $\{.2145, 2.3089, 4.4033\}$.

NOW TRY EXERCISE 33. ◀

A piano string can vibrate at more than one frequency when it is struck. It produces a complex wave that can mathematically be modeled by a sum of several pure tones. If a piano key with a frequency of f_1 is played, then the corresponding string will not only vibrate at f_1 but it will also vibrate at the higher frequencies of $2f_1, 3f_1, 4f_1, \ldots, nf_1$. f_1 is called the **fundamental frequency** of the string, and higher frequencies are called the **upper harmonics.** The human ear will hear the sum of these frequencies as one complex tone. (*Source:* Roederer, J., *Introduction to the Physics and Psychophysics of Music,* Second Edition, Springer-Verlag, 1975.)

▶ EXAMPLE 5   ANALYZING PRESSURES OF UPPER HARMONICS

Suppose that the A key above middle C is played. Its fundamental frequency is $f_1 = 440$ Hz, and its associated pressure is expressed as

$$P_1 = .002 \sin 880\pi t.$$

The string will also vibrate at

$$f_2 = 880, \quad f_3 = 1320, \quad f_4 = 1760, \quad f_5 = 2200, \ldots \text{Hz.}$$

The corresponding pressures of these upper harmonics are

$$P_2 = \frac{.002}{2} \sin 1760\pi t, \qquad P_3 = \frac{.002}{3} \sin 2640\pi t,$$

$$P_4 = \frac{.002}{4} \sin 3520\pi t, \qquad \text{and} \qquad P_5 = \frac{.002}{5} \sin 4400\pi t.$$

The graph of

$$P = P_1 + P_2 + P_3 + P_4 + P_5,$$

shown in Figure 25, is "saw-toothed."

(a) What is the maximum value of P?

(b) At what values of $t = x$ does this maximum occur over the interval $[0, .01]$?

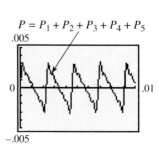

Figure 25

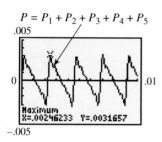

Figure 26

Solution

(a) A graphing calculator shows that the maximum value of P is approximately .00317. See Figure 26.

(b) The maximum occurs at $t = x \approx .000188, .00246, .00474, .00701,$ and .00928. Figure 26 shows how the second value is found; the others are found similarly.

NOW TRY EXERCISE 43. ◀

6.3 Exercises

Concept Check   Answer each question.

1. Suppose you are solving a trigonometric equation for solutions over the interval $[0, 2\pi)$, and your work leads to $2x = \frac{2\pi}{3}, 2\pi, \frac{8\pi}{3}$. What are the corresponding values of x?

2. Suppose you are solving a trigonometric equation for solutions over the interval $[0, 2\pi)$, and your work leads to $\frac{1}{2}x = \frac{\pi}{16}, \frac{5\pi}{12}, \frac{5\pi}{8}$. What are the corresponding values of x?

3. Suppose you are solving a trigonometric equation for solutions over the interval $[0°, 360°)$, and your work leads to $3\theta = 180°, 630°, 720°, 930°$. What are the corresponding values of θ?

4. Suppose you are solving a trigonometric equation for solutions over the interval $[0°, 360°)$, and your work leads to $\frac{1}{3}\theta = 45°, 60°, 75°, 90°$. What are the corresponding values of θ?

5. Explain what is **WRONG** with the following solution.
   Solve $\tan 2\theta = 2$ over the interval $[0, 2\pi)$.

$$\tan 2\theta = 2$$

$$\frac{\tan 2\theta}{2} = \frac{2}{2}$$

$$\tan \theta = 1$$

$$\theta = \frac{\pi}{4} \quad \text{or} \quad \theta = \frac{5\pi}{4}$$

   The solution set is $\left\{\frac{\pi}{4}, \frac{5\pi}{4}\right\}$.

6. The equation $\cot \frac{x}{2} - \csc \frac{x}{2} - 1 = 0$ has no solution over the interval $[0, 2\pi)$. Using this information, what can be said about the graph of

$$y = \cot \frac{x}{2} - \csc \frac{x}{2} - 1$$

   over this interval? Confirm your answer by actually graphing the function over the interval.

Solve each equation for exact solutions over the interval $[0, 2\pi)$. See Examples 1–4.

7. $\cos 2x = \dfrac{\sqrt{3}}{2}$

8. $\cos 2x = -\dfrac{1}{2}$

9. $\sin 3x = -1$

10. $\sin 3x = 0$

11. $3 \tan 3x = \sqrt{3}$

12. $\cot 3x = \sqrt{3}$

13. $\sqrt{2} \cos 2x = -1$

14. $2\sqrt{3} \sin 2x = \sqrt{3}$

15. $\sin \dfrac{x}{2} = \sqrt{2} - \sin \dfrac{x}{2}$

16. $\tan 4x = 0$

17. $\sin x = \sin 2x$

18. $\cos 2x - \cos x = 0$

19. $8 \sec^2 \frac{x}{2} = 4$ **20.** $\sin^2 \frac{x}{2} - 2 = 0$ **21.** $\sin \frac{x}{2} = \cos \frac{x}{2}$

22. $\sec \frac{x}{2} = \cos \frac{x}{2}$ **23.** $\cos 2x + \cos x = 0$ **24.** $\sin x \cos x = \frac{1}{4}$

Solve each equation in Exercises 25–32 for exact solutions over the interval $[0°, 360°)$. In Exercises 33–40, give all solutions. If necessary, express solutions to the nearest tenth of a degree. See Examples 1–4.

25. $\sqrt{2} \sin 3\theta - 1 = 0$ **26.** $-2 \cos 2\theta = \sqrt{3}$ **27.** $\cos \frac{\theta}{2} = 1$

28. $\sin \frac{\theta}{2} = 1$ **29.** $2\sqrt{3} \sin \frac{\theta}{2} = 3$ **30.** $2\sqrt{3} \cos \frac{\theta}{2} = -3$

31. $2 \sin \theta = 2 \cos 2\theta$ **32.** $\cos \theta - 1 = \cos 2\theta$ **33.** $1 - \sin \theta = \cos 2\theta$

34. $\sin 2\theta = 2 \cos^2 \theta$ **35.** $\csc^2 \frac{\theta}{2} = 2 \sec \theta$ **36.** $\cos \theta = \sin^2 \frac{\theta}{2}$

37. $2 - \sin 2\theta = 4 \sin 2\theta$ **38.** $4 \cos 2\theta = 8 \sin \theta \cos \theta$

39. $2 \cos^2 2\theta = 1 - \cos 2\theta$ **40.** $\sin \theta - \sin 2\theta = 0$

The following equations cannot be solved by algebraic methods. Use a graphing calculator to find all solutions over the interval $[0, 2\pi)$. Express solutions to four decimal places.

41. $2 \sin 2x - x^3 + 1 = 0$ **42.** $3 \cos \frac{x}{2} + \sqrt{x} - 2 = -\frac{1}{2}x + 2$

(Modeling)   *Solve each problem. See Example 5.*

43. *Pressure of a Plucked String*   If a string with a fundamental frequency of 110 Hz is plucked in the middle, it will vibrate at the odd harmonics of 110, 330, 550, . . . Hz but not at the even harmonics of 220, 440, 660, . . . Hz. The resulting pressure P caused by the string can be modeled by the equation

$$P = .003 \sin 220\pi t + \frac{.003}{3} \sin 660\pi t + \frac{.003}{5} \sin 1100\pi t + \frac{.003}{7} \sin 1540\pi t.$$

(*Source:* Benade, A., *Fundamentals of Musical Acoustics,* Dover Publications, 1990; Roederer, J., *Introduction to the Physics and Psychophysics of Music,* Second Edition, Springer-Verlag, 1975.)

(a) Graph P in the window $[0, .03]$ by $[-.005, .005]$.

(b) Use the graph to describe the shape of the sound wave that is produced.

(c) See **Section 6.2,** Exercise 53. At lower frequencies, the inner ear will hear a tone only when the eardrum is moving outward. Determine the times over the interval $[0, .03]$ when this will occur.

44. *Hearing Beats in Music*   Musicians sometimes tune instruments by playing the same tone on two different instruments and listening for a phenomenon known as **beats.** Beats occur when two tones vary in frequency by only a few hertz. When the two instruments are in tune, the beats disappear. The ear hears beats because the pressure slowly rises and falls as a result of this slight variation in the frequency.

This phenomenon can be seen using a graphing calculator. (*Source:* Pierce, J., *The Science of Musical Sound,* Scientific American Books, 1992.)

(a) Consider two tones with frequencies of 220 and 223 Hz and pressures $P_1 = .005 \sin 440\pi t$ and $P_2 = .005 \sin 446\pi t$, respectively. Graph the pressure $P = P_1 + P_2$ felt by an eardrum over the 1-sec interval $[.15, 1.15]$. How many beats are there in 1 sec?

(b) Repeat part (a) with frequencies of 220 and 216 Hz.

(c) Determine a simple way to find the number of beats per second if the frequency of each tone is given.

45. *Hearing Difference Tones*   Small speakers like those found in older radios and telephones often cannot vibrate slower than 200 Hz—yet 35 keys on a piano have frequencies below 200 Hz. When a musical instrument creates a tone of 110 Hz, it also creates tones at 220, 330, 440, 550, 660, . . . Hz. A small speaker cannot reproduce the 110-Hz vibration but it can reproduce the higher frequencies, which are called the upper harmonics. The low tones can still be heard because the speaker produces **difference tones** of the upper harmonics. The difference between consecutive frequencies is 110 Hz, and this difference tone will be heard by a listener. We can model this phenomenon using a graphing calculator. (*Source:* Benade, A., *Fundamentals of Musical Acoustics,* Dover Publications, 1990.)

(a) In the window $[0, .03]$ by $[-1, 1]$, graph the upper harmonics represented by the pressure

$$P = \frac{1}{2} \sin[2\pi(220)t] + \frac{1}{3} \sin[2\pi(330)t] + \frac{1}{4} \sin[2\pi(440)t].$$

(b) Estimate all t-coordinates where P is maximum.

(c) What does a person hear in addition to the frequencies of 220, 330, and 440 Hz?

(d) Graph the pressure produced by a speaker that can vibrate at 110 Hz and above.

46. *Daylight Hours in New Orleans*   The seasonal variation in length of daylight can be modeled by a sine function. For example, the daily number of hours of daylight in New Orleans is given by

$$h = \frac{35}{3} + \frac{7}{3} \sin \frac{2\pi x}{365},$$

where x is the number of days after March 21 (disregarding leap year). (*Source:* Bushaw, D., et al., *A Sourcebook of Applications of School Mathematics.* Copyright © 1980 by The Mathematical Association of America.)

(a) On what date will there be about 14 hr of daylight?

(b) What date has the least number of hours of daylight?

(c) When will there be about 10 hr of daylight?

(Modeling) Alternating Electric Current   *The study of alternating electric current requires the solutions of equations of the form*

$$i = I_{max} \sin 2\pi ft,$$

for time t in seconds, where i is instantaneous current in amperes, I_{max} is maximum current in amperes, and f is the number of cycles per second. (Source: Hannon, R. H., Basic Technical Mathematics with Calculus, W. B. Saunders Company, 1978.) Find the least positive value of t, given the following data.

47. $i = 40, I_{max} = 100, f = 60$

48. $i = 50, I_{max} = 100, f = 120$

49. $i = I_{max}, f = 60$

50. $i = \frac{1}{2} I_{max}, f = 60$

Quiz (Sections 6.1–6.3)

1. Graph $y = \cos^{-1} x$, and indicate the coordinates of three points on the graph. Give the domain and range.

2. Find the exact value of each real number y.

   (a) $y = \sin^{-1}\left(-\dfrac{\sqrt{2}}{2}\right)$ (b) $y = \tan^{-1}\sqrt{3}$ (c) $y = \sec^{-1}\left(-\dfrac{2\sqrt{3}}{3}\right)$

3. Use a calculator to give each value in decimal degrees.

   (a) $\theta = \arccos .92341853$ (b) $\theta = \cot^{-1}(-1.0886767)$

4. Give the exact value of each expression without using a calculator.

   (a) $\cos\left(\tan^{-1}\dfrac{4}{5}\right)$ (b) $\sin\left(\cos^{-1}\left(-\dfrac{1}{2}\right) + \tan^{-1}(-\sqrt{3})\right)$

Solve each equation for exact solutions over the interval $[0°, 360°)$.

5. $2\sin\theta - \sqrt{3} = 0$ 6. $\cos\theta + 1 = 2\sin^2\theta$

Solve each equation for solutions over the interval $[0, 2\pi)$.

7. $\tan^2 x - 5\tan x + 3 = 0$ 8. $3\cot 2x - \sqrt{3} = 0$

9. Solve $\cos\dfrac{x}{2} + \sqrt{3} = -\cos\dfrac{x}{2}$, giving all solutions in radians.

10. *Electromotive Force*   In an electric circuit, let

$$V = \cos 2\pi t$$

model the electromotive force in volts at t seconds. Find the least value of t where $0 \le t \le \frac{1}{2}$ for each value of V.

   (a) $V = 1$ (b) $V = .30$

6.4 Equations Involving Inverse Trigonometric Functions

Solving for *x* in Terms of *y* Using Inverse Functions ▪ Solving Inverse Trigonometric Equations

Until now, the equations in this chapter have involved trigonometric functions of angles or real numbers. Now we examine equations involving *inverse* trigonometric functions.

Solving for *x* in Terms of *y* Using Inverse Functions

▶ **EXAMPLE 1**   SOLVING AN EQUATION FOR A VARIABLE USING INVERSE NOTATION

Solve $y = 3\cos 2x$ for x.

Solution   We want $\cos 2x$ alone on one side of the equation so we can solve for $2x$, and then for x.

$$y = 3 \cos 2x \quad \boxed{\text{Our goal is to isolate } x.}$$

$$\frac{y}{3} = \cos 2x \qquad \text{Divide by 3. (Appendix A)}$$

$$2x = \arccos \frac{y}{3} \qquad \text{Definition of arccosine (Section 6.1)}$$

$$\boxed{\text{Do not multiply the argument by } \tfrac{1}{2}.}$$

$$x = \frac{1}{2} \arccos \frac{y}{3} \qquad \text{Multiply by } \tfrac{1}{2}.$$

An equivalent form of this answer is $x = \frac{1}{2} \cos^{-1} \frac{y}{3}$.

NOW TRY EXERCISE 9. ◀

Solving Inverse Trigonometric Equations

▶ **EXAMPLE 2** SOLVING AN EQUATION INVOLVING AN INVERSE TRIGONOMETRIC FUNCTION

Solve $2 \arcsin x = \pi$.

Solution First solve for $\arcsin x$, and then for x.

$$2 \arcsin x = \pi$$

$$\arcsin x = \frac{\pi}{2} \qquad \text{Divide by 2.}$$

$$x = \sin \frac{\pi}{2} \qquad \text{Definition of arcsine (Section 6.1)}$$

$$x = 1 \qquad \text{(Section 3.3)}$$

Verify that the solution satisfies the given equation. The solution set is $\{1\}$.

NOW TRY EXERCISE 25. ◀

▶ **EXAMPLE 3** SOLVING AN EQUATION INVOLVING INVERSE TRIGONOMETRIC FUNCTIONS

Solve $\cos^{-1} x = \sin^{-1} \frac{1}{2}$.

Solution Let $\sin^{-1} \frac{1}{2} = u$. Then $\sin u = \frac{1}{2}$ and for u in quadrant I, we have

$$\cos^{-1} x = \sin^{-1} \frac{1}{2}$$

$$\cos^{-1} x = u \qquad \text{Substitute.}$$

$$\cos u = x. \qquad \text{Alternative form (Section 6.1)}$$

Sketch a triangle and label it using the facts that u is in quadrant I and $\sin u = \frac{1}{2}$. See Figure 27. Since $x = \cos u$, $x = \frac{\sqrt{3}}{2}$, and the solution set is $\left\{\frac{\sqrt{3}}{2}\right\}$. *Check.*

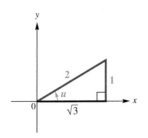

Figure 27

NOW TRY EXERCISE 33. ◀

▶ **EXAMPLE 4**   SOLVING AN INVERSE TRIGONOMETRIC EQUATION USING AN IDENTITY

Solve $\arcsin x - \arccos x = \dfrac{\pi}{6}$.

Solution   Isolate one inverse function on one side of the equation.

$$\arcsin x - \arccos x = \frac{\pi}{6}$$

$$\arcsin x = \arccos x + \frac{\pi}{6} \qquad \text{Add arccos } x. \quad (1)$$

$$x = \sin\left(\arccos x + \frac{\pi}{6}\right) \qquad \text{Definition of arcsine}$$

Let $u = \arccos x$. The arccosine function yields angles in quadrants I and II, so $0 \le u \le \pi$ by definition.

$$x = \sin\left(u + \frac{\pi}{6}\right) \qquad \text{Substitute.}$$

$$x = \sin u \cos \frac{\pi}{6} + \cos u \sin \frac{\pi}{6} \qquad \text{Sine sum identity (Section 5.4)} \quad (2)$$

From equation (1) and by the definition of the arcsine function,

$$-\frac{\pi}{2} \le \arccos x + \frac{\pi}{6} \le \frac{\pi}{2} \qquad \text{Range of arcsine is } \left[-\frac{\pi}{2}, \frac{\pi}{2}\right].$$

$$-\frac{2\pi}{3} \le \arccos x \le \frac{\pi}{3}. \qquad \text{Subtract } \frac{\pi}{6} \text{ from each part. (Appendix A)}$$

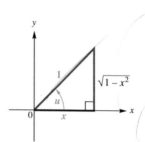

Figure 28

Since $0 \le \arccos x \le \pi$ **and** $-\frac{2\pi}{3} \le \arccos x \le \frac{\pi}{3}$, the intersection yields $0 \le \arccos x \le \frac{\pi}{3}$. This places u in quadrant I, and we can sketch the triangle in Figure 28. From this triangle we find that $\sin u = \sqrt{1 - x^2}$. Now substitute into equation (2) using $\sin u = \sqrt{1 - x^2}$, $\sin \frac{\pi}{6} = \frac{1}{2}$, $\cos \frac{\pi}{6} = \frac{\sqrt{3}}{2}$, and $\cos u = x$.

$$x = \sin u \cos \frac{\pi}{6} + \cos u \sin \frac{\pi}{6} \qquad (2)$$

$$x = \left(\sqrt{1 - x^2}\right)\frac{\sqrt{3}}{2} + x \cdot \frac{1}{2} \qquad \text{Substitute.}$$

$$2x = \left(\sqrt{1 - x^2}\right)\sqrt{3} + x \qquad \text{Multiply by 2.}$$

$$x = \left(\sqrt{3}\right)\sqrt{1 - x^2} \qquad \text{Subtract } x.$$

> Square *each* term.

$$x^2 = 3(1 - x^2) \qquad \text{Square both sides.}$$

$$x^2 = 3 - 3x^2 \qquad \text{Distributive property}$$

$$4x^2 = 3 \qquad \text{Add } 3x^2.$$

$$x^2 = \frac{3}{4} \qquad \text{Divide by 4.}$$

> Choose the positive square root; $x > 0$.

$$x = \sqrt{\frac{3}{4}} \qquad \text{Take the square root of both sides. (Appendix A)}$$

$$x = \frac{\sqrt{3}}{2} \qquad \text{Quotient rule; } \sqrt[v]{\frac{a}{b}} = \frac{\sqrt[v]{a}}{\sqrt[v]{b}}$$

To *check*, replace x with $\frac{\sqrt{3}}{2}$ in the original equation:

$$\arcsin\frac{\sqrt{3}}{2} - \arccos\frac{\sqrt{3}}{2} = \frac{\pi}{3} - \frac{\pi}{6} = \frac{\pi}{6},$$

as required. The solution set is $\left\{\frac{\sqrt{3}}{2}\right\}$.

NOW TRY EXERCISE 35. ◄

6.4 Exercises

Concept Check   *Answer each question.*

1. Which one of the following equations has solution 0?

   A. $\arctan 1 = x$ **B.** $\arccos 0 = x$ **C.** $\arcsin 0 = x$

2. Which one of the following equations has solution $\frac{\pi}{4}$?

   A. $\arcsin\frac{\sqrt{2}}{2} = x$ **B.** $\arccos\left(-\frac{\sqrt{2}}{2}\right) = x$ **C.** $\arctan\frac{\sqrt{3}}{3} = x$

3. Which one of the following equations has solution $\frac{3\pi}{4}$?

   A. $\arctan 1 = x$ **B.** $\arcsin\frac{\sqrt{2}}{2} = x$ **C.** $\arccos\left(-\frac{\sqrt{2}}{2}\right) = x$

4. Which one of the following equations has solution $-\frac{\pi}{6}$?

   A. $\arctan\frac{\sqrt{3}}{3} = x$ **B.** $\arccos\left(-\frac{1}{2}\right) = x$ **C.** $\arcsin\left(-\frac{1}{2}\right) = x$

Solve each equation for x. See Example 1.

5. $y = 5\cos x$ **6.** $4y = \sin x$ **7.** $2y = \cot 3x$

8. $6y = \frac{1}{2}\sec x$ **9.** $y = 3\tan 2x$ **10.** $y = 3\sin\frac{x}{2}$

11. $y = 6\cos\frac{x}{4}$ **12.** $y = -\sin\frac{x}{3}$ **13.** $y = -2\cos 5x$

14. $y = 3\cot 5x$ **15.** $y = \cos(x + 3)$ **16.** $y = \tan(2x - 1)$

17. $y = \sin x - 2$ **18.** $y = \cot x + 1$ **19.** $y = 2\sin x - 4$

20. $y = 4 + 3\cos x$ **21.** $y = \sqrt{2} + 3\sec 2x$ **22.** $y = 2\csc\frac{x}{2} - \sqrt{3}$

23. Refer to Exercise 17. A student attempting to solve this equation wrote as the first step $y = \sin(x - 2)$, inserting parentheses as shown. Explain why this is incorrect.

24. Explain why the equation $\sin^{-1} x = \cos^{-1} 2$ cannot have a solution. (No work is required.)

Solve each equation for exact solutions. See Examples 2 and 3.

25. $-4\arcsin x = \pi$ **26.** $6\arccos x = 5\pi$

27. $\frac{4}{3}\cos^{-1}\frac{y}{4} = \pi$ **28.** $4\pi + 4\tan^{-1} y = \pi$

29. $2 \arccos\left(\dfrac{y - \pi}{3}\right) = 2\pi$

30. $\arccos\left(y - \dfrac{\pi}{3}\right) = \dfrac{\pi}{6}$

31. $\arcsin x = \arctan \dfrac{3}{4}$

32. $\arctan x = \arccos \dfrac{5}{13}$

33. $\cos^{-1} x = \sin^{-1} \dfrac{3}{5}$

34. $\cot^{-1} x = \tan^{-1} \dfrac{4}{3}$

Solve each equation for exact solutions. See Example 4.

35. $\sin^{-1} x - \tan^{-1} 1 = -\dfrac{\pi}{4}$

36. $\sin^{-1} x + \tan^{-1} \sqrt{3} = \dfrac{2\pi}{3}$

37. $\arccos x + 2 \arcsin \dfrac{\sqrt{3}}{2} = \pi$

38. $\arccos x + 2 \arcsin \dfrac{\sqrt{3}}{2} = \dfrac{\pi}{3}$

39. $\arcsin 2x + \arccos x = \dfrac{\pi}{6}$

40. $\arcsin 2x + \arcsin x = \dfrac{\pi}{2}$

41. $\cos^{-1} x + \tan^{-1} x = \dfrac{\pi}{2}$

42. $\sin^{-1} x + \tan^{-1} x = 0$

43. Provide graphical support for the solution in Example 4 by showing that the graph of $y = \arcsin x - \arccos x - \frac{\pi}{6}$ has x-intercept $\frac{\sqrt{3}}{2} \approx .8660254$.

44. Provide graphical support for the solution in Example 4 by showing that the x-coordinate of the point of intersection of the graphs of $Y_1 = \arcsin X - \arccos X$ and $Y_2 = \frac{\pi}{6}$ is $\frac{\sqrt{3}}{2} \approx .8660254$.

The following equations cannot be solved by algebraic methods. Use a graphing calculator to find all solutions over the interval $[0, 6]$. Express solutions to four decimal places.

45. $(\arctan x)^3 - x + 2 = 0$

46. $\pi \sin^{-1}(.2x) - 3 = -\sqrt{x}$

(Modeling)   Solve each problem.

47. *Tone Heard by a Listener*   When two sources located at different positions produce the same pure tone, the human ear will often hear one sound that is equal to the sum of the individual tones. Since the sources are at different locations, they will have different phase angles ϕ. If two speakers located at different positions produce pure tones $P_1 = A_1 \sin(2\pi ft + \phi_1)$ and $P_2 = A_2 \sin(2\pi ft + \phi_2)$, where $-\frac{\pi}{4} \le \phi_1, \phi_2 \le \frac{\pi}{4}$, then the resulting tone heard by a listener can be written as $P = A \sin(2\pi ft + \phi)$, where

$$A = \sqrt{(A_1 \cos \phi_1 + A_2 \cos \phi_2)^2 + (A_1 \sin \phi_1 + A_2 \sin \phi_2)^2}$$

and $\phi = \arctan\left(\dfrac{A_1 \sin \phi_1 + A_2 \sin \phi_2}{A_1 \cos \phi_1 + A_2 \cos \phi_2}\right).$

(*Source:* Fletcher, N. and T. Rossing, *The Physics of Musical Instruments,* Second Edition, Springer-Verlag, 1998.)

(a) Calculate A and ϕ if $A_1 = .0012$, $\phi_1 = .052$, $A_2 = .004$, and $\phi_2 = .61$. Also find an expression for $P = A \sin(2\pi ft + \phi)$ if $f = 220$.

(b) Graph $Y_1 = P$ and $Y_2 = P_1 + P_2$ on the same coordinate axes over the interval $[0, .01]$. Are the two graphs the same?

48. *Tone Heard by a Listener* Repeat Exercise 47, using $A_1 = .0025$, $\phi_1 = \frac{\pi}{7}$, $A_2 = .001$, $\phi_2 = \frac{\pi}{6}$, and $f = 300$.

49. *Depth of Field* When a large-view camera is used to take a picture of an object that is not parallel to the film, the lens board should be tilted so that the planes containing the subject, the lens board, and the film intersect in a line. This gives the best "depth of field." See the figure. (*Source:* Bushaw, D., et al., *A Sourcebook of Applications of School Mathematics.* Copyright © 1980 by The Mathematical Association of America.)

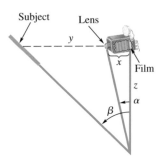

(a) Write two equations, one relating α, x, and z, and the other relating β, x, y, and z.
(b) Eliminate z from the equations in part (a) to get one equation relating α, β, x, and y.
(c) Solve the equation from part (b) for α.
(d) Solve the equation from part (b) for β.

50. *Programming Language for Inverse Functions* In Visual Basic, a widely used programming language for PCs, the only inverse trigonometric function available is arctangent. The other inverse trigonometric functions can be expressed in terms of arctangent as follows.

(a) Let $u = \arcsin x$. Solve the equation for x in terms of u.
(b) Use the result of part (a) to label the three sides of the triangle in the figure in terms of x.

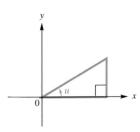

(c) Use the triangle from part (b) to write an equation for $\tan u$ in terms of x.
(d) Solve the equation from part (c) for u.

51. *Alternating Electric Current* In the study of alternating electric current, instantaneous voltage is modeled by

$$E = E_{max} \sin 2\pi ft,$$

where f is the number of cycles per second, E_{max} is the maximum voltage, and t is time in seconds.

(a) Solve the equation for t.
(b) Find the least positive value of t if $E_{max} = 12$, $E = 5$, and $f = 100$. Use a calculator.

52. *Viewing Angle of an Observer*   While visiting a museum, Marsha Langlois views a painting that is 3 ft high and hangs 6 ft above the ground. See the figure. Assume her eyes are 5 ft above the ground, and let x be the distance from the spot where she is standing to the wall displaying the painting.

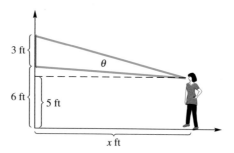

(a) Show that θ, the viewing angle subtended by the painting, is given by

$$\theta = \tan^{-1}\left(\frac{4}{x}\right) - \tan^{-1}\left(\frac{1}{x}\right).$$

(b) Find the value of x to the nearest hundredth for each value of θ.

  (i) $\theta = \dfrac{\pi}{6}$ **(ii)** $\theta = \dfrac{\pi}{8}$

(c) Find the value of θ to the nearest hundredth for each value of x.

  (i) $x = 4$ **(ii)** $x = 3$

53. *Movement of an Arm*   In the **Section 4.1** Exercises we found the equation

$$y = \frac{1}{3}\sin\frac{4\pi t}{3},$$

where t is time (in seconds) and y is the angle formed by a rhythmically moving arm.

(a) Solve the equation for t.

(b) At what time(s) does the arm form an angle of .3 radian?

54. The function $y = \sec^{-1} x$ is not found on graphing calculators. However, with some models it can be graphed as

$$y = \frac{\pi}{2} - ((x > 0) - (x < 0))\left(\frac{\pi}{2} - \tan^{-1}\left(\sqrt{(x^2 - 1)}\right)\right).$$

(This formula appears as Y_1 in the TI-83/84 Plus screen here.) Use the formula to obtain the graph of $y = \sec^{-1} x$ in the window $[-4, 4]$ by $[0, \pi]$.

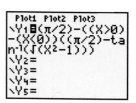

Chapter 6 Summary

KEY TERMS

6.1 one-to-one function
inverse function

NEW SYMBOLS

f^{-1}	inverse of function f	$\cot^{-1}x$ **(arccot x)**	inverse cotangent of x
$\sin^{-1}x$ **(arcsin x)**	inverse sine of x	$\sec^{-1}x$ **(arcsec x)**	inverse secant of x
$\cos^{-1}x$ **(arccos x)**	inverse cosine of x	$\csc^{-1}x$ **(arccsc x)**	inverse cosecant of x
$\tan^{-1}x$ **(arctan x)**	inverse tangent of x		

QUICK REVIEW

CONCEPTS	EXAMPLES

6.1 Inverse Circular Functions

Inverse Function	Domain	Range	
		Interval	Quadrants of the Unit Circle
$y = \sin^{-1}x$	$[-1, 1]$	$\left[-\frac{\pi}{2}, \frac{\pi}{2}\right]$	I and IV
$y = \cos^{-1}x$	$[-1, 1]$	$[0, \pi]$	I and II
$y = \tan^{-1}x$	$(-\infty, \infty)$	$\left(-\frac{\pi}{2}, \frac{\pi}{2}\right)$	I and IV
$y = \cot^{-1}x$	$(-\infty, \infty)$	$(0, \pi)$	I and II
$y = \sec^{-1}x$	$(-\infty, -1] \cup [1, \infty)$	$[0, \pi], y \neq \frac{\pi}{2}$	I and II
$y = \csc^{-1}x$	$(-\infty, -1] \cup [1, \infty)$	$\left[-\frac{\pi}{2}, \frac{\pi}{2}\right], y \neq 0$	I and IV

Evaluate $y = \cos^{-1}0$.

Write $y = \cos^{-1}0$ as $\cos y = 0$. Then $y = \frac{\pi}{2}$, because $\cos\frac{\pi}{2} = 0$ and $\frac{\pi}{2}$ is in the range of $\cos^{-1}x$.

Use a calculator to find y in radians if $y = \sec^{-1}(-3)$.

With the calculator in radian mode, enter $\sec^{-1}(-3)$ as $\cos^{-1}\left(-\frac{1}{3}\right)$ to get $y \approx 1.9106332$.

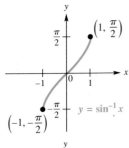

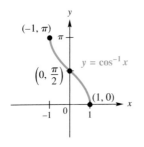

Evaluate $\sin\left(\tan^{-1}\left(-\frac{3}{4}\right)\right)$.

Let $u = \tan^{-1}\left(-\frac{3}{4}\right)$. Then $\tan u = -\frac{3}{4}$. Since $\tan u$ is negative when u is in quadrant IV, sketch a triangle as shown.

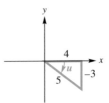

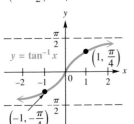

See Figure 14 on page 264 for graphs of the other inverse circular (trigonometric) functions.

We want $\sin\left(\tan^{-1}\left(-\frac{3}{4}\right)\right) = \sin u$. From the triangle,

$$\sin u = -\frac{3}{5}.$$

CONCEPTS	EXAMPLES

6.2 Trigonometric Equations I
6.3 Trigonometric Equations II

Solving a Trigonometric Equation

1. Decide whether the equation is linear or quadratic in form, so you can determine the solution method.

2. If only one trigonometric function is present, solve the equation for that function.

3. If more than one trigonometric function is present, re-arrange the equation so that one side equals 0. Then try to factor and set each factor equal to 0 to solve.

4. If the equation is quadratic in form, but not factorable, use the quadratic formula. Check that solutions are in the desired interval.

5. Try using identities to change the form of the equation. It may be helpful to square both sides of the equation first. If this is done, check for extraneous solutions.

Solve $\tan \theta + \sqrt{3} = 2\sqrt{3}$ over the interval $[0°, 360°)$. Use a linear method.

$$\tan \theta + \sqrt{3} = 2\sqrt{3}$$
$$\tan \theta = \sqrt{3}$$
$$\theta = 60°$$

Another solution over $[0°, 360°)$ is

$$\theta = 60° + 180° = 240°.$$

The solution set is $\{60°, 240°\}$.

Solve $2\cos^2 x = 1$ over the interval $[0, 2\pi)$ using a double-angle identity.

$$2\cos^2 x = 1$$
$$2\cos^2 x - 1 = 0$$
$$\cos 2x = 0 \quad \text{Cosine double-angle identity}$$
$$2x = \frac{\pi}{2}, \frac{3\pi}{2}, \frac{5\pi}{2}, \frac{7\pi}{2} \quad \begin{array}{l} 0 \le x < 2\pi, \text{ so} \\ 0 \le 2x < 4\pi. \end{array}$$
$$x = \frac{\pi}{4}, \frac{3\pi}{4}, \frac{5\pi}{4}, \frac{7\pi}{4} \quad \text{Divide by 2.}$$

The solution set is $\left\{ \frac{\pi}{4}, \frac{3\pi}{4}, \frac{5\pi}{4}, \frac{7\pi}{4} \right\}$.

6.4 Equations Involving Inverse Trigonometric Functions

We solve equations of the form $y = f(x)$ where $f(x)$ is a trigonometric function using inverse trigonometric functions.

Solve $y = 2\sin 3x$ for x.

$$y = 2\sin 3x$$
$$\frac{y}{2} = \sin 3x \qquad \text{Divide by 2.}$$
$$3x = \arcsin \frac{y}{2} \qquad \text{Definition of arcsine}$$
$$x = \frac{1}{3}\arcsin \frac{y}{2} \qquad \text{Multiply by } \tfrac{1}{3}.$$

Techniques introduced in this section also show how to solve equations that involve inverse functions.

Solve $4\tan^{-1} x = \pi$.

$$\tan^{-1} x = \frac{\pi}{4} \qquad \text{Divide by 4.}$$

$$x = \tan \frac{\pi}{4} = 1 \qquad \text{Definition of arctangent}$$

The solution set is $\{1\}$.

CHAPTER 6 ▶ Review Exercises

1. Graph the inverse sine, cosine, and tangent functions, indicating three points on each graph. Give the domain and range for each.

Concept Check Tell whether each statement is true or false. If false, tell why.

2. The ranges of the inverse tangent and inverse cotangent functions are the same.

3. It is true that $\sin \frac{11\pi}{6} = -\frac{1}{2}$, and therefore $\arcsin\left(-\frac{1}{2}\right) = \frac{11\pi}{6}$.

4. For all x, $\tan(\tan^{-1} x) = x$.

Give the exact real number value of y. Do not use a calculator.

$y = \frac{\pi}{4}$ 5. $y = \sin^{-1}\frac{\sqrt{2}}{2}$ $\frac{2\pi}{3}$ 6. $y = \arccos\left(-\frac{1}{2}\right)$ $-\frac{\pi}{3}$ 7. $y = \tan^{-1}\left(-\sqrt{3}\right)$

$-\frac{\pi}{2}$ 8. $y = \arcsin(-1)$ $\frac{3\pi}{4}$ 9. $y = \cos^{-1}\left(-\frac{\sqrt{2}}{2}\right)$ $\frac{\pi}{6}$ 10. $y = \arctan\frac{\sqrt{3}}{3}$

$\frac{2\pi}{3}$ 11. $y = \sec^{-1}(-2)$ $\frac{\pi}{3}$ 12. $y = \text{arccsc}\frac{2\sqrt{3}}{3}$ $-\frac{\pi}{4}$ 13. $y = \text{arccot}(-1)$

Give the degree measure of θ. Do not use a calculator.

14. $\theta = \arccos\frac{1}{2}$

$60°$

15. $\theta = \arcsin\left(-\frac{\sqrt{3}}{2}\right)$

$-60°$

16. $\theta = \tan^{-1} 0$

$0°$

Use a calculator to give the degree measure of θ.

17. $\theta = \arctan 1.7804675$ 18. $\theta = \sin^{-1}(-.66045320)$ 19. $\theta = \cos^{-1} .80396577$

20. $\theta = \cot^{-1} 4.5046388$ 21. $\theta = \text{arcsec } 3.4723155$ 22. $\theta = \csc^{-1} 7.4890096$

Evaluate the following without using a calculator.

23. $\cos(\arccos(-1))$ 24. $\sin\left(\arcsin\left(-\frac{\sqrt{3}}{2}\right)\right)$ 25. $\arccos\left(\cos\frac{3\pi}{4}\right)$

26. $\text{arcsec}(\sec \pi)$ 27. $\tan^{-1}\left(\tan\frac{\pi}{4}\right)$ 28. $\cos^{-1}(\cos 0)$

29. $\sin\left(\arccos\frac{3}{4}\right)$ 30. $\cos(\arctan 3)$ 31. $\cos(\csc^{-1}(-2))$

32. $\sec\left(2\sin^{-1}\left(-\frac{1}{3}\right)\right)$ 33. $\tan\left(\arcsin\frac{3}{5} + \arccos\frac{5}{7}\right)$

Write each of the following as an algebraic (nontrigonometric) expression in u, $u > 0$.

34. $\cos\left(\arctan\frac{u}{\sqrt{1-u^2}}\right)$ 35. $\tan\left(\text{arcsec}\frac{\sqrt{u^2+1}}{u}\right)$

Solve each equation for solutions over the interval $[0, 2\pi)$. If necessary, express approximate solutions to four decimal places.

36. $\sin^2 x = 1$ 37. $2 \tan x - 1 = 0$

38. $3 \sin^2 x - 5 \sin x + 2 = 0$ 39. $\tan x = \cot x$

40. $\sec^4 2x = 4$

41. $\tan^2 2x - 1 = 0$

Give all solutions for each equation.

42. $\sec \dfrac{x}{2} = \cos \dfrac{x}{2}$

43. $\cos 2x + \cos x = 0$

44. $4 \sin x \cos x = \sqrt{3}$

Solve each equation for solutions over the interval $[0°, 360°)$. If necessary, express solutions to the nearest tenth of a degree.

45. $\sin^2 \theta + 3 \sin \theta + 2 = 0$

46. $2 \tan^2 \theta = \tan \theta + 1$

47. $\sin 2\theta = \cos 2\theta + 1$

48. $2 \sin 2\theta = 1$

49. $3 \cos^2 \theta + 2 \cos \theta - 1 = 0$

50. $5 \cot^2 \theta - \cot \theta - 2 = 0$

Solve each equation in Exercises 51–57 for x.

51. $4y = 2 \sin x$

52. $y = 3 \cos \dfrac{x}{2}$

53. $2y = \tan(3x + 2)$

54. $5y = 4 \sin x - 3$

55. $\dfrac{4}{3} \arctan \dfrac{x}{2} = \pi$

56. $\arccos x = \arcsin \dfrac{2}{7}$

57. $\arccos x + \arctan 1 = \dfrac{11\pi}{12}$

58. Solve $d = 550 + 450 \cos\left(\dfrac{\pi}{50} t\right)$ for t in terms of d.

(Modeling) Solve each problem.

59. *Viewing Angle of an Observer* A 10-ft-wide chalkboard is situated 5 ft from the left wall of a classroom. See the figure. A student sitting next to the wall x feet from the front of the classroom has a viewing angle of θ radians.

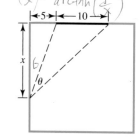

    (a) Show that the value of θ is given by the function defined by

$$f(x) = \arctan\left(\dfrac{15}{x}\right) - \arctan\left(\dfrac{5}{x}\right).$$

    (b) Graph $f(x)$ with a graphing calculator to estimate the value of x that maximizes the viewing angle.

60. *Snell's Law* Recall Snell's law from Exercises 65 and 66 of **Section 2.3:**

$$\dfrac{c_1}{c_2} = \dfrac{\sin \theta_1}{\sin \theta_2},$$

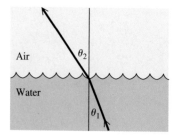

where c_1 is the speed of light in one medium, c_2 is the speed of light in a second medium, and θ_1 and θ_2 are the angles shown in the figure. Suppose a light is shining up through water into the air as in the figure. As θ_1 increases, θ_2 approaches 90°, at which point no light will emerge from the water. Assume the ratio $\frac{c_1}{c_2}$ in this case is .752. For what value of θ_1 does $\theta_2 = 90°$? This value of θ_1 is called the **critical angle** for water.

61. *Snell's Law* Refer to Exercise 60. What happens when θ_1 is greater than the critical angle?

62. *British Nautical Mile* The British nautical mile is defined as the length of a minute of arc of a meridian. Since Earth is flat at its poles, the nautical mile, in feet, is given by

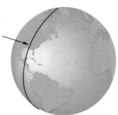

A nautical mile is the length on any of the meridians cut by a central angle of measure 1 minute.

$$L = 6077 - 31 \cos 2\theta,$$

where θ is the latitude in degrees. See the figure. (*Source:* Bushaw, D., et al., *A Sourcebook of Applications of School Mathematics.* Copyright © 1980 by The Mathematical Association of America.)

   (a) Find the latitude between $0°$ and $90°$ at which the nautical mile is 6074 ft.
   (b) At what latitude between $0°$ and $180°$ is the nautical mile 6108 ft?
   (c) In the United States, the nautical mile is defined everywhere as 6080.2 ft. At what latitude between $0°$ and $90°$ does this agree with the British nautical mile?

63. The function $y = \csc^{-1} x$ is not found on graphing calculators. However, with some models it can be graphed as

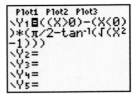

$$y = ((x > 0) - (x < 0))\left(\frac{\pi}{2} - \tan^{-1}\left(\sqrt{(x^2 - 1)}\right)\right).$$

(This formula appears as Y_1 in the TI-83/84 Plus screen here.) Use the formula to obtain the graph of $y = \csc^{-1} x$ in the window $[-4, 4]$ by $\left[-\frac{\pi}{2}, \frac{\pi}{2}\right]$.

64. **(a)** Use the graph of $y = \sin^{-1} x$ to approximate $\sin^{-1} .4$.
   (b) Use the inverse sine key of a graphing calculator to approximate $\sin^{-1} .4$.

CHAPTER 6 ▶ Test

1. Graph $y = \sin^{-1} x$, and indicate the coordinates of three points on the graph. Give the domain and range.

2. Find the exact value of y for each equation.

   (a) $y = \arccos\left(-\dfrac{1}{2}\right)$ **(b)** $y = \sin^{-1}\left(-\dfrac{\sqrt{3}}{2}\right)$

   (c) $y = \tan^{-1} 0$ **(d)** $y = \text{arcsec}(-2)$

3. Give the degree measure of θ.

   (a) $\theta = \arccos\dfrac{\sqrt{3}}{2}$ **(b)** $\theta = \tan^{-1}(-1)$

   (c) $\theta = \cot^{-1}(-1)$ **(d)** $\theta = \csc^{-1}\left(-\dfrac{2\sqrt{3}}{3}\right)$

4. Use a calculator to give each value in decimal degrees to the nearest hundredth.

   (a) $\sin^{-1}.67610476$　　　　(b) $\sec^{-1}1.0840880$　　　　(c) $\cot^{-1}(-.7125586)$

5. Find each exact value.

   (a) $\cos\left(\arcsin\dfrac{2}{3}\right)$　　　　　　　　　　(b) $\sin\left(2\cos^{-1}\dfrac{1}{3}\right)$

6. Explain why $\sin^{-1}3$ cannot be defined.

7. Explain why $\arcsin\left(\sin\dfrac{5\pi}{6}\right)\neq\dfrac{5\pi}{6}$.

8. Write $\tan(\arcsin u)$ as an algebraic (nontrigonometric) expression in u, $u>0$.

Solve each equation for solutions over the interval $[0°,360°)$. *Express approximate solutions to the nearest tenth of a degree.*

   9. $-3\sec\theta+2\sqrt{3}=0$　　　　10. $\sin^2\theta=\cos^2\theta+1$　　　　11. $\csc^2\theta-2\cot\theta=4$

Solve each equation for solutions over the interval $[0,2\pi)$. *Express approximate solutions to four decimal places.*

   12. $\cos x=\cos 2x$　　　　13. $\sqrt{2}\cos 3x-1=0$　　　　14. $\sin x\cos x=\dfrac{1}{3}$

15. Solve $\sin^2\theta=-\cos 2\theta$, giving all solutions in degrees.

16. Solve $2\sqrt{3}\sin\dfrac{x}{2}=3$, giving all solutions in radians.

17. Solve each equation for x.

   (a) $y=\cos 3x$　　　　　　　　　　(b) $\arcsin x=\arctan\dfrac{4}{3}$

18. *(Modeling) Movement of a Runner's Arm* A runner's arm swings rhythmically according to the model

$$y=\frac{\pi}{8}\cos\left[\pi\left(t-\frac{1}{3}\right)\right],$$

where y represents the angle between the actual position of the upper arm and the downward vertical position, and t represents time in seconds. At what times over the interval $[0,3)$ is the angle y equal to 0?

CHAPTER 6 ▶ Quantitative Reasoning

How can we determine the amount of oil in a submerged storage tank?

The level of oil in a storage tank buried in the ground can be found in much the same way a dipstick is used to determine the oil level in an automobile crankcase. The person in the figure on the left has lowered a calibrated rod into an oil storage tank. When the rod is removed, the reading on the rod can be used with the dimensions of the storage tank to calculate the amount of oil in the tank.

Suppose the ends of the cylindrical storage tank in the figure are circles of radius 3 ft and the cylinder is 20 ft long. Determine the volume of oil in the tank to the nearest cubic foot if the rod shows a depth of 2 ft. (*Hint:* The volume will be 20 times the area of the shaded segment of the circle shown in the figure on the right.)

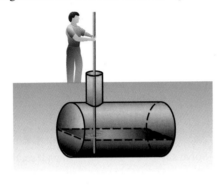

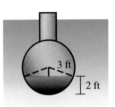

7 Applications of Trigonometry and Vectors

World technology took a giant step forward when the *Global Positioning System (GPS),* a network of more than 24 communications satellites, achieved full operational capability on July 17, 1995. GPS satellites transmit signals from 12,552 miles above Earth's surface.

The magnitude and direction of four arrows, or *vectors,* representing the signals from four GPS satellites to a receiver on Earth, are used to calculate location on the globe. Navigators use a similar calculation process with three vectors, called *triangulation,* to determine the location of a ship or an airplane. (*Source:* A. El-Rabbany, *Introduction to GPS, The Global Positioning System.*)

In this chapter, we use trigonometry and vectors to solve problems involving navigation, weight and force balancing, and distance and angle determination, among others.

7.1 Oblique Triangles and the Law of Sines

Congruency and Oblique Triangles ■ **Derivation of the Law of Sines** ■ **Solving SAA and ASA Triangles (Case 1)** ■ **Area of a Triangle**

The concepts of solving triangles developed in **Chapter 2** can be extended to *all* triangles. If any three of the six side and angle measures of a triangle are known (provided at least one measure is the length of a side), then the other three measures can be found.

Congruency and Oblique Triangles The following axioms from geometry allow us to prove that two triangles are congruent (that is, their corresponding sides and angles are equal).

CONGRUENCE AXIOMS

Side-Angle-Side (SAS)	If two sides and the included angle of one triangle are equal, respectively, to two sides and the included angle of a second triangle, then the triangles are congruent.
Angle-Side-Angle (ASA)	If two angles and the included side of one triangle are equal, respectively, to two angles and the included side of a second triangle, then the triangles are congruent.
Side-Side-Side (SSS)	If three sides of one triangle are equal, respectively, to three sides of a second triangle, then the triangles are congruent.

If a side and *any* two angles are given (SAA), the third angle is easily determined by the angle sum formula ($A + B + C = 180°$), and then the ASA axiom can be applied. Keep in mind that whenever SAS, ASA, or SSS is given, the triangle is unique. We continue to label triangles as we did earlier with right triangles: side a opposite angle A, side b opposite angle B, and side c opposite angle C.

A triangle that is not a right triangle is called an **oblique triangle.** The measures of the three sides and the three angles of a triangle can be found if at least one side and any other two measures are known. There are four possible cases.

DATA REQUIRED FOR SOLVING OBLIQUE TRIANGLES

Case 1 One side and two angles are known (SAA or ASA).

Case 2 Two sides and one angle not included between the two sides are known (SSA). This case may lead to more than one triangle.

Case 3 Two sides and the angle included between the two sides are known (SAS).

Case 4 Three sides are known (SSS).

▶ **Note** *If we know three angles of a triangle, we cannot find unique side lengths since* **AAA** *assures us only of similarity, not congruence.* For example, there are infinitely many triangles ABC with $A = 35°$, $B = 65°$, and $C = 80°$.

Case 1, discussed in this section, and Case 2, discussed in **Section 7.2,** require the *law of sines.* Cases 3 and 4, discussed in **Section 7.3,** require the *law of cosines.*

Derivation of the Law of Sines

To derive the law of sines, we start with an oblique triangle, such as the *acute triangle* in Figure 1(a) or the *obtuse triangle* in Figure 1(b). This discussion applies to both triangles. First, construct the perpendicular from B to side AC (or its extension). Let h be the length of this perpendicular. Then c is the hypotenuse of right triangle ADB, and a is the hypotenuse of right triangle BDC.

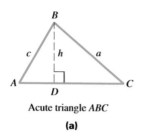

Acute triangle ABC

(a)

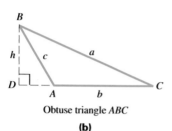

Obtuse triangle ABC

(b)

Figure 1

In triangle ADB, $\sin A = \dfrac{h}{c}$,   or $h = c \sin A.$

(Section 2.1)

In triangle BDC, $\sin C = \dfrac{h}{a}$,   or $h = a \sin C.$

Since $h = c \sin A$ and $h = a \sin C$,

$$a \sin C = c \sin A$$

$$\frac{a}{\sin A} = \frac{c}{\sin C}.$$ Divide each side by $\sin A \sin C$.

In a similar way, by constructing perpendicular lines from the other vertices, it can be shown that

$$\frac{a}{\sin A} = \frac{b}{\sin B} \quad \text{and} \quad \frac{b}{\sin B} = \frac{c}{\sin C}.$$

This discussion proves the following theorem.

LAW OF SINES

In any triangle ABC, with sides a, b, and c,

$$\frac{a}{\sin A} = \frac{b}{\sin B}, \qquad \frac{a}{\sin A} = \frac{c}{\sin C}, \qquad \text{and} \qquad \frac{b}{\sin B} = \frac{c}{\sin C}.$$

This can be written in compact form as

$$\frac{a}{\sin A} = \frac{b}{\sin B} = \frac{c}{\sin C}.$$

That is, according to the law of sines, the lengths of the sides in a triangle are proportional to the sines of the measures of the angles opposite them.

When solving for an angle, we use an alternative form of the law of sines,

$$\frac{\sin A}{a} = \frac{\sin B}{b} = \frac{\sin C}{c}.$$ Alternative form

---

▶ **Note** When using the law of sines, a good strategy is to select an equation so that the unknown variable is in the numerator and all other variables are known.

Solving SAA and ASA Triangles (Case 1)   If two angles and one side of a triangle are known (Case 1, SAA or ASA), then the law of sines can be used to solve the triangle.

▶ **EXAMPLE 1**   USING THE LAW OF SINES TO SOLVE A TRIANGLE (SAA)

Solve triangle ABC if $A = 32.0°$, $B = 81.8°$, and $a = 42.9$ cm.

Solution   Start by drawing a triangle, roughly to scale, and labeling the given parts as in Figure 2. Since the values of A, B, and a are known, use the form of the law of sines that involves these variables, and then solve for b.

> Be sure to label a sketch carefully to help set up the correct equation.

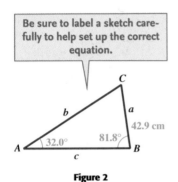

Figure 2

> Choose a form that has the unknown variable in the numerator.

$$\frac{a}{\sin A} = \frac{b}{\sin B} \qquad \text{Law of sines}$$

$$\frac{42.9}{\sin 32.0°} = \frac{b}{\sin 81.8°} \qquad \text{Substitute the given values.}$$

$$b = \frac{42.9 \sin 81.8°}{\sin 32.0°} \qquad \text{Multiply by } \sin 81.8°; \text{ rewrite.}$$

$$b \approx 80.1 \text{ cm} \qquad \text{Approximate with a calculator.}$$

To find C, use the fact that the sum of the angles of any triangle is $180°$.

$$A + B + C = 180° \qquad \text{(Section 1.2)}$$

$$C = 180° - A - B$$

$$C = 180° - 32.0° - 81.8°$$

$$C = 66.2°$$

Use the law of sines to find c. (Why does the Pythagorean theorem not apply?)

$$\frac{a}{\sin A} = \frac{c}{\sin C} \qquad \text{Law of sines}$$

$$\frac{42.9}{\sin 32.0°} = \frac{c}{\sin 66.2°} \qquad \text{Substitute.}$$

$$c = \frac{42.9 \sin 66.2°}{\sin 32.0°} \qquad \text{Multiply by } \sin 66.2°; \text{ rewrite.}$$

$$c \approx 74.1 \text{ cm} \qquad \text{Approximate with a calculator.}$$

NOW TRY EXERCISE 7. ◀

▶ **EXAMPLE 2**   USING THE LAW OF SINES IN AN APPLICATION (ASA)

Jerry Keefe wishes to measure the distance across the Big Muddy River. See Figure 3. He determines that $C = 112.90°$, $A = 31.10°$, and $b = 347.6$ ft. Find the distance a across the river.

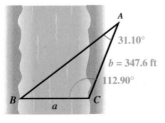

Figure 3

Solution   To use the law of sines, one side and the angle opposite it must be known. Since b is the only side whose length is given, angle B must be found before the law of sines can be used.

$$B = 180° - A - C$$

$$B = 180° - 31.10° - 112.90° = 36.00°$$

Now use the form of the law of sines involving A, B, and b to find a.

Solve for a.

$$\frac{a}{\sin A} = \frac{b}{\sin B} \qquad \text{Law of sines}$$

$$\frac{a}{\sin 31.10°} = \frac{347.6}{\sin 36.00°} \qquad \text{Substitute known values.}$$

$$a = \frac{347.6 \sin 31.10°}{\sin 36.00°} \qquad \text{Multiply by } \sin 31.10°.$$

$$a \approx 305.5 \text{ ft} \qquad \text{Use a calculator.}$$

NOW TRY EXERCISE 25. ◀

The next example involves the concept of bearing, first discussed in **Chapter 2.**

▶ **EXAMPLE 3**   USING THE LAW OF SINES IN AN APPLICATION (ASA)

Two ranger stations are on an east-west line 110 mi apart. A forest fire is located on a bearing of N 42° E from the western station at A and a bearing of N 15° E from the eastern station at B. How far is the fire from the western station?

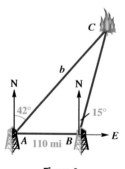

Figure 4

Solution   Figure 4 shows the two stations at points A and B and the fire at point C. Angle $BAC = 90° - 42° = 48°$, the obtuse angle at B measures $90° + 15° = 105°$, and the third angle, C, measures $180° - 105° - 48° = 27°$. We use the law of sines to find side b.

Solve for b.

$$\frac{b}{\sin B} = \frac{c}{\sin C} \qquad \text{Law of sines}$$

$$\frac{b}{\sin 105°} = \frac{110}{\sin 27°} \qquad \text{Substitute known values.}$$

$$b = \frac{110 \sin 105°}{\sin 27°} \qquad \text{Multiply by } \sin 105°.$$

$$b \approx 234 \text{ mi} \qquad \text{Use a calculator.}$$

NOW TRY EXERCISE 27. ◀

Area of a Triangle   The method used to derive the law of sines can also be used to derive a formula to find the area of a triangle. A familiar formula for the area of a triangle is $\mathcal{A} = \frac{1}{2}bh$, where $\mathcal{A}$ represents area, b base, and h height. This formula cannot always be used easily since in practice, h is often unknown. To find another formula, refer to acute triangle ABC in Figure 5(a) or obtuse triangle ABC in Figure 5(b), shown on the next page.

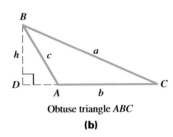

Acute triangle *ABC*
(a)

Obtuse triangle *ABC*
(b)

Figure 5

A perpendicular has been drawn from *B* to the base of the triangle (or the extension of the base). This perpendicular forms two right triangles. Using triangle *ABD*,

$$\sin A = \frac{h}{c}, \quad \text{or} \quad h = c \sin A.$$

Substituting into the formula for the area of a triangle,

$$\mathcal{A} = \frac{1}{2}bh = \frac{1}{2}bc \sin A.$$

Any other pair of sides and the angle between them could have been used.

AREA OF A TRIANGLE (SAS)

In any triangle *ABC*, the area $\mathcal{A}$ is given by the following formulas:

$$\mathcal{A} = \frac{1}{2}bc \sin A, \quad \mathcal{A} = \frac{1}{2}ab \sin C, \quad \text{and} \quad \mathcal{A} = \frac{1}{2}ac \sin B.$$

That is, the area is half the product of the lengths of two sides and the sine of the angle included between them.

▶ **Note** If the included angle measures 90°, its sine is 1, and the formula becomes the familiar $\mathcal{A} = \frac{1}{2}bh$.

▶ **EXAMPLE 4** FINDING THE AREA OF A TRIANGLE (SAS)

Find the area of triangle *ABC* in Figure 6.

Solution We are given $B = 55°\ 10'$, $a = 34.0$ ft, and $c = 42.0$ ft, so

$$\mathcal{A} = \frac{1}{2}ac \sin B = \frac{1}{2}(34.0)(42.0) \sin 55°\ 10' \approx 586 \text{ ft}^2.$$

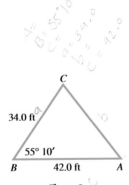

34.0 ft

55° 10′

B   42.0 ft   *A*

Figure 6

NOW TRY EXERCISE 43. ◀

▶ **EXAMPLE 5** FINDING THE AREA OF A TRIANGLE (ASA)

Find the area of triangle *ABC* if $A = 24°\ 40'$, $b = 27.3$ cm, and $C = 52°\ 40'$.

Solution Before the area formula can be used, we must find either *a* or *c*. Since the sum of the measures of the angles of any triangle is 180°,

$$B = 180° - 24°\ 40' - 52°\ 40' = 102°\ 40'.$$

We can use the law of sines to find a.

Solve for a.
$$\frac{a}{\sin A} = \frac{b}{\sin B} \qquad \text{Law of sines}$$

$$\frac{a}{\sin 24° \, 40'} = \frac{27.3}{\sin 102° \, 40'} \qquad \text{Substitute known values.}$$

$$a = \frac{27.3 \sin 24° \, 40'}{\sin 102° \, 40'} \qquad \text{Multiply by } \sin 24° \, 40'.$$

$$a \approx 11.7 \text{ cm} \qquad \text{Use a calculator.}$$

Now, we find the area.

$$\mathcal{A} = \frac{1}{2} ab \sin C = \frac{1}{2} (11.7)(27.3) \sin 52° \, 40' \approx 127 \text{ cm}^2$$

NOW TRY EXERCISE 49. ◄

▶ **Caution** Whenever possible, use given values in solving triangles or finding areas rather than values obtained in intermediate steps to avoid possible rounding errors.

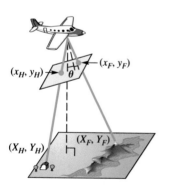

Figure 7

CONNECTIONS Aerial photographs can be used to provide coordinates of ordered pairs to determine distances on the ground. Suppose we assign coordinates as shown in Figure 7. If an object's photographic coordinates are (x, y), then its ground coordinates (X, Y) in feet can be computed using the following formulas.

$$X = \frac{(a - h)x}{f \sec \theta - y \sin \theta}, \qquad Y = \frac{(a - h)y \cos \theta}{f \sec \theta - y \sin \theta}$$

Here, f is focal length of the camera in inches, a is altitude in feet of the airplane, and h is elevation in feet of the object. Suppose that a house has photographic coordinates $(x_H, y_H) = (.9, 3.5)$ with elevation 150 ft, while a nearby forest fire has photographic coordinates $(x_F, y_F) = (2.1, -2.4)$ and is at elevation 690 ft. If the photograph was taken at 7400 ft by a camera with focal length 6 in. and tilt angle $\theta = 4.1°$, we can use these formulas to find the distance on the ground between the house and the fire. (*Source:* Moffitt, F. and E. Mikhail, *Photogrammetry,* Third Edition, Harper & Row, 1980.)

FOR DISCUSSION OR WRITING

1. Use the formulas to find the ground coordinates of the house and the fire to the nearest tenth of a foot.

2. Use the distance formula given in **Appendix B** to find the required distance on the ground to the nearest tenth of a foot.

7.1 **Exercises**

1. *Concept Check* Consider the oblique triangle *ABC*. Which one of the following proportions is *not* valid?

A. $\dfrac{a}{b} = \dfrac{\sin A}{\sin B}$ **B.** $\dfrac{a}{\sin A} = \dfrac{b}{\sin B}$

C. $\dfrac{\sin A}{a} = \dfrac{b}{\sin B}$ **D.** $\dfrac{\sin A}{a} = \dfrac{\sin B}{b}$

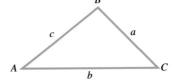

2. *Concept Check* Which two of the following situations do not provide sufficient information for solving a triangle by the law of sines?

A. You are given two angles and the side included between them.
B. You are given two angles and a side opposite one of them.
C. You are given two sides and the angle included between them.
D. You are given three sides.

Find the length of each side a. Do not use a calculator.

3.

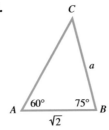

4.

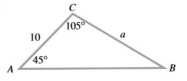

Determine the remaining sides and angles of each triangle ABC. See Example 1.

5.

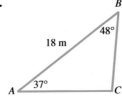

6.

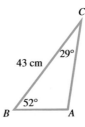

7.

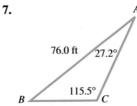

8.

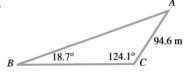

9. $A = 68.41°, B = 54.23°, a = 12.75$ ft

10. $C = 74.08°, B = 69.38°, c = 45.38$ m

11. $A = 87.2°, b = 75.9$ yd, $C = 74.3°$

12. $B = 38° \, 40', a = 19.7$ cm, $C = 91° \, 40'$

13. $B = 20° \, 50'$, $C = 103° \, 10'$, $AC = 132$ ft

14. $A = 35.3°$, $B = 52.8°$, $AC = 675$ ft

15. $A = 39.70°$, $C = 30.35°$, $b = 39.74$ m

16. $C = 71.83°$, $B = 42.57°$, $a = 2.614$ cm

17. $B = 42.88°$, $C = 102.40°$, $b = 3974$ ft

18. $A = 18.75°$, $B = 51.53°$, $c = 2798$ yd

19. $A = 39° \, 54'$, $a = 268.7$ m, $B = 42° \, 32'$

20. $C = 79° \, 18'$, $c = 39.81$ mm, $A = 32° \, 57'$

21. Explain why the law of sines cannot be used to solve a triangle if we are given the lengths of the three sides of the triangle.

22. *Concept Check* In Example 1, we asked the question, "Why does the Pythagorean theorem not apply?" Answer this question.

23. Terry Harris, a perceptive trigonometry student, makes the statement, "If we know *any* two angles and one side of a triangle, then the triangle is uniquely determined." Is this a valid statement? Explain, referring to the congruence axioms given in this section.

24. *Concept Check* If a is twice as long as b, is A necessarily twice as large as B?

Solve each problem. See Examples 2 and 3.

25. *Distance Across a River* To find the distance AB across a river, a surveyor laid off a distance $BC = 354$ m on one side of the river. It is found that $B = 112° \, 10'$ and $C = 15° \, 20'$. Find AB. See the figure.

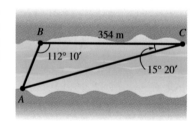

26. *Distance Across a Canyon* To determine the distance RS across a deep canyon, Rhonda lays off a distance $TR = 582$ yd. She then finds that $T = 32° \, 50'$ and $R = 102° \, 20'$. Find RS.

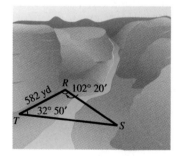

27. *Distance a Ship Travels* A ship is sailing due north. At a certain point the bearing of a lighthouse 12.5 km away is N 38.8° E. Later on, the captain notices that the bearing of the lighthouse has become S 44.2° E. How far did the ship travel between the two observations of the lighthouse?

28. *Distance Between Radio Direction Finders* Radio direction finders are placed at points A and B, which are 3.46 mi apart on an east-west line, with A west of B. From A the bearing of a certain radio transmitter is 47.7°, and from B the bearing is 302.5°. Find the distance of the transmitter from A.

29. *Measurement of a Folding Chair* A folding chair is to have a seat 12.0 in. deep with angles as shown in the figure. How far down from the seat should the crossing legs be joined? (Find x in the figure.)

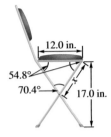

30. *Distance Across a River* Standing on one bank of a river flowing north, Mark notices a tree on the opposite bank at a bearing of 115.45°. Lisa is on the same bank as Mark, but 428.3 m away. She notices that the bearing of the tree is 45.47°. The two banks are parallel. What is the distance across the river?

31. *Angle Formed by Radii of Gears* Three gears are arranged as shown in the figure. Find angle θ.

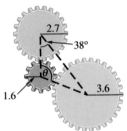

32. *Distance Between Atoms* Three atoms with atomic radii of 2.0, 3.0, and 4.5 are arranged as in the figure. Find the distance between the centers of atoms A and C.

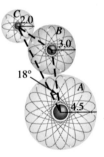

33. *Distance Between a Ship and a Lighthouse* The bearing of a lighthouse from a ship was found to be N 37° E. After the ship sailed 2.5 mi due south, the new bearing was N 25° E. Find the distance between the ship and the lighthouse at each location.

34. *Height of a Balloon* A balloonist is directly above a straight road 1.5 mi long that joins two villages. She finds that the town closer to her is at an angle of depression of 35°, and the farther town is at an angle of depression of 31°. How high above the ground is the balloon?

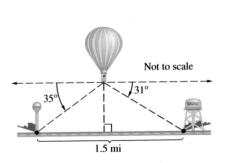

35. *Distance to the Moon*   Since the moon is a relatively close celestial object, its distance can be measured directly by taking two different photographs at precisely the same time from two different locations. The moon will have a different angle of elevation at each location. On April 29, 1976, at 11:35 A.M., the lunar angles of elevation during a partial solar eclipse at Bochum in upper Germany and at Donaueschingen in lower Germany were measured as 52.6997° and 52.7430°, respectively. The two cities are 398 km apart. Calculate the distance to the moon from Bochum on this day, and compare it with the actual value of 406,000 km. Disregard the curvature of Earth in this calculation. (*Source:* Scholosser, W., T. Schmidt-Kaler, and E. Milone, *Challenges of Astronomy,* Springer-Verlag, 1991.)

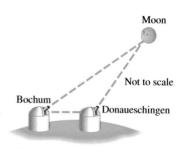

36. *Ground Distances Measured by Aerial Photography*   The distance covered by an aerial photograph is determined by both the focal length of the camera and the tilt of the camera from the perpendicular to the ground. Although the tilt is usually small, both archaeological and Canadian photographs often use larger tilts. A camera lens with a 12-in. focal length will have an angular coverage of 60°. If an aerial photograph is taken with this camera tilted $\theta = 35°$ at an altitude of 5000 ft, calculate the ground distance d to the nearest foot that will be shown in this photograph. (*Source:* Brooks, R. and D. Johannes, *Phytoarchaeology,* Dioscorides Press, 1990; Moffitt, F. and E. Mikhail, *Photogrammetry,* Third Edition, Harper & Row, 1980.)

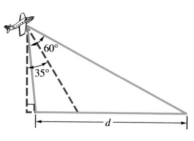

37. *Ground Distances Measured by Aerial Photography*   Refer to Exercise 36. A camera lens with a 6-in. focal length has an angular coverage of 86°. Suppose an aerial photograph is taken vertically with no tilt at an altitude of 3500 ft over ground with an increasing slope of 5°, as shown in the figure. Calculate the ground distance CB that would appear in the resulting photograph. (*Source:* Moffitt, F. and E. Mikhail, *Photogrammetry,* Third Edition, Harper & Row, 1980.)

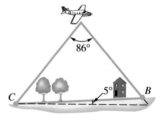

38. *Ground Distances Measured by Aerial Photography*   Repeat Exercise 37 if the camera lens has an 8.25-in. focal length with an angular coverage of 72°.

Find the area of each triangle using the formula $\mathcal{A} = \frac{1}{2}bh$, and then verify that the formula $\mathcal{A} = \frac{1}{2}ab \sin C$ gives the same result.

39.

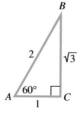

40.

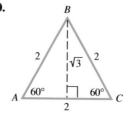

41.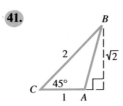

42.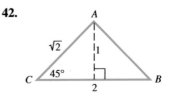

Find the area of each triangle ABC. See Examples 4 and 5.

43. $A = 42.5°$, $b = 13.6$ m, $c = 10.1$ m

44. $C = 72.2°$, $b = 43.8$ ft, $a = 35.1$ ft

45. $B = 124.5°$, $a = 30.4$ cm, $c = 28.4$ cm

46. $C = 142.7°$, $a = 21.9$ km, $b = 24.6$ km

47. $A = 56.80°$, $b = 32.67$ in., $c = 52.89$ in.

48. $A = 34.97°$, $b = 35.29$ m, $c = 28.67$ m

49. $A = 30.50°$, $b = 13.00$ cm, $C = 112.60°$

50. $A = 59.80°$, $b = 15.00$ m, $C = 53.10°$

Solve each problem.

51. *Area of a Metal Plate*   A painter is going to apply a special coating to a triangular metal plate on a new building. Two sides measure 16.1 m and 15.2 m. She knows that the angle between these sides is 125°. What is the area of the surface she plans to cover with the coating?

52. *Area of a Triangular Lot*   A real estate agent wants to find the area of a triangular lot. A surveyor takes measurements and finds that two sides are 52.1 m and 21.3 m, and the angle between them is 42.2°. What is the area of the triangular lot?

53. *Triangle Inscribed in a Circle*   For a triangle inscribed in a circle of radius r, each law of sines ratio $\frac{a}{\sin A}$, $\frac{b}{\sin B}$, and $\frac{c}{\sin C}$ has value $2r$. The circle in the figure has diameter 1. What are the values of a, b, and c? (*Note:* This result provides an alternative way to define the sine function for angles between 0° and 180°. It was used nearly 2000 yr ago by the mathematician Ptolemy to construct one of the earliest trigonometric tables.)

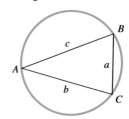

54. *Theorem of Ptolemy*   The following theorem is also attributed to Ptolemy: *In a quadrilateral inscribed in a circle, the product of the diagonals is equal to the sum of the products of the opposite sides.* (*Source:* Eves, H., *An Introduction to the History of Mathematics,* Sixth Edition, Saunders College Publishing, 1990.) The circle in the figure has diameter 1. Explain why the lengths of the line segments are as shown, and then apply Ptolemy's theorem to derive the formula for the sine of the sum of two angles.

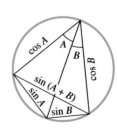

55. Several of the exercises on right triangle applications involved a figure similar to the one shown here, in which angles α and β and the length of line segment AB are known, and the length of side CD is to be determined. Use the law of sines to obtain x in terms of α, β, and d.

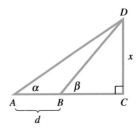

7.2 The Ambiguous Case of the Law of Sines

Description of the Ambiguous Case ▪ Solving SSA Triangles (Case 2) ▪ Analyzing Data for Possible Number of Triangles

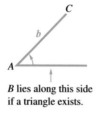

B lies along this side
if a triangle exists.

Figure 8

Description of the Ambiguous Case   We used the law of sines to solve triangles involving Case 1, SAA or ASA, in **Section 7.1.** If we are given the lengths of two sides and the angle opposite one of them (Case 2, SSA), then zero, one, or two such triangles may exist. (There is no SSA axiom.)

Suppose we know the measure of acute angle *A* of triangle *ABC*, the length of side *a*, and the length of side *b*, as shown in Figure 8. Now we must draw the side of length *a* opposite angle *A*. The table shows possible outcomes. This situation (SSA) is called the **ambiguous case** of the law of sines.

If angle *A* is acute, there are four possible outcomes.

Number of Triangles	Sketch	Applying Law of Sines Leads to
0		$\sin B > 1$, $a < h < b$ $a + b < AB$
1		$\sin B = 1$, $a = h$ and $h < b$
1		$0 < \sin B < 1$, $a \geq b$
2		$0 < \sin B_2 < 1$, $h < a < b$

If angle *A* is obtuse, there are two possible outcomes.

Number of Triangles	Sketch	Applying Law of Sines Leads to
0		$\sin B \geq 1$, $a \leq b$
1		$0 < \sin B < 1$, $a > b$

The following basic facts should be kept in mind to determine which situation applies.

APPLYING THE LAW OF SINES

1. For any angle θ of a triangle, $0 < \sin \theta \leq 1$. If $\sin \theta = 1$, then $\theta = 90°$ and the triangle is a right triangle.
2. $\sin \theta = \sin(180° - \theta)$ (Supplementary angles have the same sine value.)
3. The smallest angle is opposite the shortest side, the largest angle is opposite the longest side, and the middle-valued angle is opposite the intermediate side (assuming the triangle has sides that are all of different lengths).

Solving SSA Triangles (Case 2)

▶ **EXAMPLE 1** SOLVING THE AMBIGUOUS CASE (NO SUCH TRIANGLE)

Solve triangle ABC if $B = 55° \, 40'$, $b = 8.94$ m, and $a = 25.1$ m.

Solution Since we are given B, b, and a, we can use the law of sines to find A.

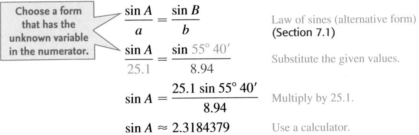

Choose a form that has the unknown variable in the numerator.

$$\frac{\sin A}{a} = \frac{\sin B}{b}$$ Law of sines (alternative form) (Section 7.1)

$$\frac{\sin A}{25.1} = \frac{\sin 55° \, 40'}{8.94}$$ Substitute the given values.

$$\sin A = \frac{25.1 \sin 55° \, 40'}{8.94}$$ Multiply by 25.1.

$$\sin A \approx 2.3184379$$ Use a calculator.

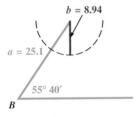

Since $\sin A$ cannot be greater than 1, there can be no such angle A and thus, no triangle with the given information. An attempt to sketch such a triangle leads to the situation shown in Figure 9.

Figure 9

NOW TRY EXERCISE 17. ◀

▶ **Note** In the ambiguous case, we are given two sides and an angle opposite one of the sides (SSA). For example, suppose b, c, and angle C are given. This situation represents the ambiguous case because angle C is opposite side c.

▶ **EXAMPLE 2** SOLVING THE AMBIGUOUS CASE (TWO TRIANGLES)

Solve triangle ABC if $A = 55.3°$, $a = 22.8$ ft, and $b = 24.9$ ft.

Solution To begin, use the law of sines to find angle B.

$$\frac{\sin A}{a} = \frac{\sin B}{b}$$ ◁ Solve for sin B.

$$\frac{\sin 55.3°}{22.8} = \frac{\sin B}{24.9}$$ Substitute known values.

$$\sin B = \frac{24.9 \sin 55.3°}{22.8}$$ Multiply by 24.9; rewrite.

$$\sin B \approx .8978678$$ Use a calculator.

There are two angles B between $0°$ and $180°$ that satisfy this condition. Since $\sin B \approx .8978678$, to the nearest tenth one value of B is

$$B_1 = 63.9°. \quad \text{Use the inverse sine function. (Section 6.1)}$$

Supplementary angles have the same sine value, so another *possible* value of B is

$$B_2 = 180° - 63.9° = 116.1°. \quad \text{(Section 1.1)}$$

To see if $B_2 = 116.1°$ is a valid possibility, add $116.1°$ to the measure of A, $55.3°$. Since $116.1° + 55.3° = 171.4°$, and this sum is less than $180°$, it is a valid angle measure for this triangle.

Now separately solve triangles AB_1C_1 and AB_2C_2 shown in Figure 10. Begin with AB_1C_1. Find C_1 first.

$$C_1 = 180° - A - B_1 \quad \text{(Section 1.2)}$$
$$C_1 = 180° - 55.3° - 63.9°$$
$$C_1 = 60.8°$$

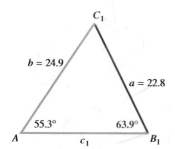

Now, use the law of sines to find c_1.

$$\frac{a}{\sin A} = \frac{c_1}{\sin C_1} \quad \boxed{\text{Solve for } c_1.}$$
$$\frac{22.8}{\sin 55.3°} = \frac{c_1}{\sin 60.8°}$$
$$c_1 = \frac{22.8 \sin 60.8°}{\sin 55.3°}$$
$$c_1 \approx 24.2 \text{ ft}$$

To solve triangle AB_2C_2, first find C_2.

$$C_2 = 180° - A - B_2$$
$$C_2 = 180° - 55.3° - 116.1°$$
$$C_2 = 8.6°$$

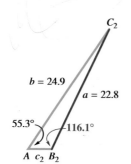

Figure 10

Use the law of sines to find c_2.

$$\frac{a}{\sin A} = \frac{c_2}{\sin C_2} \quad \boxed{\text{Solve for } c_2.}$$
$$\frac{22.8}{\sin 55.3°} = \frac{c_2}{\sin 8.6°}$$
$$c_2 = \frac{22.8 \sin 8.6°}{\sin 55.3°}$$
$$c_2 \approx 4.15 \text{ ft}$$

$$\boxed{\text{NOW TRY EXERCISE 25.} \blacktriangleleft}$$

The ambiguous case results in zero, one, or two triangles. The following guidelines can be used to determine how many triangles there are.

NUMBER OF TRIANGLES SATISFYING THE AMBIGUOUS CASE (SSA)

Let sides a and b and angle A be given in triangle ABC. (The law of sines can be used to calculate the value of $\sin B$.)

1. If applying the law of sines results in an equation having $\sin B > 1$, then *no triangle* satisfies the given conditions.
2. If $\sin B = 1$, then *one triangle* satisfies the given conditions and $B = 90°$.
3. If $0 < \sin B < 1$, then either *one or two triangles* satisfy the given conditions.
    (a) If $\sin B = k$, then let $B_1 = \sin^{-1} k$ and use B_1 for B in the first triangle.
    (b) Let $B_2 = 180° - B_1$. If $A + B_2 < 180°$, then a second triangle exists. In this case, use B_2 for B in the second triangle.

▶ **EXAMPLE 3**   SOLVING THE AMBIGUOUS CASE (ONE TRIANGLE)

Solve triangle ABC, given $A = 43.5°$, $a = 10.7$ in., and $c = 7.2$ in.

Solution   To find angle C, use an alternative form of the law of sines.

$$\frac{\sin C}{c} = \frac{\sin A}{a}$$

$$\frac{\sin C}{7.2} = \frac{\sin 43.5°}{10.7}$$

$$\sin C = \frac{7.2 \sin 43.5°}{10.7} \approx .46319186$$

$$C \approx 27.6° \quad \text{Use the inverse sine function.}$$

There is another angle C that has sine value .46319186; it is $C = 180° - 27.6° = 152.4°$. However, notice in the given information that $c < a$, meaning that in the triangle, angle C must have measure *less than* angle A. Notice also that when we add this obtuse value to the given angle $A = 43.5°$, we obtain

$$152.4° + 43.5° = 195.9°,$$

which is greater than 180°. So either of these approaches shows that there can be only one triangle. See Figure 11. Then

$$B = 180° - 27.6° - 43.5° = 108.9°,$$

and we can find side b with the law of sines.

$$\frac{b}{\sin B} = \frac{a}{\sin A}$$

$$\frac{b}{\sin 108.9°} = \frac{10.7}{\sin 43.5°}$$

$$b = \frac{10.7 \sin 108.9°}{\sin 43.5°}$$

$$b \approx 14.7 \text{ in.}$$

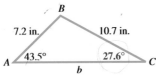

Figure 11

NOW TRY EXERCISE 21. ◀

Analyzing Data for Possible Number of Triangles

▶ **EXAMPLE 4**   ANALYZING DATA INVOLVING AN OBTUSE ANGLE

Without using the law of sines, explain why $A = 104°$, $a = 26.8$ m, and $b = 31.3$ m cannot be valid for a triangle ABC.

Solution   Since A is an obtuse angle, it is the largest angle and so the longest side of the triangle must be a. However, we are given $b > a$; thus, $B > A$, which is impossible if A is obtuse. Therefore, no such triangle ABC exists.

NOW TRY EXERCISE 33. ◀

7.2 Exercises

1. *Concept Check*   Which one of the following sets of data does not determine a unique triangle?
   A. $A = 40°, B = 60°, C = 80°$ B. $a = 5, b = 12, c = 13$
   C. $a = 3, b = 7, C = 50°$ D. $a = 2, b = 2, c = 2$

2. *Concept Check*   Which one of the following sets of data determines a unique triangle?
   A. $A = 50°, B = 50°, C = 80°$ B. $a = 3, b = 5, c = 20$
   C. $A = 40°, B = 20°, C = 30°$ D. $a = 7, b = 24, c = 25$

Concept Check   *In each figure, a line of length h is to be drawn from the given point to the positive x-axis in order to form a triangle. For what value(s) of h can you draw the following?*

(a) *two triangles*   **(b)** *exactly one triangle*   **(c)** *no triangle*

3.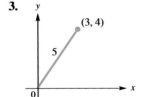

4.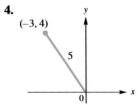

Determine the number of triangles ABC possible with the given parts. See Examples 1–4.

5. $a = 50, b = 26, A = 95°$ **6.** $b = 60, a = 82, B = 100°$

7. $a = 31, b = 26, B = 48°$ **8.** $a = 35, b = 30, A = 40°$

9. $a = 50, b = 61, A = 58°$ **10.** $B = 54°, c = 28, b = 23$

Find each angle B. Do not use a calculator.

11.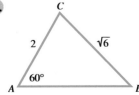

12.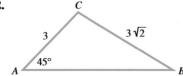

Find the unknown angles in triangle ABC for each triangle that exists. See Examples 1–3.

13. $A = 29.7°, b = 41.5$ ft, $a = 27.2$ ft

14. $B = 48.2°, a = 890$ cm, $b = 697$ cm

15. $C = 41° 20', b = 25.9$ m, $c = 38.4$ m

16. $B = 48° 50', a = 3850$ in., $b = 4730$ in.

17. $B = 74.3°, a = 859$ m, $b = 783$ m

18. $C = 82.2°, a = 10.9$ km, $c = 7.62$ km

19. $A = 142.13°, b = 5.432$ ft, $a = 7.297$ ft

20. $B = 113.72°, a = 189.6$ yd, $b = 243.8$ yd

Solve each triangle ABC that exists. See Examples 1–3.

21. $A = 42.5°, a = 15.6$ ft, $b = 8.14$ ft   **22.** $C = 52.3°, a = 32.5$ yd, $c = 59.8$ yd

23. $B = 72.2°, b = 78.3$ m, $c = 145$ m   **24.** $C = 68.5°, c = 258$ cm, $b = 386$ cm

25. $A = 38° 40', a = 9.72$ km, $b = 11.8$ km

26. $C = 29° 50', a = 8.61$ m, $c = 5.21$ m

27. $A = 96.80°, b = 3.589$ ft, $a = 5.818$ ft

28. $C = 88.70°, b = 56.87$ yd, $c = 112.4$ yd

29. $B = 39.68°, a = 29.81$ m, $b = 23.76$ m

30. $A = 51.20°, c = 7986$ cm, $a = 7208$ cm

31. Apply the law of sines to the following: $a = \sqrt{5}, c = 2\sqrt{5}, A = 30°$. What is the value of sin C? What is the measure of C? Based on its angle measures, what kind of triangle is triangle ABC?

32. Explain the condition that must exist to determine that there is no triangle satisfying the given values of a, b, and B, once the value of sin B is found.

33. Without using the law of sines, explain why no triangle ABC exists satisfying $A = 103° 20', a = 14.6$ ft, $b = 20.4$ ft.

34. Apply the law of sines to the data given in Example 4. Describe in your own words what happens when you try to find the measure of angle B using a calculator.

Use the law of sines to solve each problem.

35. *Distance Between Inaccessible Points*   To find the distance between a point X and an inaccessible point Z, a line segment XY is constructed. It is found that $XY = 960$ m, angle $XYZ = 43° 30'$, and angle $YZX = 95° 30'$. Find the distance between X and Z to the nearest meter.

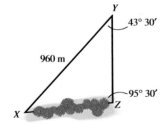

36. *Height of an Antenna Tower*   The angle of elevation from the top of a building 45.0 ft high to the top of a nearby antenna tower is $15° 20'$. From the base of the building, the angle of elevation of the tower is $29° 30'$. Find the height of the tower.

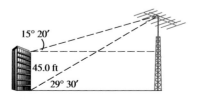

37. *Height of a Building*   A flagpole 95.0 ft tall is on the top of a building. From a point on level ground, the angle of elevation of the top of the flagpole is $35.0°$, while the angle of elevation of the bottom of the flagpole is $26.0°$. Find the height of the building.

38. *Flight Path of a Plane* A pilot flies her plane on a heading of 35° 00′ from point X to point Y, 400 mi from X. Then she turns and flies on a heading of 145° 00′ to point Z, which is 400 mi from her starting point X. What is the heading of Z from X, and what is the distance YZ?

Use the law of sines to prove that each statement is true for any triangle ABC, with corresponding sides a, b, and c.

39. $\dfrac{a+b}{b} = \dfrac{\sin A + \sin B}{\sin B}$

40. $\dfrac{a-b}{a+b} = \dfrac{\sin A - \sin B}{\sin A + \sin B}$

7.3 The Law of Cosines

Derivation of the Law of Cosines ▪ Solving SAS and SSS Triangles (Cases 3 and 4) ▪ Heron's Formula for the Area of a Triangle

As mentioned in **Section 7.1,** if we are given two sides and the included angle (Case 3) or three sides (Case 4) of a triangle, then a unique triangle is determined. These are the SAS and SSS cases, respectively. Both cases require using the *law of cosines*. Remember the following property of triangles when applying the law of cosines to solve a triangle.

TRIANGLE SIDE LENGTH RESTRICTION

In any triangle, the sum of the lengths of any two sides must be greater than the length of the remaining side.

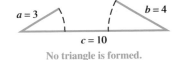

$a = 3$ $b = 4$

$c = 10$

No triangle is formed.

Figure 12

For example, it would be impossible to construct a triangle with sides of lengths 3, 4, and 10. See Figure 12.

Derivation of the Law of Cosines To derive the law of cosines, let ABC be any oblique triangle. Choose a coordinate system so that vertex B is at the origin and side BC is along the positive x-axis. See Figure 13.

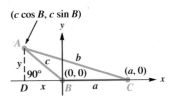

$(c \cos B, c \sin B)$

Figure 13

Let (x, y) be the coordinates of vertex A of the triangle. Verify that for angle B, whether obtuse or acute,

$$\sin B = \frac{y}{c} \qquad \text{and} \qquad \cos B = \frac{x}{c}. \qquad \text{(Section 2.2)}$$

$$y = c \sin B \qquad \text{and} \qquad x = c \cos B.$$

Here x is negative when B is obtuse.

Thus, the coordinates of point A become $(c \cos B, c \sin B)$.

Point C in Figure 13 has coordinates $(a, 0)$, and AC has length b. Since point A has coordinates $(c \cos B, c \sin B)$, by the distance formula,

$$b = \sqrt{(c \cos B - a)^2 + (c \sin B - 0)^2} \qquad \text{(Appendix B)}$$

$$b^2 = (c \cos B - a)^2 + (c \sin B)^2 \qquad \text{Square both sides.}$$

$$= (c^2 \cos^2 B - 2ac \cos B + a^2) + c^2 \sin^2 B \qquad \text{Multiply;}$$
$$(x - y)^2 = x^2 - 2xy + y^2$$

Remember the middle term.

$$= a^2 + c^2(\cos^2 B + \sin^2 B) - 2ac \cos B \qquad \text{Properties of real numbers}$$

$$= a^2 + c^2(1) - 2ac \cos B \qquad \text{Fundamental identity (Section 5.1)}$$

$$b^2 = a^2 + c^2 - 2ac \cos B.$$

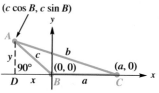

$(c \cos B, c \sin B)$

Figure 13 (repeated)

This result is one form of the law of cosines. In our work, we could just as easily have placed A or C at the origin. This would have given the same result, but with the variables rearranged.

LAW OF COSINES

In any triangle ABC, with sides a, b, and c,

$$a^2 = b^2 + c^2 - 2bc \cos A,$$
$$b^2 = a^2 + c^2 - 2ac \cos B,$$
$$c^2 = a^2 + b^2 - 2ab \cos C.$$

That is, according to the law of cosines, the square of a side of a triangle is equal to the sum of the squares of the other two sides, minus twice the product of those two sides and the cosine of the angle included between them.

▶ **Note** If we let $C = 90°$ in the third form of the law of cosines, then $\cos C = \cos 90° = 0$, and the formula becomes $c^2 = a^2 + b^2$, the Pythagorean theorem (Appendix B). The Pythagorean theorem is a special case of the law of cosines.

Solving SAS and SSS Triangles (Cases 3 and 4)

▶ **EXAMPLE 1**   USING THE LAW OF COSINES IN AN APPLICATION (SAS)

A surveyor wishes to find the distance between two inaccessible points A and B on opposite sides of a lake. While standing at point C, she finds that $AC = 259$ m, $BC = 423$ m, and angle ACB measures $132° \, 40'$. Find the distance AB. See Figure 14.

Solution   The law of cosines can be used here since we know the lengths of two sides of the triangle and the measure of the included angle.

$$AB^2 = 259^2 + 423^2 - 2(259)(423) \cos 132° \, 40'$$

$$AB^2 \approx 394{,}510.6 \quad \text{Use a calculator.}$$

$$AB \approx 628 \qquad \text{Take the square root of each side. (Appendix A)}$$

The distance between the points is approximately 628 m.

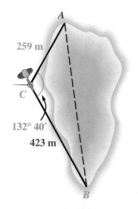

259 m

C

132° 40'

423 m

B

Figure 14

NOW TRY EXERCISE 39. ◀

▶ **EXAMPLE 2**   USING THE LAW OF COSINES TO SOLVE A TRIANGLE (SAS)

Solve triangle ABC if $A = 42.3°$, $b = 12.9$ m, and $c = 15.4$ m.

Solution   See Figure 15. We start by finding a with the law of cosines.

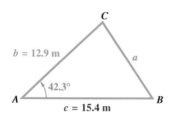

Figure 15

$$a^2 = b^2 + c^2 - 2bc \cos A \qquad \text{Law of cosines}$$
$$a^2 = 12.9^2 + 15.4^2 - 2(12.9)(15.4)\cos 42.3° \qquad \text{Substitute.}$$
$$a^2 \approx 109.7 \qquad \text{Use a calculator.}$$
$$a \approx 10.47 \text{ m} \qquad \text{Take square roots.}$$

Of the two remaining angles B and C, B must be the smaller since it is opposite the shorter of the two sides b and c. Therefore, B cannot be obtuse.

$$\frac{\sin A}{a} = \frac{\sin B}{b} \qquad \text{Law of sines (Section 7.1)}$$
$$\frac{\sin 42.3°}{10.47} = \frac{\sin B}{12.9} \qquad \text{Substitute.}$$
$$\sin B = \frac{12.9 \sin 42.3°}{10.47} \qquad \text{Multiply by 12.9; rewrite.}$$
$$B \approx 56.0° \qquad \text{Use the inverse sine function. (Section 6.1)}$$

The easiest way to find C is to subtract the measures of A and B from $180°$.

$$C = 180° - 42.3° - 56.0° = 81.7° \quad \text{(Section 1.2)}$$

NOW TRY EXERCISE 19. ◀

▶ **Caution**   Had we used the law of sines to find C rather than B in Example 2, we would not have known whether C is equal to $81.7°$ or its supplement, $98.3°$.

▶ **EXAMPLE 3**   USING THE LAW OF COSINES TO SOLVE A TRIANGLE (SSS)

Solve triangle ABC if $a = 9.47$ ft, $b = 15.9$ ft, and $c = 21.1$ ft.

Solution   We can use the law of cosines to solve for any angle of the triangle. We solve for C, the largest angle. We will know that C is obtuse if $\cos C < 0$.

$$c^2 = a^2 + b^2 - 2ab \cos C \qquad \text{Law of cosines}$$
$$\cos C = \frac{a^2 + b^2 - c^2}{2ab} \qquad \text{Solve for } \cos C.$$
$$\cos C = \frac{9.47^2 + 15.9^2 - 21.1^2}{2(9.47)(15.9)} \qquad \text{Substitute.}$$
$$\cos C \approx -.34109402 \qquad \text{Use a calculator.}$$
$$C \approx 109.9° \qquad \text{Use the inverse cosine function.}$$
$$\text{(Section 6.1)}$$

We can use either the law of sines or the law of cosines to find $B \approx 45.1°$. (Verify this.) Since $A = 180° - B - C$, we obtain $A \approx 25.0°$.

NOW TRY EXERCISE 23. ◀

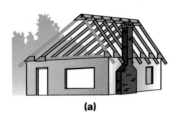

(a)

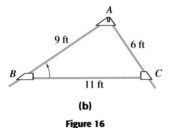

(b)

Figure 16

Trusses are frequently used to support roofs on buildings, as illustrated in Figure 16(a). The simplest type of roof truss is a triangle, as shown in Figure 16(b). One basic task when constructing a roof truss is to cut the ends of the rafters so that the roof has the correct slope. (*Source:* Riley, W., L. Sturges, and D. Morris, *Statics and Mechanics of Materials,* John Wiley and Sons, 1995.)

▶ **EXAMPLE 4**   DESIGNING A ROOF TRUSS (SSS)

Find angle B for the truss shown in Figure 16(b).

Solution $b^2 = a^2 + c^2 - 2ac \cos B$ Law of cosines

$$\cos B = \frac{a^2 + c^2 - b^2}{2ac}$$ Solve for $\cos B$.

$$\cos B = \frac{11^2 + 9^2 - 6^2}{2(11)(9)}$$ Let $a = 11$, $b = 6$, and $c = 9$.

$$\cos B \approx .83838384$$ Use a calculator.

$$B \approx 33°$$ Use the inverse cosine function.

NOW TRY EXERCISE 45. ◀

Four possible cases can occur when solving an oblique triangle, as summarized in the following table. In all four cases, it is assumed that the given information actually produces a triangle.

Oblique Triangle	Suggested Procedure for Solving
Case 1: One side and two angles are known. **(SAA or ASA)**	**Step 1** Find the remaining angle using the angle sum formula ($A + B + C = 180°$). **Step 2** Find the remaining sides using the law of sines.
Case 2: Two sides and one angle (not included between the two sides) are known. **(SSA)**	*This is the ambiguous case; there may be no triangle, one triangle, or two triangles.* **Step 1** Find an angle using the law of sines. **Step 2** Find the remaining angle using the angle sum formula. **Step 3** Find the remaining side using the law of sines. *If two triangles exist, repeat Steps 2 and 3.*
Case 3: Two sides and the included angle are known. **(SAS)**	**Step 1** Find the third side using the law of cosines. **Step 2** Find the smaller of the two remaining angles using the law of sines. **Step 3** Find the remaining angle using the angle sum formula.
Case 4: Three sides are known. **(SSS)**	**Step 1** Find the largest angle using the law of cosines. **Step 2** Find either remaining angle using the law of sines. **Step 3** Find the remaining angle using the angle sum formula.

Heron of Alexandria

Heron's Formula for the Area of a Triangle

The law of cosines can be used to derive a formula for the area of a triangle given the lengths of the three sides. This formula is known as **Heron's formula,** named after the Greek mathematician Heron of Alexandria, who lived around A.D. 75. It is found in his work *Metrica*. Heron's formula can be used for the case SSS.

HERON'S AREA FORMULA (SSS)

If a triangle has sides of lengths a, b, and c, with **semiperimeter**

$$s = \frac{1}{2}(a + b + c),$$

then the area of the triangle is

$$\mathcal{A} = \sqrt{s(s - a)(s - b)(s - c)}.$$

That is, according to Heron's formula, the area of a triangle is the square root of the product of four factors: (1) the semiperimeter, (2) the semiperimeter minus the first side, (3) the semiperimeter minus the second side, and (4) the semiperimeter minus the third side. A derivation of Heron's formula is given in the Connections at the end of this section.

▶ **EXAMPLE 5** USING HERON'S FORMULA TO FIND AN AREA (SSS)

The distance "as the crow flies" from Los Angeles to New York is 2451 mi, from New York to Montreal is 331 mi, and from Montreal to Los Angeles is 2427 mi. What is the area of the triangular region having these three cities as vertices? (Ignore the curvature of Earth.)

Solution In Figure 17 we let $a = 2451$, $b = 331$, and $c = 2427$.

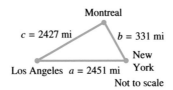

Montreal

$c = 2427$ mi

$b = 331$ mi

New York

Los Angeles $a = 2451$ mi

Not to scale

Figure 17

The semiperimeter s is

$$s = \frac{1}{2}(2451 + 331 + 2427) = 2604.5.$$

Using Heron's formula, the area $\mathcal{A}$ is

$$\mathcal{A} = \sqrt{s(s - a)(s - b)(s - c)}$$

> Don't forget the factor s.

$$\mathcal{A} = \sqrt{2604.5(2604.5 - 2451)(2604.5 - 331)(2604.5 - 2427)}$$

$$\mathcal{A} \approx 401{,}700 \text{ mi}^2.$$

NOW TRY EXERCISE 75. ◀

44. *Distance Between a Ship and a Submarine* From an airplane flying over the ocean, the angle of depression to a submarine lying under the surface is 24° 10′. At the same moment, the angle of depression from the airplane to a battleship is 17° 30′. (See the figure.) The distance from the airplane to the battleship is 5120 ft. Find the distance between the battleship and the submarine. (Assume the airplane, submarine, and battleship are in a vertical plane.)

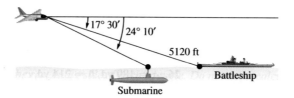

45. *Truss Construction* A triangular truss is shown in the figure. Find angle θ.

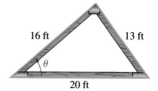

46. *Distance Between Points on a Crane* A crane with a counterweight is shown in the figure. Find the horizontal distance between points A and B to the nearest foot.

47. *Distance Between a Beam and Cables* A weight is supported by cables attached to both ends of a balance beam, as shown in the figure. What angles are formed between the beam and the cables?

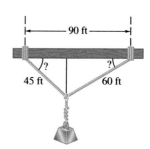

48. *Length of a Tunnel* To measure the distance through a mountain for a proposed tunnel, a point C is chosen that can be reached from each end of the tunnel. (See the figure.) If $AC = 3800$ m, $BC = 2900$ m, and angle $C = 110°$, find the length of the tunnel.

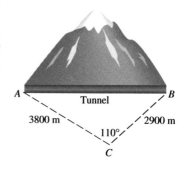

49. *Distance on a Baseball Diamond*   A baseball diamond is a square, 90.0 ft on a side, with home plate and the three bases as vertices. The pitcher's position is 60.5 ft from home plate. Find the distance from the pitcher's position to each of the bases.

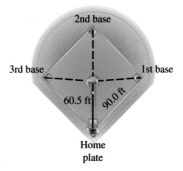

50. *Distance Between Ends of the Vietnam Memorial*   The Vietnam Veterans' Memorial in Washington, D.C., is V-shaped with equal sides of length 246.75 ft. The angle between these sides measures 125° 12′. Find the distance between the ends of the two sides. (*Source:* Pamphlet obtained at Vietnam Veterans' Memorial.)

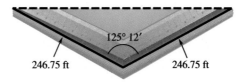

51. *Distance Between a Ship and a Point*   Starting at point A, a ship sails 18.5 km on a bearing of 189°, then turns and sails 47.8 km on a bearing of 317°. Find the distance of the ship from point A.

52. *Distance Between Two Factories*   Two factories blow their whistles at exactly 5:00. A man hears the two blasts at 3 sec and 6 sec after 5:00, respectively. The angle between his lines of sight to the two factories is 42.2°. If sound travels 344 m per sec, how far apart are the factories?

53. *Bearing of One Town to Another*   Two towns 21 mi apart are separated by a dense forest. (See the figure.) To travel from town A to town B, a person must go 17 mi on a bearing of 325°, then turn and continue for 9 mi to reach town B. Find the bearing of B from A.

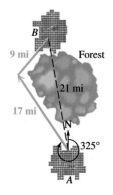

54. *Distance Traveled by a Plane*   An airplane flies 180 mi from point X at a bearing of 125°, and then turns and flies at a bearing of 230° for 100 mi. How far is the plane from point X?

55. *Measurement Using Triangulation* Surveyors are often confronted with obstacles, such as trees, when measuring the boundary of a lot. One technique used to obtain an accurate measurement is the so-called **triangulation method. In** this technique, a triangle is constructed around the obstacle and one angle and two sides of the triangle are measured. Use this technique to find the length of the property line (the straight line between the two markers) in the figure. (*Source:* Kavanagh, B., *Surveying Principles and Applications,* Sixth Edition, Prentice-Hall, 2003.)

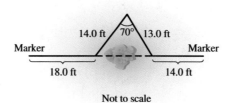

Not to scale

56. *Path of a Ship* A ship sailing due east in the North Atlantic has been warned to change course to avoid icebergs. The captain turns and sails on a bearing of 62°, then changes course again to a bearing of 115° until the ship reaches its original course. (See the figure.) How much farther did the ship have to travel to avoid the icebergs?

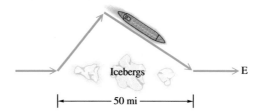

57. *Angle in a Parallelogram* A parallelogram has sides of length 25.9 cm and 32.5 cm. The longer diagonal has length 57.8 cm. Find the angle opposite the longer diagonal.

58. *Distance Between an Airplane and a Mountain* A person in a plane flying straight north observes a mountain at a bearing of 24.1°. At that time, the plane is 7.92 km from the mountain. A short time later, the bearing to the mountain becomes 32.7°. How far is the airplane from the mountain when the second bearing is taken?

59. *Layout for a Playhouse* The layout for a child's playhouse has the dimensions given in the figure. Find x.

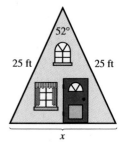

60. *Distance Between Two Towns* To find the distance between two small towns, an electronic distance measuring (EDM) instrument is placed on a hill from which both towns are visible. The distance to each town from the EDM and the angle between the two lines of sight are measured. (See the figure.) Find the distance between the towns.

Find the measure of each angle θ to two decimal places.

61.

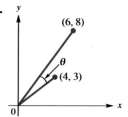

62.

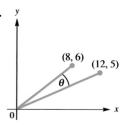

Find the exact area of each triangle using the formula $\mathcal{A} = \frac{1}{2}bh$, and then verify that Heron's formula gives the same result.

63.

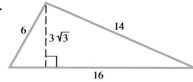

64.

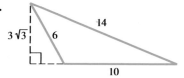

Find the area of each triangle ABC. See Example 5.

65. $a = 12$ m, $b = 16$ m, $c = 25$ m

66. $a = 22$ in., $b = 45$ in., $c = 31$ in.

67. $a = 154$ cm, $b = 179$ cm, $c = 183$ cm

68. $a = 25.4$ yd, $b = 38.2$ yd, $c = 19.8$ yd

69. $a = 76.3$ ft, $b = 109$ ft, $c = 98.8$ ft

70. $a = 15.89$ in., $b = 21.74$ in., $c = 10.92$ in.

Volcano Movement   *To help predict eruptions from the volcano Mauna Loa on the island of Hawaii, scientists keep track of the volcano's movement by using a "super triangle" with vertices on the three volcanoes shown on the map at the right. (For example, in one year, Mauna Loa moved 6 in., a result of increasing internal pressure.) Refer to the map to work Exercises 71 and 72.*

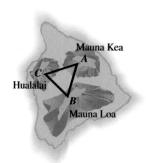

71. $AB = 22.47928$ mi, $AC = 28.14276$ mi, $A = 58.56989°$;   find BC

72. $AB = 22.47928$ mi, $BC = 25.24983$ mi, $A = 58.56989°$;   find B

Solve each problem. See Example 5.

73. *Perfect Triangles*   A **perfect triangle** is a triangle whose sides have whole number lengths and whose area is numerically equal to its perimeter. Show that the triangle with sides of length 9, 10, and 17 is perfect.

74. *Heron Triangles*   A **Heron triangle** is a triangle having integer sides and area. Show that each of the following is a Heron triangle.

(a) $a = 11, b = 13, c = 20$ (b) $a = 13, b = 14, c = 15$
(c) $a = 7, b = 15, c = 20$ (d) $a = 9, b = 10, c = 17$

75. *Area of the Bermuda Triangle* Find the area of the Bermuda Triangle if the sides of the triangle have approximate lengths 850 mi, 925 mi, and 1300 mi.

76. *Required Amount of Paint* A painter needs to cover a triangular region 75 m by 68 m by 85 m. A can of paint covers 75 m² of area. How many cans (to the next higher number of cans) will be needed?

77. Consider triangle *ABC* shown here.

    (a) Use the law of sines to find candidates for the value of angle *C*. Round angle measures to the nearest tenth of a degree.

    (b) Rework part (a) using the law of cosines.

    (c) Why is the law of cosines a better method in this case?

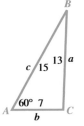

78. Show that the measure of angle *A* is twice the measure of angle *B*. (*Hint:* Use the law of cosines to find cos *A* and cos *B*, and then show that $\cos A = 2\cos^2 B - 1$.)

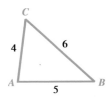

79. Let point *D* on side *AB* of triangle *ABC* be such that *CD* bisects angle *C*. Show that

$$\frac{AD}{DB} = \frac{b}{a}.$$

80. In addition to the law of sines and the law of cosines, there is a **law of tangents.** In any triangle *ABC*,

$$\frac{\tan\frac{1}{2}(A - B)}{\tan\frac{1}{2}(A + B)} = \frac{a - b}{a + b}.$$

Verify this law for the triangle *ABC* with $a = 2$, $b = 2\sqrt{3}$, $A = 30°$, and $B = 60°$.

RELATING CONCEPTS

For individual or collaborative investigation
(Exercises 81–85)

We have introduced two new formulas for the area of a triangle in this chapter. You should now be able to find the area $\mathcal{A}$ of a triangle using one of three formulas:

(a) $\mathcal{A} = \frac{1}{2}bh$

(b) $\mathcal{A} = \frac{1}{2}ab \sin C$ $\left(\text{or } \mathcal{A} = \frac{1}{2}ac \sin B \text{ or } \mathcal{A} = \frac{1}{2}bc \sin A\right)$

(c) $\mathcal{A} = \sqrt{s(s - a)(s - b)(s - c)}$ (Heron's formula).

(continued)

Another area formula can be used when the coordinates of the vertices of a triangle are given. If the vertices are the ordered pairs (x_1, y_1), (x_2, y_2), and (x_3, y_3), then the following is valid:

(d) $\mathcal{A} = \dfrac{1}{2} \left| (x_1 y_2 - y_1 x_2 + x_2 y_3 - y_2 x_3 + x_3 y_1 - y_3 x_1) \right|.$

Work Exercises 81–85 in order, *showing that the various formulas all lead to the same area.*

81. Draw a triangle with vertices $A(2, 5)$, $B(-1, 3)$, and $C(4, 0)$.

82. Use the distance formula to find the lengths of the sides a, b, and c.

83. Find the area of triangle ABC using formula (b). (First use the law of cosines to find the measure of an angle.)

84. Find the area of triangle ABC using formula (c) (that is, Heron's formula).

85. Find the area of triangle ABC using new formula (d).

CHAPTER 7 ▶   Quiz (Sections 7.1–7.3)

Find the indicated part of each triangle ABC.

1. Find A if $B = 30.6°$, $b = 7.42$ in., and $c = 4.54$ in.

2. Find a if $A = 144°$, $c = 135$ m, and $b = 75.0$ m.

3. Find C if $a = 28.4$ ft, $b = 16.9$ ft, and $c = 21.2$ ft.

4. Find the area of the triangle shown here.

5. Find the area of triangle ABC if $a = 19.5$ km, $b = 21.0$ km, and $c = 22.5$ km.

6. For triangle ABC with $c = 345$, $a = 534$, and $C = 25.4°$, there are two possible values for angle A. What are they?

7. Solve triangle ABC if $c = 326$, $A = 111°$, and $B = 41°$.

8. *Height of a Balloon*   The angles of elevation of a hot air balloon from two observation points X and Y on level ground are $42° \, 10'$ and $23° \, 30'$, respectively. As shown in the figure, points X, Y, and Z are in the same vertical plane and points X and Y are 12.2 mi apart. Approximate the height of the balloon to the nearest tenth of a mile.

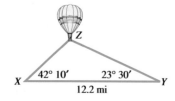

7.4 Vectors, Operations, and the Dot Product

Basic Terminology ▪ **Algebraic Interpretation of Vectors** ▪ **Operations with Vectors** ▪ **Dot Product and the Angle Between Vectors**

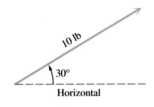

Figure 18

Basic Terminology Many quantities involve magnitudes, such as 45 lb or 60 mph. These quantities are called **scalars** and can be represented by real numbers. Other quantities, called **vector quantities,** involve both magnitude and direction. Typical vector quantities are velocity, acceleration, and force. For example, traveling 50 mph *east* represents a vector quantity.

A vector quantity is often represented with a directed line segment (a segment that uses an arrowhead to indicate direction), called a **vector.** The length of the vector represents the **magnitude** of the vector quantity. The direction of the vector, indicated by the arrowhead, represents the direction of the quantity. For example, the vector in Figure 18 represents a force of 10 lb applied at an angle of 30° from the horizontal.

The symbol for a vector is often printed in boldface type. When writing vectors by hand, it is customary to use an arrow over the letter or letters. Thus **OP** and $\overrightarrow{OP}$ both represent the vector **OP**. Vectors may be named with either one lowercase or uppercase letter, or two uppercase letters. When two letters are used, the first indicates the **initial point** and the second indicates the **terminal point** of the vector. Knowing these points gives the direction of the vector. For example, vectors **OP** and **PO** in Figure 19 are not the same vector. They have the same magnitude, but *opposite* directions. The magnitude of vector **OP** is written |**OP**|.

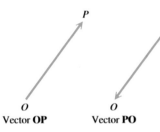

Figure 19

Two vectors are equal if and only if they have the same direction and the same magnitude. In Figure 20, vectors **A** and **B** are equal, as are vectors **C** and **D**. As Figure 20 shows, equal vectors need not coincide, but they must be parallel. Vectors **A** and **E** are unequal because they do not have the same direction, while **A** ≠ **F** because they have different magnitudes.

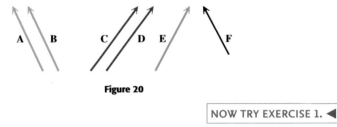

Figure 20

NOW TRY EXERCISE 1. ◀

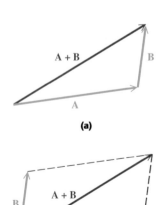

(a)

(b)

Figure 21

To find the sum of two vectors **A** and **B**, we place the initial point of vector **B** at the terminal point of vector **A**, as shown in Figure 21(a). The vector with the same initial point as **A** and the same terminal point as **B** is the sum **A** + **B**. The sum of two vectors is also a vector.

Another way to find the sum of two vectors is to use the **parallelogram rule.** Place vectors **A** and **B** so that their initial points coincide, as in Figure 21(b). Then, complete a parallelogram that has **A** and **B** as two sides. The diagonal of the parallelogram with the same initial point as **A** and **B** is the sum **A** + **B** found by the definition. Compare Figures 21(a) and (b). Parallelograms can be used to show that vector **B** + **A** is the same as vector **A** + **B**, or that **A** + **B** = **B** + **A**, so vector addition is commutative. The vector sum **A** + **B** is called the **resultant** of vectors **A** and **B**.

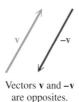

Vectors **v** and **−v**
are opposites.

Figure 22

For every vector **v** there is a vector **−v** that has the same magnitude as **v** but opposite direction. Vector **−v** is called the **opposite** of **v**. (See Figure 22.) The sum of **v** and **−v** has magnitude 0 and is called the **zero vector.** As with real numbers, to subtract vector **B** from vector **A**, find the vector sum **A** + (**−B**). (See Figure 23.)

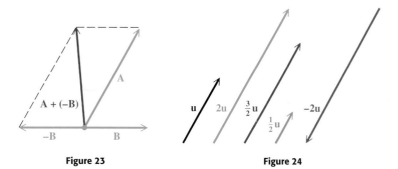

Figure 23   **Figure 24**

The **scalar product** of a real number (or scalar) k and a vector **u** is the vector $k \cdot$ **u**, which has magnitude $|k|$ times the magnitude of **u**. As suggested by Figure 24, the vector $k \cdot$ **u** has the same direction as **u** if $k > 0$, and opposite direction if $k < 0$.

NOW TRY EXERCISES 5, 7, 9, AND 11. ◀

Algebraic Interpretation of Vectors

We now consider vectors in a rectangular coordinate system. A vector with its initial point at the origin is called a **position vector.** A position vector **u** with its endpoint at the point (a, b) is written $\langle a, b \rangle$, so

$$\mathbf{u} = \langle a, b \rangle.$$

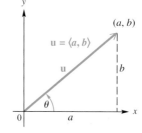

Figure 25

This means that every vector in the real plane corresponds to an ordered pair of real numbers. Thus, geometrically a vector is a directed line segment; algebraically, it is an ordered pair. The numbers a and b are the **horizontal component** and **vertical component,** respectively, of vector **u**. Figure 25 shows the vector $\mathbf{u} = \langle a, b \rangle$. The positive angle between the x-axis and a position vector is the **direction angle** for the vector. In Figure 25, θ is the direction angle for vector **u**.

From Figure 25, we can see that the magnitude and direction of a vector are related to its horizontal and vertical components.

MAGNITUDE AND DIRECTION ANGLE OF A VECTOR $\langle a, b \rangle$

The magnitude (length) of vector $\mathbf{u} = \langle a, b \rangle$ is given by

$$|\mathbf{u}| = \sqrt{a^2 + b^2}.$$

The direction angle θ satisfies $\tan \theta = \frac{b}{a}$, where $a \neq 0$.

▶ **EXAMPLE 1** FINDING MAGNITUDE AND DIRECTION ANGLE

Find the magnitude and direction angle for $\mathbf{u} = \langle 3, -2 \rangle$.

Algebraic Solution

The magnitude is $|\mathbf{u}| = \sqrt{3^2 + (-2)^2} = \sqrt{13}$. To find the direction angle θ, start with $\tan \theta = \frac{b}{a} = \frac{-2}{3} = -\frac{2}{3}$. Vector $\mathbf{u}$ has a positive horizontal component and a negative vertical component, placing the position vector in quadrant IV. A calculator gives $\tan^{-1}\left(-\frac{2}{3}\right) \approx -33.7°$. Adding 360° yields the direction angle $\theta \approx 326.3°$. See Figure 26.

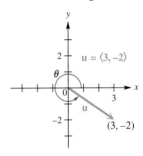

Figure 26

Graphing Calculator Solution

A calculator returns the magnitude and direction angle, given the horizontal and vertical components. An approximation for $\sqrt{13}$ is given, and the direction angle has a measure with least possible absolute value. We must add 360° to the value of θ to obtain the positive direction angle. See Figure 27.

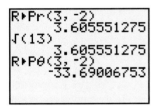

Figure 27

For more information, see your owner's manual or the graphing calculator manual that accompanies this text.

NOW TRY EXERCISE 33. ◀

HORIZONTAL AND VERTICAL COMPONENTS

The horizontal and vertical components, respectively, of a vector $\mathbf{u}$ having magnitude $|\mathbf{u}|$ and direction angle θ are given by

$$a = |\mathbf{u}| \cos \theta \quad \text{and} \quad b = |\mathbf{u}| \sin \theta.$$

That is, $\mathbf{u} = \langle a, b \rangle = \langle |\mathbf{u}| \cos \theta, |\mathbf{u}| \sin \theta \rangle$.

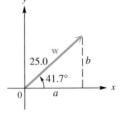

Figure 28

▶ **EXAMPLE 2** FINDING HORIZONTAL AND VERTICAL COMPONENTS

Vector $\mathbf{w}$ in Figure 28 has magnitude 25.0 and direction angle 41.7°. Find the horizontal and vertical components.

Algebraic Solution

Use the two formulas in the box, with $|\mathbf{w}| = 25.0$ and $\theta = 41.7°$.

$$a = 25.0 \cos 41.7° \qquad b = 25.0 \sin 41.7°$$
$$a \approx 18.7 \qquad b \approx 16.6$$

Therefore, $\mathbf{w} = \langle 18.7, 16.6 \rangle$. The horizontal component is 18.7, and the vertical component is 16.6 (rounded to the nearest tenth).

Graphing Calculator Solution

See Figure 29. The results support the algebraic solution.

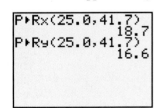

Figure 29

NOW TRY EXERCISE 37. ◀

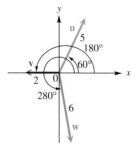

Figure 30

▶ **EXAMPLE 3**   **WRITING VECTORS IN THE FORM** $\langle a, b \rangle$

Write each vector in Figure 30 in the form $\langle a, b \rangle$.

Solution

$$\mathbf{u} = \langle 5 \cos 60°, 5 \sin 60° \rangle = \left\langle 5 \cdot \frac{1}{2}, 5 \cdot \frac{\sqrt{3}}{2} \right\rangle = \left\langle \frac{5}{2}, \frac{5\sqrt{3}}{2} \right\rangle$$

$$\mathbf{v} = \langle 2 \cos 180°, 2 \sin 180° \rangle = \langle 2(-1), 2(0) \rangle = \langle -2, 0 \rangle$$

$$\mathbf{w} = \langle 6 \cos 280°, 6 \sin 280° \rangle \approx \langle 1.0419, -5.9088 \rangle \quad \text{Use a calculator.}$$

NOW TRY EXERCISE 43. ◀

The following properties of parallelograms are helpful when studying vectors.

PROPERTIES OF PARALLELOGRAMS

1. A parallelogram is a quadrilateral whose opposite sides are parallel.
2. The opposite sides and opposite angles of a parallelogram are equal, and adjacent angles of a parallelogram are supplementary.
3. The diagonals of a parallelogram bisect each other, but do not necessarily bisect the angles of the parallelogram.

▶ **EXAMPLE 4**   **FINDING THE MAGNITUDE OF A RESULTANT**

Two forces of 15 and 22 newtons act on a point in the plane. (A **newton** is a unit of force that equals .225 lb.) If the angle between the forces is 100°, find the magnitude of the resultant force.

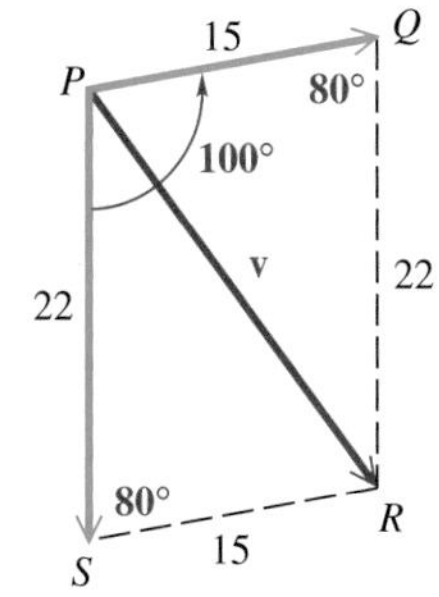

Solution   As shown in Figure 31, a parallelogram that has the forces as adjacent sides can be formed. The angles of the parallelogram adjacent to angle P measure 80°, since adjacent angles of a parallelogram are supplementary. Opposite sides of the parallelogram are equal in length. The resultant force divides the parallelogram into two triangles. Use the law of cosines with either triangle.

$$|\mathbf{v}|^2 = 15^2 + 22^2 - 2(15)(22) \cos 80° \quad \text{Law of cosines (Section 7.3)}$$
$$\approx 225 + 484 - 115$$
$$|\mathbf{v}|^2 \approx 594$$
$$|\mathbf{v}| \approx 24 \qquad\qquad \text{Take square roots. (Appendix A)}$$

To the nearest unit, the magnitude of the resultant force is 24 newtons.

Figure 31

NOW TRY EXERCISE 49. ◀

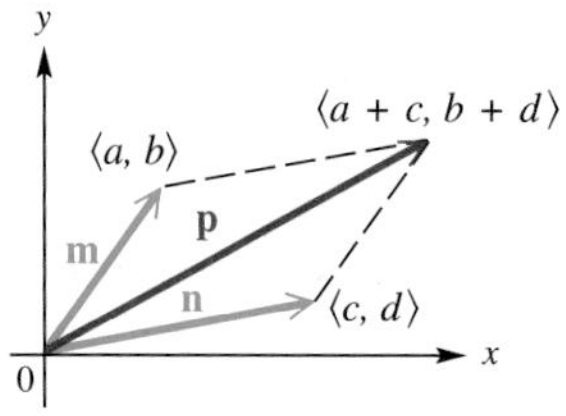

Figure 32

Operations with Vectors   In Figure 32, $\mathbf{m} = \langle a, b \rangle$, $\mathbf{n} = \langle c, d \rangle$, and $\mathbf{p} = \langle a + c, b + d \rangle$. Using geometry, we can show that the endpoints of the three vectors and the origin form a parallelogram. Since a diagonal of this parallelogram gives the resultant of $\mathbf{m}$ and $\mathbf{n}$, we have $\mathbf{p} = \mathbf{m} + \mathbf{n}$ or

$$\langle a + c, b + d \rangle = \langle a, b \rangle + \langle c, d \rangle.$$

Similarly, we could verify the following vector operations.

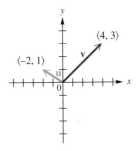

Figure 33

VECTOR OPERATIONS

For any real numbers a, b, c, d, and k,

$$\langle a, b \rangle + \langle c, d \rangle = \langle a + c, b + d \rangle$$

$$k \cdot \langle a, b \rangle = \langle ka, kb \rangle.$$

If $\mathbf{a} = \langle a_1, a_2 \rangle$, then $-\mathbf{a} = \langle -a_1, -a_2 \rangle$.

$$\langle a, b \rangle - \langle c, d \rangle = \langle a, b \rangle + (-\langle c, d \rangle) = \langle a - c, b - d \rangle$$

▶ **EXAMPLE 5** PERFORMING VECTOR OPERATIONS

Let $\mathbf{u} = \langle -2, 1 \rangle$ and $\mathbf{v} = \langle 4, 3 \rangle$. (See Figure 33.) Find the following: **(a)** $\mathbf{u} + \mathbf{v}$, **(b)** $-2\mathbf{u}$, **(c)** $4\mathbf{u} - 3\mathbf{v}$.

Algebraic Solution

(a) $\mathbf{u} + \mathbf{v} = \langle -2, 1 \rangle + \langle 4, 3 \rangle$

$\qquad = \langle -2 + 4, 1 + 3 \rangle$

$\qquad = \langle 2, 4 \rangle$

(b) $-2\mathbf{u} = -2 \cdot \langle -2, 1 \rangle$

$\qquad = \langle -2(-2), -2(1) \rangle$

$\qquad = \langle 4, -2 \rangle$

(c) $4\mathbf{u} - 3\mathbf{v} = 4 \cdot \langle -2, 1 \rangle - 3 \cdot \langle 4, 3 \rangle$

$\qquad = \langle -8, 4 \rangle - \langle 12, 9 \rangle$

$\qquad = \langle -8 - 12, 4 - 9 \rangle$

$\qquad = \langle -20, -5 \rangle$

Graphing Calculator Solution

Vector arithmetic can be performed with a graphing calculator, as shown in Figure 34.

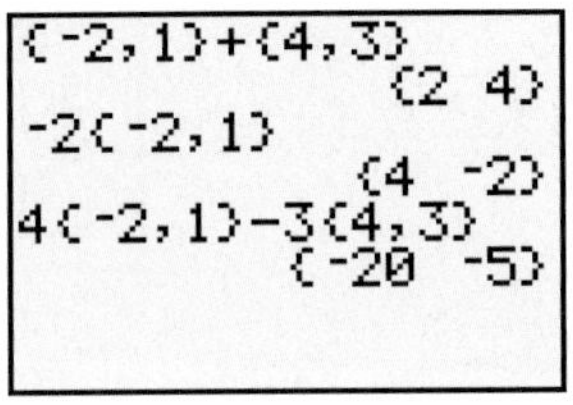

```
{-2,1)+{4,3)
                {2 4)
-2{-2,1)
                {4 -2)
4{-2,1)-3{4,3)
             {-20 -5)
```

Figure 34

NOW TRY EXERCISES 59, 61, AND 63. ◀

A **unit vector** is a vector that has magnitude 1. Two very useful unit vectors are defined as follows and shown in Figure 35(a).

$$\mathbf{i} = \langle 1, 0 \rangle \qquad \mathbf{j} = \langle 0, 1 \rangle$$

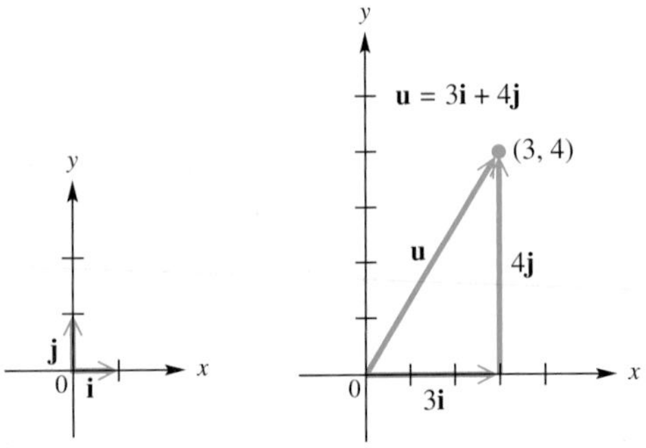

(a)

(b)

$\mathbf{u} = 3\mathbf{i} + 4\mathbf{j}$

Figure 35

With the unit vectors $\mathbf{i}$ and $\mathbf{j}$, we can express any other vector $\langle a, b \rangle$ in the form $a\mathbf{i} + b\mathbf{j}$, as shown in Figure 35(b), where $\langle 3, 4 \rangle = 3\mathbf{i} + 4\mathbf{j}$. The vector operations previously given can be restated, using $a\mathbf{i} + b\mathbf{j}$ notation.

i, j FORM FOR VECTORS

If $\mathbf{v} = \langle a, b \rangle$, then $\mathbf{v} = a\mathbf{i} + b\mathbf{j}$, where $\mathbf{i} = \langle 1, 0 \rangle$ and $\mathbf{j} = \langle 0, 1 \rangle$.

Dot Product and the Angle Between Vectors

The *dot product* of two vectors is a real number, not a vector. It is also known as the *inner product*. Dot products are used to determine the angle between two vectors, derive geometric theorems, and solve physics problems.

DOT PRODUCT

The **dot product** of the two vectors $\mathbf{u} = \langle a, b \rangle$ and $\mathbf{v} = \langle c, d \rangle$ is denoted $\mathbf{u} \cdot \mathbf{v}$, read "$\mathbf{u}$ dot $\mathbf{v}$," and given by

$$\mathbf{u} \cdot \mathbf{v} = ac + bd.$$

That is, the dot product of two vectors is the sum of the product of their first components and the product of their second components.

▶ **EXAMPLE 6** FINDING DOT PRODUCTS

Find each dot product.

(a) $\langle 2, 3 \rangle \cdot \langle 4, -1 \rangle$ **(b)** $\langle 6, 4 \rangle \cdot \langle -2, 3 \rangle$

Solution

(a) $\langle 2, 3 \rangle \cdot \langle 4, -1 \rangle = 2(4) + 3(-1) = 5$

(b) $\langle 6, 4 \rangle \cdot \langle -2, 3 \rangle = 6(-2) + 4(3) = 0$

NOW TRY EXERCISE 71. ◀

The following properties of dot products are easily verified.

PROPERTIES OF THE DOT PRODUCT

For all vectors $\mathbf{u}$, $\mathbf{v}$, and $\mathbf{w}$ and real numbers k,

(a) $\mathbf{u} \cdot \mathbf{v} = \mathbf{v} \cdot \mathbf{u}$ **(b)** $\mathbf{u} \cdot (\mathbf{v} + \mathbf{w}) = \mathbf{u} \cdot \mathbf{v} + \mathbf{u} \cdot \mathbf{w}$

(c) $(\mathbf{u} + \mathbf{v}) \cdot \mathbf{w} = \mathbf{u} \cdot \mathbf{w} + \mathbf{v} \cdot \mathbf{w}$ **(d)** $(k\mathbf{u}) \cdot \mathbf{v} = k(\mathbf{u} \cdot \mathbf{v}) = \mathbf{u} \cdot (k\mathbf{v})$

(e) $\mathbf{0} \cdot \mathbf{u} = 0$ **(f)** $\mathbf{u} \cdot \mathbf{u} = |\mathbf{u}|^2.$

For example, to prove the first part of (d), we let $\mathbf{u} = \langle a, b \rangle$ and $\mathbf{v} = \langle c, d \rangle$.

$$(k\mathbf{u}) \cdot \mathbf{v} = (k\langle a, b \rangle) \cdot \langle c, d \rangle = \langle ka, kb \rangle \cdot \langle c, d \rangle$$
$$= kac + kbd = k(ac + bd)$$
$$= k(\langle a, b \rangle \cdot \langle c, d \rangle) = k(\mathbf{u} \cdot \mathbf{v})$$

The proofs of the remaining properties are similar.

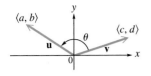

Figure 36

The dot product of two vectors can be positive, 0, or negative. A geometric interpretation of the dot product explains when each of these cases occurs. This interpretation involves the angle between the two vectors. Consider the vectors $\mathbf{u} = \langle a, b \rangle$ and $\mathbf{v} = \langle c, d \rangle$, as shown in Figure 36. The **angle θ between u and v** is defined to be the angle having the two vectors as its sides for which $0° \leq \theta \leq 180°$. The following theorem relates the dot product to the angle between the vectors. Its proof is outlined in Exercise 32 in **Section 7.5.**

GEOMETRIC INTERPRETATION OF DOT PRODUCT

If θ is the angle between the two nonzero vectors $\mathbf{u}$ and $\mathbf{v}$, where $0° \leq \theta \leq 180°$, then

$$\mathbf{u} \cdot \mathbf{v} = |\mathbf{u}||\mathbf{v}|\cos\theta \quad \text{or, equivalently,} \quad \cos\theta = \frac{\mathbf{u} \cdot \mathbf{v}}{|\mathbf{u}||\mathbf{v}|}.$$

▶ **EXAMPLE 7** FINDING THE ANGLE BETWEEN TWO VECTORS

Find the angle θ between the two vectors $\mathbf{u} = \langle 3, 4 \rangle$ and $\mathbf{v} = \langle 2, 1 \rangle$.

Solution By the geometric interpretation,

$$\cos\theta = \frac{\mathbf{u} \cdot \mathbf{v}}{|\mathbf{u}||\mathbf{v}|} = \frac{\langle 3, 4 \rangle \cdot \langle 2, 1 \rangle}{|\langle 3, 4 \rangle||\langle 2, 1 \rangle|}$$

$$= \frac{3(2) + 4(1)}{\sqrt{9 + 16} \cdot \sqrt{4 + 1}}$$

$$= \frac{10}{5\sqrt{5}} \approx .894427191.$$

Therefore, $\theta \approx \cos^{-1} .894427191 \approx 26.57°$. (Section 6.1)

NOW TRY EXERCISE 77. ◀

For angles θ between 0° and 180°, $\cos\theta$ is positive, 0, or negative when θ is less than, equal to, or greater than 90°, respectively. Therefore, the dot product is positive, 0, or negative according to this table.

Dot Product	Angle Between Vectors
Positive	Acute
0	Right
Negative	Obtuse

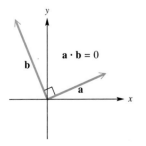

Figure 37

▶ **Note** If $\mathbf{a} \cdot \mathbf{b} = 0$ for two nonzero vectors $\mathbf{a}$ and $\mathbf{b}$, then $\cos\theta = 0$ and $\theta = 90°$. Thus, $\mathbf{a}$ and $\mathbf{b}$ are perpendicular or **orthogonal vectors.** See Figure 37.

NOW TRY EXERCISES 87 AND 89. ◀

7.4 Exercises

Concept Check   Exercises 1–4 refer to the vectors **m**-**t** *at the right.*

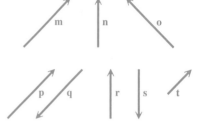

1. Name all pairs of vectors that appear to be equal.

2. Name all pairs of vectors that are opposites.

3. Name all pairs of vectors where the first is a scalar multiple of the other, with the scalar positive.

4. Name all pairs of vectors where the first is a scalar multiple of the other, with the scalar negative.

Concept Check   Refer to vectors **a**-**h** *below. Make a copy or a sketch of each vector, and then draw a sketch to represent each vector in Exercises 5–16. For example, find* **a** + **e** *by placing* **a** *and* **e** *so that their initial points coincide. Then use the parallelogram rule to find the resultant, as shown in the figure on the right.*

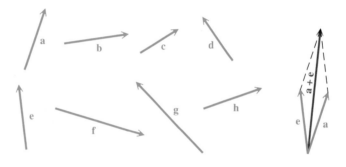

5. −**b**	6. −**g**	7. 3**a**	8. 2**h**
9. **a** + **b**	10. **h** + **g**	11. **a** − **c**	12. **d** − **e**
13. **a** + (**b** + **c**)	14. (**a** + **b**) + **c**	15. **c** + **d**	16. **d** + **c**

17. From the results of Exercises 13 and 14, does it appear that vector addition is associative?

18. From the results of Exercises 15 and 16, does it appear that vector addition is commutative?

In Exercises 19–24, use the figure to find each vector: **(a) a** + **b (b) a** − **b (c)** −**a**. *Use* ⟨*x, y*⟩ *notation as in Example 3.*

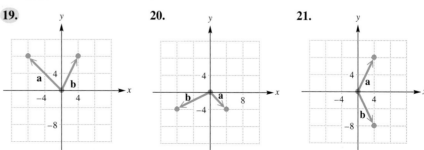

22. **23.** **24.**

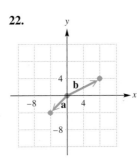 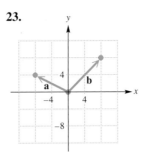 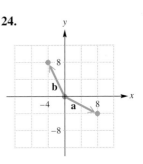

Given vectors **a** *and* **b**, *find:* **(a)** 2**a** **(b)** 2**a** + 3**b** **(c)** **b** − 3**a**.

25. **a** = 2**i**, **b** = **i** + **j**   **26.** **a** = −**i** + 2**j**, **b** = **i** − **j**

27. **a** = ⟨−1, 2⟩, **b** = ⟨3, 0⟩   **28.** **a** = ⟨−2, −1⟩, **b** = ⟨−3, 2⟩

For each pair of vectors **u** *and* **w** *with angle θ between them, sketch the resultant.*

29. |**u**| = 12, |**w**| = 20, θ = 27°   **30.** |**u**| = 8, |**w**| = 12, θ = 20°

31. |**u**| = 20, |**w**| = 30, θ = 30°   **32.** |**u**| = 50, |**w**| = 70, θ = 40°

Find the magnitude and direction angle for **u**. *See Example 1.*

33. ⟨15, −8⟩   **34.** ⟨−7, 24⟩

35. ⟨−4, 4√3⟩   **36.** ⟨8√2, −8√2⟩

For each of the following, vector **v** *has the given magnitude and direction. Find the magnitudes of the horizontal and vertical components of* **v**, *if α is the direction angle of* **v** *from the horizontal. See Example 2.*

37. α = 20°, |**v**| = 50   **38.** α = 50°, |**v**| = 26

39. α = 35° 50′, |**v**| = 47.8   **40.** α = 27° 30′, |**v**| = 15.4

41. α = 128.5°, |**v**| = 198   **42.** α = 146.3°, |**v**| = 238

Write each vector in the form ⟨a, b⟩. *See Example 3.*

43. **44.** **45.**

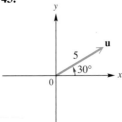 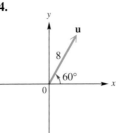 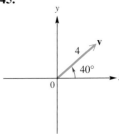

46. **47.** **48.**

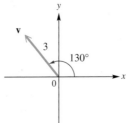 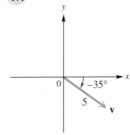 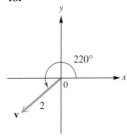

Two forces act at a point in the plane. The angle between the two forces is given. Find the magnitude of the resultant force. See Example 4.

49. forces of 250 and 450 newtons, forming an angle of 85°

50. forces of 19 and 32 newtons, forming an angle of 118°

51. forces of 116 and 139 lb, forming an angle of 140° 50′

52. forces of 37.8 and 53.7 lb, forming an angle of 68.5°

Use the parallelogram rule to find the magnitude of the resultant force for the two forces shown in each figure. Round answers to the nearest tenth.

53.

40 lb

40°

60 lb

54.

85 lb

65°

102 lb

55.

15 lb

110°

25 lb

56.

1500 lb

140°

2000 lb

57. *Concept Check*   If $\mathbf{u} = \langle a, b \rangle$ and $\mathbf{v} = \langle c, d \rangle$, what is $\mathbf{u} + \mathbf{v}$?

58. Explain how addition of vectors is similar to addition of complex numbers.

Given $\mathbf{u} = \langle -2, 5 \rangle$ and $\mathbf{v} = \langle 4, 3 \rangle$, find the following. See Example 5.

59. $\mathbf{u} + \mathbf{v}$ **60.** $\mathbf{u} - \mathbf{v}$ **61.** $-4\mathbf{u}$ **62.** $-5\mathbf{v}$

63. $3\mathbf{u} - 6\mathbf{v}$ **64.** $-2\mathbf{u} + 4\mathbf{v}$ **65.** $\mathbf{u} + \mathbf{v} - 3\mathbf{u}$ **66.** $2\mathbf{u} + \mathbf{v} - 6\mathbf{v}$

Write each vector in the form $a\mathbf{i} + b\mathbf{j}$. See Figure 35(b).

67. $\langle -5, 8 \rangle$ **68.** $\langle 6, -3 \rangle$ **69.** $\langle 2, 0 \rangle$ **70.** $\langle 0, -4 \rangle$

Find the dot product for each pair of vectors. See Example 6.

71. $\langle 6, -1 \rangle, \langle 2, 5 \rangle$ **72.** $\langle -3, 8 \rangle, \langle 7, -5 \rangle$ **73.** $\langle 2, -3 \rangle, \langle 6, 5 \rangle$

74. $\langle 1, 2 \rangle, \langle 3, -1 \rangle$ **75.** $4\mathbf{i}, 5\mathbf{i} - 9\mathbf{j}$ **76.** $2\mathbf{i} + 4\mathbf{j}, -\mathbf{j}$

Find the angle between each pair of vectors. See Example 7.

77. $\langle 2, 1 \rangle, \langle -3, 1 \rangle$ **78.** $\langle 1, 7 \rangle, \langle 1, 1 \rangle$ **79.** $\langle 1, 2 \rangle, \langle -6, 3 \rangle$

80. $\langle 4, 0 \rangle, \langle 2, 2 \rangle$ **81.** $3\mathbf{i} + 4\mathbf{j}, \mathbf{j}$ **82.** $-5\mathbf{i} + 12\mathbf{j}, 3\mathbf{i} + 2\mathbf{j}$

Let $\mathbf{u} = \langle -2, 1 \rangle$, $\mathbf{v} = \langle 3, 4 \rangle$, and $\mathbf{w} = \langle -5, 12 \rangle$. Evaluate each expression.

83. $(3\mathbf{u}) \cdot \mathbf{v}$ **84.** $\mathbf{u} \cdot (\mathbf{v} - \mathbf{w})$ **85.** $\mathbf{u} \cdot \mathbf{v} - \mathbf{u} \cdot \mathbf{w}$ **86.** $\mathbf{u} \cdot (3\mathbf{v})$

Determine whether each pair of vectors is orthogonal. See Figure 37.

87. $\langle 1, 2 \rangle, \langle -6, 3 \rangle$ **88.** $\langle 3, 4 \rangle, \langle 6, 8 \rangle$

89. $\langle 1, 0 \rangle, \langle \sqrt{2}, 0 \rangle$ **90.** $\langle 1, 1 \rangle, \langle 1, -1 \rangle$

91. $\sqrt{5}\mathbf{i} - 2\mathbf{j}, -5\mathbf{i} + 2\sqrt{5}\mathbf{j}$ **92.** $-4\mathbf{i} + 3\mathbf{j}, 8\mathbf{i} - 6\mathbf{j}$

RELATING CONCEPTS

For individual or collaborative investigation
(Exercises 93–98)

Consider the two vectors **u** *and* **v** *shown. Assume all values are exact.* **Work Exercises 93–98 in order.**

93. Use trigonometry alone (without using vector notation) to find the magnitude and direction angle of **u** + **v**. Use the law of cosines and the law of sines in your work.

94. Find the horizontal and vertical components of **u**, using your calculator.

95. Find the horizontal and vertical components of **v**, using your calculator.

96. Find the horizontal and vertical components of **u** + **v** by adding the results you obtained in Exercises 94 and 95.

97. Use your calculator to find the magnitude and direction angle of the vector **u** + **v**.

98. Compare your answers in Exercises 93 and 97. What do you notice? Which method of solution do you prefer?

7.5 Applications of Vectors

The Equilibrant ▪ Incline Applications ▪ Navigation Applications

The Equilibrant   The previous section covered methods for finding the resultant of two vectors. Sometimes it is necessary to find a vector that will counterbalance the resultant. This opposite vector is called the **equilibrant;** that is, the equilibrant of vector **u** is the vector −**u**.

▶ **EXAMPLE 1**   FINDING THE MAGNITUDE AND DIRECTION OF AN EQUILIBRANT

Find the magnitude of the equilibrant of forces of 48 newtons and 60 newtons acting on a point A, if the angle between the forces is 50°. Then find the angle between the equilibrant and the 48-newton force.

Solution

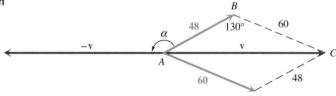

Figure 38

In Figure 38, the equilibrant is −**v**. The magnitude of **v**, and hence of −**v**, is found by using triangle ABC and the law of cosines.

$$|\mathbf{v}|^2 = 48^2 + 60^2 - 2(48)(60)\cos 130°$$ Law of cosines **(Section 7.3)**

$$|\mathbf{v}|^2 \approx 9606.5$$ Use a calculator.

$$|\mathbf{v}| \approx 98 \text{ newtons}$$ Two significant digits **(Section 2.4)**

The required angle, labeled α in Figure 38, can be found by subtracting angle *CAB* from 180°. Use the law of sines to find angle *CAB*.

$$\frac{98}{\sin 130°} = \frac{60}{\sin CAB} \qquad \text{Law of sines (Section 7.1)}$$

$$\sin CAB \approx .46900680$$

$$CAB \approx 28° \qquad \text{Use the inverse sine function. (Section 6.1)}$$

Finally, $\alpha \approx 180° - 28° = 152°$.

NOW TRY EXERCISE 1. ◀

Incline Applications   We can use vectors to solve incline problems.

▶ **EXAMPLE 2**   FINDING A REQUIRED FORCE

Find the force required to keep a 50-lb wagon from sliding down a ramp inclined at 20° to the horizontal. (Assume there is no friction.)

Solution   In Figure 39, the vertical 50-lb force **BA** represents the force of gravity. It is the sum of vectors **BC** and $-$**AC**. The vector **BC** represents the force with which the weight pushes against the ramp. Vector **BF** represents the force that would pull the weight up the ramp. Since vectors **BF** and **AC** are equal, $|\mathbf{AC}|$ gives the magnitude of the required force.

Vectors **BF** and **AC** are parallel, so angle *EBD* equals angle *A*. Since angle *BDE* and angle *C* are right angles, triangles *CBA* and *DEB* have two corresponding angles equal and, thus, are similar triangles. Therefore, angle *ABC* equals angle *E*, which is 20°. From right triangle *ABC*,

$$\sin 20° = \frac{|\mathbf{AC}|}{50} \qquad \text{(Section 2.1)}$$

$$|\mathbf{AC}| = 50 \sin 20° \approx 17.$$

Approximately a 17-lb force will keep the wagon from sliding down the ramp.

NOW TRY EXERCISE 9. ◀

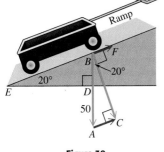

Figure 39

▶ **EXAMPLE 3**   FINDING AN INCLINE ANGLE

A force of 16.0 lb is required to hold a 40.0 lb lawn mower on an incline. What angle does the incline make with the horizontal?

Solution   Figure 40 illustrates the situation. Consider right triangle *ABC*. Angle *B* equals angle θ, the magnitude of vector **BA** represents the weight of the mower, and vector **AC** equals vector **BE**, which represents the force required to hold the mower on the incline. From the figure,

$$\sin B = \frac{16}{40} = .4$$

$$B \approx 23.5782°. \quad \text{Use the inverse sine function.}$$

Therefore, the hill makes an angle of about 23.6° with the horizontal.

NOW TRY EXERCISE 11. ◀

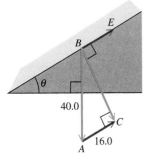

Figure 40

Navigation Applications Problems involving bearing (defined in **Section 2.5**) can also be worked with vectors.

▶ **EXAMPLE 4** APPLYING VECTORS TO A NAVIGATION PROBLEM

A ship leaves port on a bearing of 28.0° and travels 8.20 mi. The ship then turns due east and travels 4.30 mi. How far is the ship from port? What is its bearing from port?

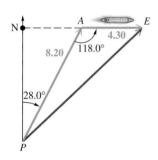

Figure 41

Solution In Figure 41, vectors **PA** and **AE** represent the ship's path. The magnitude and bearing of the resultant **PE** can be found as follows. Triangle *PNA* is a right triangle, so angle $NAP = 90° - 28.0° = 62.0°$. Then angle $PAE = 180° - 62.0° = 118.0°$. Use the law of cosines to find $|\mathbf{PE}|$, the magnitude of vector **PE**.

$$|\mathbf{PE}|^2 = 8.20^2 + 4.30^2 - 2(8.20)(4.30)\cos 118.0° \qquad \text{Law of cosines}$$

$$|\mathbf{PE}|^2 \approx 118.84 \qquad \text{Approximate.}$$

$$|\mathbf{PE}| \approx 10.9 \qquad \text{Square root property (Appendix A)}$$

The ship is about 10.9 mi from port.

To find the bearing of the ship from port, first find angle *APE*. Use the law of sines.

$$\frac{\sin APE}{4.30} = \frac{\sin 118.0°}{10.9} \qquad \text{Law of sines}$$

$$\sin APE = \frac{4.30 \sin 118.0°}{10.9} \qquad \text{Multiply by 4.30.}$$

$$APE \approx 20.4° \qquad \text{Use the inverse sine function.}$$

Now add 20.4° to 28.0° to find that the bearing is 48.4°.

NOW TRY EXERCISE 15. ◀

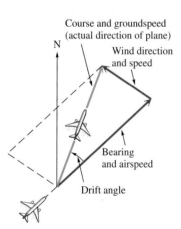

Figure 42

In air navigation, the **airspeed** of a plane is its speed relative to the air, while the **groundspeed** is its speed relative to the ground. Because of wind, these two speeds are usually different. The groundspeed of the plane is represented by the vector sum of the airspeed and windspeed vectors. See Figure 42.

▶ **EXAMPLE 5** APPLYING VECTORS TO A NAVIGATION PROBLEM

A plane with an airspeed of 192 mph is headed on a bearing of 121°. A north wind is blowing (from north to south) at 15.9 mph. Find the groundspeed and the actual bearing of the plane.

Solution In Figure 43, the groundspeed is represented by $|\mathbf{x}|$. We must find angle α to determine the bearing, which will be $121° + \alpha$. From Figure 43, angle BCO = angle AOC, which measures 121°. Find $|\mathbf{x}|$ by the law of cosines.

$$|\mathbf{x}|^2 = 192^2 + 15.9^2 - 2(192)(15.9)\cos 121°$$

$$|\mathbf{x}|^2 \approx 40{,}261$$

$$|\mathbf{x}| \approx 200.7, \quad \text{or} \quad \text{about 201 mph}$$

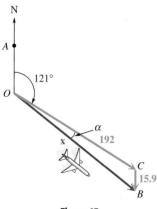

Figure 43

Now find α by using the law of sines. Use the value of $|\mathbf{x}|$ before rounding.

$$\frac{\sin \alpha}{15.9} = \frac{\sin 121°}{200.7}$$

$$\sin \alpha \approx .0679$$

$$\alpha \approx 3.89°$$

To the nearest degree, α is 4°. The groundspeed is about 201 mph on a bearing of $121° + 4° = 125°$.

NOW TRY EXERCISE 25. ◀

7.5 Exercises

Solve each problem. See Examples 1–3.

1. *Direction and Magnitude of an Equilibrant* Two tugboats are pulling a disabled speedboat into port with forces of 1240 lb and 1480 lb. The angle between these forces is 28.2°. Find the direction and magnitude of the equilibrant.

2. *Direction and Magnitude of an Equilibrant* Two rescue vessels are pulling a broken-down motorboat toward a boathouse with forces of 840 lb and 960 lb. The angle between these forces is 24.5°. Find the direction and magnitude of the equilibrant.

3. *Angle Between Forces* Two forces of 692 newtons and 423 newtons act at a point. The resultant force is 786 newtons. Find the angle between the forces.

4. *Angle Between Forces* Two forces of 128 lb and 253 lb act at a point. The equilibrant force is 320 lb. Find the angle between the forces.

5. *Magnitudes of Forces* A force of 176 lb makes an angle of 78° 50′ with a second force. The resultant of the two forces makes an angle of 41° 10′ with the first force. Find the magnitudes of the second force and of the resultant.

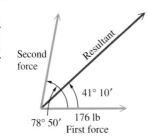

6. *Magnitudes of Forces* A force of 28.7 lb makes an angle of 42° 10′ with a second force. The resultant of the two forces makes an angle of 32° 40′ with the first force. Find the magnitudes of the second force and of the resultant.

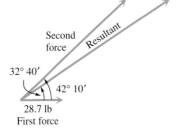

7. *Angle of a Hill Slope* A force of 25 lb is required to hold an 80-lb crate on a hill. What angle does the hill make with the horizontal?

8. *Force Needed to Keep a Car Parked* Find the force required to keep a 3000-lb car parked on a hill that makes an angle of 15° with the horizontal.

9. *Force Needed for a Monolith*   To build the pyramids in Egypt, it is believed that giant causeways were constructed to transport the building materials to the site. One such causeway is said to have been 3000 ft long, with a slope of about 2.3°. How much force would be required to hold a 60-ton monolith on this causeway?

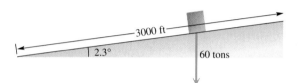

3000 ft

2.3°

60 tons

10. *Weight of a Box*   Two people are carrying a box. One person exerts a force of 150 lb at an angle of 62.4° with the horizontal. The other person exerts a force of 114 lb at an angle of 54.9°. Find the weight of the box.

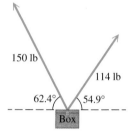

150 lb

114 lb

62.4°   54.9°

Box

11. *Incline Angle*   A force of 18.0 lb is required to hold a 60.0-lb stump grinder on an incline. What angle does the incline make with the horizontal?

12. *Incline Angle*   A force of 30.0 lb is required to hold an 80.0-lb pressure washer on an incline. What angle does the incline make with the horizontal?

13. *Weight of a Crate and Tension of a Rope*   A crate is supported by two ropes. One rope makes an angle of 46° 20′ with the horizontal and has a tension of 89.6 lb on it. The other rope is horizontal. Find the weight of the crate and the tension in the horizontal rope.

14. *Angles Between Forces*   Three forces acting at a point are in equilibrium. The forces are 980 lb, 760 lb, and 1220 lb. Find the angles between the directions of the forces to the nearest tenth of a degree. (*Hint:* Arrange the forces to form the sides of a triangle.)

Solve each problem. See Examples 4 and 5.

15. *Distance and Bearing of a Ship*   A ship leaves port on a bearing of 34.0° and travels 10.4 mi. The ship then turns due east and travels 4.6 mi. How far is the ship from port, and what is its bearing from port?

16. *Distance and Bearing of a Luxury Liner*   A luxury liner leaves port on a bearing of 110.0° and travels 8.8 mi. It then turns due west and travels 2.4 mi. How far is the liner from port, and what is its bearing from port?

17. *Distance of a Ship from Its Starting Point*   Starting at point *A*, a ship sails 18.5 km on a bearing of 189°, then turns and sails 47.8 km on a bearing of 317°. Find the distance of the ship from point *A*.

18. *Distance of a Ship from Its Starting Point*   Starting at point *X*, a ship sails 15.5 km on a bearing of 200°, then turns and sails 2.4 km on a bearing of 320°. Find the distance of the ship from point *X*.

19. *Distance and Direction of a Motorboat*   A motorboat sets out in the direction N 80° 00′ E. The speed of the boat in still water is 20.0 mph. If the current is flowing directly south, and the actual direction of the motorboat is due east, find the speed of the current and the actual speed of the motorboat.

20. *Path Traveled by a Plane*   The aircraft carrier *Tallahassee* is traveling at sea on a steady course with a bearing of 30° at 32 mph. Patrol planes on the carrier have enough fuel for 2.6 hr of flight when traveling at a speed of 520 mph. One of the pilots takes off on a bearing of 338° and then turns and heads in a straight line, so as to be able to catch the carrier and land on the deck at the exact instant that his fuel runs out. If the pilot left at 2 P.M., at what time did he turn to head for the carrier?

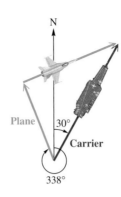

21. *Bearing and Groundspeed of a Plane*   An airline route from San Francisco to Honolulu is on a bearing of 233.0°. A jet flying at 450 mph on that bearing runs into a wind blowing at 39.0 mph from a direction of 114.0°. Find the resulting bearing and groundspeed of the plane.

22. *Movement of a Motorboat*   Suppose you would like to cross a 132-ft-wide river in a motorboat. Assume that the motorboat can travel at 7.0 mph relative to the water and that the current is flowing west at the rate of 3.0 mph. The bearing θ is chosen so that the motorboat will land at a point exactly across from the starting point.

  (a) At what speed will the motorboat be traveling relative to the banks?

  (b) How long will it take for the motorboat to make the crossing?

  (c) What is the measure of angle θ?

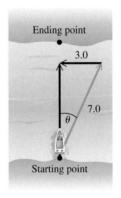

23. *Airspeed and Groundspeed*   A pilot wants to fly on a bearing of 74.9°. By flying due east, he finds that a 42.0-mph wind, blowing from the south, puts him on course. Find the airspeed and the groundspeed.

24. *Bearing of a Plane*   A plane flies 650 mph on a bearing of 175.3°. A 25-mph wind, from a direction of 266.6°, blows against the plane. Find the resulting bearing of the plane.

25. *Bearing and Groundspeed of a Plane*   A pilot is flying at 190.0 mph. He wants his flight path to be on a bearing of 64° 30′. A wind is blowing from the south at 35.0 mph. Find the bearing he should fly, and find the plane's groundspeed.

26. *Bearing and Groundspeed of a Plane*   A pilot is flying at 168 mph. She wants her flight path to be on a bearing of 57° 40′. A wind is blowing from the south at 27.1 mph. Find the bearing the pilot should fly, and find the plane's groundspeed.

27. *Bearing and Airspeed of a Plane*   What bearing and airspeed are required for a plane to fly 400 mi due north in 2.5 hr if the wind is blowing from a direction of 328° at 11 mph?

28. *Groundspeed and Bearing of a Plane*   A plane is headed due south with an airspeed of 192 mph. A wind from a direction of 78.0° is blowing at 23.0 mph. Find the groundspeed and resulting bearing of the plane.

29. *Groundspeed and Bearing of a Plane*   An airplane is headed on a bearing of 174° at an airspeed of 240 km per hr. A 30 km per hr wind is blowing from a direction of 245°. Find the groundspeed and resulting bearing of the plane.

30. *Velocity of a Star* The space velocity **v** of a star relative to the sun can be expressed as the resultant vector of two perpendicular vectors—the radial velocity $\mathbf{v}_r$ and the tangential velocity $\mathbf{v}_t$, where $\mathbf{v} = \mathbf{v}_r + \mathbf{v}_t$. If a star is located near the sun and its space velocity is large, then its motion across the sky will also be large. Barnard's Star is a relatively close star with a distance of 35 trillion mi from the sun. It moves across the sky through an angle of 10.34″ per year, which is the largest motion of any known star. Its radial velocity is $\mathbf{v}_r = 67$ mi per sec toward the sun. (*Source:* Zeilik, M., S. Gregory, and E. Smith, *Introductory Astronomy and Astrophysics,* Second Edition, Saunders College Publishing, 1998; Acker, A. and C. Jaschek, *Astronomical Methods and Calculations,* John Wiley and Sons, 1986.)

Not to scale

(a) Approximate the tangential velocity $\mathbf{v}_t$ of Barnard's Star. (*Hint:* Use the arc length formula $s = r\theta$.)

(b) Compute the magnitude of **v**.

31. *(Modeling) Measuring Rainfall* Suppose that vector **R** models the amount of rainfall in inches and the direction it falls, and vector **A** models the area in square inches and orientation of the opening of a rain gauge, as illustrated in the figure. The total volume V of water collected in the rain gauge is given by $V = |\mathbf{R} \cdot \mathbf{A}|$. This formula calculates the volume of water collected even if the wind is blowing the rain in a slanted direction or the rain gauge is not exactly vertical. Let $\mathbf{R} = \mathbf{i} - 2\mathbf{j}$ and $\mathbf{A} = .5\mathbf{i} + \mathbf{j}$.

(a) Find $|\mathbf{R}|$ and $|\mathbf{A}|$. Interpret your results.

(b) Calculate V and interpret this result.

(c) For the rain gauge to collect the maximum amount of water, what should be true about vectors **R** and **A**?

32. *The Dot Product* In the figure at the right, $\mathbf{a} = \langle a_1, a_2 \rangle$, $\mathbf{b} = \langle b_1, b_2 \rangle$, and $\mathbf{a} - \mathbf{b} = \langle a_1 - b_1, a_2 - b_2 \rangle$. Apply the law of cosines to the triangle and derive the equation

$$\mathbf{a} \cdot \mathbf{b} = |\mathbf{a}||\mathbf{b}| \cos \theta.$$

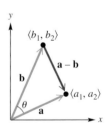

Summary Exercises on Applications of Trigonometry and Vectors

These summary exercises provide practice with applications that involve solving triangles and using vectors.

1. *Wires Supporting a Flagpole* A flagpole stands vertically on a hillside that makes an angle of 20° with the horizontal. Two supporting wires are attached as shown in the figure. What are the lengths of the supporting wires?

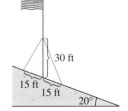

2. *Distance between Points on a Softball Field*   The pitcher's mound on a regulation softball field is 46 ft from home plate. The distance between the bases is 60 ft, as shown in the figure. How far from third base (point *T*) is the pitcher's mound (point *M*)? Give your answer to the nearest foot.

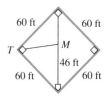

3. *Distance Between a Pin and a Rod*   A slider crank mechanism is shown in the figure. Find the distance between the wrist pin *W* and the connecting rod center *C*.

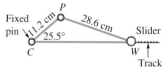

4. *Distance Between Two Lighthouses*   Two lighthouses are located on a north-south line. From lighthouse *A*, the bearing of a ship 3742 m away is 129° 43′. From lighthouse *B*, the bearing of a ship is 39° 43′. Find the distance between the lighthouses.

5. *Hot-Air Balloon*   A hot-air balloon is rising straight up at the speed of 15.0 ft per sec. Then a wind starts blowing horizontally at 5.00 ft per sec. What will the new speed of the balloon be and what angle with the horizontal will the balloon's path make?

6. *Playing on a Swing*   Mary is playing with her daughter Brittany on a swing. Starting from rest, Mary pulls the swing through an angle of 40° and holds it briefly before releasing the swing. If Brittany weighs 50 lb, what horizontal force, to the nearest pound, must Mary apply while holding the swing?

7. *Height of an Airplane*   Two observation points *A* and *B* are 950 ft apart. From these points the angles of elevation of an airplane are 52° and 57°. (See the figure.) Find the height of the airplane.

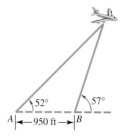

8. *Wind and Vectors*   A wind can be described by **v** = 6**i** + 8**j**, where vector **j** points north and represents a south wind of 1 mph.

   (a) What is the speed of the wind?   **(b)** Find 3**v**. Interpret the result.
   (c) Interpret the wind if it switches to **u** = −8**i** + 8**j**.

9. *Property Survey*   A surveyor reported the following data about a piece of property: "The property is triangular in shape, with dimensions as shown in the figure." Use the law of sines to see whether such a piece of property could exist.

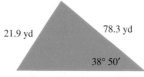

Can such a triangle exist?

10. *Property Survey*   A second triangular piece of property has dimensions as shown. This time it turns out that the surveyor did not consider every possible case. Use the law of sines to show why.

Chapter 7 Summary

KEY TERMS

7.1 Side-Angle-Side
    (SAS)
    Angle-Side-Angle
    (ASA)
    Side-Side-Side (SSS)
    oblique triangle
    Side-Angle-Angle
    (SAA)
7.2 ambiguous case

7.3 semiperimeter
7.4 scalar
    vector quantity
    vector
    magnitude
    initial point
    terminal point
    parallelogram rule

resultant
opposite (of a vector)
zero vector
scalar product
position vector
horizontal component
vertical component
direction angle

unit vector
dot product
angle between two
    vectors
orthogonal vectors
7.5 equilibrant
    airspeed
    groundspeed

NEW SYMBOLS

OP or $\overrightarrow{\textbf{OP}}$ vector **OP**
$|\textbf{OP}|$ magnitude of vector **OP**

$\langle \textbf{\textit{a}},\textbf{\textit{b}} \rangle$ position vector
i, j   unit vectors

QUICK REVIEW

CONCEPTS

EXAMPLES

7.1 Oblique Triangles and the Law of Sines

Law of Sines
In any triangle ABC, with sides a, b, and c,

$$\frac{a}{\sin A} = \frac{b}{\sin B}, \quad \frac{a}{\sin A} = \frac{c}{\sin C}, \quad \text{and} \quad \frac{b}{\sin B} = \frac{c}{\sin C}.$$

In triangle ABC, find c if $A = 44°$, $C = 62°$, and $a = 12.00$ units. Then find its area.

$$\frac{a}{\sin A} = \frac{c}{\sin C}$$

$$\frac{12.00}{\sin 44°} = \frac{c}{\sin 62°}$$

$$c = \frac{12.00 \sin 62°}{\sin 44°} \approx 15.25 \text{ units}$$

Area of a Triangle
In any triangle ABC, the area is half the product of the lengths of two sides and the sine of the angle between them.

$$\mathscr{A} = \frac{1}{2}bc \sin A, \quad \mathscr{A} = \frac{1}{2}ab \sin C, \quad \mathscr{A} = \frac{1}{2}ac \sin B$$

For triangle ABC above,

$$\mathscr{A} = \frac{1}{2}ac \sin B$$

$$= \frac{1}{2}(12.00)(15.25)\sin 74° \quad B = 180° - 44° - 62°$$

$$\approx 87.96 \text{ sq units.}$$

CONCEPTS	EXAMPLES

7.2 The Ambiguous Case of the Law of Sines

Ambiguous Case

If we are given the lengths of two sides and the angle opposite one of them, for example, A, a, and b in triangle ABC, then it is possible that zero, one, or two such triangles exist. If A is acute, h is the altitude from C, and

1. $a < h < b$, then there is no triangle.
2. $a = h$ and $h < b$, then there is one triangle (a right triangle).
3. $a \geq b$, then there is one triangle.
4. $h < a < b$, then there are two triangles.

If A is obtuse and

1. $a \leq b$, then there is no triangle.
2. $a > b$, then there is one triangle.

See the table on page 313 that illustrates the possible outcomes.

Solve triangle ABC, given $A = 44.5°$, $a = 11.0$ in., and $c = 7.0$ in.

Find angle C.

$$\frac{\sin C}{7.0} = \frac{\sin 44.5°}{11.0}$$

$$\sin C \approx .4460$$

$$C \approx 26.5°$$

Another angle with this sine value is

$$180° - 26.5° = 153.5°.$$

However, $153.5° + 44.5° > 180°$, so there is only one triangle.

$$B = 180° - 44.5° - 26.5°$$

$$B = 109°$$

Using the law of sines again,

$$b \approx 14.8 \text{ in.}$$

7.3 The Law of Cosines

Law of Cosines

In any triangle ABC, with sides a, b, and c,

$$a^2 = b^2 + c^2 - 2bc \cos A$$

$$b^2 = a^2 + c^2 - 2ac \cos B$$

$$c^2 = a^2 + b^2 - 2ab \cos C.$$

In triangle ABC, find C if $a = 11$ units, $b = 13$ units, and $c = 20$ units. Then find its area.

$$c^2 = a^2 + b^2 - 2ab \cos C$$

$$20^2 = 11^2 + 13^2 - 2(11)(13) \cos C$$

$$400 = 121 + 169 - 286 \cos C$$

$$\frac{400 - 121 - 169}{-286} = \cos C$$

$$C = \cos^{-1}\left(\frac{400 - 121 - 169}{-286}\right)$$

$$C \approx 112.6°$$

Heron's Area Formula

If a triangle has sides of lengths a, b, and c, with semiperimeter

$$s = \frac{1}{2}(a + b + c),$$

then the area of the triangle is

$$\mathcal{A} = \sqrt{s(s - a)(s - b)(s - c)}.$$

The semiperimeter s is

$$s = \frac{1}{2}(11 + 13 + 20) = 22,$$

so

$$\mathcal{A} = \sqrt{22(22 - 11)(22 - 13)(22 - 20)} = 66 \text{ sq units.}$$

(continued)

CONCEPTS

EXAMPLES

7.4 Vectors, Operations, and the Dot Product

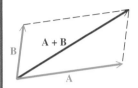

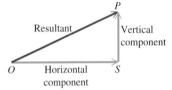

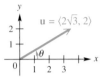

Magnitude and Direction Angle of a Vector

The magnitude (length) of vector $\mathbf{u} = \langle a, b \rangle$ is given by

$$|\mathbf{u}| = \sqrt{a^2 + b^2}.$$

$$|\mathbf{u}| = \sqrt{(2\sqrt{3})^2 + 2^2} = \sqrt{16} = 4$$

The direction angle θ satisfies $\tan \theta = \frac{b}{a}$, where $a \neq 0$.

Since $\tan \theta = \frac{2}{2\sqrt{3}} = \frac{1}{\sqrt{3}} \cdot \frac{\sqrt{3}}{\sqrt{3}} = \frac{\sqrt{3}}{3}$, it follows that $\theta = 30°$.

Vector Operations

For any real numbers a, b, c, d, and k,

$$\langle a, b \rangle + \langle c, d \rangle = \langle a + c, b + d \rangle$$

$$k \cdot \langle a, b \rangle = \langle ka, kb \rangle.$$

If $\mathbf{a} = \langle a_1, a_2 \rangle$, then $-\mathbf{a} = \langle -a_1, -a_2 \rangle$.

$$\langle a, b \rangle - \langle c, d \rangle = \langle a, b \rangle + (-\langle c, d \rangle) = \langle a - c, b - d \rangle.$$

If $\mathbf{u} = \langle x, y \rangle$ has direction angle θ, then

$$\mathbf{u} = \langle |\mathbf{u}| \cos \theta, |\mathbf{u}| \sin \theta \rangle.$$

$$\langle 4, 6 \rangle + \langle -8, 3 \rangle = \langle -4, 9 \rangle$$

$$5\langle -2, 1 \rangle = \langle -10, 5 \rangle$$

$$-\langle -9, 6 \rangle = \langle 9, -6 \rangle$$

$$\langle 4, 6 \rangle - \langle -8, 3 \rangle = \langle 12, 3 \rangle$$

For $\mathbf{u}$ defined above,

$$\mathbf{u} = \langle 4 \cos 30°, 4 \sin 30° \rangle$$

$$= \langle 2\sqrt{3}, 2 \rangle$$

i, j Form for Vectors

If $\mathbf{v} = \langle a, b \rangle$, then $\mathbf{v} = a\mathbf{i} + b\mathbf{j}$, where $\mathbf{i} = \langle 1, 0 \rangle$ and $\mathbf{j} = \langle 0, 1 \rangle$.

and

$$\mathbf{u} = 2\sqrt{3}\mathbf{i} + 2\mathbf{j}.$$

Dot Product

The dot product of the two vectors $\mathbf{u} = \langle a, b \rangle$ and $\mathbf{v} = \langle c, d \rangle$, denoted $\mathbf{u} \cdot \mathbf{v}$, is given by

$$\mathbf{u} \cdot \mathbf{v} = ac + bd.$$

$$\langle 2, 1 \rangle \cdot \langle 5, -2 \rangle = 2 \cdot 5 + 1(-2) = 8$$

If θ is the angle between $\mathbf{u}$ and $\mathbf{v}$, where $0° \leq \theta \leq 180°$, then

$$\mathbf{u} \cdot \mathbf{v} = |\mathbf{u}||\mathbf{v}| \cos \theta, \quad \text{or} \quad \cos \theta = \frac{\mathbf{u} \cdot \mathbf{v}}{|\mathbf{u}||\mathbf{v}|}.$$

Find the angle θ between $\mathbf{u} = \langle 3, 1 \rangle$ and $\mathbf{v} = \langle 2, -3 \rangle$.

$$\cos \theta = \frac{\langle 3, 1 \rangle \cdot \langle 2, -3 \rangle}{\sqrt{3^2 + 1^2} \cdot \sqrt{2^2 + (-3)^2}}$$

$$\cos \theta = \frac{6 + (-3)}{\sqrt{10} \cdot \sqrt{13}}$$

$$\cos \theta = \frac{3}{\sqrt{130}}$$

$$\theta = \cos^{-1} \frac{3}{\sqrt{130}} \approx 74.7°$$

CHAPTER 7 ▶   Review Exercises

Use the law of sines to find the indicated part of each triangle ABC.

1. Find b if $C = 74.2°$, $c = 96.3$ m, $B = 39.5°$.

2. Find B if $A = 129.7°$, $a = 127$ ft, $b = 69.8$ ft.

3. Find B if $C = 51.3°$, $c = 68.3$ m, $b = 58.2$ m.

4. Find b if $a = 165$ m, $A = 100.2°$, $B = 25.0°$.

5. Find A if $B = 39° 50'$, $b = 268$ m, $a = 340$ m.

6. Find A if $C = 79° 20'$, $c = 97.4$ mm, $a = 75.3$ mm.

7. If we are given a, A, and C in a triangle ABC, does the possibility of the ambiguous case exist? If not, explain why.

8. Can triangle ABC exist if $a = 4.7$, $b = 2.3$, and $c = 7.0$? If not, explain why. Answer this question without using trigonometry.

9. Given $a = 10$ and $B = 30°$, determine the values of b for which A has

   (a) exactly one value   (b) two possible values   (c) no value.

10. Explain why there can be no triangle ABC satisfying $A = 140°$, $a = 5$, and $b = 7$.

Use the law of cosines to find the indicated part of each triangle ABC.

11. Find A if $a = 86.14$ in., $b = 253.2$ in., $c = 241.9$ in.

12. Find b if $B = 120.7°$, $a = 127$ ft, $c = 69.8$ ft.

13. Find a if $A = 51° 20'$, $c = 68.3$ m, $b = 58.2$ m.

14. Find B if $a = 14.8$ m, $b = 19.7$ m, $c = 31.8$ m.

15. Find a if $A = 60°$, $b = 5.0$ cm, $c = 21$ cm.

16. Find A if $a = 13$ ft, $b = 17$ ft, $c = 8$ ft.

Solve each triangle ABC having the given information.

17. $A = 25.2°$, $a = 6.92$ yd, $b = 4.82$ yd

18. $A = 61.7°$, $a = 78.9$ m, $b = 86.4$ m

19. $a = 27.6$ cm, $b = 19.8$ cm, $C = 42° 30'$

20. $a = 94.6$ yd, $b = 123$ yd, $c = 109$ yd

Find the area of each triangle ABC with the given information.

21. $b = 840.6$ m, $c = 715.9$ m, $A = 149.3°$

22. $a = 6.90$ ft, $b = 10.2$ ft, $C = 35° 10'$

23. $a = .913$ km, $b = .816$ km, $c = .582$ km

24. $a = 43$ m, $b = 32$ m, $c = 51$ m

Solve each problem.

25. *Distance Across a Canyon* To measure the distance AB across a canyon for a power line, a surveyor measures angles B and C and the distance BC, as shown in the figure. What is the distance from A to B?

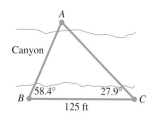

26. *Length of a Brace* A banner on an 8.0-ft pole is to be mounted on a building at an angle of 115°, as shown in the figure. Find the length of the brace.

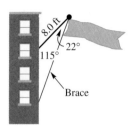

27. *Height of a Tree* A tree leans at an angle of 8.0° from the vertical. From a point 7.0 m from the bottom of the tree, the angle of elevation to the top of the tree is 68°. How tall is the tree?

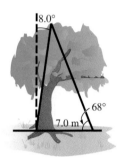

28. *Hanging Sculpture* A hanging sculpture in an art gallery is to be hung with two wires of lengths 15.0 ft and 12.2 ft so that the angle between them is 70.3°. How far apart should the ends of the wire be placed on the ceiling?

29. *Height of a Tree* A hill makes an angle of 14.3° with the horizontal. From the base of the hill, the angle of elevation to the top of a tree on top of the hill is 27.2°. The distance along the hill from the base to the tree is 212 ft. Find the height of the tree.

30. *Pipeline Position* A pipeline is to run between points *A* and *B*, which are separated by a protected wetlands area. To avoid the wetlands, the pipe will run from point *A* to *C* and then to *B*. The distances involved are $AB = 150$ km, $AC = 102$ km, and $BC = 135$ km. What angle should be used at point *C*?

31. *Distance Between Two Boats* Two boats leave a dock together. Each travels in a straight line. The angle between their courses measures 54° 10′. One boat travels 36.2 km per hr, and the other travels 45.6 km per hr. How far apart will they be after 3 hr?

32. *Distance from a Ship to a Lighthouse* A ship sailing parallel to shore sights a lighthouse at an angle of 30° from its direction of travel. After the ship travels 2.0 mi farther, the angle has increased to 55°. At that time, how far is the ship from the lighthouse?

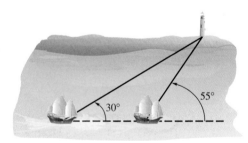

33. *Area of a Triangle* Find the area of the triangle shown in the figure using Heron's area formula.

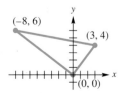

34. Show that the triangle in Exercise 33 is a right triangle. Then use the formula $\mathcal{A} = \frac{1}{2} ac \sin B$, with $B = 90°$, to find the area.

In Exercises 35 and 36, use the given vectors to sketch the following.

35. $\mathbf{a} - \mathbf{b}$

36. $\mathbf{a} + 3\mathbf{c}$

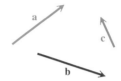

37. *Concept Check*   Decide whether each statement is true or false.

(a) Opposite angles of a parallelogram are equal.

(b) A diagonal of a parallelogram must bisect two angles of the parallelogram.

Given two forces and the angle between them, find the magnitude of the resultant force.

38. forces of 142 and 215 newtons, forming an angle of 112°

39.

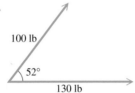

Vector $\mathbf{v}$ *has the given magnitude and direction angle. Find the magnitudes of the horizontal and vertical components of* $\mathbf{v}$.

40. $|\mathbf{v}| = 50$, $\theta = 45°$

(Give exact values.)

41. $|\mathbf{v}| = 964$, $\theta = 154° \, 20'$

Find the magnitude and direction angle for $\mathbf{u}$ *rounded to the nearest tenth.*

42. $\mathbf{u} = \langle 21, -20 \rangle$

43. $\mathbf{u} = \langle -9, 12 \rangle$

44. Let $\mathbf{v} = 2\mathbf{i} - \mathbf{j}$ and $\mathbf{u} = -3\mathbf{i} + 2\mathbf{j}$. Express each in terms of $\mathbf{i}$ and $\mathbf{j}$.

(a) $2\mathbf{v} + \mathbf{u}$ (b) $2\mathbf{v}$ (c) $\mathbf{v} - 3\mathbf{u}$

45. Let $\mathbf{a} = \langle 3, -2 \rangle$ and $\mathbf{b} = \langle -1, 3 \rangle$. Find $\mathbf{a} \cdot \mathbf{b}$ and the angle between $\mathbf{a}$ and $\mathbf{b}$, rounded to the nearest tenth of a degree.

Find the vector of magnitude 1 *having the same direction angle as the given vector.*

46. $\mathbf{u} = \langle -4, 3 \rangle$

47. $\mathbf{u} = \langle 5, 12 \rangle$

Solve each problem.

48. *Force Placed on a Barge*   One rope pulls a barge directly east with a force of 100 newtons. Another rope pulls the barge to the northeast with a force of 200 newtons. Find the resultant force acting on the barge, to the nearest unit, and the angle between the resultant and the first rope, to the nearest tenth.

49. *Weight of a Sled and Passenger* Paula and Steve are pulling their daughter Jessie on a sled. Steve pulls with a force of 18 lb at an angle of 10°. Paula pulls with a force of 12 lb at an angle of 15°. Find the magnitude of the resultant force on Jessie and the sled.

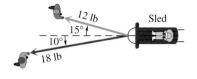

50. *Angle of a Hill* A 186-lb force just keeps a 2800-lb car from rolling down a hill. What angle does the hill make with the horizontal?

51. *Direction and Speed of a Plane* A plane has an airspeed of 520 mph. The pilot wishes to fly on a bearing of 310°. A wind of 37 mph is blowing from a bearing of 212°. What direction should the pilot fly, and what will be her actual speed?

52. *Speed and Direction of a Boat* A boat travels 15 km per hr in still water. The boat is traveling across a large river, on a bearing of 130°. The current in the river, coming from the west, has a speed of 7 km per hr. Find the resulting speed of the boat and its resulting direction of travel.

53. *Control Points* To obtain accurate aerial photographs, ground control must determine the coordinates of **control points** located on the ground that can be identified in the photographs. Using these known control points, the orientation and scale of each photograph can be found. Then, unknown positions and distances can easily be determined. The figure shows three consecutive control points *A*, *B*, and *C*.

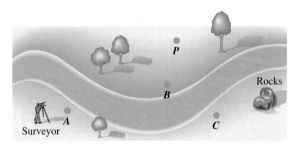

A surveyor measures a baseline distance of 92.1300 ft from *B* to an arbitrary point *P*. Angles *BAP* and *BCP* are found to be 2° 22′ 47″ and 5° 13′ 11″, respectively. Then, angles *APB* and *CPB* are determined to be 63° 4′ 25″ and 74° 19′ 49″, respectively. Determine the distance between control points *A* and *B* and between *B* and *C*. (*Source:* Moffitt, F. and E. Mikhail, *Photogrammetry,* Third Edition, Harper & Row, 1980.)

The following identities involve all six parts of a triangle ABC and are thus useful for checking answers.

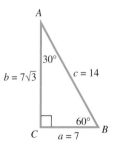

$$\frac{a + b}{c} = \frac{\cos \frac{1}{2}(A - B)}{\sin \frac{1}{2} C}$$ Newton's formula

$$\frac{a - b}{c} = \frac{\sin \frac{1}{2}(A - B)}{\cos \frac{1}{2} C}$$ Mollweide's formula

54. Apply Newton's formula to the given triangle to verify the accuracy of the information.

55. Apply Mollweide's formula to the given triangle to verify the accuracy of the information.

CHAPTER 7 ▶ Test

Find the indicated part of each triangle ABC.

1. Find C if $A = 25.2°$, $a = 6.92$ yd, and $b = 4.82$ yd.
2. Find c if $C = 118°$, $a = 75.0$ km, and $b = 131$ km.
3. Find B if $a = 17.3$ ft, $b = 22.6$ ft, $c = 29.8$ ft.
4. Find the area of triangle ABC if $a = 14$, $b = 30$, and $c = 40$.
5. Find the area of triangle XYZ shown here.

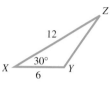

6. Given $a = 10$ and $B = 150°$ in triangle ABC, determine the values of b for which A has
   (a) exactly one value   (b) two possible values   (c) no value.

Solve each triangle ABC.

7. $A = 60°$, $b = 30$ m, $c = 45$ m
8. $b = 1075$ in., $c = 785$ in., $C = 38° 30'$
9. Find the magnitude and the direction angle for the vector shown in the figure.

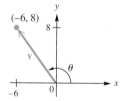

10. Use the given vectors to sketch $\mathbf{a} + \mathbf{b}$.

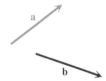

11. For the vectors $\mathbf{u} = \langle -1, 3 \rangle$ and $\mathbf{v} = \langle 2, -6 \rangle$, find each of the following.
    (a) $\mathbf{u} + \mathbf{v}$ (b) $-3\mathbf{v}$ (c) $\mathbf{u} \cdot \mathbf{v}$ (d) $|\mathbf{u}|$

Solve each problem.

12. *Height of a Balloon*   The angles of elevation of a balloon from two points A and B on level ground are $24° 50'$ and $47° 20'$, respectively. As shown in the figure, points A, B, and C are in the same vertical plane and points A and B are 8.4 mi apart. Approximate the height of the balloon above the ground to the nearest tenth of a mile.

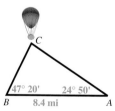

13. *Horizontal and Vertical Components*   Find the horizontal and vertical components of the vector with magnitude 569 that is inclined 127.5° from the horizontal. Give your answer in the form $\langle a, b \rangle$.

14. *Radio Direction Finders*   Radio direction finders are placed at points A and B, which are 3.46 mi apart on an east-west line, with A west of B. From A, the bearing of a certain illegal pirate radio transmitter is 48°, and from B the bearing is 302°. Find the distance between the transmitter and A to the nearest hundredth of a mile.

15. *Height of a Tree* A tree leans at an angle of 8.0°
from the vertical, as shown in the figure. From a
point 8.0 m from the bottom of the tree, the angle of
elevation to the top of the tree is 66°. How tall is the
tree?

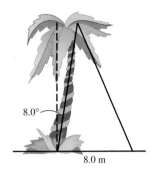

8.0°

8.0 m

16. *Walking Dogs on Leashes* While Michael is
walking his two dogs, Duke and Prince, they
reach a corner and must wait for a WALK sign.
Michael is holding the two leashes in the same
hand, and the dogs are pulling on their leashes
at the angles and forces shown in the figure.
Find the magnitude of the force (to the nearest
tenth of a pound) Michael must apply to restrain
the dogs.

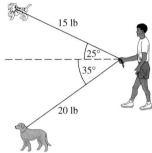

15 lb

25°
35°

20 lb

CHAPTER 7 ▶ Quantitative Reasoning

Just how much does the U.S. flag "show its colors"?

The flag of the United States includes the colors red, white, and blue. Which color is predominant? Clearly the answer is either red or white. (It can be shown that only 18.73% of
the total area is blue.) (*Source:* Banks, R., *Slicing Pizzas, Racing Turtles, and Further
Adventures in Applied Mathematics,* Princeton University Press, 1999.)

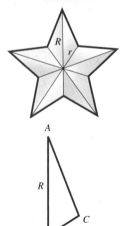

1. Let R denote the radius of the circumscribing circle of a five-pointed star appearing
on the American flag. The star can be decomposed into ten congruent triangles. In the
figure, r is the radius of the circumscribing circle of the pentagon in the interior of the
star. Show that the area of a star is

$$A = \left[5\frac{\sin A \sin B}{\sin(A + B)} \right] R^2.$$

(*Hint:* $\sin C = \sin[180° - (A + B)] = \sin(A + B)$.)

2. Angles A and B have values 18° and 36°, respectively. Express the area of a star in
terms of its radius, R.

3. To determine whether red or white is predominant, we must know the measurements
of the flag. Consider a flag of width 10 in., length 19 in., length of each upper stripe
11.4 in., and radius R of the circumscribing circle of each star .308 in. The thirteen
stripes consist of six matching pairs of red and white stripes and one additional red,
upper stripe. Therefore, we must compare the area of a red, upper stripe with the total
area of the 50 white stars.

   (a) Compute the area of the red, upper stripe.
   (b) Compute the total area of the 50 white stars.
   (c) Which color occupies the greatest area on the flag?

8

Complex Numbers, Polar Equations, and Parametric Equations

High-resolution computer graphics and *complex numbers* make it possible to produce beautiful shapes called *fractals*. Benoit B. Mandelbrot first used the term *fractal* in 1975. At its basic level, a fractal is a unique, enchanting geometric figure with an endless self-similarity property. A fractal image repeats itself infinitely with ever-decreasing dimensions. If you look at smaller and smaller portions of a fractal image, you will continue to see the whole—much like looking into two parallel mirrors that are facing each other.

The fractal called *Newton's basins of attraction for the cube roots of unity* is discussed in Exercise 50, Section 8.4. (*Source:* Crownover, R., *Introduction to Fractals and Chaos,* Jones and Bartlett, 1995; Lauwerier, H., *Fractals,* Princeton University Press, 1991.)

8.1 Complex Numbers

Basic Concepts of Complex Numbers ▪ Complex Solutions of Equations ▪ Operations on Complex Numbers

Basic Concepts of Complex Numbers The set of real numbers does not include all numbers needed in mathematics. For example, there is no real number solution of the equation

$$x^2 = -1,$$

since -1 has no real square root. Square roots of negative numbers were not incorporated into an integrated number system until the 16th century. They were then used as solutions of equations and, later in the 18th century, in surveying. Today, such numbers are used extensively in science and engineering.

To extend the real number system to include numbers such as $\sqrt{-1}$, the number i is defined to have the following property.

$$i^2 = -1$$

Thus, $i = \sqrt{-1}$. The number i is called the **imaginary unit.** Numbers of the form $a + bi$, where a and b are real numbers, are called **complex numbers.** In the complex number $a + bi$, a is the **real part** and b is the **imaginary part.***

The relationships among the various sets of numbers are shown in Figure 1.

▼ **LOOKING AHEAD TO CALCULUS**

The letters j and k are also used to represent $\sqrt{-1}$ in calculus and some applications (electronics, for example).

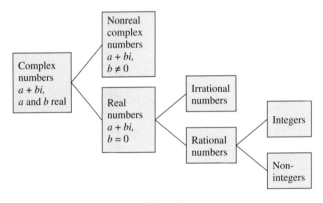

Figure 1

Two complex numbers $a + bi$ and $c + di$ are equal provided that their real parts are equal and their imaginary parts are equal; that is,

$$a + bi = c + di \quad \text{if and only if} \quad a = c \quad \text{and} \quad b = d.$$

For a complex number $a + bi$, if $b = 0$, then $a + bi = a$, which is a real number. Thus, the set of real numbers is a subset of the set of complex numbers. If $a = 0$ and $b \neq 0$, the complex number is said to be a **pure imaginary number.** For example, $3i$ is a pure imaginary number. A pure imaginary number, or a number such as $7 + 2i$ with $a \neq 0$ and $b \neq 0$, is a **nonreal complex number.** A complex number written in the form $a + bi$ (or $a + ib$) is in

---

*In some texts, the term bi is defined to be the imaginary part.

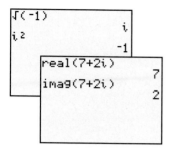

The calculator is in complex number mode.

Figure 2

standard form. $\left(\text{The form } a + ib \text{ is used to write expressions such as } i\sqrt{5}, \text{ since } \sqrt{5}i \text{ could be mistaken for } \sqrt{5i}.\right)$

NOW TRY EXERCISES 9, 11, AND 13. ◀

Some graphing calculators, such as the TI-83/84 Plus, are capable of working with complex numbers, as seen in Figure 2. The top screen supports the definition of i. The bottom screen shows how the calculator returns the real and imaginary parts of $7 + 2i$. ■

For a positive real number a, $\sqrt{-a}$ is defined as follows.

THE EXPRESSION $\sqrt{-a}$

If $a > 0$, then $$\sqrt{-a} = i\sqrt{a}.$$

▶ **EXAMPLE 1** WRITING $\sqrt{-a}$ AS $i\sqrt{a}$

Write as the product of a real number and i, using the definition of $\sqrt{-a}$.

(a) $\sqrt{-16}$ **(b)** $\sqrt{-70}$ **(c)** $\sqrt{-48}$

Solution

(a) $\sqrt{-16} = i\sqrt{16} = 4i$ **(b)** $\sqrt{-70} = i\sqrt{70}$

(c) $\sqrt{-48} = i\sqrt{48} = i\sqrt{16 \cdot 3} = 4i\sqrt{3}$ Product rule for radicals; $\sqrt[n]{ab} = \sqrt[n]{a} \cdot \sqrt[n]{b}$

NOW TRY EXERCISES 17, 19, AND 21. ◀

Complex Solutions of Equations

▶ **EXAMPLE 2** SOLVING QUADRATIC EQUATIONS FOR COMPLEX SOLUTIONS

Solve each equation.

(a) $x^2 = -9$ **(b)** $x^2 + 24 = 0$

Solution

(a) Take the square root on both sides, remembering that we must find both roots, indicated by the $\pm$ sign.

$$x^2 = -9$$

$$x = \pm\sqrt{-9} \quad \text{Square root property (Appendix A)}$$

$$x = \pm i\sqrt{9} \quad \sqrt{-a} = i\sqrt{a}$$

$$x = \pm 3i \quad \sqrt{9} = 3$$

Take *both* square roots.

The solution set is $\{\pm 3i\}$.

(b) $x^2 + 24 = 0$

$$x^2 = -24 \qquad \text{Subtract 24.}$$

$$x = \pm\sqrt{-24} \qquad \text{Square root property}$$

$$x = \pm i\sqrt{24} \qquad \sqrt{-a} = i\sqrt{a}$$

$$x = \pm i\sqrt{4 \cdot 6} = \pm 2i\sqrt{6} \qquad \text{Product rule for radicals}$$

The solution set is $\left\{\pm 2i\sqrt{6}\right\}$.

NOW TRY EXERCISES 25 AND 27. ◀

▶ **EXAMPLE 3**   **SOLVING A QUADRATIC EQUATION FOR COMPLEX SOLUTIONS**

Solve $9x^2 + 5 = 6x$.

Solution   Write the equation in standard form, $9x^2 - 6x + 5 = 0$. Then use the quadratic formula.

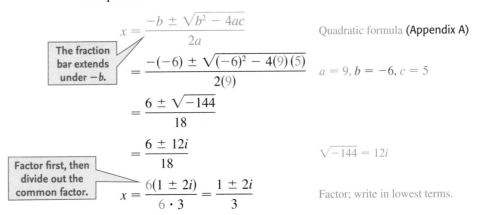

$$x = \frac{-b \pm \sqrt{b^2 - 4ac}}{2a} \qquad \text{Quadratic formula (Appendix A)}$$

The fraction bar extends under $-b$.

$$= \frac{-(-6) \pm \sqrt{(-6)^2 - 4(9)(5)}}{2(9)} \qquad a = 9, b = -6, c = 5$$

$$= \frac{6 \pm \sqrt{-144}}{18}$$

$$= \frac{6 \pm 12i}{18} \qquad \sqrt{-144} = 12i$$

Factor first, then divide out the common factor.

$$x = \frac{6(1 \pm 2i)}{6 \cdot 3} = \frac{1 \pm 2i}{3} \qquad \text{Factor; write in lowest terms.}$$

The solution set is $\left\{\frac{1}{3} \pm \frac{2}{3}i\right\}$.

NOW TRY EXERCISE 29. ◀

Operations on Complex Numbers   Products or quotients with negative radicands are simplified by first rewriting $\sqrt{-a}$ as $i\sqrt{a}$ for a positive number a. Then the properties of real numbers are applied, together with the fact that $i^2 = -1$.

▶ **Caution** *When working with negative radicands, use the definition* $\sqrt{-a} = i\sqrt{a}$ *before using any of the other rules for radicals.* In particular, the rule $\sqrt{c} \cdot \sqrt{d} = \sqrt{cd}$ is valid only when c and d are *not* both negative. For example,

$$\sqrt{(-4)(-9)} = \sqrt{36} = 6,$$

while

$$\sqrt{-4} \cdot \sqrt{-9} = 2i(3i) = 6i^2 = -6,$$

so

$$\sqrt{-4} \cdot \sqrt{-9} \neq \sqrt{(-4)(-9)}.$$

▶ EXAMPLE 4   FINDING PRODUCTS AND QUOTIENTS INVOLVING NEGATIVE RADICANDS

Multiply or divide, as indicated. Simplify each answer.

(a) $\sqrt{-7} \cdot \sqrt{-7}$ **(b)** $\sqrt{-6} \cdot \sqrt{-10}$ **(c)** $\dfrac{\sqrt{-20}}{\sqrt{-2}}$ **(d)** $\dfrac{\sqrt{-48}}{\sqrt{24}}$

Solution

(a) $\sqrt{-7} \cdot \sqrt{-7} = i\sqrt{7} \cdot i\sqrt{7}$

First write all square roots in terms of i.

$= i^2 \cdot (\sqrt{7})^2$

$= -1 \cdot 7 \qquad i^2 = -1$

$= -7$

(b) $\sqrt{-6} \cdot \sqrt{-10} = i\sqrt{6} \cdot i\sqrt{10}$

$= i^2 \cdot \sqrt{60}$

$= -1\sqrt{4 \cdot 15}$

$= -1 \cdot 2\sqrt{15}$

$= -2\sqrt{15}$

(c) $\dfrac{\sqrt{-20}}{\sqrt{-2}} = \dfrac{i\sqrt{20}}{i\sqrt{2}} = \sqrt{\dfrac{20}{2}} = \sqrt{10}$ Quotient rule for radicals; $\dfrac{\sqrt[n]{a}}{\sqrt[n]{b}} = \sqrt[n]{\dfrac{a}{b}}$

(d) $\dfrac{\sqrt{-48}}{\sqrt{24}} = \dfrac{i\sqrt{48}}{\sqrt{24}} = i\sqrt{\dfrac{48}{24}} = i\sqrt{2}$

NOW TRY EXERCISES 37, 39, 41, AND 43. ◀

▶ EXAMPLE 5   SIMPLIFYING A QUOTIENT INVOLVING A NEGATIVE RADICAND

Write $\dfrac{-8 + \sqrt{-128}}{4}$ in standard form $a + bi$.

Solution

$$\dfrac{-8 + \sqrt{-128}}{4} = \dfrac{-8 + \sqrt{-64 \cdot 2}}{4}$$

$$= \dfrac{-8 + 8i\sqrt{2}}{4} \qquad \sqrt{-64} = 8i$$

Be sure to factor before simplifying.

$$= \dfrac{4(-2 + 2i\sqrt{2})}{4} \qquad \text{Factor.}$$

$$= -2 + 2i\sqrt{2} \qquad \text{Lowest terms}$$

NOW TRY EXERCISE 49. ◀

With the definitions $i^2 = -1$ and $\sqrt{-a} = i\sqrt{a}$ for $a > 0$, all properties of real numbers are extended to complex numbers. As a result, complex numbers are added, subtracted, multiplied, and divided using the definitions on the following pages and real number properties.

ADDITION AND SUBTRACTION OF COMPLEX NUMBERS

For complex numbers $a + bi$ and $c + di$,

$$(a + bi) + (c + di) = (a + c) + (b + d)i$$

and

$$(a + bi) - (c + di) = (a - c) + (b - d)i.$$

That is, to add or subtract complex numbers, add or subtract the real parts and add or subtract the imaginary parts.

▶ **EXAMPLE 6** ADDING AND SUBTRACTING COMPLEX NUMBERS

Find each sum or difference.

(a) $(3 - 4i) + (-2 + 6i)$ **(b)** $(-9 + 7i) + (3 - 15i)$

(c) $(-4 + 3i) - (6 - 7i)$ **(d)** $(12 - 5i) - (8 - 3i) + (-4 + 2i)$

Solution

(a) $(3 - 4i) + (-2 + 6i) = \overbrace{[3 + (-2)]}^{\text{Add real parts.}} + \overbrace{[-4 + 6]}^{\text{Add imaginary parts.}}i$ Commutative, associative, distributive properties

$$= 1 + 2i$$

(b) $(-9 + 7i) + (3 - 15i) = -6 - 8i$

(c) $(-4 + 3i) - (6 - 7i) = (-4 - 6) + [3 - (-7)]i$

$$= -10 + 10i$$

(d) $(12 - 5i) - (8 - 3i) + (-4 + 2i) = (12 - 8 - 4) + (-5 + 3 + 2)i$

$$= 0, \quad \text{or} \quad 0 + 0i$$

NOW TRY EXERCISES 55 AND 57. ◀

The product of two complex numbers is found by multiplying as if the numbers were binomials and using the fact that $i^2 = -1$, as follows.

$$(a + bi)(c + di) = ac + adi + bic + bidi \qquad \text{FOIL (multiply, First, Outer, Inner, Last terms)}$$

$$= ac + adi + bci + bdi^2$$

$$= ac + (ad + bc)i + bd(-1) \qquad \text{Distributive property; } i^2 = -1$$

$$= (ac - bd) + (ad + bc)i$$

MULTIPLICATION OF COMPLEX NUMBERS

For complex numbers $a + bi$ and $c + di$,

$$(a + bi)(c + di) = (ac - bd) + (ad + bc)i.$$

This definition is not practical in routine calculations. To find a given product, it is easier just to multiply as with binomials.

▶ **EXAMPLE 7** MULTIPLYING COMPLEX NUMBERS

Find each product.

(a) $(2 - 3i)(3 + 4i)$ **(b)** $(4 + 3i)^2$

(c) $(2 + i)(-2 - i)$ **(d)** $(6 + 5i)(6 - 5i)$

Solution

(a) $(2 - 3i)(3 + 4i) = 2(3) + 2(4i) - 3i(3) - 3i(4i)$ FOIL

$$= 6 + 8i - 9i - 12i^2$$

$$= 6 - i - 12(-1) \qquad\qquad i^2 = -1$$

$$= 18 - i$$

(b) $(4 + 3i)^2 = 4^2 + 2(4)(3i) + (3i)^2$ Square of a binomial;

$(x + y)^2 = x^2 + 2xy + y^2$

> **Remember to add twice the product of the two terms.**

$$= 16 + 24i + 9i^2$$

$$= 16 + 24i + 9(-1) \qquad i^2 = -1$$

$$= 7 + 24i$$

(c) $(2 + i)(-2 - i) = -4 - 2i - 2i - i^2$ FOIL

$$= -4 - 4i - (-1) \qquad \text{Combine terms; } i^2 = -1$$

$$= -4 - 4i + 1$$

$$= -3 - 4i \qquad\qquad \text{Standard form}$$

(d) $(6 + 5i)(6 - 5i) = 6^2 - (5i)^2$ Product of the sum and difference of two terms; $(x + y)(x - y) = x^2 - y^2$

$$= 36 - 25(-1) \qquad i^2 = -1$$

$$= 36 + 25$$

$$= 61, \quad \text{or} \quad 61 + 0i \quad \text{Standard form}$$

```
(2-3i)(3+4i)
               18-i
(4+3i)²
              7+24i
(6+5i)(6-5i)
                  61
```

This screen shows how the TI–83/84 Plus displays the results found in Example 7(a), (b), and (d).

> **NOW TRY EXERCISES 63, 67, AND 71.** ◀

Powers of i can be simplified using the facts

$$i^2 = -1 \quad \text{and} \quad i^4 = (i^2)^2 = (-1)^2 = 1.$$

▶ **EXAMPLE 8** SIMPLIFYING POWERS OF i

Simplify each power of i.

(a) i^{15} **(b)** i^{-3} **(c)** $\dfrac{1}{i^{-13}}$

Solution

(a) Since $i^2 = -1$ and $i^4 = 1$, write the given power as a product involving i^2 or i^4. For example,

$$i^3 = i^2 \cdot i = (-1) \cdot i = -i.$$

Alternatively, using i^4 and i^3 to rewrite i^{15} gives

$$i^{15} = i^{12} \cdot i^3 = (i^4)^3 \cdot i^3 = 1^3(-i) = -i.$$

(b) $i^{-3} = i^{-4} \cdot i = (i^4)^{-1} \cdot i = (1)^{-1} \cdot i = i$

(c) $\dfrac{1}{i^{-13}} = i^{13} = i^{12} \cdot i = (i^4)^3 \cdot i = 1^3 \cdot i = i$

NOW TRY EXERCISES 81, 89, AND 91. ◄

We can use the method of Example 8 to construct the following list of powers of i.

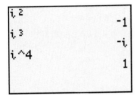

Powers of i can be found on the TI–83/84 Plus calculator.

POWERS OF i			
$i^1 = i$	$i^5 = i$	$i^9 = i$	
$i^2 = -1$	$i^6 = -1$	$i^{10} = -1$	
$i^3 = -i$	$i^7 = -i$	$i^{11} = -i$	
$i^4 = 1$	$i^8 = 1$	$i^{12} = 1,$	and so on.

Example 7(d) showed that $(6 + 5i)(6 - 5i) = 61$. The numbers $6 + 5i$ and $6 - 5i$ differ only in the sign of their imaginary parts, and are called **complex conjugates**. *The product of a complex number $a + bi$ and its conjugate $a - bi$ is always a real number, $a^2 + b^2$.* This product is the sum of the squares of the real and imaginary parts.

PROPERTY OF COMPLEX CONJUGATES

For real numbers a and b,
$$(a + bi)(a - bi) = a^2 + b^2.$$

To find the quotient of two complex numbers in standard form, we multiply both the numerator and the denominator by the complex conjugate of the denominator.

► **EXAMPLE 9** DIVIDING COMPLEX NUMBERS

Write each quotient in standard form $a + bi$.

(a) $\dfrac{3 + 2i}{5 - i}$

(b) $\dfrac{3}{i}$

Solution

(a) $\dfrac{3 + 2i}{5 - i} = \dfrac{(3 + 2i)(5 + i)}{(5 - i)(5 + i)}$ Multiply by the complex conjugate of the denominator in both the numerator and the denominator.

$ = \dfrac{15 + 3i + 10i + 2i^2}{25 - i^2}$ Multiply.

$ = \dfrac{13 + 13i}{26}$ $i^2 = -1$

$ = \dfrac{13}{26} + \dfrac{13i}{26}$ $\dfrac{a + bi}{c} = \dfrac{a}{c} + \dfrac{bi}{c}$

$ = \dfrac{1}{2} + \dfrac{1}{2}i$ Lowest terms; standard form

To *check* this answer, show that

$$\left(\frac{1}{2} + \frac{1}{2}i\right)(5 - i) = 3 + 2i.$$ Quotient × Divisor = Dividend

(b) $\dfrac{3}{i} = \dfrac{3(-i)}{i(-i)}$ −*i* is the conjugate of *i*.

$$= \frac{-3i}{-i^2}$$

$$= \frac{-3i}{1}$$ $-i^2 = -(-1) = 1$

$$= -3i, \quad \text{or} \quad 0 - 3i$$ Standard form

This screen supports the results in Example 9. Notice that the answer to part (a) must be interpreted $\frac{1}{2} + \frac{1}{2}i$, *not* $\frac{1}{2} + \frac{1}{2i}$.

NOW TRY EXERCISES 95 AND 101. ◀

8.1 Exercises

Concept Check   *Determine whether each statement is* true *or* false. *If it is false, tell why.*

1. Every real number is a complex number.

2. No real number is a pure imaginary number.

3. Every pure imaginary number is a complex number.

4. A number can be both real and complex.

5. There is no real number that is a complex number.

6. A complex number might not be a pure imaginary number.

Identify each number as real, complex, pure imaginary, *or* nonreal complex. *(More than one of these descriptions may apply.)*

7. -4 **8.** 0 **9.** $13i$ **10.** $-7i$ **11.** $5 + i$

12. $-6 - 2i$ **13.** π **14.** $\sqrt{24}$ **15.** $\sqrt{-25}$ **16.** $\sqrt{-36}$

Write each number as the product of a real number and i. See Example 1.

17. $\sqrt{-25}$ **18.** $\sqrt{-36}$ **19.** $\sqrt{-10}$ **20.** $\sqrt{-15}$

21. $\sqrt{-288}$ **22.** $\sqrt{-500}$ **23.** $-\sqrt{-18}$ **24.** $-\sqrt{-80}$

Solve each quadratic equation and express all nonreal complex solutions in terms of i. See Examples 2 and 3.

25. $x^2 = -16$ **26.** $x^2 = -36$ **27.** $x^2 + 12 = 0$

28. $x^2 + 48 = 0$ **29.** $3x^2 + 2 = -4x$ **30.** $2x^2 + 3x = -2$

31. $x^2 - 6x + 14 = 0$ **32.** $x^2 + 4x + 11 = 0$ **33.** $4(x^2 - x) = -7$

34. $3(3x^2 - 2x) = -7$ **35.** $x^2 + 1 = -x$ **36.** $x^2 + 2 = 2x$

Multiply or divide, as indicated. Simplify each answer. See Example 4.

37. $\sqrt{-13} \cdot \sqrt{-13}$ **38.** $\sqrt{-17} \cdot \sqrt{-17}$ **39.** $\sqrt{-3} \cdot \sqrt{-8}$

40. $\sqrt{-5} \cdot \sqrt{-15}$ **41.** $\dfrac{\sqrt{-30}}{\sqrt{-10}}$ **42.** $\dfrac{\sqrt{-70}}{\sqrt{-7}}$

43. $\dfrac{\sqrt{-24}}{\sqrt{8}}$

44. $\dfrac{\sqrt{-54}}{\sqrt{27}}$

45. $\dfrac{\sqrt{-10}}{\sqrt{-40}}$

46. $\dfrac{\sqrt{-40}}{\sqrt{20}}$

47. $\dfrac{\sqrt{-6} \cdot \sqrt{-2}}{\sqrt{3}}$

48. $\dfrac{\sqrt{-12} \cdot \sqrt{-6}}{\sqrt{8}}$

Write each number in standard form a + bi. See Example 5.

49. $\dfrac{-6 - \sqrt{-24}}{2}$

50. $\dfrac{-9 - \sqrt{-18}}{3}$

51. $\dfrac{10 + \sqrt{-200}}{5}$

52. $\dfrac{20 + \sqrt{-8}}{2}$

53. $\dfrac{-3 + \sqrt{-18}}{24}$

54. $\dfrac{-5 + \sqrt{-50}}{10}$

Find each sum or difference. Write the answer in standard form. See Example 6.

55. $(3 + 2i) + (9 - 3i)$

56. $(4 - i) + (8 + 5i)$

57. $(-2 + 4i) - (-4 + 4i)$

58. $(-3 + 2i) - (-4 + 2i)$

59. $(2 - 5i) - (3 + 4i) - (-1 - 9i)$

60. $(-4 - i) - (2 + 3i) + (6 + 4i)$

61. $-i - 2 - (6 - 4i) - (5 - 2i)$

62. $3 - (4 - i) - 4i + (-2 + 5i)$

Find each product. Write the answer in standard form. See Example 7.

63. $(2 + i)(3 - 2i)$

64. $(-2 + 3i)(4 - 2i)$

65. $(2 + 4i)(-1 + 3i)$

66. $(1 + 3i)(2 - 5i)$

67. $(3 - 2i)^2$

68. $(2 + i)^2$

69. $(3 + i)(3 - i)$

70. $(5 + i)(5 - i)$

71. $(-2 - 3i)(-2 + 3i)$

72. $(6 - 4i)(6 + 4i)$

73. $(\sqrt{6} + i)(\sqrt{6} - i)$

74. $(\sqrt{2} - 4i)(\sqrt{2} + 4i)$

75. $i(3 - 4i)(3 + 4i)$

76. $i(2 + 7i)(2 - 7i)$

77. $3i(2 - i)^2$

78. $-5i(4 - 3i)^2$

79. $(2 + i)(2 - i)(4 + 3i)$

80. $(3 - i)(3 + i)(2 - 6i)$

Simplify each power of i. See Example 8.

81. i^{25}

82. i^{29}

83. i^{22}

84. i^{26}

85. i^{23}

86. i^{27}

87. i^{32}

88. i^{40}

89. i^{-13}

90. i^{-14}

91. $\dfrac{1}{i^{-11}}$

92. $\dfrac{1}{i^{-12}}$

93. Suppose that your friend, Leah Goldberg, tells you that she has discovered a method of simplifying a positive power of i. "Just divide the exponent by 4," she says, "and then look at the remainder. Then refer to the table of powers of i in this section. The large power of i is equal to the power indicated by the remainder. And if the remainder is 0, the result is $i^0 = 1$." Explain why her method works.

94. Explain why the following method of simplifying i^{-42} works.

$$i^{-42} = i^{-42} \cdot i^{44} = i^{-42+44} = i^2 = -1$$

Find each quotient. Write the answer in standard form a + bi. See Example 9.

95. $\dfrac{6 + 2i}{1 + 2i}$

96. $\dfrac{14 + 5i}{3 + 2i}$

97. $\dfrac{2 - i}{2 + i}$

98. $\dfrac{4 - 3i}{4 + 3i}$

99. $\dfrac{1 - 3i}{1 + i}$

100. $\dfrac{-3 + 4i}{2 - i}$

101. $\dfrac{-5}{i}$

102. $\dfrac{-6}{i}$

103. $\dfrac{8}{-i}$

104. $\dfrac{12}{-i}$

105. $\dfrac{2}{3i}$

106. $\dfrac{5}{9i}$

107. Show that $\dfrac{\sqrt{2}}{2} + \dfrac{\sqrt{2}}{2}i$ is a square root of i.

108. Show that $\dfrac{\sqrt{3}}{2} + \dfrac{1}{2}i$ is a cube root of i.

109. Evaluate $3z - z^2$ if $z = 3 - 2i$.

110. Evaluate $-2z + z^3$ if $z = -6i$.

(*Modeling*) *Alternating Current*   *Complex numbers are used to describe current, I, voltage, E, and impedance, Z (the opposition to current). These three quantities are related by the equation E = IZ. Thus, if any two of these quantities are known, the third can be found. In each problem, solve the equation E = IZ for the missing variable.*

111. $I = 8 + 6i, \quad Z = 6 + 3i$

112. $I = 10 + 6i, \quad Z = 8 + 5i$

113. $I = 7 + 5i, \quad E = 28 + 54i$

114. $E = 35 + 55i, \quad Z = 6 + 4i$

(*Modeling*) *Impedance*   **Impedance** *is a measure of the opposition to the flow of alternating electrical current found in common electrical outlets. It consists of two parts called* **resistance** *and* **reactance**. *Resistance occurs when a lightbulb is turned on, while reactance is produced when electricity passes through a coil of wire like that found in electric motors. Impedance Z in ohms (Ω) can be expressed as a complex number, where the real part represents resistance and the imaginary part represents reactance.*

For example, if the resistive part is 3 ohms and the reactive part is 4 ohms, then the impedance could be described by the complex number Z = 3 + 4i. In the series circuit shown in the figure, the total impedance will be the sum of the individual impedances. (*Source:* Wilcox, G. and C. Hesselberth, *Electricity for Engineering Technology,* Allyn & Bacon, 1970.)

115. The circuit contains two light bulbs and two electric motors. Assuming that the light bulbs are pure resistive and the motors are pure reactive, find the total impedance in this circuit and express it in the form $Z = a + bi$.

116. The phase angle θ measures the phase difference between the voltage and the current in an electrical circuit. θ (in degrees) can be determined by the equation $\tan \theta = \frac{b}{a}$. Find θ for this circuit.

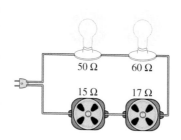

8.2 Trigonometric (Polar) Form of Complex Numbers

The Complex Plane and Vector Representation ▪ Trigonometric (Polar) Form ▪ Converting Between Rectangular and Trigonometric (Polar) Forms ▪ An Application of Complex Numbers to Fractals

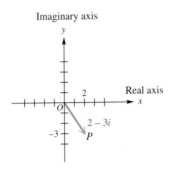

Figure 3

The Complex Plane and Vector Representation   Unlike real numbers, complex numbers cannot be ordered. One way to organize and illustrate them is by using a graph. To graph a complex number such as $2 - 3i$, we modify the familiar coordinate system by calling the horizontal axis the **real axis** and the vertical axis the **imaginary axis.** Then complex numbers can be graphed in this **complex plane,** as shown in Figure 3. Each complex number $a + bi$ determines a unique position vector with initial point $(0, 0)$ and terminal point (a, b). This relationship shows the close connection between vectors and complex numbers.

> ▶ **Note** This geometric representation is the reason that $a + bi$ is called the **rectangular form** of a complex number. (*Rectangular form* is also called *standard form.*)

Figure 4

Recall that the sum of the two complex numbers $4 + i$ and $1 + 3i$ is

$$(4 + i) + (1 + 3i) = 5 + 4i. \quad \text{(Section 8.1)}$$

Graphically, the sum of two complex numbers is represented by the vector that is the resultant of the vectors corresponding to the two numbers, as shown in Figure 4.

▶ **EXAMPLE 1**   EXPRESSING THE SUM OF COMPLEX NUMBERS GRAPHICALLY

Find the sum of $6 - 2i$ and $-4 - 3i$. Graph both complex numbers and their resultant.

Solution   The sum is found by adding the two numbers.

$$(6 - 2i) + (-4 - 3i) = 2 - 5i$$

The graphs are shown in Figure 5.

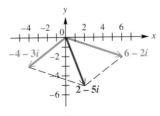

Figure 5

NOW TRY EXERCISE 13. ◀

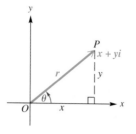

Figure 6

Trigonometric (Polar) Form Figure 6 shows the complex number $x + yi$ that corresponds to a vector **OP** with direction angle θ and magnitude r. The following relationships among x, y, r, and θ can be verified from Figure 6.

RELATIONSHIPS AMONG $x, y, r,$ AND θ

$$x = r \cos \theta \qquad\qquad y = r \sin \theta$$

$$r = \sqrt{x^2 + y^2} \qquad \tan \theta = \frac{y}{x}, \quad \text{if } x \neq 0$$

Substituting $x = r \cos \theta$ and $y = r \sin \theta$ into $x + yi$ gives

$$x + yi = r \cos \theta + (r \sin \theta)i$$
$$= r(\cos \theta + i \sin \theta). \quad \text{Factor out } r.$$

TRIGONOMETRIC (POLAR) FORM OF A COMPLEX NUMBER

The expression

$$r(\cos \theta + i \sin \theta)$$

is called the **trigonometric form** (or **polar form**) of the complex number $x + yi$. The expression $\cos \theta + i \sin \theta$ is sometimes abbreviated **cis θ**. Using this notation, $r(\cos \theta + i \sin \theta)$ is written r **cis θ.**

The number r is the **absolute value** (or **modulus**) of $x + yi$, and θ is the **argument** of $x + yi$. In this section, we choose the value of θ in the interval $[0°, 360°)$. However, any angle coterminal with θ also could serve as the argument.

▶ **EXAMPLE 2** CONVERTING FROM TRIGONOMETRIC FORM TO RECTANGULAR FORM

Express $2(\cos 300° + i \sin 300°)$ in rectangular form.

Algebraic Solution

From **Chapter 2,** $\cos 300° = \frac{1}{2}$ and $\sin 300° = -\frac{\sqrt{3}}{2}$, so

$$2(\cos 300° + i \sin 300°) = 2\left(\frac{1}{2} - i\frac{\sqrt{3}}{2}\right)$$
$$= 1 - i\sqrt{3}.$$

Notice that the real part is positive and the imaginary part is negative; this is consistent with $300°$ being a quadrant IV angle.

Graphing Calculator Solution

Figure 7 confirms the algebraic solution.

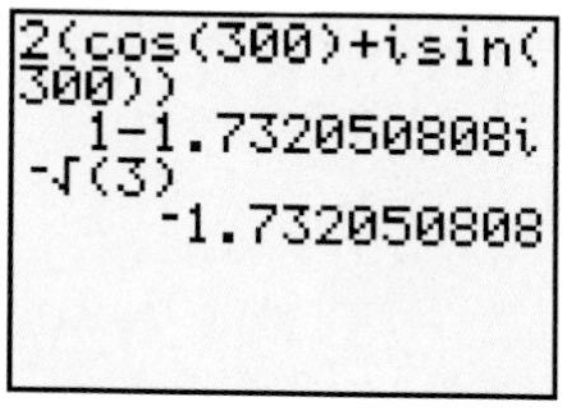

```
2(cos(300)+isin(
300))
       1-1.732050808i
-√(3)
        -1.732050808
```

The imaginary part is an approximation for $-\sqrt{3}$.

Figure 7

NOW TRY EXERCISE 29. ◀

Converting Between Rectangular and Trigonometric (Polar)
Forms To convert from rectangular form to trigonometric form, we use the following procedure.

CONVERTING FROM RECTANGULAR FORM TO TRIGONOMETRIC FORM

Step 1 Sketch a graph of the number $x + yi$ in the complex plane.

Step 2 Find r by using the equation $r = \sqrt{x^2 + y^2}$.

Step 3 Find θ by using the equation $\tan \theta = \frac{y}{x}$, $x \neq 0$, choosing the quadrant indicated in Step 1.

▶ **Caution** Errors often occur in Step 3. *Be sure to choose the correct quadrant for θ by referring to the graph sketched in Step 1.*

▶ **EXAMPLE 3** CONVERTING FROM RECTANGULAR FORM TO TRIGONOMETRIC FORM

Write each complex number in trigonometric form.

(a) $-\sqrt{3} + i$ (Use radian measure.)   **(b)** $-3i$ (Use degree measure.)

Solution

(a) We start by sketching the graph of $-\sqrt{3} + i$ in the complex plane, as shown in Figure 8. Since $x = -\sqrt{3}$ and $y = 1$,

$$r = \sqrt{x^2 + y^2} = \sqrt{\left(-\sqrt{3}\right)^2 + 1^2} = \sqrt{3 + 1} = 2,$$

and $\tan \theta = \dfrac{y}{x} = \dfrac{1}{-\sqrt{3}} = -\dfrac{1}{\sqrt{3}} \cdot \dfrac{\sqrt{3}}{\sqrt{3}} = -\dfrac{\sqrt{3}}{3}.$

Rationalize the denominator.

Since $\tan \theta = -\dfrac{\sqrt{3}}{3}$, the reference angle for θ in radians is $\frac{\pi}{6}$. From the graph, we see that θ is in quadrant II, so $\theta = \pi - \frac{\pi}{6} = \frac{5\pi}{6}$. Therefore,

$$-\sqrt{3} + i = 2\left(\cos \frac{5\pi}{6} + i \sin \frac{5\pi}{6}\right) = 2 \operatorname{cis} \frac{5\pi}{6}.$$

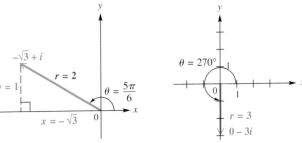

Figure 8   Figure 9

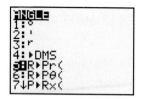

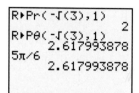

Choices 5 and 6 in the top screen show how to convert from rectangular (x, y) form to trigonometric form. The calculator is in radian mode. The results agree with our algebraic results in Example 3(a).

(b) The sketch of $-3i$ is shown in Figure 9 on the preceding page. Since $-3i = 0 - 3i$, we have $x = 0$ and $y = -3$ and

$$r = \sqrt{0^2 + (-3)^2} = \sqrt{0 + 9} = \sqrt{9} = 3.$$

We cannot find θ by using $\tan \theta = \frac{y}{x}$, because $x = 0$. From the graph, a value for θ is $270°$. In trigonometric form,

$$-3i = 3(\cos 270° + i \sin 270°) = 3 \text{ cis } 270°.$$

NOW TRY EXERCISES 41 AND 47. ◀

▶ **Note** In Example 3, we gave answers in both forms: $r(\cos \theta + i \sin \theta)$ and r cis θ. These forms will be used interchangeably from now on.

▶ **EXAMPLE 4** CONVERTING BETWEEN TRIGONOMETRIC AND RECTANGULAR FORMS USING CALCULATOR APPROXIMATIONS

Write each complex number in its alternative form, using calculator approximations as necessary.

(a) $6(\cos 115° + i \sin 115°)$ **(b)** $5 - 4i$

Solution

(a) Since $115°$ does not have a special angle as a reference angle, we cannot find exact values for $\cos 115°$ and $\sin 115°$. Use a calculator set in degree mode to find $\cos 115° \approx -.4226182617$ and $\sin 115° \approx .906307787$. Therefore, in rectangular form,

$$6(\cos 115° + i \sin 115°) \approx 6(-.4226182617 + .906307787i)$$
$$\approx -2.5357 + 5.4378i.$$

(b) A sketch of $5 - 4i$ shows that θ must be in quadrant IV. See Figure 10. Here

$$r = \sqrt{5^2 + (-4)^2} = \sqrt{41} \qquad \text{and} \qquad \tan \theta = -\frac{4}{5}.$$

Use a calculator to find that one measure of θ is $-38.66°$. In order to express θ in the interval $[0, 360°)$, we find $\theta = 360° - 38.66° = 321.34°$. Use these results to get

$$5 - 4i = \sqrt{41} \text{ cis } 321.34°.$$

NOW TRY EXERCISES 53 AND 57. ◀

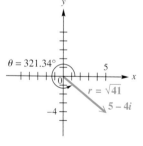

Figure 10

An Application of Complex Numbers to Fractals

As mentioned in the chapter opener, a **fractal** is a unique, enchanting geometric figure with an endless self-similarity property.

▶ **EXAMPLE 5**   **DECIDING WHETHER A COMPLEX NUMBER IS IN THE JULIA SET**

The fractal called the **Julia set** is shown in Figure 11. To determine if a complex number $z = a + bi$ is in this Julia set, perform the following sequence of calculations. Repeatedly compute the values of

$$z^2 - 1, \quad (z^2 - 1)^2 - 1, \quad [(z^2 - 1)^2 - 1]^2 - 1, \quad \ldots.$$

If the absolute values of any of the resulting complex numbers exceed 2, then the complex number z is not in the Julia set. Otherwise z is part of this set and the point (a, b) should be shaded in the graph.

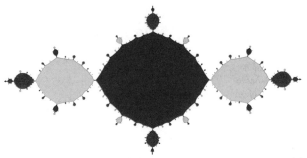

Source: Figure from Crownover, R., *Introduction to Fractals and Chaos.* Copyright © 1995. Boston: Jones and Bartlett Publishers. Reprinted with permission.

Figure 11

Determine whether each number belongs to the Julia set.

(a) $z = 0 + 0i$ **(b)** $z = 1 + 1i$

Solution

(a) Here
$$z = 0 + 0i = 0,$$
$$z^2 - 1 = 0^2 - 1 = -1,$$
$$(z^2 - 1)^2 - 1 = (-1)^2 - 1 = 0,$$
$$[(z^2 - 1)^2 - 1]^2 - 1 = 0^2 - 1 = -1,$$

and so on. We see that the calculations repeat as $0, -1, 0, -1,$ and so on. The absolute values are either 0 or 1, which do not exceed 2, so $0 + 0i$ is in the Julia set and the point $(0, 0)$ is part of the graph.

(b) We have $z^2 - 1 = (1 + i)^2 - 1 = (1 + 2i + i^2) - 1 = -1 + 2i$. The absolute value is $\sqrt{(-1)^2 + 2^2} = \sqrt{5}$. Since $\sqrt{5}$ is greater than 2, $1 + 1i$ is not in the Julia set and $(1, 1)$ is not part of the graph.

NOW TRY EXERCISE 63. ◀

8.2   Exercises

1. *Concept Check*   The absolute value of a complex number represents the _____ of the vector representing it in the complex plane.

2. *Concept Check*   What is the geometric interpretation of the argument of a complex number?

Graph each complex number. See Example 1.

3. $-3 + 2i$ **4.** $6 - 5i$ **5.** $\sqrt{2} + \sqrt{2}i$ **6.** $2 - 2i\sqrt{3}$

7. $-4i$ **8.** $3i$ **9.** -8 **10.** 2

Concept Check   *Give the rectangular form of the complex number represented in each graph.*

11. 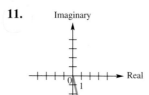   **12.**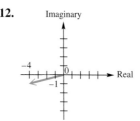

Find the sum of each pair of complex numbers. In Exercises 13–16, graph both complex numbers and their resultant. See Example 1.

13. $4 - 3i, -1 + 2i$ **14.** $2 + 3i, -4 - i$ **15.** $5 - 6i, -5 + 3i$

16. $7 - 3i, -4 + 3i$ **17.** $-3, 3i$ **18.** $6, -2i$

19. $2 + 6i, -2i$ **20.** $4 - 2i, 5$ **21.** $7 + 6i, 3i$

22. $-5 - 8i, -1$ **23.** $\dfrac{1}{2} + \dfrac{2}{3}i, \dfrac{2}{3} + \dfrac{1}{2}i$ **24.** $-\dfrac{1}{5} + \dfrac{2}{7}i, \dfrac{3}{7} - \dfrac{3}{4}i$

Write each complex number in rectangular form. See Example 2.

25. $2(\cos 45° + i \sin 45°)$ **26.** $4(\cos 60° + i \sin 60°)$

27. $10(\cos 90° + i \sin 90°)$ **28.** $8(\cos 270° + i \sin 270°)$

29. $4(\cos 240° + i \sin 240°)$ **30.** $2(\cos 330° + i \sin 330°)$

31. $3 \operatorname{cis} 150°$ **32.** $2 \operatorname{cis} 30°$

33. $5 \operatorname{cis} 300°$ **34.** $6 \operatorname{cis} 135°$

35. $\sqrt{2} \operatorname{cis} 225°$ **36.** $\sqrt{3} \operatorname{cis} 315°$

37. $4(\cos(-30°) + i \sin(-30°))$ **38.** $\sqrt{2}(\cos(-60°) + i \sin(-60°))$

Write each complex number in trigonometric form $r(\cos \theta + i \sin \theta)$, with θ in the interval $[0°, 360°)$. See Example 3.

39. $-3 - 3i\sqrt{3}$ **40.** $1 + i\sqrt{3}$ **41.** $\sqrt{3} - i$

42. $4\sqrt{3} + 4i$ **43.** $-5 - 5i$ **44.** $-2 + 2i$

45. $2 + 2i$ **46.** $4 + 4i$ **47.** $5i$

48. $-2i$ **49.** -4 **50.** 7

Perform each conversion, using a calculator to approximate answers as necessary. See Example 4.

	Rectangular Form	**Trigonometric Form**
51.	$2 + 3i$	_____
52.	_____	$\cos 35° + i \sin 35°$
53.	_____	$3(\cos 250° + i \sin 250°)$
54.	$-4 + i$	_____
55.	$12i$	_____
56.	_____	$3 \operatorname{cis} 180°$
57.	$3 + 5i$	_____
58.	_____	$\operatorname{cis} 110.5°$

Concept Check The complex number z, where z = x + yi, can be graphed in the plane as (x, y). Describe the graphs of all complex numbers z satisfying the conditions in Exercises 59–62.

59. The absolute value of z is 1.

60. The real and imaginary parts of z are equal.

61. The real part of z is 1.

62. The imaginary part of z is 1.

Julia Set Refer to Example 5 to solve Exercises 63 and 64.

63. Is $z = -.2i$ in the Julia set?

64. The graph of the Julia set in Figure 11 appears to be symmetric with respect to both the x-axis and y-axis. Complete the following to show that this is true.

  (a) Show that complex conjugates have the same absolute value.
  (b) Compute $z_1^2 - 1$ and $z_2^2 - 1$, where $z_1 = a + bi$ and $z_2 = a - bi$.
  (c) Discuss why if (a, b) is in the Julia set then so is $(a, -b)$.
  (d) Conclude that the graph of the Julia set must be symmetric with respect to the x-axis.
  (e) Using a similar argument, show that the Julia set must also be symmetric with respect to the y-axis.

In Exercises 65 and 66, suppose z = r(cos θ + i sin θ).

65. Use vectors to show that the conjugate of z is

$$r[\cos(360° - \theta) + i\sin(360° - \theta)], \quad \text{or} \quad r(\cos\theta - i\sin\theta).$$

66. Use vectors to show that $-z = r[\cos(\theta + \pi) + i\sin(\theta + \pi)]$.

Concept Check In Exercises 67–69, identify the geometric condition (A, B, or C) that implies the situation.

  A. *The corresponding vectors have opposite directions.*
  B. *The terminal points of the vectors corresponding to a + bi and c + di lie on a horizontal line.*
  C. *The corresponding vectors have the same direction.*

67. The difference between two nonreal complex numbers $a + bi$ and $c + di$ is a real number.

68. The absolute value of the sum of two complex numbers $a + bi$ and $c + di$ is equal to the sum of their absolute values.

69. The absolute value of the difference of two complex numbers $a + bi$ and $c + di$ is equal to the sum of their absolute values.

70. Show that z and iz have the same absolute value. How are the graphs of these two numbers related?

8.3 The Product and Quotient Theorems

Products of Complex Numbers in Trigonometric Form ▪ Quotients of Complex Numbers in Trigonometric Form

Products of Complex Numbers in Trigonometric Form Using the FOIL method to multiply complex numbers in rectangular form, we find the product of $1 + i\sqrt{3}$ and $-2\sqrt{3} + 2i$ as follows.

$$\left(1 + i\sqrt{3}\right)\left(-2\sqrt{3} + 2i\right) = -2\sqrt{3} + 2i - 2i(3) + 2i^2\sqrt{3} \quad \text{(Section 8.1)}$$

$$= -2\sqrt{3} + 2i - 6i - 2\sqrt{3} \qquad i^2 = -1$$

$$= -4\sqrt{3} - 4i \qquad \text{Combine terms.}$$

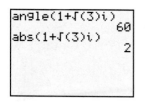

With the TI–83/84 Plus calculator in complex and degree modes, the MATH menu can be used to find the angle and the magnitude (absolute value) of the vector that corresponds to a given complex number.

We can also find this same product by first converting the complex numbers $1 + i\sqrt{3}$ and $-2\sqrt{3} + 2i$ to trigonometric form. Using the method explained in the preceding section,

$$1 + i\sqrt{3} = 2(\cos 60° + i \sin 60°) \qquad \text{(Section 8.2)}$$

and $\qquad -2\sqrt{3} + 2i = 4(\cos 150° + i \sin 150°).$

If we multiply the trigonometric forms, and if we use identities for the cosine and the sine of the sum of two angles, then the result is

$$[2(\cos 60° + i \sin 60°)][4(\cos 150° + i \sin 150°)]$$
$$= 2 \cdot 4(\cos 60° \cdot \cos 150° + i \sin 60° \cdot \cos 150°$$
$$+ \; i \cos 60° \cdot \sin 150° + i^2 \sin 60° \cdot \sin 150°) \quad \text{(Section 5.3)}$$
$$= 8[(\cos 60° \cdot \cos 150° - \sin 60° \cdot \sin 150°)$$
$$+ \; i(\sin 60° \cdot \cos 150° + \cos 60° \cdot \sin 150°)]$$
$$= 8[\cos(60° + 150°) + i \sin(60° + 150°)]$$
$$= 8(\cos 210° + i \sin 210°).$$

The absolute value of the product, 8, is equal to the product of the absolute values of the factors, $2 \cdot 4$, and the argument of the product, 210°, is equal to the sum of the arguments of the factors, 60° + 150°.

As we would expect, the product obtained when multiplying by the first method is the rectangular form of the product obtained when multiplying by the second method.

$$8(\cos 210° + i \sin 210°) = 8\left(-\frac{\sqrt{3}}{2} - \frac{1}{2}i\right)$$
$$= -4\sqrt{3} - 4i$$

We can generalize this work in the following *product theorem.*

PRODUCT THEOREM

If $r_1(\cos \theta_1 + i \sin \theta_1)$ and $r_2(\cos \theta_2 + i \sin \theta_2)$ are any two complex numbers, then

$$[r_1(\cos \theta_1 + i \sin \theta_1)] \cdot [r_2(\cos \theta_2 + i \sin \theta_2)]$$
$$= r_1 r_2[\cos(\theta_1 + \theta_2) + i \sin(\theta_1 + \theta_2)].$$

In compact form, this is written

$$(r_1 \text{ cis } \theta_1)(r_2 \text{ cis } \theta_2) = r_1 r_2 \text{ cis}(\theta_1 + \theta_2).$$

That is, to multiply complex numbers in trigonometric form, multiply their absolute values and add their arguments.

▶ **EXAMPLE 1** USING THE PRODUCT THEOREM

Find the product of $3(\cos 45° + i \sin 45°)$ and $2(\cos 135° + i \sin 135°)$. Write the result in rectangular form.

Solution

$$[3(\cos 45° + i \sin 45°)][2(\cos 135° + i \sin 135°)]$$

$$= 3 \cdot 2[\cos(45° + 135°) + i \sin(45° + 135°)] \quad \text{Product theorem}$$

$$= 6(\cos 180° + i \sin 180°) \quad \text{Multiply and add.}$$

$$= 6(-1 + i \cdot 0)$$

$$= 6(-1) = -6 \quad \text{Rectangular form}$$

NOW TRY EXERCISE 7. ◀

Quotients of Complex Numbers in Trigonometric Form The

rectangular form of the quotient of the complex numbers $1 + i\sqrt{3}$ and $-2\sqrt{3} + 2i$ is

$$\frac{1 + i\sqrt{3}}{-2\sqrt{3} + 2i} = \frac{\left(1 + i\sqrt{3}\right)\left(-2\sqrt{3} - 2i\right)}{\left(-2\sqrt{3} + 2i\right)\left(-2\sqrt{3} - 2i\right)} \quad \begin{array}{l} \text{Multiply by the conjugate of the} \\ \text{denominator. (Section 8.1)} \end{array}$$

$$= \frac{-2\sqrt{3} - 2i - 6i - 2i^2\sqrt{3}}{12 - 4i^2} \quad \begin{array}{l} \text{FOIL;} \\ (x + y)(x - y) = x^2 - y^2 \end{array}$$

$$= \frac{-8i}{16} = -\frac{1}{2}i. \quad \text{Simplify.}$$

Writing $1 + i\sqrt{3}$, $-2\sqrt{3} + 2i$, and $-\frac{1}{2}i$ in trigonometric form gives

$$1 + i\sqrt{3} = 2(\cos 60° + i \sin 60°),$$

$$-2\sqrt{3} + 2i = 4(\cos 150° + i \sin 150°),$$

and

$$-\frac{1}{2}i = \frac{1}{2}[\cos(-90°) + i \sin(-90°)].$$

The absolute value of the quotient, $\frac{1}{2}$, is the quotient of the two absolute values, $\frac{2}{4} = \frac{1}{2}$. The argument of the quotient, $-90°$, is the difference of the two arguments, $60° - 150° = -90°$. Generalizing leads to the *quotient theorem*.

QUOTIENT THEOREM

If $r_1(\cos \theta_1 + i \sin \theta_1)$ and $r_2(\cos \theta_2 + i \sin \theta_2)$ are any two complex numbers, where $r_2(\cos \theta_2 + i \sin \theta_2) \neq 0$, then

$$\frac{r_1(\cos \theta_1 + i \sin \theta_1)}{r_2(\cos \theta_2 + i \sin \theta_2)} = \frac{r_1}{r_2}[\cos(\theta_1 - \theta_2) + i \sin(\theta_1 - \theta_2)].$$

In compact form, this is written

$$\frac{r_1 \text{ cis } \theta_1}{r_2 \text{ cis } \theta_2} = \frac{r_1}{r_2} \text{cis}(\theta_1 - \theta_2).$$

That is, to divide complex numbers in trigonometric form, divide their absolute values and subtract their arguments.

▶ **EXAMPLE 2**   USING THE QUOTIENT THEOREM

Find the quotient $\dfrac{10 \text{ cis}(-60°)}{5 \text{ cis } 150°}$. Write the result in rectangular form.

Solution

$$\dfrac{10 \text{ cis}(-60°)}{5 \text{ cis } 150°} = \dfrac{10}{5} \text{ cis}(-60° - 150°) \qquad \text{Quotient theorem}$$

$$= 2 \text{ cis}(-210°) \qquad \text{Divide and subtract.}$$

$$= 2[\cos(-210°) + i \sin(-210°)] \quad \text{Rewrite.}$$

$$= 2\left[-\dfrac{\sqrt{3}}{2} + i\left(\dfrac{1}{2}\right)\right] \qquad \begin{array}{l} \cos(-210°) = -\frac{\sqrt{3}}{2}; \\ \sin(-210°) = \frac{1}{2} \text{ (Section 2.2)} \end{array}$$

$$= -\sqrt{3} + i \qquad \text{Rectangular form}$$

NOW TRY EXERCISE 17. ◀

▶ **EXAMPLE 3**   USING THE PRODUCT AND QUOTIENT THEOREMS WITH A CALCULATOR

Use a calculator to find the following. Write the results in rectangular form.

(a) $(9.3 \text{ cis } 125.2°)(2.7 \text{ cis } 49.8°)$

(b) $\dfrac{10.42\left(\cos \frac{3\pi}{4} + i \sin \frac{3\pi}{4}\right)}{5.21\left(\cos \frac{\pi}{5} + i \sin \frac{\pi}{5}\right)}$

Solution

(a) $(9.3 \text{ cis } 125.2°)(2.7 \text{ cis } 49.8°) = 9.3(2.7) \text{ cis}(125.2° + 49.8°)$

Product theorem

$$= 25.11 \text{ cis } 175°$$

$$= 25.11(\cos 175° + i \sin 175°)$$

$$\approx 25.11[-.99619470 + i(.08715574)]$$

$$\approx -25.0144 + 2.1885i$$

(b) $\dfrac{10.42\left(\cos \frac{3\pi}{4} + i \sin \frac{3\pi}{4}\right)}{5.21\left(\cos \frac{\pi}{5} + i \sin \frac{\pi}{5}\right)} = \dfrac{10.42}{5.21}\left[\cos\left(\dfrac{3\pi}{4} - \dfrac{\pi}{5}\right) + i \sin\left(\dfrac{3\pi}{4} - \dfrac{\pi}{5}\right)\right]$

Quotient theorem

$$= 2\left(\cos \dfrac{11\pi}{20} + i \sin \dfrac{11\pi}{20}\right)$$

$$\approx -.3129 + 1.9754i$$

NOW TRY EXERCISES 27 AND 29. ◀

8.3 Exercises

Concept Check *Fill in the blanks with the correct responses.*

1. When multiplying two complex numbers in trigonometric form, we _____ their absolute values and _____ their arguments.

2. When dividing two complex numbers in trigonometric form, we _____ their absolute values and _____ their arguments.

Find each product and write it in rectangular form. See Example 1.

3. $[3(\cos 60° + i \sin 60°)][2(\cos 90° + i \sin 90°)]$

4. $[4(\cos 30° + i \sin 30°)][5(\cos 120° + i \sin 120°)]$

5. $[4(\cos 60° + i \sin 60°)][6(\cos 330° + i \sin 330°)]$

6. $[8(\cos 300° + i \sin 300°)][5(\cos 120° + i \sin 120°)]$

7. $[2(\cos 45° + i \sin 45°)][2(\cos 225° + i \sin 225°)]$

8. $[8(\cos 210° + i \sin 210°)][2(\cos 330° + i \sin 330°)]$

9. $\left(\sqrt{3} \text{ cis } 45°\right)\left(\sqrt{3} \text{ cis } 225°\right)$

10. $(6 \text{ cis } 120°)[5 \text{ cis}(-30°)]$

11. $(5 \text{ cis } 90°)(3 \text{ cis } 45°)$

12. $\left(\sqrt{2} \text{ cis } 300°\right)\left(\sqrt{2} \text{ cis } 270°\right)$

Find each quotient and write it in rectangular form. In Exercises 19–24, first convert the numerator and the denominator to trigonometric form. See Example 2.

13. $\dfrac{4(\cos 120° + i \sin 120°)}{2(\cos 150° + i \sin 150°)}$

14. $\dfrac{24(\cos 150° + i \sin 150°)}{2(\cos 30° + i \sin 30°)}$

15. $\dfrac{10(\cos 230° + i \sin 230°)}{5(\cos 50° + i \sin 50°)}$

16. $\dfrac{12(\cos 293° + i \sin 293°)}{6(\cos 23° + i \sin 23°)}$

17. $\dfrac{3 \text{ cis } 305°}{9 \text{ cis } 65°}$

18. $\dfrac{16 \text{ cis } 310°}{8 \text{ cis } 70°}$

19. $\dfrac{8}{\sqrt{3} + i}$

20. $\dfrac{2i}{-1 - i\sqrt{3}}$

21. $\dfrac{-i}{1 + i}$

22. $\dfrac{1}{2 - 2i}$

23. $\dfrac{2\sqrt{6} - 2i\sqrt{2}}{\sqrt{2} - i\sqrt{6}}$

24. $\dfrac{4 + 4i}{2 - 2i}$

Use a calculator to perform the indicated operations, expressing real and imaginary parts to four decimal places. Give answers in rectangular form. See Example 3.

25. $[2.5(\cos 35° + i \sin 35°)][3.0(\cos 50° + i \sin 50°)]$

26. $[4.6(\cos 12° + i \sin 12°)][2.0(\cos 13° + i \sin 13°)]$

27. $(12 \text{ cis } 18.5°)(3 \text{ cis } 12.5°)$

28. $(4 \text{ cis } 19.25°)(7 \text{ cis } 41.75°)$

29. $\dfrac{45(\cos 127° + i \sin 127°)}{22.5(\cos 43° + i \sin 43°)}$

30. $\dfrac{30(\cos 130° + i \sin 130°)}{10(\cos 21° + i \sin 21°)}$

31. $\left[2 \text{ cis } \dfrac{5\pi}{9}\right]^2$

32. $\left[24.3 \text{ cis } \dfrac{7\pi}{12}\right]^2$

RELATING CONCEPTS

For individual or collaborative investigation
(Exercises 33–39)

Consider the complex numbers

$$w = -1 + i \quad \text{and} \quad z = -1 - i.$$

Work Exercises 33–39 in order.

33. Multiply w and z using their rectangular forms and the FOIL method from **Section 8.1.** Leave the product in rectangular form.

34. Find the trigonometric forms of w and z.

35. Multiply w and z using their trigonometric forms and the method described in this section.

36. Use the result of Exercise 35 to find the rectangular form of wz. How does this compare to your result in Exercise 33?

37. Find the quotient $\frac{w}{z}$ using their rectangular forms and multiplying both the numerator and the denominator by the conjugate of the denominator. Leave the quotient in rectangular form.

38. Use the trigonometric forms of w and z, found in Exercise 34, to divide w by z using the method described in this section.

39. Use the result of Exercise 38 to find the rectangular form of $\frac{w}{z}$. How does this compare to your result in Exercise 37?

---

40. Notice that $(r \operatorname{cis} \theta)^2 = (r \operatorname{cis} \theta)(r \operatorname{cis} \theta) = r^2 \operatorname{cis}(\theta + \theta) = r^2 \operatorname{cis} 2\theta$. State in your own words how we can square a complex number in trigonometric form. (In the next section, we will develop this idea more fully.)

41. Without actually performing the operations, state why the following products are the same.

$$[2(\cos 45° + i \sin 45°)] \cdot [5(\cos 90° + i \sin 90°)]$$

and $\quad \{2[\cos(-315°) + i \sin(-315°)]\} \cdot \{5[\cos(-270°) + i \sin(-270°)]\}$

42. Show that $\frac{1}{z} = \frac{1}{r}(\cos \theta - i \sin \theta)$, where $z = r(\cos \theta + i \sin \theta)$.

(Modeling) Solve each problem.

43. *Electrical Current* The alternating current in an electric inductor is $I = \frac{E}{Z}$ amperes, where E is voltage and $Z = R + X_L i$ is impedance. If $E = 8(\cos 20° + i \sin 20°)$, $R = 6$, and $X_L = 3$, find the current. Give the answer in rectangular form, with real and imaginary parts to the nearest hundredth.

44. *Electrical Current* The current I in a circuit with voltage E, resistance R, capacitive reactance X_c, and inductive reactance X_L is

$$I = \frac{E}{R + (X_L - X_c)i}.$$

Find I if $E = 12(\cos 25° + i \sin 25°)$, $R = 3$, $X_L = 4$, and $X_c = 6$. Give the answer in rectangular form, with real and imaginary parts to the nearest tenth.

(Modeling) Impedance *In the parallel electrical circuit shown in the figure, the impedance Z can be calculated using the equation*

$$Z = \cfrac{1}{\cfrac{1}{Z_1} + \cfrac{1}{Z_2}},$$

where Z_1 and Z_2 are the impedances for the branches of the circuit.

45. If $Z_1 = 50 + 25i$ and $Z_2 = 60 + 20i$, calculate Z.

46. Determine the angle θ for the value of Z found in Exercise 45.

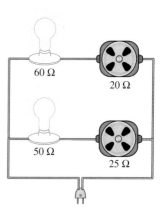

60 Ω

20 Ω

50 Ω

25 Ω

8.4 De Moivre's Theorem; Powers and Roots of Complex Numbers

Powers of Complex Numbers (De Moivre's Theorem) ▪ **Roots of Complex Numbers**

Powers of Complex Numbers (De Moivre's Theorem) In the previous section, we studied the product theorem for complex numbers in trigonometric form. Because raising a number to a positive integer power is a repeated application of the product rule, it would seem likely that a theorem for finding powers of complex numbers exists. For example,

$$[r(\cos \theta + i \sin \theta)]^2 = [r(\cos \theta + i \sin \theta)][r(\cos \theta + i \sin \theta)]$$
$$= r \cdot r[\cos(\theta + \theta) + i \sin(\theta + \theta)]$$
$$= r^2(\cos 2\theta + i \sin 2\theta).$$

In the same way,

$$[r(\cos \theta + i \sin \theta)]^3 = r^3(\cos 3\theta + i \sin 3\theta).$$

These results suggest the following theorem for positive integer values of *n*. Although the theorem is stated and can be proved for all *n*, we use it only for positive integer values of *n* and their reciprocals.

DE MOIVRE'S THEOREM

If $r(\cos \theta + i \sin \theta)$ is a complex number, and if *n* is any real number, then

$$[r(\cos \theta + i \sin \theta)]^n = r^n(\cos n\theta + i \sin n\theta).$$

In compact form, this is written

$$[r \text{ cis } \theta]^n = r^n(\text{cis } n\theta).$$

Abraham De Moivre
(1667–1754)

This theorem is named after the French expatriate friend of Isaac Newton, Abraham De Moivre, although he never explicitly stated it.

▶ **EXAMPLE 1** FINDING A POWER OF A COMPLEX NUMBER

Find $\left(1 + i\sqrt{3}\right)^8$ and express the result in rectangular form.

Solution First write $1 + i\sqrt{3}$ in trigonometric form.

$$1 + i\sqrt{3} = 2(\cos 60° + i \sin 60°) \quad \text{(Section 8.2)}$$

Now, apply De Moivre's theorem.

$$
\begin{aligned}
\left(1 + i\sqrt{3}\right)^8 &= [2(\cos 60° + i \sin 60°)]^8 \\
&= 2^8[\cos(8 \cdot 60°) + i \sin(8 \cdot 60°)] \quad \text{De Moivre's theorem} \\
&= 256(\cos 480° + i \sin 480°) \\
&= 256(\cos 120° + i \sin 120°) \quad \text{480° and 120° are coterminal.} \\
&\qquad\qquad\qquad\qquad\qquad\qquad \text{(Section 1.1)} \\
&= 256\left(-\frac{1}{2} + i\frac{\sqrt{3}}{2}\right) \quad \cos 120° = -\tfrac{1}{2}; \sin 120° = \tfrac{\sqrt{3}}{2} \\
&\qquad\qquad\qquad\qquad\qquad\qquad \text{(Section 2.2)} \\
&= -128 + 128i\sqrt{3} \quad \text{Rectangular form}
\end{aligned}
$$

NOW TRY EXERCISE 7. ◀

Roots of Complex Numbers Every nonzero complex number has exactly n distinct complex nth roots. De Moivre's theorem can be extended to find all nth roots of a complex number.

nth ROOT

For a positive integer n, the complex number $a + bi$ is an **nth root** of the complex number $x + yi$ if

$$(a + bi)^n = x + yi.$$

To find the three complex cube roots of $8(\cos 135° + i \sin 135°)$, for example, look for a complex number, say $r(\cos \alpha + i \sin \alpha)$, that will satisfy

$$[r(\cos \alpha + i \sin \alpha)]^3 = 8(\cos 135° + i \sin 135°).$$

By De Moivre's theorem, this equation becomes

$$r^3(\cos 3\alpha + i \sin 3\alpha) = 8(\cos 135° + i \sin 135°).$$

Set $r^3 = 8$ and $\cos 3\alpha + i \sin 3\alpha = \cos 135° + i \sin 135°$, to satisfy this equation. The first of these conditions implies that $r = 2$, and the second implies that

$$\cos 3\alpha = \cos 135° \quad \text{and} \quad \sin 3\alpha = \sin 135°.$$

For these equations to be satisfied, 3α must represent an angle that is coterminal with $135°$. Therefore, we must have

$$3\alpha = 135° + 360° \cdot k, \quad k \text{ any integer}$$

or

$$\alpha = \frac{135° + 360° \cdot k}{3}, \quad k \text{ any integer.}$$

Now, let k take on the integer values 0, 1, and 2.

$$\text{If } k = 0, \text{ then} \qquad \alpha = \frac{135° + 0°}{3} = 45°.$$

$$\text{If } k = 1, \text{ then} \qquad \alpha = \frac{135° + 360°}{3} = \frac{495°}{3} = 165°.$$

$$\text{If } k = 2, \text{ then} \qquad \alpha = \frac{135° + 720°}{3} = \frac{855°}{3} = 285°.$$

In the same way, $\alpha = 405°$ when $k = 3$. But note that $405° = 45° + 360°$ so $\sin 405° = \sin 45°$ and $\cos 405° = \cos 45°$. Similarly, if $k = 4$, $\alpha = 525°$, which has the same sine and cosine values as $165°$. To continue with larger values of k would just be repeating solutions already found. Therefore, all of the cube roots (three of them) can be found by letting $k = 0$, 1, or 2.

$$\text{When } k = 0, \text{ the root is} \qquad 2(\cos 45° + i \sin 45°).$$
$$\text{When } k = 1, \text{ the root is} \qquad 2(\cos 165° + i \sin 165°).$$
$$\text{When } k = 2, \text{ the root is} \qquad 2(\cos 285° + i \sin 285°).$$

In summary, we see that $2(\cos 45° + i \sin 45°)$, $2(\cos 165° + i \sin 165°)$, and $2(\cos 285° + i \sin 285°)$ are the three cube roots of $8(\cos 135° + i \sin 135°)$.

Generalizing our results, we state the following theorem.

---

*n*th ROOT THEOREM

If n is any positive integer, r is a positive real number, and θ is in degrees, then the nonzero complex number $r(\cos \theta + i \sin \theta)$ has exactly n distinct nth roots, given by

$$\sqrt[n]{r}(\cos \alpha + i \sin \alpha) \qquad \text{or} \qquad \sqrt[n]{r} \text{ cis } \alpha,$$

where

$$\alpha = \frac{\theta + 360° \cdot k}{n}, \quad \text{or} \quad \alpha = \frac{\theta}{n} + \frac{360° \cdot k}{n}, \qquad k = 0, 1, 2, \ldots, n - 1.$$

▶ **Note** In the statement of the nth root theorem, if θ is in radians, then

$$\alpha = \frac{\theta + 2\pi k}{n}, \qquad \text{or} \qquad \alpha = \frac{\theta}{n} + \frac{2\pi k}{n}.$$

▶ **EXAMPLE 2** FINDING COMPLEX ROOTS

Find the two square roots of $4i$. Write the roots in rectangular form.

Solution First write $4i$ in trigonometric form as

$$4i = 4\left(\cos\frac{\pi}{2} + i\sin\frac{\pi}{2}\right).$$

Here $r = 4$ and $\theta = \frac{\pi}{2}$. The square roots have absolute value $\sqrt{4} = 2$ and arguments as follows.

$$\alpha = \frac{\frac{\pi}{2}}{2} + \frac{2\pi k}{2} = \frac{\pi}{4} + \pi k \quad \longleftarrow \boxed{\begin{array}{c}\text{Be careful}\\\text{simplifying here.}\end{array}}$$

Since there are two square roots, let $k = 0$ and 1.

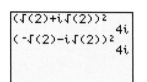

If $k = 0$, then $\qquad \alpha = \dfrac{\pi}{4} + \pi \cdot 0 = \dfrac{\pi}{4}$.

If $k = 1$, then $\qquad \alpha = \dfrac{\pi}{4} + \pi \cdot 1 = \dfrac{5\pi}{4}$.

This screen confirms the result of Example 2.

Using these values for α, the square roots are $2 \operatorname{cis}\frac{\pi}{4}$ and $2 \operatorname{cis}\frac{5\pi}{4}$, which can be written in rectangular form as $\sqrt{2} + i\sqrt{2}$ and $-\sqrt{2} - i\sqrt{2}$.

NOW TRY EXERCISE 17(a). ◀

▶ **EXAMPLE 3** FINDING COMPLEX ROOTS

Find all fourth roots of $-8 + 8i\sqrt{3}$. Write the roots in rectangular form.

Solution $\qquad -8 + 8i\sqrt{3} = 16 \operatorname{cis} 120°$ Write in trigonometric form.

Here $r = 16$ and $\theta = 120°$. The fourth roots of this number have absolute value $\sqrt[4]{16} = 2$ and arguments as follows.

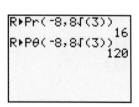

$$\alpha = \frac{120°}{4} + \frac{360° \cdot k}{4} = 30° + 90° \cdot k$$

Degree mode

This screen shows how a calculator finds r and θ for the number in Example 3.

Since there are four fourth roots, let $k = 0, 1, 2,$ and 3.

If $k = 0$, then $\qquad \alpha = 30° + 90° \cdot 0 = 30°$.

If $k = 1$, then $\qquad \alpha = 30° + 90° \cdot 1 = 120°$.

If $k = 2$, then $\qquad \alpha = 30° + 90° \cdot 2 = 210°$.

If $k = 3$, then $\qquad \alpha = 30° + 90° \cdot 3 = 300°$.

Using these angles, the fourth roots are

$$2 \operatorname{cis} 30°, \qquad 2 \operatorname{cis} 120°, \qquad 2 \operatorname{cis} 210°, \qquad \text{and} \qquad 2 \operatorname{cis} 300°.$$

These four roots can be written in rectangular form as

$$\sqrt{3} + i, \qquad -1 + i\sqrt{3}, \qquad -\sqrt{3} - i, \qquad \text{and} \qquad 1 - i\sqrt{3}.$$

The graphs of these roots lie on a circle with center at the origin and radius 2. See Figure 12. The roots are equally spaced about the circle, 90° apart.

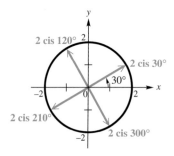

Figure 12

NOW TRY EXERCISES 23(a) AND (b). ◀

▶ **EXAMPLE 4**   **SOLVING AN EQUATION BY FINDING COMPLEX ROOTS**

Find all complex number solutions of $x^5 - 1 = 0$. Graph them as vectors in the complex plane.

Solution   Write the equation as

$$x^5 - 1 = 0, \quad \text{or} \quad x^5 = 1.$$

While there is only one real number solution, 1, there are five complex number solutions. To find these solutions, first write 1 in trigonometric form as

$$1 = 1 + 0i = 1(\cos 0° + i \sin 0°).$$

The absolute value of the fifth roots is $\sqrt[5]{1} = 1$, and the arguments are given by

$$0° + 72° \cdot k, \quad k = 0, 1, 2, 3, \text{ and } 4.$$

By using these arguments, the fifth roots are

$$1(\cos 0° + i \sin 0°), \qquad k = 0$$
$$1(\cos 72° + i \sin 72°), \qquad k = 1$$
$$1(\cos 144° + i \sin 144°), \quad k = 2$$
$$1(\cos 216° + i \sin 216°), \quad k = 3$$

and $\qquad\qquad 1(\cos 288° + i \sin 288°). \quad k = 4$

The solution set of the equation can be written as

$$\{\text{cis } 0°, \text{cis } 72°, \text{cis } 144°, \text{cis } 216°, \text{cis } 288°\}.$$

The first of these roots equals 1; the others cannot easily be expressed in rectangular form but can be approximated with a calculator.

The tips of the arrows representing the five fifth roots all lie on a unit circle and are equally spaced around it every 72°, as shown in Figure 13.

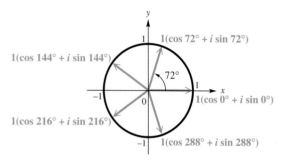

Figure 13

NOW TRY EXERCISE 35. ◀

8.4 Exercises

Find each power. Write each answer in rectangular form. See Example 1.

1. $[3(\cos 30° + i \sin 30°)]^3$

2. $[2(\cos 135° + i \sin 135°)]^4$

3. $(\cos 45° + i \sin 45°)^8$

4. $[2(\cos 120° + i \sin 120°)]^3$

5. $[3 \text{ cis } 100°]^3$

6. $[3 \text{ cis } 40°]^3$

7. $(\sqrt{3} + i)^5$

8. $(2 - 2i\sqrt{3})^4$

9. $(2\sqrt{2} - 2i\sqrt{2})^6$

10. $\left(\dfrac{\sqrt{2}}{2} - \dfrac{\sqrt{2}}{2}i\right)^8$

11. $(-2 - 2i)^5$

12. $(-1 + i)^7$

In Exercises 13–24, (a) find all cube roots of each complex number. Leave answers in trigonometric form. (b) Graph each cube root as a vector in the complex plane. See Examples 2 and 3.

13. $\cos 0° + i \sin 0°$

14. $\cos 90° + i \sin 90°$

15. $8 \text{ cis } 60°$

16. $27 \text{ cis } 300°$

17. $-8i$

18. $27i$

19. -64

20. 27

21. $1 + i\sqrt{3}$

22. $2 - 2i\sqrt{3}$

23. $-2\sqrt{3} + 2i$

24. $\sqrt{3} - i$

Find and graph all specified roots of 1.

25. second (square)

26. fourth

27. sixth

Find and graph all specified roots of i.

28. second (square)

29. third (cube)

30. fourth

Find all complex number solutions of each equation. Leave answers in trigonometric form. See Example 4.

31. $x^3 - 1 = 0$

32. $x^3 + 1 = 0$

33. $x^3 + i = 0$

34. $x^4 + i = 0$

35. $x^3 - 8 = 0$

36. $x^3 + 27 = 0$

37. $x^4 + 1 = 0$

38. $x^4 + 16 = 0$

39. $x^4 - i = 0$

40. $x^5 - i = 0$

41. $x^3 - (4 + 4i\sqrt{3}) = 0$

42. $x^4 - (8 + 8i\sqrt{3}) = 0$

43. Solve the equation $x^3 - 1 = 0$ by factoring the left side as the difference of two cubes and setting each factor equal to 0. Apply the quadratic formula as needed. Then compare your solutions to those of Exercise 31.

44. Solve the equation $x^3 + 27 = 0$ by factoring the left side as the sum of two cubes and setting each factor equal to 0. Apply the quadratic formula as needed. Then compare your solutions to those of Exercise 36.

RELATING CONCEPTS

For individual or collaborative investigation
(Exercises 45–48)

*In **Chapter 5** we derived identities, or formulas, for* $\cos 2\theta$ *and* $\sin 2\theta$. *These identities can also be derived using De Moivre's theorem.* **Work Exercises 45–48 in order, to see how this is done.**

45. De Moivre's theorem states that $(\cos \theta + i \sin \theta)^2 = $ _____.

46. Expand the left side of the equation in Exercise 45 as a binomial and collect terms to write the left side in the form $a + bi$.

47. Use the result of Exercise 46 to obtain the double-angle formula for cosine.

48. Repeat Exercise 47, but find the double-angle formula for sine.

Solve each problem.

49. *Mandelbrot Set* The fractal called the **Mandelbrot set** is shown in the figure. To determine if a complex number $z = a + bi$ is in this set, perform the following sequence of calculations. Repeatedly compute

$$z, \quad z^2 + z, \quad (z^2 + z)^2 + z,$$
$$[(z^2 + z)^2 + z]^2 + z, \ldots.$$

In a manner analogous to the Julia set, the complex number z does not belong to the Mandelbrot set if any of the resulting absolute values exceed 2. Otherwise z is in the set and the point (a, b) should be shaded in the graph. Determine whether or not the following numbers belong to the Mandelbrot set. (*Source:* Lauwerier, H., *Fractals,* Princeton University Press, 1991.)

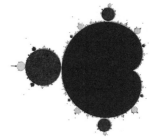

Source: Figure from Crownover, R., *Introduction to Fractals and Chaos.* Copyright © 1995. Boston: Jones and Bartlett Publishers. Reprinted with permission.

(**a**) $z = 0 + 0i$ (**b**) $z = 1 - 1i$ (**c**) $z = -.5i$

50. *Basins of Attraction* The fractal shown in the figure is the solution to Cayley's problem of determining the basins of attraction for the cube roots of unity. The three cube roots of unity are

$$w_1 = 1, \quad w_2 = -\frac{1}{2} + \frac{\sqrt{3}}{2}i,$$

and $\quad w_3 = -\frac{1}{2} - \frac{\sqrt{3}}{2}i.$

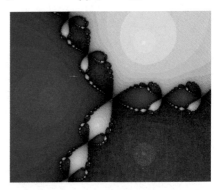

This fractal can be generated by repeatedly evaluating the function defined by

$$f(z) = \frac{2z^3 + 1}{3z^2},$$

where z is a complex number. One begins by picking $z_1 = a + bi$ and then successively computing $z_2 = f(z_1)$, $z_3 = f(z_2)$, $z_4 = f(z_3)$, ... If the resulting values of $f(z)$ approach w_1, color the pixel at (a, b) red. If it approaches w_2, color it blue, and if it approaches w_3, color it yellow. If this process continues for a large number of different z_1, the fractal in the figure will appear. Determine the appropriate color of the pixel for each value of z_1. (*Source:* Crownover, R., *Introduction to Fractals and Chaos,* Jones and Bartlett Publishers, 1995.)

(a) $z_1 = i$ **(b)** $z_1 = 2 + i$ **(c)** $z_1 = -1 - i$

51. The screens here illustrate how a pentagon can be graphed using a graphing calculator. Note that a pentagon has five sides, and the T-step is $\frac{360}{5} = 72$. The display at the bottom of the graph screen indicates that one fifth root of 1 is $1 + 0i = 1$. Use this technique to find all fifth roots of 1, and express the real and imaginary parts in decimal form.

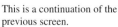
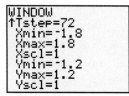
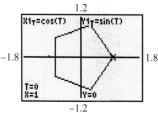

This is a continuation of the previous screen.

The calculator is in parametric, degree, and connected graph modes.

52. Use the method of Exercise 51 to find the first three of the ten 10th roots of 1.

53. One of the three cube roots of a complex number is $2 + 2i\sqrt{3}$. Determine the rectangular form of its other two cube roots.

Use a calculator to find all solutions of each equation in rectangular form.

54. $x^2 + 2 - i = 0$ **55.** $x^2 - 3 + 2i = 0$

56. $x^3 + 4 - 5i = 0$ **57.** $x^5 + 2 + 3i = 0$

58. *Concept Check* How many complex 64th roots does 1 have? How many are real? How many are not?

59. *Concept Check* *True* or *false*: Every real number must have two distinct real square roots.

60. *Concept Check* *True* or *false*: Some real numbers have three real cube roots.

61. Show that if z is an nth root of 1, then so is $\frac{1}{z}$.

62. Explain why a real number can have only one real cube root.

63. Explain why the n nth roots of 1 are equally spaced around the unit circle.

64. Refer to Figure 13. A regular pentagon can be created by joining the tips of the arrows. Explain how you can use this principle to create a regular octagon.

1. Multiply or divide as indicated. Simplify each answer.

   (a) $\sqrt{-24} \cdot \sqrt{-3}$ **(b)** $\dfrac{\sqrt{-8}}{\sqrt{72}}$

2. For the complex numbers

$$w = 3 + 5i \quad \text{and} \quad z = -4 + i,$$

   find each of the following in rectangular form.

   (a) $w + z$ (and give a geometric representation)

   (b) $w - z$ **(c)** wz **(d)** $\dfrac{w}{z}$

3. Express each of the following in rectangular form.

   (a) $(1 - i)^3$ **(b)** i^{33}

4. Solve $3x^2 - x + 4 = 0$ over the complex number system.

5. Write each complex number in trigonometric (polar) form, where $0° \le \theta < 360°$.

   (a) $-4i$ **(b)** $1 - i\sqrt{3}$ **(c)** $-3 - i$

6. Write each complex number in rectangular form.

   (a) $4(\cos 60° + i \sin 60°)$ **(b)** $5 \operatorname{cis} 130°$

   (c) $7(\cos 270° + i \sin 270°)$

7. For the complex numbers

$$w = 12(\cos 80° + i \sin 80°) \quad \text{and} \quad z = 3(\cos 50° + i \sin 50°),$$

   find each of the following in the form specified.

   (a) wz (trigonometric form)   **(b)** $\dfrac{w}{z}$ (rectangular form)

   (c) z^3 (rectangular form)

8. Find the four complex fourth roots of -16. Express them in both trigonometric and rectangular forms.

8.5 Polar Equations and Graphs

Polar Coordinate System ▪ **Graphs of Polar Equations** ▪ **Converting from Polar to Rectangular Equations** ▪ **Classifying Polar Equations**

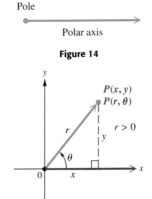

Pole

Polar axis

Figure 14

Figure 15

Polar Coordinate System   We have been using the rectangular coordinate system to graph equations. The **polar coordinate system** is based on a point, called the **pole,** and a ray, called the **polar axis.** The polar axis is usually drawn in the direction of the positive x-axis, as shown in Figure 14.

In Figure 15, the pole has been placed at the origin of a rectangular coordinate system so that the polar axis coincides with the positive x-axis. Point P has rectangular coordinates (x, y). Point P can also be located by giving the directed angle θ from the positive x-axis to ray OP and the *directed* distance r from the pole to point P. The ordered pair (r, θ) gives the **polar coordinates** of point P. If $r > 0$ then point P lies on the terminal side of θ, and if $r < 0$ then point P lies on the ray pointing in the opposite direction of the terminal side of θ, a distance $|r|$ from the pole. See Figure 16 on the next page.

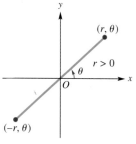

Figure 16

Using Figure 15 and trigonometry, we can establish the following relationships between rectangular and polar coordinates.

RECTANGULAR AND POLAR COORDINATES

If a point has rectangular coordinates (x, y) and polar coordinates (r, θ), then these coordinates are related as follows.

$$x = r \cos \theta \qquad\qquad y = r \sin \theta$$

$$r^2 = x^2 + y^2 \qquad \tan \theta = \frac{y}{x}, \quad \text{if } x \neq 0$$

▶ **EXAMPLE 1** PLOTTING POINTS WITH POLAR COORDINATES

Plot each point by hand in the polar coordinate system. Then, determine the rectangular coordinates of each point.

(a) $P(2, 30°)$ **(b)** $Q\left(-4, \dfrac{2\pi}{3}\right)$ **(c)** $R\left(5, -\dfrac{\pi}{4}\right)$

Solution

(a) In this case, $r = 2$ and $\theta = 30°$, so the point P is located 2 units from the origin in the positive direction on a ray making a 30° angle with the polar axis, as shown in Figure 17 below.

        Using the conversion equations, we find the rectangular coordinates as follows.

$$
\begin{array}{l|l}
x = r \cos \theta & y = r \sin \theta \\
x = 2 \cos 30° & y = 2 \sin 30° \\
x = 2\left(\dfrac{\sqrt{3}}{2}\right) = \sqrt{3} & y = 2\left(\dfrac{1}{2}\right) = 1 \quad \text{(Section 2.1)}
\end{array}
$$

The rectangular coordinates are $\left(\sqrt{3}, 1\right)$.

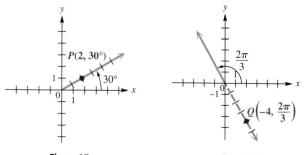

     Figure 17   **Figure 18**

(b) Since r is *negative*, Q is 4 units in the *opposite* direction from the pole on an extension of the $\frac{2\pi}{3}$ ray. See Figure 18. The rectangular coordinates are

$$x = -4 \cos \frac{2\pi}{3} = -4\left(-\frac{1}{2}\right) = 2$$

and $y = -4 \sin \dfrac{2\pi}{3} = -4\left(\dfrac{\sqrt{3}}{2}\right) = -2\sqrt{3}.$

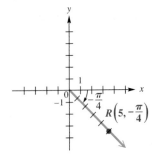

Figure 19

(c) Point R is shown in Figure 19. Since θ is negative, the angle is measured in the clockwise direction. Furthermore, we have

$$x = 5 \cos\left(-\frac{\pi}{4}\right) = \frac{5\sqrt{2}}{2} \quad \text{and} \quad y = 5 \sin\left(-\frac{\pi}{4}\right) = -\frac{5\sqrt{2}}{2}.$$

NOW TRY EXERCISES 3(a), (c), 5(a), (c), AND 11(a), (c). ◀

While a given point in the plane can have only one pair of rectangular coordinates, this same point can have an infinite number of pairs of polar coordinates. For example, $(2, 30°)$ locates the same point as

$$(2, 390°), \quad (2, -330°), \quad \text{and} \quad (-2, 210°).$$

▼ **LOOKING AHEAD TO CALCULUS**
Techniques studied in calculus associated with derivatives and integrals provide methods of finding slopes of tangent lines to polar curves, areas bounded by such curves, and lengths of their arcs.

▶ **EXAMPLE 2** **GIVING ALTERNATIVE FORMS FOR COORDINATES OF A POINT**

(a) Give three other pairs of polar coordinates for the point $P(3, 140°)$.

(b) Determine two pairs of polar coordinates for the point with rectangular coordinates $(-1, 1)$.

Solution

(a) Three pairs that could be used for the point are $(3, -220°)$, $(-3, 320°)$, and $(-3, -40°)$. See Figure 20.

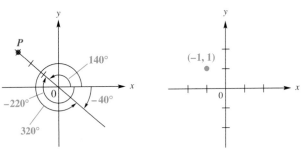

Figure 20   **Figure 21**

(b) As shown in Figure 21, the point $(-1, 1)$ lies in the second quadrant. Since $\tan \theta = \frac{1}{-1} = -1$, one possible value for θ is $135°$. Also,

$$r = \sqrt{x^2 + y^2} = \sqrt{(-1)^2 + 1^2} = \sqrt{2}.$$

Two pairs of polar coordinates are $\left(\sqrt{2}, 135°\right)$ and $\left(-\sqrt{2}, 315°\right)$.

NOW TRY EXERCISES 3(b), 5(b), 11(b), AND 13. ◀

Graphs of Polar Equations

Equations in x and y are called **rectangular (or Cartesian) equations.** An equation in which r and θ are the variables is a **polar equation.**

$$r = 3 \sin \theta, \qquad r = 2 + \cos \theta, \qquad r = \theta \quad \text{Polar equations}$$

While the rectangular forms of lines and circles are the ones most often encountered, they can also be defined in terms of polar coordinates. The polar equation of the line $ax + by = c$ can be derived as follows.

$$ax + by = c \quad \text{Rectangular equation}$$

$$a(r \cos \theta) + b(r \sin \theta) = c \quad \text{Convert from rectangular to polar coordinates.}$$

$$r(a \cos \theta + b \sin \theta) = c \quad \text{Factor out } r.$$

This is the polar equation of $ax + by = c$. $\longrightarrow \quad r = \dfrac{c}{a \cos \theta + b \sin \theta}$

For the circle $x^2 + y^2 = a^2$, we have

$$x^2 + y^2 = a^2$$

$$\left(\sqrt{x^2 + y^2}\right)^2 = a^2 \qquad \left(\sqrt{k}\right)^2 = k$$

$$r^2 = a^2 \qquad \sqrt{x^2 + y^2} = r$$

These are polar equations of $x^2 + y^2 = a^2$. $\longrightarrow \quad \boldsymbol{r = \pm a.} \quad r$ can be negative in polar coordinates.

▶ EXAMPLE 3 **EXAMINING POLAR AND RECTANGULAR EQUATIONS OF LINES AND CIRCLES**

For each rectangular equation, give the equivalent polar equation and sketch its graph.

(a) $y = x - 3$
(b) $x^2 + y^2 = 4$

Solution

(a) This is the equation of a line. Rewrite $y = x - 3$ as

$$x - y = 3. \quad \text{Standard form } ax + by = c$$

Using the general form for the polar equation of a line,

$$r = \dfrac{c}{a \cos \theta + b \sin \theta}$$

$$r = \dfrac{3}{1 \cdot \cos \theta + (-1) \cdot \sin \theta} \qquad a = 1, b = -1, c = 3$$

$$r = \dfrac{3}{\cos \theta - \sin \theta}.$$

A traditional graph is shown in Figure 22(a), and a calculator graph is shown in Figure 22(b).

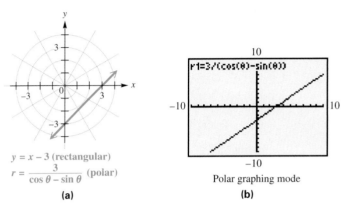

$y = x - 3$ (rectangular)

$r = \dfrac{3}{\cos\theta - \sin\theta}$ (polar)

(a)

Polar graphing mode

(b)

Figure 22

(b) The graph of $x^2 + y^2 = 4$ is a circle with center at the origin and radius 2.

$$x^2 + y^2 = 4 \qquad \text{(Appendix B)}$$

$$\left(\sqrt{x^2 + y^2}\right)^2 = (\pm 2)^2$$

$$\sqrt{x^2 + y^2} = 2 \qquad \text{or} \qquad \sqrt{x^2 + y^2} = -2$$

$$r = 2 \qquad \text{or} \qquad r = -2$$

(In polar coordinates, we may have $r < 0$.) The graphs of $r = 2$ and $r = -2$ coincide. See the graphs in Figure 23.

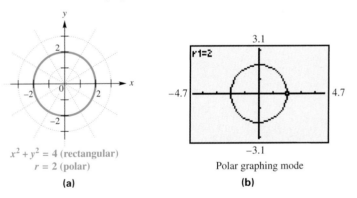

$x^2 + y^2 = 4$ (rectangular)

$r = 2$ (polar)

(a)

Polar graphing mode

(b)

Figure 23

NOW TRY EXERCISES 23 AND 25. ◀

To graph polar equations, evaluate r for various values of θ until a pattern appears, and then join the points with a smooth curve.

A graphing calculator can be used to graph an equation in the form $r = f(\theta)$. (See the screens in Example 3.) Refer to your owner's manual to see how your model handles polar graphs. As always, you must set an appropriate window and choose the correct angle mode (radians or degrees). You will need to decide on maximum and minimum values of θ. Keep in mind the periods of the functions, so a complete set of ordered pairs is generated. ■

▶ **EXAMPLE 4** GRAPHING A POLAR EQUATION (CARDIOID)

Graph $r = 1 + \cos \theta$.

Algebraic Solution

To graph this equation, find some ordered pairs (as in the table). Once the pattern of values of r becomes clear, it is not necessary to find more ordered pairs. That is why we stopped with the ordered pair $(1.9, 330°)$ in the table.

θ	$\cos \theta$	$r = 1 + \cos \theta$	θ	$\cos \theta$	$r = 1 + \cos \theta$
0°	1	2	135°	−.7	.3
30°	.9	1.9	150°	−.9	.1
45°	.7	1.7	180°	−1	0
60°	.5	1.5	270°	0	1
90°	0	1	315°	.7	1.7
120°	−.5	.5	330°	.9	1.9

Connect the points in order—from $(2, 0°)$ to $(1.9, 30°)$ to $(1.7, 45°)$ and so on. See Figure 24. This curve is called a **cardioid** because of its heart shape. The curve has been graphed on a **polar grid.**

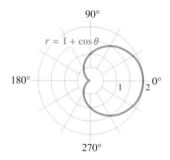

Figure 24

Graphing Calculator Solution

We choose degree mode and graph values of θ in the interval $[0°, 360°]$. The screens in Figure 25(a) show the choices needed to generate the graph in Figure 25(b).

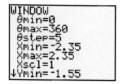

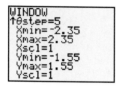

This is a continuation of the previous screen.

(a)

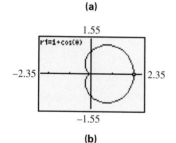

(b)

Figure 25

NOW TRY EXERCISE 41. ◀

▶ **EXAMPLE 5** GRAPHING A POLAR EQUATION (ROSE)

Graph $r = 3 \cos 2\theta$.

Solution Because of the argument 2θ, the graph requires a larger number of points than when the argument is just θ. A few ordered pairs are given in the table. You should complete the table similarly through the first 180° so that 2θ has values up to 360°.

θ	0°	15°	30°	45°	60°	75°	90°
2θ	0°	30°	60°	90°	120°	150°	180°
$\cos 2\theta$	1	.9	.5	0	−.5	−.9	−1
$r = 3 \cos 2\theta$	3	2.6	1.5	0	−1.5	−2.6	−3

Plotting the points from the table in order gives the graph, called a **four-leaved rose**. Notice in Figure 26(a) on the next page how the graph is developed with a continuous curve, beginning with the upper half of the right horizontal leaf and ending with the lower half of that leaf. As the graph is traced, the curve goes through the pole four times. This can actually be seen as a calculator graphs the curve. See Figure 26(b).

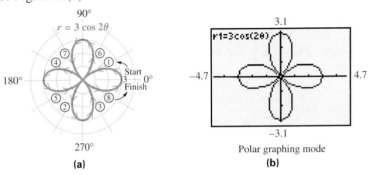

(a)   **(b)**

Figure 26

NOW TRY EXERCISE 45. ◀

The equation $r = 3 \cos 2\theta$ in Example 5 has a graph that belongs to a family of curves called **roses**. The graphs of $r = a \sin n\theta$ and $r = a \cos n\theta$ are roses, with n leaves if n is odd, and $2n$ leaves if n is even. The value of a determines the length of the leaves.

▶ **EXAMPLE 6** **GRAPHING A POLAR EQUATION (LEMNISCATE)**

Graph $r^2 = \cos 2\theta$.

Algebraic Solution

Complete a table of ordered pairs, and sketch the graph, as in Figure 27. The point $(-1, 0°)$, with r negative, may be plotted as $(1, 180°)$. Also, $(-.7, 30°)$ may be plotted as $(.7, 210°)$, and so on.

Values of θ for $45° < \theta < 135°$ are not included in the table because the corresponding values of $\cos 2\theta$ are negative (quadrants II and III) and so do not have real square roots. Values of θ larger than $180°$ give 2θ larger than $360°$ and would repeat the points already found. This curve is called a **lemniscate**.

θ	0°	30°	45°	135°	150°	180°
2θ	0°	60°	90°	270°	300°	360°
$\cos 2\theta$	1	.5	0	0	.5	1
$r = \pm\sqrt{\cos 2\theta}$	±1	±.7	0	0	±.7	±1

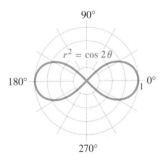

Figure 27

Graphing Calculator Solution

To graph $r^2 = \cos 2\theta$ with a graphing calculator, let

$$r_1 = \sqrt{\cos 2\theta}$$

and $$r_2 = -\sqrt{\cos 2\theta}.$$

See Figures 28(a) and (b).

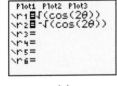

(a)

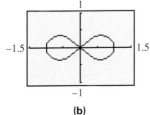

(b)

Figure 28

NOW TRY EXERCISE 47. ◀

▶ EXAMPLE 7   GRAPHING A POLAR EQUATION (SPIRAL OF ARCHIMEDES)

Graph $r = 2\theta$ (θ measured in radians).

Solution   Some ordered pairs are shown in the table. Since $r = 2\theta$, rather than a trigonometric function of θ, we must also consider negative values of θ. Radian measures have been rounded. The graph in Figure 29(a), is called a **spiral of Archimedes.** Figure 29(b) shows a calculator graph of this spiral.

θ (radians)	$r = 2\theta$	θ (radians)	$r = 2\theta$
$-\pi$	-6.3	$\frac{\pi}{3}$	2.1
$-\frac{\pi}{2}$	-3.1	$\frac{\pi}{2}$	3.1
$-\frac{\pi}{4}$	-1.6	π	6.3
0	0	$\frac{3\pi}{2}$	9.4
$\frac{\pi}{6}$	1	2π	12.6

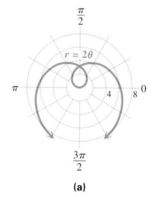

(a)

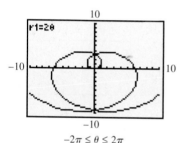

$-2\pi \le \theta \le 2\pi$
More of the spiral can be seen in this calculator graph.

(b)

Figure 29

NOW TRY EXERCISE 63. ◀

Converting from Polar to Rectangular Equations

▶ EXAMPLE 8   CONVERTING A POLAR EQUATION TO A RECTANGULAR EQUATION

Convert the equation $r = \frac{4}{1 + \sin \theta}$ to rectangular coordinates, and graph.

Solution

$$r = \frac{4}{1 + \sin \theta} \qquad \text{Polar equation}$$

$$r + r \sin \theta = 4 \qquad \text{Multiply by } 1 + \sin \theta.$$

$$\sqrt{x^2 + y^2} + y = 4 \qquad \text{Let } r = \sqrt{x^2 + y^2} \text{ and } r \sin \theta = y.$$

$$\sqrt{x^2 + y^2} = 4 - y \qquad \text{Subtract } y.$$

$$x^2 + y^2 = (4 - y)^2 \qquad \text{Square each side.}$$

$$x^2 + y^2 = 16 - 8y + y^2 \qquad \text{Expand the right side.}$$

$$x^2 = -8y + 16 \qquad \text{Subtract } y^2.$$

$$x^2 = -8(y - 2) \qquad \text{Rectangular equation}$$

The final equation represents a parabola and is graphed in Figure 30.

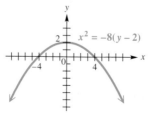

Figure 30

NOW TRY EXERCISE 55. ◀

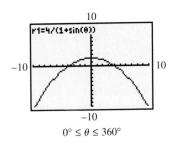

$0° \leq \theta \leq 360°$

Figure 31

The conversion in Example 8 is not necessary when using a graphing calculator. Figure 31 shows the graph of $r = \frac{4}{1 + \sin \theta}$, graphed directly with the calculator in polar mode. ■

Classifying Polar Equations   The following table summarizes some common polar graphs and forms of their equations. (In addition to circles, lemniscates, and roses, we include **limaçons.** Cardioids are a special case of limaçons, where $\left| \frac{a}{b} \right| = 1$.)

Circles and Lemniscates			
Circles		Lemniscates	
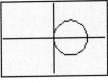	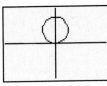	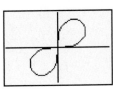	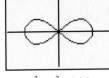
$r = a \cos \theta$	$r = a \sin \theta$	$r^2 = a^2 \sin 2\theta$	$r^2 = a^2 \cos 2\theta$

Limaçons			
$r = a \pm b \sin \theta$ or $r = a \pm b \cos \theta$			
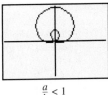	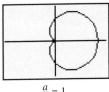	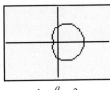	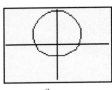
$\frac{a}{b} < 1$	$\frac{a}{b} = 1$	$1 < \frac{a}{b} < 2$	$\frac{a}{b} \geq 2$

Rose Curves			
$2n$ leaves if n is even, $n \geq 2$		n leaves if n is odd	
$n = 2$	$n = 4$	$n = 3$	$n = 5$
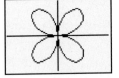	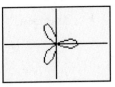		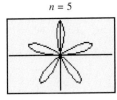
$r = a \sin n\theta$	$r = a \cos n\theta$	$r = a \cos n\theta$	$r = a \sin n\theta$

▶ **Note** Some other polar curves are the **cissoid, kappa curve, conchoid, trisectrix, cruciform, strophoid,** and **lituus.** Refer to older textbooks on analytic geometry to investigate them.

8.5 Exercises

1. *Concept Check* For each point given in polar coordinates, state the quadrant in which the point lies if it is graphed in a rectangular coordinate system.

   (a) $(5, 135°)$ **(b)** $(2, 60°)$ **(c)** $(6, -30°)$ **(d)** $(4.6, 213°)$

2. *Concept Check* For each point given in polar coordinates, state the axis on which the point lies if it is graphed in a rectangular coordinate system. Also state whether it is on the positive portion or the negative portion of the axis. (For example, $(5, 0°)$ lies on the positive x-axis.)

   (a) $(7, 360°)$ **(b)** $(4, 180°)$ **(c)** $(2, -90°)$ **(d)** $(8, 450°)$

For each pair of polar coordinates, (a) plot the point, (b) give two other pairs of polar coordinates for the point, and (c) give the rectangular coordinates for the point. See Examples 1 and 2.

3. $(1, 45°)$ 4. $(3, 120°)$ 5. $(-2, 135°)$ 6. $(-4, 30°)$

7. $(5, -60°)$ 8. $(2, -45°)$ 9. $(-3, -210°)$ 10. $(-1, -120°)$

11. $\left(3, \dfrac{5\pi}{3}\right)$ 12. $\left(4, \dfrac{3\pi}{2}\right)$

For each pair of rectangular coordinates, (a) plot the point and (b) give two pairs of polar coordinates for the point, where $0° \leq \theta < 360°$. See Example 2(b).

13. $(1, -1)$ 14. $(1, 1)$ 15. $(0, 3)$ 16. $(0, -3)$

17. $\left(\sqrt{2}, \sqrt{2}\right)$ 18. $\left(-\sqrt{2}, \sqrt{2}\right)$ 19. $\left(\dfrac{\sqrt{3}}{2}, \dfrac{3}{2}\right)$ 20. $\left(-\dfrac{\sqrt{3}}{2}, -\dfrac{1}{2}\right)$

21. $(3, 0)$ 22. $(-2, 0)$

For each rectangular equation, give its equivalent polar equation and sketch its graph. See Example 3.

23. $x - y = 4$ 24. $x + y = -7$ 25. $x^2 + y^2 = 16$

26. $x^2 + y^2 = 9$ 27. $2x + y = 5$ 28. $3x - 2y = 6$

RELATING CONCEPTS

For individual or collaborative investigation
(Exercises 29–36)

In rectangular coordinates, the graph of $ax + by = c$ is a horizontal line if $a = 0$ or a vertical line if $b = 0$. **Work Exercises 29–36 in order,** *to determine the general forms of polar equations for horizontal and vertical lines.*

29. Begin with the equation $y = k$, whose graph is a horizontal line. Make a trigonometric substitution for y using r and θ.

30. Solve the equation in Exercise 29 for r.

31. Rewrite the equation in Exercise 30 using the appropriate reciprocal function.

32. Sketch the graph of $r = 3 \csc \theta$. What is the corresponding rectangular equation?

33. Begin with the equation $x = k$, whose graph is a vertical line. Make a trigonometric substitution for x using r and θ.

34. Solve the equation in Exercise 33 for r.

35. Rewrite the equation in Exercise 34 using the appropriate reciprocal function.

36. Sketch the graph of $r = 3 \sec \theta$. What is the corresponding rectangular equation?

Concept Check In Exercises 37–40, match each equation with its polar graph from choices A–D.

37. $r = 3$ **38.** $r = \cos 3\theta$ **39.** $r = \cos 2\theta$ **40.** $r = \dfrac{2}{\cos \theta + \sin \theta}$

A.

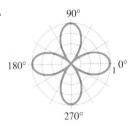

B.

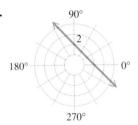

C.

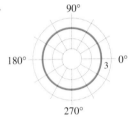

D.

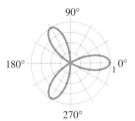

Give a complete graph of each polar equation. In Exercises 41–50, also identify the type of polar graph. See Examples 4–6.

41. $r = 2 + 2 \cos \theta$

42. $r = 8 + 6 \cos \theta$

43. $r = 3 + \cos \theta$

44. $r = 2 - \cos \theta$

45. $r = 4 \cos 2\theta$

46. $r = 3 \cos 5\theta$

47. $r^2 = 4 \cos 2\theta$

48. $r^2 = 4 \sin 2\theta$

49. $r = 4 - 4 \cos \theta$

50. $r = 6 - 3 \cos \theta$

51. $r = 2 \sin \theta \tan \theta$
(This is a **cissoid**.)

52. $r = \dfrac{\cos 2\theta}{\cos \theta}$
(This is a **cissoid with a loop**.)

For each equation, find an equivalent equation in rectangular coordinates and graph. See Example 8.

53. $r = 2 \sin \theta$

54. $r = 2 \cos \theta$

55. $r = \dfrac{2}{1 - \cos \theta}$

56. $r = \dfrac{3}{1 - \sin \theta}$

57. $r = -2 \cos \theta - 2 \sin \theta$

58. $r = \dfrac{3}{4 \cos \theta - \sin \theta}$

59. $r = 2 \sec \theta$

60. $r = -5 \csc \theta$

61. $r = \dfrac{2}{\cos \theta + \sin \theta}$

62. $r = \dfrac{2}{2 \cos \theta + \sin \theta}$

63. Graph $r = \theta$, a spiral of Archimedes. (See Example 7.) Use both positive and non-positive values for θ.

64. Use a graphing calculator window of $[-1250, 1250]$ by $[-1250, 1250]$, in degree mode, to graph more of $r = 2\theta$ (a spiral of Archimedes) than what is shown in Figure 29. Use $-1250° \le \theta \le 1250°$.

65. Find the polar equation of the line that passes through the points $(1, 0°)$ and $(2, 90°)$.

66. Explain how to plot a point (r, θ) in polar coordinates, if $r < 0$.

Concept Check   *The polar graphs in this section exhibit symmetry. (See **Appendix D**.) Visualize an xy-plane superimposed on the polar coordinate system, with the pole at the origin and the polar axis on the positive x-axis. Then a polar graph may be symmetric with respect to the x-axis (the polar axis), the y-axis $\left(\text{the line } \theta = \frac{\pi}{2}\right)$, or the origin (the pole). Use this information to work Exercises 67 and 68.*

67. Complete the missing ordered pairs in the graphs below.

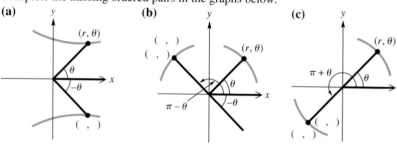

(a)   **(b)**   **(c)**

68. Based on your results in Exercise 67, fill in the blanks with the correct responses.
   (a) The graph of $r = f(\theta)$ is symmetric with respect to the polar axis if substitution of _____ for θ leads to an equivalent equation.
   (b) The graph of $r = f(\theta)$ is symmetric with respect to the vertical line $\theta = \frac{\pi}{2}$ if substitution of _____ for θ leads to an equivalent equation.
   (c) Alternatively, the graph of $r = f(\theta)$ is symmetric with respect to the vertical line $\theta = \frac{\pi}{2}$ if substitution of _____ for r and _____ for θ leads to an equivalent equation.
   (d) The graph of $r = f(\theta)$ is symmetric with respect to the pole if substitution of _____ for r leads to an equivalent equation.
   (e) Alternatively, the graph of $r = f(\theta)$ is symmetric with respect to the pole if substitution of _____ for θ leads to an equivalent equation.
   (f) In general, the completed statements in parts (a)–(e) mean that the graphs of polar equations of the form $r = a \pm b \cos \theta$ (where a may be 0) are symmetric with respect to _____.
   (g) In general, the completed statements in parts (a)–(e) mean that the graphs of polar equations of the form $r = a \pm b \sin \theta$ (where a may be 0) are symmetric with respect to _____.

The graph of $r = a\theta$ in polar coordinates is an example of the spiral of Archimedes. With your calculator set to radian mode, use the given value of a and interval of θ to graph the spiral in the window specified.

69. $a = 1, 0 \le \theta \le 4\pi, [-15, 15]$ by $[-15, 15]$

70. $a = 2, -4\pi \le \theta \le 4\pi, [-30, 30]$ by $[-30, 30]$

71. $a = 1.5, -4\pi \le \theta \le 4\pi, [-20, 20]$ by $[-20, 20]$

72. $a = -1, 0 \le \theta \le 12\pi, [-40, 40]$ by $[-40, 40]$

...—

Find the polar coordinates of the points of intersection of the given curves for the specified interval of θ.

73. $r = 4 \sin \theta, r = 1 + 2 \sin \theta; 0 \le \theta < 2\pi$

74. $r = 3, r = 2 + 2 \cos \theta; 0° \le \theta < 360°$

75. $r = 2 + \sin \theta, r = 2 + \cos \theta; 0 \le \theta < 2\pi$

76. $r = \sin 2\theta, r = \sqrt{2} \cos \theta; 0 \le \theta < \pi$

(Modeling) Solve each problem.

77. *Orbits of Satellites* The polar equation

$$r = \frac{a(1 - e^2)}{1 + e \cos \theta}$$

can be used to graph the orbits of the satellites of our sun, where a is the average distance in astronomical units from the sun and e is a constant called the **eccentricity**. The sun will be located at the pole. The table lists the values of a and e.

Satellite	a	e
Mercury	.39	.206
Venus	.78	.007
Earth	1.00	.017
Mars	1.52	.093
Jupiter	5.20	.048
Saturn	9.54	.056
Uranus	19.20	.047
Neptune	30.10	.009
Pluto	39.40	.249

Source: Karttunen, H., P. Kröger, H. Oja, M. Putannen, and K. Donners (Editors), *Fundamental Astronomy, 4th edition,* Springer-Verlag, 2003; Zeilik, M., S. Gregory, and E. Smith, *Introductory Astronomy and Astrophysics,* Saunders College Publishers, 1992.

(a) Graph the orbits of the four closest satellites on the same polar grid. Choose a viewing window that results in a graph with nearly circular orbits.

(b) Plot the orbits of Earth, Jupiter, Uranus, and Pluto on the same polar grid. How does Earth's distance from the sun compare to the others' distances from the sun?

(c) Use graphing to determine whether or not Pluto is always farthest from the sun.

78. *Radio Towers and Broadcasting Patterns* Many times radio stations do not broadcast in all directions with the same intensity. To avoid interference with an existing station to the north, a new station may be licensed to broadcast only east and west. To create an east-west signal, two radio towers are sometimes used, as illustrated in the figure. Locations where the radio signal is received correspond to the interior of the curve defined by

$$r^2 = 40{,}000 \cos 2\theta,$$

where the polar axis (or positive x-axis) points east.

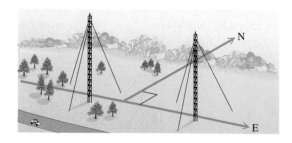

(a) Graph $r^2 = 40{,}000 \cos 2\theta$ for $0° \le \theta \le 360°$, where units are in miles. Assuming the radio towers are located near the pole, use the graph to describe the regions where the signal can be received and where the signal cannot be received.

(b) Suppose a radio signal pattern is given by $r^2 = 22{,}500 \sin 2\theta$. Graph this pattern and interpret the results.

8.6 Parametric Equations, Graphs, and Applications

Basic Concepts ▪ **Parametric Graphs and Their Rectangular Equivalents** ▪ **The Cycloid** ▪ **Applications of Parametric Equations**

Basic Concepts   Throughout this text, we have graphed sets of ordered pairs of real numbers that correspond to a function of the form $y = f(x)$ or $r = g(\theta)$. Another way to determine a set of ordered pairs involves two functions f and g defined by $x = f(t)$ and $y = g(t)$, where t is a real number in some interval I. Each value of t leads to a corresponding x-value and a corresponding y-value, and thus to an ordered pair (x, y).

PARAMETRIC EQUATIONS OF A PLANE CURVE

A **plane curve** is a set of points (x, y) such that $x = f(t)$, $y = g(t)$, and f and g are both defined on an interval I. The equations $x = f(t)$ and $y = g(t)$ are **parametric equations** with **parameter t.**

Graphing calculators are capable of graphing plane curves defined by parametric equations. The calculator must be set in parametric mode, and the window requires intervals for the parameter t, as well as for x and y. ▪

Parametric Graphs and Their Rectangular Equivalents

▶ **EXAMPLE 1**   GRAPHING A PLANE CURVE DEFINED PARAMETRICALLY

Let $x = t^2$ and $y = 2t + 3$, for t in $[-3, 3]$. Graph the set of ordered pairs (x, y).

Algebraic Solution

Make a table of corresponding values of t, x, and y over the domain of t. Then plot the points as shown in Figure 32. The graph is a portion of a parabola with horizontal axis $y = 3$. The arrowheads indicate the direction the curve traces as t increases.

t	x	y
-3	9	-3
-2	4	-1
-1	1	1
0	0	3
1	1	5
2	4	7
3	9	9

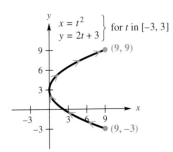

Figure 32

Graphing Calculator Solution

We set the parameters of the TI-83/84 Plus as shown in the top two screens to obtain the bottom screen in Figure 33.

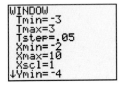

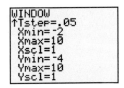

This is a continuation of the previous screen.

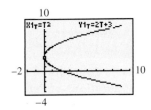

Figure 33

NOW TRY EXERCISE 5(a). ◀

▶ **EXAMPLE 2**   FINDING AN EQUIVALENT RECTANGULAR EQUATION

Find a rectangular equation for the plane curve of Example 1 defined as follows:

$$x = t^2, \quad y = 2t + 3, \qquad \text{for } t \text{ in } [-3, 3].$$

Solution   To eliminate the parameter t, solve either equation for t. Here, only the second equation, $y = 2t + 3$, leads to a unique solution for t, so we choose it.

$$y = 2t + 3 \qquad \text{Solve for } t. \text{ (Appendix A)}$$
$$2t = y - 3$$
$$t = \frac{y - 3}{2}$$

Now substitute this result into the first equation to get

$$x = t^2 = \left(\frac{y - 3}{2}\right)^2 = \frac{(y - 3)^2}{4}, \qquad \text{or} \qquad 4x = (y - 3)^2.$$

This is the equation of a horizontal parabola opening to the right, which agrees with the graph given in Figure 32. Because t is in $[-3, 3]$, x is in $[0, 9]$ and y is in $[-3, 9]$. The rectangular equation must be given with its restricted domain as

$$4x = (y - 3)^2, \qquad \text{for } x \text{ in } [0, 9].$$

NOW TRY EXERCISE 5(b). ◀

▶ **EXAMPLE 3** GRAPHING A PLANE CURVE DEFINED PARAMETRICALLY

Graph the plane curve defined by $x = 2 \sin t$, $y = 3 \cos t$, for t in $[0, 2\pi]$.

Solution To convert to a rectangular equation, it is not productive here to solve either equation for t. Instead, we use the fact that $\sin^2 t + \cos^2 t = 1$ to apply another approach. Square both sides of each equation; solve one for $\sin^2 t$, the other for $\cos^2 t$.

$x = 2 \sin t$	$y = 3 \cos t$ Given equations
$x^2 = 4 \sin^2 t$	$y^2 = 9 \cos^2 t$ Square both sides.
$\dfrac{x^2}{4} = \sin^2 t$	$\dfrac{y^2}{9} = \cos^2 t$ Divide.

Now add corresponding sides of the two equations.

$$\frac{x^2}{4} + \frac{y^2}{9} = \sin^2 t + \cos^2 t$$

$$\frac{x^2}{4} + \frac{y^2}{9} = 1 \qquad \sin^2 t + \cos^2 t = 1 \text{ (Section 5.1)}$$

This is the equation of an **ellipse.** See Figure 34 for traditional and calculator graphs. (Ellipses are covered in more detail in college algebra courses.)

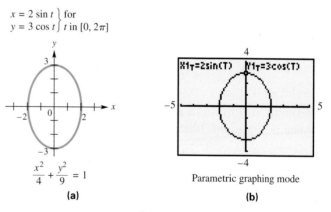

$x = 2 \sin t$ for
$y = 3 \cos t$ t in $[0, 2\pi]$

$$\frac{x^2}{4} + \frac{y^2}{9} = 1$$

(a)

X1ᴛ=2sin(T)   Y1ᴛ=3cos(T)

Parametric graphing mode

(b)

Figure 34

NOW TRY EXERCISE 27. ◀

Parametric representations of a curve are not unique. In fact, there are infinitely many parametric representations of a given curve. If the curve can be described by a rectangular equation $y = f(x)$, with domain X, then one simple parametric representation is

$$x = t, \quad y = f(t), \qquad \text{for } t \text{ in } X.$$

▶ **EXAMPLE 4** FINDING ALTERNATIVE PARAMETRIC EQUATION FORMS

Give two parametric representations for the equation of the parabola

$$y = (x - 2)^2 + 1.$$

Solution The simplest choice is to let

$$x = t, \quad y = (t - 2)^2 + 1, \qquad \text{for } t \text{ in } (-\infty, \infty).$$

Another choice, which leads to a simpler equation for y, is

$$x = t + 2, \quad y = t^2 + 1, \qquad \text{for } t \text{ in } (-\infty, \infty).$$

NOW TRY EXERCISE 29. ◀

▶ **Note** Sometimes trigonometric functions are desirable. One choice in Example 4 might be

$$x = 2 + \tan t, \quad y = \sec^2 t, \qquad \text{for } t \text{ in } \left(-\frac{\pi}{2}, \frac{\pi}{2}\right).$$

The Cycloid The path traced by a fixed point on the circumference of a circle rolling along a line is called a *cycloid*. A **cycloid** is defined by

$$x = at - a \sin t, \quad y = a - a \cos t, \qquad \text{for } t \text{ in } (-\infty, \infty).$$

The cycloid is a special case of a curve traced out by a point at a given distance from the center of a circle as the circle rolls along a straight line. Such a curve is called a **trochoid. Bezier curves** are used in manufacturing, and **Conchoids of Nicodemes** are so named because the shape of their outer branches resembles a conch shell. Other examples are **hypocycloids, epicycloids, the witch of Agnesi, swallowtail catastrophe curves,** and **Lissajou figures.** (*Source:* Stewart, J., *Calculus*, Fifth Edition, Brooks/Cole Publishing Co., 2003.)

▶ **EXAMPLE 5** GRAPHING A CYCLOID

Graph the cycloid $x = t - \sin t$, $y = 1 - \cos t$, for t in $[0, 2\pi]$.

Algebraic Solution

There is no simple way to find a rectangular equation for the cycloid from its parametric equations. Instead, begin with a table of values.

t	0	$\frac{\pi}{4}$	$\frac{\pi}{2}$	π	$\frac{3\pi}{2}$	2π
x	0	.08	.6	π	5.7	2π
y	0	.3	1	2	1	0

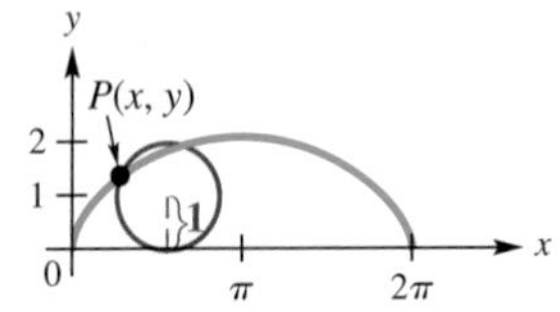

Figure 35

Plotting the ordered pairs (x, y) from the table of values leads to the portion of the graph in Figure 35 from 0 to 2π.

Graphing Calculator Solution

It is easier to graph a cycloid with a graphing calculator in parametric mode than with traditional methods. See Figure 36.

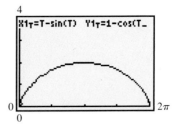

Figure 36

NOW TRY EXERCISE 33. ◀

Figure 37

The cycloid has an interesting physical property. If a flexible cord or wire goes through points P and Q as in Figure 37, and a bead is allowed to slide due to the force of gravity without friction along this path from P to Q, the path that requires the shortest time takes the shape of the graph of an inverted cycloid.

Applications of Parametric Equations

Parametric equations are used to simulate motion. If a ball is thrown with a velocity of v feet per second at an angle θ with the horizontal, its flight can be modeled by the parametric equations

$$x = (v \cos \theta)t \quad \text{and} \quad y = (v \sin \theta)t - 16t^2 + h,$$

where t is in seconds and h is the ball's initial height in feet above the ground. The term $-16t^2$ occurs because gravity is pulling downward. See Figure 38. These equations ignore air resistance.

▼ LOOKING AHEAD TO CALCULUS

At any time t, the velocity of an object is given by the vector $\mathbf{v} = \langle f'(t), g'(t) \rangle$. The object's speed at time t is

$$|\mathbf{v}| = \sqrt{(f'(t))^2 + (g'(t))^2}.$$

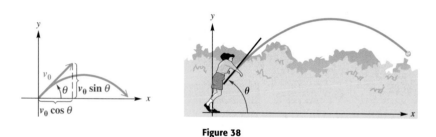

Figure 38

▶ **EXAMPLE 6** SIMULATING MOTION WITH PARAMETRIC EQUATIONS

Three golf balls are hit simultaneously into the air at 132 ft per sec (90 mph) at angles of 30°, 50°, and 70° with the horizontal.

(a) Assuming the ground is level, determine graphically which ball travels the farthest. Estimate this distance.

(b) Which ball reaches the greatest height? Estimate this height.

Solution

(a) The three sets of parametric equations determined by the three golf balls are as follows since $h = 0$.

$$x_1 = (132 \cos 30°)t, \quad y_1 = (132 \sin 30°)t - 16t^2$$
$$x_2 = (132 \cos 50°)t, \quad y_2 = (132 \sin 50°)t - 16t^2$$
$$x_3 = (132 \cos 70°)t, \quad y_3 = (132 \sin 70°)t - 16t^2$$

The graphs of the three sets of parametric equations are shown on the next page in Figure 39(a), where $0 \le t \le 9$. From the graph in Figure 39(b), we can see that the ball hit at 50° travels the farthest distance. Using the TRACE feature of the TI-83/84 Plus, we estimate this distance to be about 540 ft.

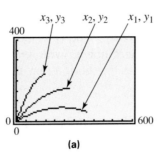

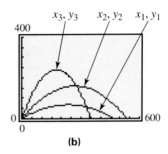

(a)   **(b)**

Figure 39

(b) Again, use the TRACE feature to find that the ball hit at 70° reaches the greatest height, about 240 ft.

NOW TRY EXERCISE 39. ◀

▶ **Note** The TI-83/84 Plus graphing calculator allows the user to view the graphing of more than one equation either *sequentially* or *simultaneously*. By choosing the latter, the three balls in Figure 39 can be viewed in flight at the same time.

▶ **EXAMPLE 7**   **EXAMINING PARAMETRIC EQUATIONS OF FLIGHT**

Jack Lukas launches a small rocket from a table that is 3.36 ft above the ground. Its initial velocity is 64 ft per sec, and it is launched at an angle of 30° with respect to the ground. Find the rectangular equation that models its path. What type of path does the rocket follow?

Solution   The path of the rocket is defined by the parametric equations

$$x = (64 \cos 30°)t \qquad \text{and} \qquad y = (64 \sin 30°)t - 16t^2 + 3.36$$

or, equivalently,

$$x = 32\sqrt{3}t \qquad \text{and} \qquad y = -16t^2 + 32t + 3.36.$$

From $x = 32\sqrt{3}t$, we obtain

$$t = \frac{x}{32\sqrt{3}}.$$

Substituting into the other parametric equation for t yields

$$y = -16\left(\frac{x}{32\sqrt{3}}\right)^2 + 32\left(\frac{x}{32\sqrt{3}}\right) + 3.36$$

$$y = -\frac{1}{192}x^2 + \frac{\sqrt{3}}{3}x + 3.36. \qquad \text{Simplify.}$$

Because this equation defines a parabola, the rocket follows a parabolic path.

NOW TRY EXERCISE 43(a). ◀

▶ EXAMPLE 8   ANALYZING THE PATH OF A PROJECTILE

Determine the total flight time and the horizontal distance traveled by the rocket in Example 7.

Algebraic Solution

The equation $y = -16t^2 + 32t + 3.36$ tells the vertical position of the rocket at time t. We need to determine those values of t for which $y = 0$ since these values correspond to the rocket at ground level. This yields

$$0 = -16t^2 + 32t + 3.36.$$

Using the quadratic formula, the solutions are $t = -.1$ or $t = 2.1$. Since t represents time, $t = -.1$ is an unacceptable answer. Therefore, the flight time is 2.1 sec.

The rocket was in the air for 2.1 sec, so we can use $t = 2.1$ and the parametric equation that models the horizontal position, $x = 32\sqrt{3}t$, to obtain

$$x = 32\sqrt{3}(2.1) \approx 116.4 \text{ ft.}$$

Graphing Calculator Solution

Figure 40 shows that when T = 2.1, the horizontal distance X covered is approximately 116.4 ft, which agrees with the algebraic solution.

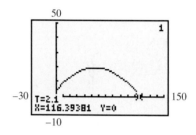

Figure 40

NOW TRY EXERCISE 43(b). ◀

8.6 Exercises

Concept Check   *Match the ordered pair from Column II with the pair of parametric equations in Column I on whose graph the point lies. In each case, consider the given value of t.*

I	II
1. $x = 3t + 6, y = -2t + 4;$ $t = 2$	**A.** $(5, 25)$
2. $x = \cos t, y = \sin t;$ $t = \dfrac{\pi}{4}$	**B.** $(7, 2)$
	C. $(12, 0)$
3. $x = t, y = t^2;$ $t = 5$	**D.** $\left(\dfrac{\sqrt{2}}{2}, \dfrac{\sqrt{2}}{2}\right)$
4. $x = t^2 + 3, y = t^2 - 2;$ $t = 2$	

*For each plane curve **(a)** graph the curve, and **(b)** find a rectangular equation for the curve. See Examples 1 and 2.*

5. $x = t + 2, y = t^2,$ for t in $[-1, 1]$

6. $x = 2t, y = t + 1,$ for t in $[-2, 3]$

7. $x = \sqrt{t}, y = 3t - 4,$ for t in $[0, 4]$

8. $x = t^2, y = \sqrt{t},$ for t in $[0, 4]$

9. $x = t^3 + 1, y = t^3 - 1,$ for t in $(-\infty, \infty)$

10. $x = 2t - 1, y = t^2 + 2,$ for t in $(-\infty, \infty)$

11. $x = 2 \sin t, y = 2 \cos t,$ for t in $[0, 2\pi]$

12. $x = \sqrt{5} \sin t, y = \sqrt{3} \cos t,$ for t in $[0, 2\pi]$

13. $x = 3 \tan t, y = 2 \sec t,$ for t in $\left(-\dfrac{\pi}{2}, \dfrac{\pi}{2}\right)$

14. $x = \cot t, y = \csc t,$ for t in $(0, \pi)$

15. $x = \sin t, y = \csc t,$ for t in $(0, \pi)$

16. $x = \tan t, y = \cot t,$ for t in $\left(0, \dfrac{\pi}{2}\right)$

17. $x = t, y = \sqrt{t^2 + 2},$ for t in $(-\infty, \infty)$

18. $x = \sqrt{t}, y = t^2 - 1,$ for t in $[0, \infty)$

19. $x = 2 + \sin t, y = 1 + \cos t,$ for t in $[0, 2\pi]$

20. $x = 1 + 2 \sin t, y = 2 + 3 \cos t,$ for t in $[0, 2\pi]$

21. $x = t + 2, y = \dfrac{1}{t + 2},$ for $t \neq -2$

22. $x = t - 3, y = \dfrac{2}{t - 3},$ for $t \neq 3$

23. $x = t + 2, y = t - 4,$ for t in $(-\infty, \infty)$

24. $x = t^2 + 2, y = t^2 - 4,$ for t in $(-\infty, \infty)$

Graph each plane curve defined by the parametric equations for t in $[0, 2\pi]$. Then find a rectangular equation for the plane curve. See Example 3.

25. $x = 3 \cos t, y = 3 \sin t$ **26.** $x = 2 \cos t, y = 2 \sin t$

27. $x = 3 \sin t, y = 2 \cos t$ **28.** $x = 4 \sin t, y = 3 \cos t$

Give two parametric representations for the equation of each parabola. See Example 4.

29. $y = (x + 3)^2 - 1$ **30.** $y = (x + 4)^2 + 2$

31. $y = x^2 - 2x + 3$ **32.** $y = x^2 - 4x + 6$

Graph each cycloid defined by the given equations for t in the specified interval. See Example 5.

33. $x = 2t - 2 \sin t, y = 2 - 2 \cos t,$ for t in $[0, 4\pi]$

34. $x = t - \sin t, y = 1 - \cos t,$ for t in $[0, 4\pi]$

*Lissajous Figures   The screen shown here is an example of a **Lissajous figure**. Lissajous figures occur in electronics and may be used to find the frequency of an unknown voltage. Graph each Lissajous figure for t in $[0, 6.5]$ in the window $[-6, 6]$ by $[-4, 4]$.*

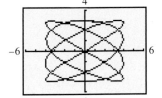

35. $x = 2 \cos t, y = 3 \sin 2t$

36. $x = 3 \cos 2t, y = 3 \sin 3t$

37. $x = 3 \sin 4t, y = 3 \cos 3t$

38. $x = 4 \sin 4t, y = 3 \sin 5t$

(Modeling)   In Exercises 39–42, do the following.

(a) *Determine the parametric equations that model the path of the projectile.*
(b) *Determine the rectangular equation that models the path of the projectile.*
(c) *Determine approximately how long the projectile is in flight and the horizontal distance covered.*

See Examples 6–8.

39. *Flight of a Model Rocket*   A model rocket is launched from the ground with velocity 48 ft per sec at an angle of 60° with respect to the ground.

40. *Flight of a Golf Ball*   Tiger is playing golf. He hits a golf ball from the ground at an angle of 60° with respect to the ground at velocity 150 ft per sec.

41. *Flight of a Softball*   Sally hits a softball when it is 2 ft above the ground. The ball leaves her bat at an angle of 20° with respect to the ground at velocity 88 ft per sec.

42. *Flight of a Baseball*   Pronk hits a baseball when it is 2.5 ft above the ground. The ball leaves his bat at an angle of 29° from the horizontal with velocity 136 ft per sec.

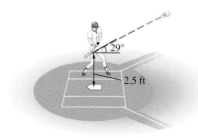

(Modeling)   Solve each problem. See Examples 7 and 8.

43. *Path of a Rocket*   A rocket is launched from the top of an 8-ft ladder. Its initial velocity is 128 ft per sec, and it is launched at an angle of 60° with respect to the ground.

   (a) Find the rectangular equation that models its path. What type of path does the rocket follow?
   (b) Determine the total flight time and the horizontal distance the rocket travels.

44. *Simulating Gravity on the Moon*   If an object is thrown on the moon, then the parametric equations of flight are

$$x = (v \cos \theta)t \qquad \text{and} \qquad y = (v \sin \theta)t - 2.66t^2 + h.$$

Estimate the distance that a golf ball hit at 88 ft per sec (60 mph) at an angle of 45° with the horizontal travels on the moon if the moon's surface is level.

45. *Flight of a Baseball*   A baseball is hit from a height of 3 ft at a 60° angle above the horizontal. Its initial velocity is 64 ft per sec.

   (a) Write parametric equations that model the flight of the baseball.
   (b) Determine the horizontal distance traveled by the ball in the air. Assume that the ground is level.
   (c) What is the maximum height of the baseball? At that time, how far has the ball traveled horizontally?
   (d) Would the ball clear a 5-ft-high fence that is 100 ft from the batter?

(Modeling) Path of a Projectile In Exercises 46 and 47, a projectile has been launched from the ground with initial velocity 88 ft per sec. You are supplied with the parametric equations modeling the path of the projectile.

(a) *Graph the parametric equations.*

(b) *Approximate θ, the angle the projectile makes with the horizontal at launch, to the nearest tenth of a degree.*

(c) *Based on your answer to part (b), write parametric equations for the projectile using the cosine and sine functions.*

46. $x = 82.69295063t$, $y = -16t^2 + 30.09777261t$

47. $x = 56.56530965t$, $y = -16t^2 + 67.41191099t$

48. Give two parametric representations of the line through the point (x_1, y_1) with slope m.

49. Give two parametric representations of the parabola $y = a(x - h)^2 + k$.

50. Give a parametric representation of the rectangular equation $\frac{x^2}{a^2} - \frac{y^2}{b^2} = 1$.

51. Give a parametric representation of the rectangular equation $\frac{x^2}{a^2} + \frac{y^2}{b^2} = 1$.

52. The spiral of Archimedes has polar equation $r = a\theta$, where $r^2 = x^2 + y^2$. Show that a parametric representation of the spiral of Archimedes is

$$x = a\theta \cos \theta, \quad y = a\theta \sin \theta, \qquad \text{for } \theta \text{ in } (-\infty, \infty).$$

53. Show that the **hyperbolic spiral** $r\theta = a$, where $r^2 = x^2 + y^2$, is given parametrically by

$$x = \frac{a \cos \theta}{\theta}, \quad y = \frac{a \sin \theta}{\theta}, \qquad \text{for } \theta \text{ in } (-\infty, 0) \cup (0, \infty).$$

54. The parametric equations $x = \cos t$, $y = \sin t$, for t in $[0, 2\pi]$ and the parametric equations $x = \cos t$, $y = -\sin t$, for t in $[0, 2\pi]$ both have the unit circle as their graph. However, in one case the circle is traced out clockwise (as t moves from 0 to 2π) and in the other case the circle is traced out counterclockwise. For which pair of equations is the circle traced out in the clockwise direction?

Concept Check Consider the parametric equations $x = f(t)$, $y = g(t)$, for t in $[a, b]$, with $c > 0, d > 0$.

55. How is the graph affected if the equation $x = f(t)$ is replaced by $x = c + f(t)$?

56. How is the graph affected if the equation $y = g(t)$ is replaced by $y = d + g(t)$?

Chapter 8 Summary

KEY TERMS

8.1 imaginary unit
complex number
real part
imaginary part
pure imaginary number
nonreal complex
number
standard form
complex conjugates
8.2 real axis
imaginary axis

complex plane
rectangular form of a
complex number
trigonometric (polar)
form of a complex
number
absolute value
(modulus)
argument
8.4 nth root of a complex
number

8.5 polar coordinate
system
pole
polar axis
polar coordinates
rectangular (Cartesian)
equation
polar equation
cardioid
polar grid
rose curve

lemniscate
spiral of Archimedes
limaçon
8.6 plane curve
parametric equations
of a plane curve
parameter
cycloid

NEW SYMBOLS

i imaginary unit

$a + bi$ complex number

QUICK REVIEW

CONCEPTS	EXAMPLES

8.1 Complex Numbers

Definition of i

$$i^2 = -1 \quad \text{or} \quad i = \sqrt{-1}$$

Definition of $\sqrt{-a}$
For $a > 0$,

$$\sqrt{-a} = i\sqrt{a}.$$

$$\sqrt{-4} = 2i$$
$$\sqrt{-12} = i\sqrt{12} = 2i\sqrt{3}$$

Adding and Subtracting Complex Numbers
Add or subtract the real parts and add or subtract the imaginary parts.

$$(2 + 3i) + (3 + i) - (2 - i)$$
$$= (2 + 3 - 2) + (3 + 1 + 1)i$$
$$= 3 + 5i$$

Multiplying and Dividing Complex Numbers
Multiply complex numbers as with binomials, and use the fact that $i^2 = -1$.

$$(6 + i)(3 - 2i) = 18 - 12i + 3i - 2i^2 \quad \text{FOIL}$$
$$= (18 + 2) + (-12 + 3)i \quad i^2 = -1$$
$$= 20 - 9i$$

Divide complex numbers by multiplying the numerator and denominator by the complex conjugate of the denominator.

$$\frac{3 + i}{1 + i} = \frac{(3 + i)(1 - i)}{(1 + i)(1 - i)} = \frac{3 - 3i + i - i^2}{1 - i^2}$$
$$= \frac{4 - 2i}{2} = \frac{2(2 - i)}{2} = 2 - i$$

(continued)

CONCEPTS	EXAMPLES

8.2 Trigonometric (Polar) Form of Complex Numbers

Trigonometric (Polar) Form of Complex Numbers

If the complex number $x + yi$ corresponds to the vector with direction angle θ and magnitude r, then

$$x = r \cos \theta \qquad\qquad y = r \sin \theta$$

$$r = \sqrt{x^2 + y^2} \qquad \tan \theta = \frac{y}{x}, \quad \text{if } x \neq 0.$$

The expression

$$r(\cos \theta + i \sin \theta) \qquad \text{or} \qquad r \operatorname{cis} \theta$$

is the trigonometric form (or polar form) of $x + yi$.

Write $2(\cos 60° + i \sin 60°)$ in rectangular form.

$$2(\cos 60° + i \sin 60°) = 2\left(\frac{1}{2} + i \cdot \frac{\sqrt{3}}{2}\right) = 1 + i\sqrt{3}$$

Write $-\sqrt{2} + i\sqrt{2}$ in trigonometric form.

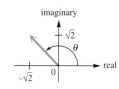

$$r = \sqrt{(-\sqrt{2})^2 + (\sqrt{2})^2} = 2$$

$\tan \theta = -1$ and θ is in quadrant II, so $\theta = 180° - 45° = 135°$. Therefore,

$$-\sqrt{2} + i\sqrt{2} = 2 \operatorname{cis} 135°.$$

8.3 The Product and Quotient Theorems

Product and Quotient Theorems

For any two complex numbers $r_1(\cos \theta_1 + i \sin \theta_1)$ and $r_2(\cos \theta_2 + i \sin \theta_2)$,

$$[r_1(\cos \theta_1 + i \sin \theta_1)] \cdot [r_2(\cos \theta_2 + i \sin \theta_2)]$$
$$= r_1 r_2[\cos(\theta_1 + \theta_2) + i \sin(\theta_1 + \theta_2)]$$

and

$$\frac{r_1(\cos \theta_1 + i \sin \theta_1)}{r_2(\cos \theta_2 + i \sin \theta_2)} = \frac{r_1}{r_2}[\cos(\theta_1 - \theta_2) + i \sin(\theta_1 - \theta_2)],$$

where $r_2 \operatorname{cis} \theta_2 \neq 0$.

If $\quad z_1 = 4(\cos 135° + i \sin 135°)$

and $\quad z_2 = 2(\cos 45° + i \sin 45°)$, then

$$z_1 z_2 = 8(\cos 180° + i \sin 180°)$$
$$= 8(-1 + i \cdot 0) = -8,$$

and

$$\frac{z_1}{z_2} = 2(\cos 90° + i \sin 90°)$$
$$= 2(0 + i \cdot 1) = 2i.$$

8.4 De Moivre's Theorem; Powers and Roots of Complex Numbers

De Moivre's Theorem

$$[r(\cos \theta + i \sin \theta)]^n = r^n(\cos n\theta + i \sin n\theta)$$

nth Root Theorem

If n is any positive integer, r is a positive real number, and θ is in degrees, then the nonzero complex number $r(\cos \theta + i \sin \theta)$ has exactly n distinct nth roots, given by

$$\sqrt[n]{r}(\cos \alpha + i \sin \alpha), \qquad \text{or} \qquad \sqrt[n]{r} \operatorname{cis} \alpha,$$

where

$$\alpha = \frac{\theta + 360° \cdot k}{n}, \qquad \text{or} \qquad \alpha = \frac{\theta}{n} + \frac{360° \cdot k}{n},$$

$k = 0, 1, 2, \ldots, n - 1.$

Let $z = 4(\cos 180° + i \sin 180°)$. Find z^3 and the square roots of z.

$$z^3 = 4^3(\cos 3 \cdot 180° + i \sin 3 \cdot 180°)$$
$$= 64(\cos 540° + i \sin 540°)$$
$$= 64(-1 + i \cdot 0)$$
$$= -64$$

For the given z, $r = 4$ and $\theta = 180°$. Its square roots are

$$\sqrt{4}\left(\cos \frac{180°}{2} + i \sin \frac{180°}{2}\right) = 2(0 + i \cdot 1) = 2i$$

and $\sqrt{4}\left(\cos \dfrac{180° + 360°}{2} + i \sin \dfrac{180° + 360°}{2}\right)$

$$= 2(0 + i(-1)) = -2i.$$

CONCEPTS | EXAMPLES

8.5 Polar Equations and Graphs

Rectangular and Polar Coordinates
The following relationships hold between the point (x, y) in the rectangular coordinate plane and the same point (r, θ) in the polar coordinate plane.

$$x = r \cos \theta \qquad y = r \sin \theta$$
$$r^2 = x^2 + y^2 \qquad \tan \theta = \frac{y}{x}, \quad \text{if } x \neq 0$$

Find the rectangular coordinates for the point $(5, 60°)$ in polar coordinates.

$$x = 5 \cos 60° = 5\left(\frac{1}{2}\right) = \frac{5}{2}$$
$$y = 5 \sin 60° = 5\left(\frac{\sqrt{3}}{2}\right) = \frac{5\sqrt{3}}{2}$$

The rectangular coordinates are $\left(\frac{5}{2}, \frac{5\sqrt{3}}{2}\right)$.

Find polar coordinates for $(-1, -1)$ in rectangular coordinates.

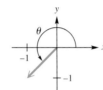

$$r = \sqrt{(-1)^2 + (-1)^2} = \sqrt{2}$$

$\tan \theta = 1$ and θ is in quadrant III, so $\theta = 225°$.

One pair of polar coordinates for $(-1, -1)$ is $\left(\sqrt{2}, 225°\right)$.

Polar Graphs
Examples of polar graphs include lines, circles, limaçons, lemniscates, roses, and spirals. (See page 400.)

Graph $r = 4 \cos 2\theta$.

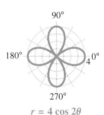

$r = 4 \cos 2\theta$

8.6 Parametric Equations, Graphs, and Applications

Plane Curve
A **plane curve** is a set of points (x, y) such that $x = f(t)$, $y = g(t)$, and f and g are both defined on an interval I. The equations $x = f(t)$ and $y = g(t)$ are **parametric equations** with **parameter t.**

Graph $x = 2 - \sin t$, $y = \cos t - 1$, for $0 \leq t \leq 2\pi$.

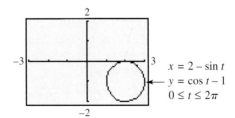

$x = 2 - \sin t$
$y = \cos t - 1$
$0 \leq t \leq 2\pi$

(continued)

CONCEPTS	EXAMPLES
Flight of an Object	Joe kicks a football from the ground at an angle of 45°

Flight of an Object

If an object has an initial velocity v, initial height h, and travels so that its initial angle of elevation is θ, then its flight after t seconds is modeled by the parametric equations

$$x = (v \cos \theta)t \quad \text{and} \quad y = (v \sin \theta)t - 16t^2 + h.$$

Joe kicks a football from the ground at an angle of 45° with a velocity of 48 ft per sec. Give the parametric equations that model the path of the football and the distance it travels before hitting the ground.

$$x = (48 \cos 45°)t = 24\sqrt{2}\,t$$
$$y = (48 \sin 45°)t - 16t^2 = 24\sqrt{2}\,t - 16t^2$$

When the ball hits the ground, $y = 0$.

$$24\sqrt{2}\,t - 16t^2 = 0$$
$$8t\left(3\sqrt{2} - 2t\right) = 0$$
$$t = 0 \quad \text{or} \quad t = \frac{3\sqrt{2}}{2}$$

(Reject)

The distance it travels is $x = 24\sqrt{2}\left(\dfrac{3\sqrt{2}}{2}\right) = 72$ ft.

CHAPTER 8 ▶ Review Exercises

Write as a multiple of i.

1. $\sqrt{-9}$

2. $\sqrt{-12}$

Solve each quadratic equation.

3. $x^2 = -81$

4. $x(2x + 3) = -4$

Perform each operation. Write answers in rectangular form.

5. $(1 - i) - (3 + 4i) + 2i$

6. $(2 - 5i) + (9 - 10i) - 3$

7. $(6 - 5i) + (2 + 7i) - (3 - 2i)$

8. $(4 - 2i) - (6 + 5i) - (3 - i)$

9. $(3 + 5i)(8 - i)$

10. $(4 - i)(5 + 2i)$

11. $(2 + 6i)^2$

12. $(6 - 3i)^2$

13. $(1 - i)^3$

14. $(2 + i)^3$

15. $\dfrac{25 - 19i}{5 + 3i}$

16. $\dfrac{2 - 5i}{1 + i}$

17. $\dfrac{2 + i}{1 - 5i}$

18. $\dfrac{3 + 2i}{i}$

19. i^{53}

20. i^{-41}

Perform each operation. Write answers in rectangular form.

21. $[5(\cos 90° + i \sin 90°)][6(\cos 180° + i \sin 180°)]$

22. $[3 \operatorname{cis} 135°][2 \operatorname{cis} 105°]$

23. $\dfrac{2(\cos 60° + i \sin 60°)}{8(\cos 300° + i \sin 300°)}$

24. $\dfrac{4 \text{ cis } 270°}{2 \text{ cis } 90°}$

25. $\left(\sqrt{3} + i\right)^3$

26. $(2 - 2i)^5$

27. $(\cos 100° + i \sin 100°)^6$

28. *Concept Check*   The vector representing a real number will lie on the _____-axis in the complex plane.

Graph each complex number as a vector.

29. $5i$

30. $-4 + 2i$

31. $3 - 3i\sqrt{3}$

32. Find and graph the resultant of $7 + 3i$ and $-2 + i$.

Perform each conversion, using a calculator to approximate answers as necessary.

	Rectangular Form	Trigonometric Form
33.	$-2 + 2i$	_____
34.	_____	$3(\cos 90° + i \sin 90°)$
35.	_____	$2(\cos 225° + i \sin 225°)$
36.	$-4 + 4i\sqrt{3}$	_____
37.	$1 - i$	_____
38.	_____	$4 \text{ cis } 240°$
39.	$-4i$	_____

Concept Check   The complex number z, where z = x + yi, can be graphed in the plane as (x, y). Describe the graphs of all complex numbers z satisfying the conditions in Exercises 40 and 41.

40. The absolute value of z is 2.

41. The imaginary part of z is the negative of the real part of z.

Find all roots as indicated. Express them in trigonometric form.

42. the fifth roots of $-2 + 2i$

43. the cube roots of $1 - i$

44. *Concept Check*   How many real fifth roots does -32 have?

45. *Concept Check*   How many real sixth roots does -64 have?

Solve each equation. Leave answers in trigonometric form.

46. $x^3 + 125 = 0$

47. $x^4 + 16 = 0$

48. $x^2 + i = 0$

49. Convert $\left(-1, \sqrt{3}\right)$ to polar coordinates, with $0° \le \theta < 360°$ and $r > 0$.

50. Convert $(5, 315°)$ to rectangular coordinates.

51. *Concept Check*   If a point lies on an axis in the rectangular plane, then what kind of angle must θ be if (r, θ) represents the point in polar coordinates?

52. *Concept Check*   What will the graph of $r = k$ be, for $k > 0$?

Identify and graph each polar equation for θ in $[0°, 360°)$.

53. $r = 4 \cos \theta$

54. $r = -1 + \cos \theta$

55. $r = 2 \sin 4\theta$

56. $r = \dfrac{2}{2 \cos \theta - \sin \theta}$

Find an equivalent equation in rectangular coordinates.

57. $r = \dfrac{3}{1 + \cos \theta}$ **58.** $r = \sin \theta + \cos \theta$ **59.** $r = 2$

Find an equivalent equation in polar coordinates.

60. $y = x$ **61.** $y = x^2$

In Exercises 62–65, identify the geometric symmetry (A, B, or C) that the graph will possess.

    A. *symmetry with respect to the origin*
    B. *symmetry with respect to the y-axis*
    C. *symmetry with respect to the x-axis*

62. Whenever (r, θ) is on the graph, then so is $(-r, -\theta)$.

63. Whenever (r, θ) is on the graph, then so is $(-r, \theta)$.

64. Whenever (r, θ) is on the graph, then so is $(r, -\theta)$.

65. Whenever (r, θ) is on the graph, then so is $(r, \pi - \theta)$.

In Exercises 66–69, find a polar equation having the given graph.

66.

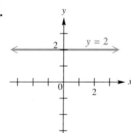

67.

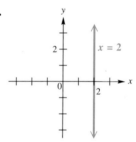

68.

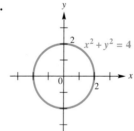

69.

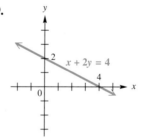

70. Show that the distance between (r_1, θ_1) and (r_2, θ_2) in polar coordinates is given by

$$d = \sqrt{r_1^2 + r_2^2 - 2r_1 r_2 \cos(\theta_1 - \theta_2)}.$$

71. Graph the plane curve defined by the parametric equations $x = t + \cos t$, $y = \sin t$, for t in $[0, 2\pi]$.

Find a rectangular equation for each plane curve with the given parametric equations.

72. $x = 3t + 2, y = t - 1$,   for t in $[-5, 5]$

73. $x = \sqrt{t - 1}, y = \sqrt{t}$,   for t in $[1, \infty)$

74. $x = t^2 + 5, y = \dfrac{1}{t^2 + 1}$,   for t in $(-\infty, \infty)$

75. $x = 5 \tan t, y = 3 \sec t$,   for t in $\left(-\dfrac{\pi}{2}, \dfrac{\pi}{2} \right)$

76. $x = \cos 2t,\ y = \sin t,$ for t in $(-\pi, \pi)$

77. Find a pair of parametric equations whose graph is the circle having center $(3, 4)$ and passing through the origin.

78. *Mandelbrot Set* Follow the steps in Exercise 64 of **Section 8.2** to show that the graph of the Mandelbrot set in Exercise 49 of **Section 8.4** is symmetric with respect to the x-axis.

79. *Flight of a Baseball* Albert hits a baseball when it is 3.2 ft above the ground. It leaves the bat with velocity 118 ft per sec at an angle of 27° with respect to the ground.

  (a) Determine the parametric equations that model the path of the baseball.

  (b) Determine the rectangular equation that models the path of the baseball.

  (c) Determine approximately how long the projectile is in flight and the horizontal distance covered.

CHAPTER 8 ▶ Test

1. Multiply or divide as indicated. Simplify each answer.

  (a) $\sqrt{-8} \cdot \sqrt{-6}$ **(b)** $\dfrac{\sqrt{-2}}{\sqrt{8}}$ **(c)** $\dfrac{\sqrt{-20}}{\sqrt{-180}}$

2. For the complex numbers $w = 2 - 4i$ and $z = 5 + i$, find each of the following in rectangular form.

  (a) $w + z$ (and give a geometric representation)   **(b)** $w - z$ **(c)** wz **(d)** $\dfrac{w}{z}$

3. Express each of the following in rectangular form.

  (a) i^{15} **(b)** $(1 + i)^2$

4. Solve $2x^2 - x + 4 = 0$ over the complex number system.

5. Write each complex number in trigonometric (polar) form, where $0° \le \theta < 360°$.

  (a) $3i$ **(b)** $1 + 2i$ **(c)** $-1 - i\sqrt{3}$

6. Write each complex number in rectangular form.

  (a) $3(\cos 30° + i \sin 30°)$ **(b)** $4 \operatorname{cis} 40°$ **(c)** $3(\cos 90° + i \sin 90°)$

7. For the complex numbers $w = 8(\cos 40° + i \sin 40°)$ and $z = 2(\cos 10° + i \sin 10°)$, find each of the following in the form specified.

  (a) wz (trigonometric form)   **(b)** $\dfrac{w}{z}$ (rectangular form)   **(c)** z^3 (rectangular form)

8. Find the four complex fourth roots of $-16i$. Express them in trigonometric form.

9. Convert the given rectangular coordinates to polar coordinates. Give two pairs of polar coordinates for each point.

  (a) $(0, 5)$ **(b)** $(-2, -2)$

10. Convert the given polar coordinates to rectangular coordinates.

(a) $(3, 315°)$ **(b)** $(-4, 90°)$

Identify and graph each polar equation for θ in $[0°, 360°)$.

11. $r = 1 - \cos\theta$ **12.** $r = 3\cos 3\theta$

13. Convert each polar equation to a rectangular equation, and sketch its graph.

(a) $r = \dfrac{4}{2\sin\theta - \cos\theta}$ **(b)** $r = 6$

Graph each pair of parametric equations.

14. $x = 4t - 3, y = t^2,$ for t in $[-3, 4]$

15. $x = 2\cos 2t, y = 2\sin 2t,$ for t in $[0, 2\pi]$

16. *Julia Set* Consider the complex number $z = -1 + i$. Compute the value of $z^2 - 1$, and show that its absolute value exceeds 2, indicating that $-1 + i$ is not in the Julia set.

CHAPTER 8 ▶ **Quantitative Reasoning**

Lake Tahoe

How Rugged Is Your Coastline?*

An interesting feature of coastlines is that their ruggedness is independent of the distance from which they are viewed. From an airplane, we see irregularities as bays, peninsulas, river mouths, and so on. On foot, we see each rock outcropping and creek that makes the coastline appear more rugged. An ant sees every pebble as a mountain to be scaled.

An interesting result of this phenomenon is that the total distance you travel along a coastline is dependent on the size of the steps you take. The closer (or smaller) you are, the smaller your steps will be. This means you will have more obstacles in your way, which results in a longer distance to travel. The more rugged the coastline, the longer it will be. In theory, this means that if you could take small enough steps on a rugged enough coastline, the length of the coastline would approach infinity. This is related to the study of fractals.

From a mathematical perspective, we can say that the number of steps needed (y) varies inversely with some power of the size of the steps taken (x):

$$y = \frac{k}{x^n}.$$

Each coastline will have different values for k and n, depending on its ruggedness. For a particular map of Lake Tahoe, these values are $k = 23.5$ and $n = 1.153$. Use the equation to determine how long the coastline would be if you "walked" with the given step sizes. Remember that the equation gives you the number of steps, so multiply that value by the step size to get the total length. Round answers to the nearest tenth of an inch.

1. 6 in.   **2.** .1 in.   **3.** .01 in.

---

*This material is based on an idea presented by Lori Lambertson, of the Nueva School and the Exploratorium in San Francisco.

Appendices

Equations and Inequalities

Equations ▪ Solving Linear Equations ▪ Solving Quadratic Equations ▪ Inequalities ▪ Solving Linear Inequalities ▪ Interval Notation ▪ Three-Part Inequalities

Equations   Recall from algebra that an **equation** is a statement that two expressions are equal.

$$x + 2 = 9, \qquad 11x = 5x + 6x, \qquad x^2 - 2x - 1 = 0 \qquad \text{Equations}$$

To *solve* an equation means to find all numbers that make the equation a true statement. These numbers are called **solutions** or **roots** of the equation. A number that is a solution of an equation is said to *satisfy* the equation, and the solutions of an equation make up its **solution set.** Equations with the same solution set are **equivalent equations.** For example, $x = 4$, $x + 1 = 5$, and $6x + 3 = 27$ are equivalent equations because they have the same solution set, $\{4\}$. However, the equations $x^2 = 9$ and $x = 3$ are not equivalent, since the first has solution set $\{-3, 3\}$ while the solution set of the second is $\{3\}$.

To solve an equation, we use the **addition and multiplication properties of equality.** For real numbers a, b, and c:

> **If $a = b$, then $a + c = b + c$;**
>
> **If $a = b$ and $c \neq 0$, then $ac = bc$.**

That is, the same number may be added to (or subtracted from) both sides of an equation and both sides of an equation may be multiplied (or divided) by the same nonzero number, without changing the solution set.

Solving Linear Equations   A **linear equation in one variable** is an equation that can be written in the form

$$ax + b = 0,$$

where a and b are real numbers with $a \neq 0$. A linear equation is also called a **first-degree equation** since the greatest degree of the variable is one.

$$3x + \sqrt{2} = 0, \qquad \frac{3}{4}x = 12, \qquad .5(x + 3) = 2x - 6 \qquad \text{Linear equations}$$

$$\sqrt{x} + 2 = 5, \qquad \frac{1}{x} = -8, \qquad x^2 + 3x + .2 = 0 \qquad \text{Nonlinear equations}$$

▶ **EXAMPLE 1**   SOLVING A LINEAR EQUATION

Solve $3(2x - 4) = 7 - (x + 5)$.

Solution $3(2x - 4) = 7 - (x + 5)$ Be careful with signs.

$$3(2x) + 3(-4) = 7 - x - 5 \qquad \text{Distributive property}$$

$$6x - 12 = 7 - x - 5 \qquad \text{Multiply.}$$

$$6x - 12 = 2 - x \qquad \text{Combine terms.}$$

423

$$6x - 12 + x = 2 - x + x \qquad \text{Add } x \text{ to each side.}$$

$$7x - 12 = 2 \qquad \text{Combine terms.}$$

$$7x - 12 + 12 = 2 + 12 \qquad \text{Add 12 to each side.}$$

$$7x = 14 \qquad \text{Combine terms.}$$

$$\frac{7x}{7} = \frac{14}{7} \qquad \text{Divide each side by 7.}$$

$$x = 2$$

Check: $\quad 3(2x - 4) = 7 - (x + 5) \qquad$ Original equation

$$3(2 \cdot 2 - 4) = 7 - (2 + 5) \qquad ? \quad \text{Let } x = 2.$$

$$3(4 - 4) = 7 - (7) \qquad ?$$

$$0 = 0 \qquad \text{True}$$

> A check of the solution is recommended.

Since replacing x with 2 results in a true statement, 2 is a solution of the given equation. The solution set is $\{2\}$.

NOW TRY EXERCISE 9. ◀

▶ **EXAMPLE 2** CLEARING FRACTIONS BEFORE SOLVING A LINEAR EQUATION

Solve $\dfrac{2t + 4}{3} + \dfrac{1}{2} t = \dfrac{1}{4} t - \dfrac{7}{3}$.

Solution $\qquad \dfrac{2t + 4}{3} + \dfrac{1}{2} t = \dfrac{1}{4} t - \dfrac{7}{3}$

> Distribute to *all* terms within the parentheses.

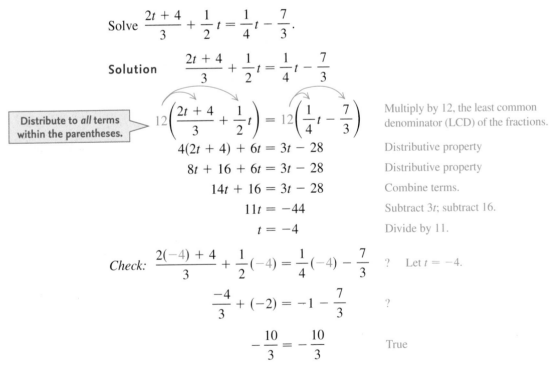

$$12 \left(\frac{2t + 4}{3} + \frac{1}{2} t \right) = 12 \left(\frac{1}{4} t - \frac{7}{3} \right) \qquad \begin{array}{l} \text{Multiply by 12, the least common} \\ \text{denominator (LCD) of the fractions.} \end{array}$$

$$4(2t + 4) + 6t = 3t - 28 \qquad \text{Distributive property}$$

$$8t + 16 + 6t = 3t - 28 \qquad \text{Distributive property}$$

$$14t + 16 = 3t - 28 \qquad \text{Combine terms.}$$

$$11t = -44 \qquad \text{Subtract } 3t; \text{ subtract 16.}$$

$$t = -4 \qquad \text{Divide by 11.}$$

Check: $\dfrac{2(-4) + 4}{3} + \dfrac{1}{2}(-4) = \dfrac{1}{4}(-4) - \dfrac{7}{3} \qquad ? \quad \text{Let } t = -4.$

$$\frac{-4}{3} + (-2) = -1 - \frac{7}{3} \qquad ?$$

$$-\frac{10}{3} = -\frac{10}{3} \qquad \text{True}$$

The solution set is $\{-4\}$.

NOW TRY EXERCISE 17. ◀

An equation satisfied by every number that is a meaningful replacement for the variable is called an **identity**. The equation $3(x + 1) = 3x + 3$ is an example of an identity. An equation that is satisfied by some numbers but not others, such as $2x = 4$, is called a **conditional equation.** The equations in Examples 1 and 2 are conditional equations. An equation that has no solution, such as $x = x + 1$, is called a **contradiction.**

▶ **EXAMPLE 3**   IDENTIFYING TYPES OF EQUATIONS

Decide whether each equation is an *identity,* a *conditional equation,* or a *contradiction.* Give the solution set.

(a) $-2(x + 4) + 3x = x - 8$ **(b)** $5x - 4 = 11$ **(c)** $3(3x - 1) = 9x + 7$

Solution

(a) $-2(x + 4) + 3x = x - 8$

$\quad -2x - 8 + 3x = x - 8$ Distributive property

$\quad\quad\quad x - 8 = x - 8$ Combine terms.

$\quad\quad\quad\quad 0 = 0$ Subtract x; add 8.

When a *true* statement such as $0 = 0$ results, the equation is an identity, and the solution set is {**all real numbers**}.

(b) $5x - 4 = 11$

$\quad 5x = 15$ Add 4.

$\quad x = 3$ Divide by 5.

This is a conditional equation, and its solution set is {3}.

(c) $3(3x - 1) = 9x + 7$

$\quad 9x - 3 = 9x + 7$ Distributive property

$\quad -3 = 7$ Subtract $9x$.

When a *false* statement such as $-3 = 7$ results, the equation is a contradiction, and the solution set is the **empty set** or **null set,** symbolized $\emptyset$.

NOW TRY EXERCISES 23, 25, AND 27. ◀

Solving Quadratic Equations   An equation that can be written in the form

$$ax^2 + bx + c = 0,$$

where a, b, and c are real numbers with $a \neq 0$, is a **quadratic equation.** The given form is called **standard form.** A quadratic equation is a **second-degree equation**—that is, an equation with a squared variable term and no terms of greater degree.

$\quad x^2 = 25, \quad 4x^2 + 4x - 5 = 0, \quad 3x^2 = 4x - 8$ Quadratic equations

Factoring, the simplest method of solving a quadratic equation, depends on the **zero-factor property.**

If a and b are complex numbers with $ab = 0$,

then $a = 0$ or $b = 0$ or both.

▶ **EXAMPLE 4** USING THE ZERO-FACTOR PROPERTY

Solve $6x^2 + 7x = 3$.

Solution

Don't factor out x here.	$\longmapsto$ $6x^2 + 7x = 3$	

$$6x^2 + 7x - 3 = 0 \qquad \text{Standard form}$$

$$(3x - 1)(2x + 3) = 0 \qquad \text{Factor.}$$

$$3x - 1 = 0 \quad \text{or} \quad 2x + 3 = 0 \qquad \text{Zero-factor property}$$

$$3x = 1 \quad \text{or} \quad 2x = -3 \qquad \text{Solve each equation.}$$

$$x = \frac{1}{3} \quad \text{or} \qquad x = -\frac{3}{2}$$

Check: $\qquad\qquad 6x^2 + 7x = 3 \quad$ Original equation

$$6\left(\frac{1}{3}\right)^2 + 7\left(\frac{1}{3}\right) = 3 \quad \text{Let } x = \tfrac{1}{3}. \qquad 6\left(-\frac{3}{2}\right)^2 + 7\left(-\frac{3}{2}\right) = 3 \quad \text{Let } x = -\tfrac{3}{2}.$$

$$\frac{6}{9} + \frac{7}{3} = 3 \quad ? \qquad\qquad\qquad \frac{54}{4} - \frac{21}{2} = 3 \quad ?$$

$$3 = 3 \quad \text{True} \qquad\qquad\qquad\qquad 3 = 3 \quad \text{True}$$

Both values check, since true statements result. The solution set is $\left\{\frac{1}{3}, -\frac{3}{2}\right\}$.

NOW TRY EXERCISE 37. ◀

A quadratic equation of the form $x^2 = k$ can be solved by the **square root property.**

$$\textbf{If } x^2 = k, \textbf{ then } x = \sqrt{k} \textbf{ or } x = -\sqrt{k}.$$

That is, the solution set of $x^2 = k$ is $\left\{\sqrt{k}, -\sqrt{k}\right\}$, which may be abbreviated $\left\{\pm\sqrt{k}\right\}$.

▶ **EXAMPLE 5** USING THE SQUARE ROOT PROPERTY

Solve each quadratic equation.

(a) $x^2 = 17$

(b) $(x - 4)^2 = 12$

Solution

(a) By the square root property, the solution set of $x^2 = 17$ is $\left\{\pm\sqrt{17}\right\}$.

(b) Use a generalization of the square root property.

$$(x - 4)^2 = 12$$

$$x - 4 = \pm\sqrt{12} \qquad \text{Generalized square root property}$$

$$x = 4 \pm \sqrt{12} \qquad \text{Add 4.}$$

$$x = 4 \pm 2\sqrt{3} \qquad \sqrt{12} = \sqrt{4 \cdot 3} = 2\sqrt{3}$$

Check: $\left(4 + 2\sqrt{3} - 4\right)^2 = \left(2\sqrt{3}\right)^2 = 12$

$\left(4 - 2\sqrt{3} - 4\right)^2 = \left(-2\sqrt{3}\right)^2 = 12$

The solution set is $\left\{4 \pm 2\sqrt{3}\right\}$.

NOW TRY EXERCISES 39 AND 43. ◀

Any quadratic equation can be solved by the **quadratic formula,** which says that the solutions of the quadratic equation $ax^2 + bx + c = 0$, where $a \neq 0$, are

$$x = \frac{-b \pm \sqrt{b^2 - 4ac}}{2a}.$$

▶ EXAMPLE 6   USING THE QUADRATIC FORMULA (REAL SOLUTIONS)

Solve $x^2 - 4x = -2$.

Solution $x^2 - 4x + 2 = 0$ Write in standard form.

Here $a = 1$, $b = -4$, and $c = 2$.

$$x = \frac{-b \pm \sqrt{b^2 - 4ac}}{2a}$$ Quadratic formula

$$= \frac{-(-4) \pm \sqrt{(-4)^2 - 4(1)(2)}}{2(1)}$$ $a = 1, b = -4, c = 2$

The fraction bar extends *under* $-b$.

$$= \frac{4 \pm \sqrt{16 - 8}}{2}$$

$$= \frac{4 \pm 2\sqrt{2}}{2}$$ $\sqrt{16-8} = \sqrt{8} = \sqrt{4 \cdot 2} = 2\sqrt{2}$

Factor first, then divide. $$= \frac{2(2 \pm \sqrt{2})}{2}$$ Factor out 2 in the numerator.

$$= 2 \pm \sqrt{2}$$ Lowest terms

The solution set is $\{2 \pm \sqrt{2}\}$.

NOW TRY EXERCISE 51. ◀

Inequalities   An **inequality** says that one expression is greater than, greater than or equal to, less than, or less than or equal to, another. As with equations, a value of the variable for which the inequality is true is a solution of the inequality; the set of all solutions is the solution set of the inequality. Two inequalities with the same solution set are equivalent. Inequalities are solved with the **properties of inequality.** For real numbers a, b, and c:

1. **If $a < b$, then $a + c < b + c$,**
2. **If $a < b$ and if $c > 0$, then $ac < bc$,**
3. **If $a < b$ and if $c < 0$, then $ac > bc$.**

Replacing $<$ with $>$, $\leq$, or $\geq$ results in similar properties. (Restrictions on c remain the same.) Multiplication may be replaced by division in properties 2 and 3. *Always remember to reverse the direction of the inequality symbol when multiplying or dividing by a negative number.*

Solving Linear Inequalities   A **linear inequality in one variable** is an inequality that can be written in the form

$$ax + b > 0,$$

where a and b are real numbers with $a \neq 0$. (Any of the symbols $\geq$, $<$, or $\leq$ may also be used.)

▶ EXAMPLE 7   SOLVING A LINEAR INEQUALITY

Solve $-3x + 5 > -7$.

Solution $-3x + 5 - 5 > -7 - 5$ Subtract 5 from each side.

$$-3x > -12$$

> **Don't forget to reverse the symbol here.**

$$\frac{-3x}{-3} < \frac{-12}{-3}$$ Divide by -3; reverse the direction of the inequality symbol when multiplying or dividing by a negative number.

$$x < 4$$

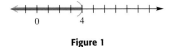

Figure 1

The original inequality is satisfied by any real number less than 4. The solution set can be written using **set-builder notation** as $\{x \mid x < 4\}$, which is read "the set of all x such that x is less than 4." A graph of the solution set is shown in Figure 1, where the parenthesis is used to show that 4 itself does not belong to the solution set.   ◀

Interval Notation   The solution set for the inequality in Example 7, $\{x \mid x < 4\}$, is an example of an **interval.** We use a simplified notation, called **interval notation,** to write intervals. With this notation, we write the interval in Example 7 as $(-\infty, 4)$. The symbol $-\infty$ does not represent an actual number; it is used to show that the interval includes all real numbers less than 4. The interval $(-\infty, 4)$ is an example of an **open interval,** since the endpoint, 4, is not part of the interval. A **closed interval** includes both endpoints. A square bracket is used to show that a number *is* part of the graph, and a parenthesis is used to indicate that a number *is not* part of the graph. In the table that follows, we assume that $a < b$.

Type of Interval	Set	Interval Notation	Graph
Open interval	$\{x \mid x > a\}$	(a, ∞)	
	$\{x \mid a < x < b\}$	(a, b)	
	$\{x \mid x < b\}$	$(-\infty, b)$	
Other intervals	$\{x \mid x \geq a\}$	$[a, \infty)$	
	$\{x \mid a < x \leq b\}$	$(a, b]$	
	$\{x \mid a \leq x < b\}$	$[a, b)$	
	$\{x \mid x \leq b\}$	$(-\infty, b]$	
Closed interval	$\{x \mid a \leq x \leq b\}$	$[a, b]$	
Disjoint interval	$\{x \mid x < a \text{ or } x > b\}$	$(-\infty, a) \cup (b, \infty)$	
All real numbers	$\{x \mid x \text{ is a real number}\}$	$(-\infty, \infty)$	

NOW TRY EXERCISES 59 AND 63. ◀

Three-Part Inequalities The inequality $-2 < 5 + 3x < 20$ in the next example says that $5 + 3x$ is *between* -2 and 20. This inequality is solved using an extension of the properties of inequality given earlier, working with all three expressions at the same time.

▶ **EXAMPLE 8**   SOLVING A THREE-PART INEQUALITY

Solve $-2 < 5 + 3x < 20$.

Solution

$$-2 < 5 + 3x < 20$$

$$-2 - 5 < 5 + 3x - 5 < 20 - 5 \quad \text{Subtract 5 from each part.}$$

$$-7 < 3x < 15$$

$$\frac{-7}{3} < \frac{3x}{3} < \frac{15}{3} \quad \text{Divide each part by 3.}$$

$$-\frac{7}{3} < x < 5$$

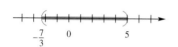

Figure 2

The solution set, graphed in Figure 2, is the interval $\left(-\frac{7}{3}, 5\right)$.

NOW TRY EXERCISE 73. ◀

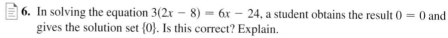

Appendix A   Exercises

Concept Check   In Exercises 1–4, decide whether each statement is true *or* false.

1. The solution set of $2x + 7 = x - 1$ is $\{-8\}$.

2. The equation $5(x - 10) = 5x - 50$ is an example of an identity.

3. The equations $x^2 = 4$ and $x + 2 = 4$ are equivalent equations.

4. It is possible for a linear equation to have exactly two solutions.

5. *Concept Check*   Which one is not a linear equation?

    A. $5x + 7(x - 1) = -3x$ **B.** $9x^2 - 4x + 3 = 0$

    C. $7x + 8x = 13x$ **D.** $.04x - .08x = .40$

6. In solving the equation $3(2x - 8) = 6x - 24$, a student obtains the result $0 = 0$ and gives the solution set $\{0\}$. Is this correct? Explain.

Solve each equation. See Examples 1 and 2.

7. $5x + 4 = 3x - 4$ **8.** $9x + 11 = 7x + 1$

9. $6(3x - 1) = 8 - (10x - 14)$ **10.** $4(-2x + 1) = 6 - (2x - 4)$

11. $\dfrac{5}{6}x - 2x + \dfrac{4}{3} = \dfrac{5}{3}$ **12.** $\dfrac{7}{4} + \dfrac{1}{5}x - \dfrac{3}{2} = \dfrac{4}{5}x$

13. $3x + 5 - 5(x + 1) = 6x + 7$ **14.** $5(x + 3) + 4x - 3 = -(2x - 4) + 2$

15. $2[x - (4 + 2x) + 3] = 2x + 2$ **16.** $4[2x - (3 - x) + 5] = -7x - 2$

17. $\dfrac{1}{14}(3x - 2) = \dfrac{x + 10}{10}$ **18.** $\dfrac{1}{15}(2x + 5) = \dfrac{x + 2}{9}$

19. $.2x - .5 = .1x + 7$

20. $.01x + 3.1 = 2.03x - 2.96$

21. $-4(2x - 6) + 8x = 5x + 24 + x$

22. $-8(3x + 4) + 6x = 4(x - 8) + 4x$

Decide whether each equation is an identity, *a* conditional equation, *or a* contradiction. *Give the solution set. See Example 3.*

23. $4(2x + 7) = 2x + 22 + 3(2x + 2)$

24. $\frac{1}{2}(6x + 20) = x + 4 + 2(x + 3)$

25. $2(x - 8) = 3x - 16$

26. $-8(x + 3) = -8x - 5(x + 1)$

27. $4(x + 7) = 2(x + 12) + 2(x + 1)$

28. $-6(2x + 1) - 3(x - 4) = -15x + 1$

29. *Concept Check* Match the equation in Column I with its solution(s) in Column II.

I	II

(a) $x^2 = 25$ **(b)** $x^2 - 5 = 0$ **A.** $\pm2\sqrt{5}$ **B.** 5

(c) $x^2 = 20$ **(d)** $x - 5 = 0$ **C.** $\pm\sqrt{5}$ **D.** -5

(e) $x + 5 = 0$ **E.** ±5

Concept Check *Answer each question.*

30. Which one of the following equations is set up for direct use of the zero-factor property? Solve it.

  A. $3x^2 - 17x - 6 = 0$

  B. $(2x + 5)^2 = 7$

  C. $x^2 + x = 12$

  D. $(3x - 1)(x - 7) = 0$

31. Which one of the following equations is set up for direct use of the square root property? Solve it.

  A. $3x^2 - 17x - 6 = 0$

  B. $(2x + 5)^2 = 7$

  C. $x^2 + x = 12$

  D. $(3x - 1)(x - 7) = 0$

32. Only one of the following equations is set up so that the values of a, b, and c can be determined immediately for direct use of the quadratic formula. Which one is it? Solve it.

  A. $3x^2 - 17x - 6 = 0$

  B. $(2x + 5)^2 = 7$

  C. $x^2 + x = 12$

  D. $(3x - 1)(x - 7) = 0$

Solve each equation by the zero-factor property. See Example 4.

33. $x^2 - 5x + 6 = 0$

34. $x^2 + 2x - 8 = 0$

35. $5x^2 - 3x - 2 = 0$

36. $2x^2 - x - 15 = 0$

37. $-4x^2 + x = -3$

38. $-6x^2 + 7x = -10$

Solve each equation by the square root property. See Example 5.

39. $x^2 = 16$

40. $x^2 = 25$

41. $27 - x^2 = 0$

42. $48 - x^2 = 0$

43. $(x + 5)^2 = 40$

44. $(x - 7)^2 = 24$

45. $(3x - 1)^2 = 12$

46. $(4x + 1)^2 = 20$

Solve each equation using the quadratic formula. See Example 6.

47. $x^2 - 4x + 3 = 0$

48. $x^2 - 7x + 12 = 0$

49. $x^2 - x - 1 = 0$

50. $x^2 - 3x - 2 = 0$

51. $x^2 - 6x = -7$

52. $x^2 - 4x = -1$

53. $\frac{1}{2}x^2 + \frac{1}{4}x - 3 = 0$

54. $\frac{2}{3}x^2 + \frac{1}{4}x = 3$

55. $.2x^2 + .4x - .3 = 0$

56. $.1x^2 - .1x = .3$

57. $(4x - 1)(x + 2) = 4x$

58. $(3x + 2)(x - 1) = 3x$

59. *Concept Check*   Match the inequality in each exercise in Column I with its equivalent interval notation in Column II.

I

(a) $x < -6$

(b) $x \le 6$

(c) $-2 < x \le 6$

(d) $-3 \le x \le 3$

(e) $x \ge -6$

(f) $6 \le x$

(g)

(h)

(i)

(j)

II

A. $(-2, 6]$

B. $[-2, 6)$

C. $(-\infty, -6]$

D. $[6, \infty)$

E. $(-\infty, -3) \cup (3, \infty)$

F. $(-\infty, -6)$

G. $(0, 8)$

H. $[-3, 3]$

I. $[-6, \infty)$

J. $(-\infty, 6]$

60. *Concept Check*   The three-part inequality $a < x < b$ means "a is less than x and x is less than b." Which one of the following inequalities is not satisfied by some real number x?

A. $-3 < x < 10$ **B.** $0 < x < 6$

C. $-3 < x < -1$ **D.** $-8 < x < -10$

Solve each inequality. Write each solution set in interval notation, and graph it. See Example 7.

61. $2x + 8 \le 16$ **62.** $3x - 8 \le 7$

63. $-2x - 2 \le 1 + x$ **64.** $-4x + 3 \ge -2 + x$

65. $2(x + 5) + 1 \ge 5 + 3x$ **66.** $6x - (2x + 3) \ge 4x - 5$

67. $8x - 3x + 2 < 2(x + 7)$ **68.** $2 - 4x + 5(x - 1) < -6(x - 2)$

69. $\dfrac{4x + 7}{-3} \le 2x + 5$ **70.** $\dfrac{2x - 5}{-8} \le 1 - x$

71. $\dfrac{1}{3}x + \dfrac{2}{5}x - \dfrac{1}{2}(x + 3) \le \dfrac{1}{10}$ **72.** $-\dfrac{2}{3}x - \dfrac{1}{6}x + \dfrac{2}{3}(x + 1) \le \dfrac{4}{3}$

Solve each inequality. Write each solution set in interval notation, and graph it. See Example 8.

73. $-5 < 5 + 2x < 11$ **74.** $-7 < 2 + 3x < 5$

75. $10 \le 2x + 4 \le 16$ **76.** $-6 \le 6x + 3 \le 21$

77. $-11 > -3x + 1 > -17$ **78.** $2 > -6x + 3 > -3$

79. $-4 \le \dfrac{x + 1}{2} \le 5$ **80.** $-5 \le \dfrac{x - 3}{3} \le 1$

B   Graphs of Equations

The Rectangular Coordinate System ▪ **The Pythagorean Theorem and the Distance Formula** ▪ **The Midpoint Formula** ▪ **Graphing Equations** ▪ **Circles**

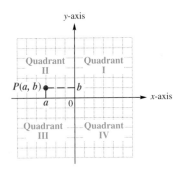

Figure 1

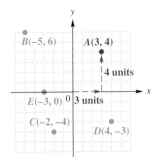

Figure 2

The Rectangular Coordinate System   Recall from algebra that each point in a plane corresponds to an **ordered pair,** two numbers written inside parentheses, such as (3, 4), in which the order of the numbers is important. We graph ordered pairs of real numbers by using two perpendicular number lines, one horizontal and one vertical, that intersect at their zero-points. This point of intersection is called the **origin.** The horizontal line is called the **x-axis,** and the vertical line is called the **y-axis.** Starting at the origin, on the x-axis the positive numbers go to the right and the negative numbers go to the left. The y-axis has positive numbers going up and negative numbers going down.

The x-axis and y-axis together make up a **rectangular coordinate system,** or **Cartesian coordinate system** (named for one of its coinventors, René Descartes; the other coinventor was Pierre de Fermat). The plane into which the coordinate system is introduced is the **coordinate plane,** or **xy-plane.** The x-axis and y-axis divide the plane into four regions, or **quadrants,** labeled as shown in Figure 1. The points on the x-axis and y-axis belong to no quadrant.

Each point P in the xy-plane corresponds to a unique ordered pair (a, b) of real numbers. The numbers a and b are the **coordinates** of point P. (See Figure 1.) To locate on the xy-plane the point corresponding to the ordered pair $(3, 4)$, for example, start at the origin, move 3 units in the positive x-direction, and then move 4 units in the positive y-direction. See Figure 2, where point A corresponds to the ordered pair $(3, 4)$.

The Pythagorean Theorem and the Distance Formula   The distance between any two points in a plane can be found by using a formula derived from the **Pythagorean Theorem.**

PYTHAGOREAN THEOREM

In a right triangle, the sum of the squares of the lengths of the legs is equal to the square of the length of the hypotenuse.

$$a^2 + b^2 = c^2$$

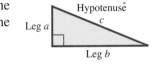

To find the distance between two points (x_1, y_1) and (x_2, y_2), draw the line segment connecting the points, as shown in Figure 3. Complete a right triangle by drawing a line through (x_1, y_1) parallel to the x-axis and a line through (x_2, y_2) parallel to the y-axis. The ordered pair at the right angle of this triangle is (x_2, y_1).

The horizontal side of the right triangle in Figure 3 has length $x_2 - x_1$, while the vertical side has length $y_2 - y_1$. If d represents the distance between the two original points, then by the Pythagorean theorem,

$$d^2 = (x_2 - x_1)^2 + (y_2 - y_1)^2.$$

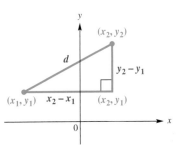

Figure 3

Solving for d, we obtain the **distance formula.**

Suppose that $P(x_1, y_1)$ and $Q(x_2, y_2)$ are two points in a coordinate plane. Then the distance between P and Q, written $d(P, Q)$, is given by

$$d(P,Q) = \sqrt{(x_2 - x_1)^2 + (y_2 - y_1)^2}.$$

That is, the distance between two points in a coordinate plane is the square root of the sum of the square of the difference between their x-coordinates and the square of the difference between their y-coordinates.

▶ EXAMPLE 1   USING THE DISTANCE FORMULA

Find the distance between $P(-8, 4)$ and $Q(3, -2)$.

Solution   According to the distance formula,

$$d(P,Q) = \sqrt{[3 - (-8)]^2 + (-2 - 4)^2} \quad {\scriptstyle x_1 = -8, \, y_1 = 4, \, x_2 = 3, \, y_2 = -2}$$

$$= \sqrt{11^2 + (-6)^2}$$

Be careful when subtracting a negative number.

$$= \sqrt{121 + 36}$$

$$= \sqrt{157}.$$

NOW TRY EXERCISE 25(a). ◀

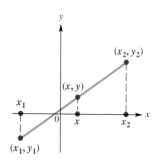

Figure 4

The Midpoint Formula   The **midpoint formula** is used to find the co-ordinates of the midpoint of a line segment. (Recall that the midpoint of a line segment is equidistant from the endpoints of the segment.) To develop the mid-point formula, let (x_1, y_1) and (x_2, y_2) be any two distinct points in a plane. (Although Figure 4 shows $x_1 < x_2$, no particular order is required.) Let (x, y) be the midpoint of the segment connecting (x_1, y_1) and (x_2, y_2). Draw vertical lines from each of the three points to the x-axis, as shown in Figure 4.

Since (x, y) is the midpoint of the line segment connecting (x_1, y_1) and (x_2, y_2), the distance between x and x_1 equals the distance between x and x_2, so

$$x_2 - x = x - x_1$$

$$x_2 + x_1 = 2x \qquad {\scriptstyle \text{Add } x; \text{ add } x_1. \text{ (Appendix A)}}$$

$$x = \frac{x_1 + x_2}{2}. \qquad {\scriptstyle \text{Divide by 2; rewrite.}}$$

Similarly, the y-coordinate is $\dfrac{y_1 + y_2}{2}$, yielding the following formula.

The midpoint of the line segment with endpoints (x_1, y_1) and (x_2, y_2) has coordinates

$$\left(\frac{x_1 + x_2}{2}, \frac{y_1 + y_2}{2} \right).$$

That is, the *x*-coordinate of the midpoint of a line segment is the *average* of the *x*-coordinates of the segment's endpoints, and the *y*-coordinate is the *average* of the *y*-coordinates of the segment's endpoints.

▶ EXAMPLE 2 USING THE MIDPOINT FORMULA

Find the coordinates of the midpoint M of the segment with endpoints $(8, -4)$ and $(-6, 1)$.

Solution The coordinates of M are

$$\left(\frac{8 + (-6)}{2}, \frac{-4 + 1}{2}\right) = \left(1, -\frac{3}{2}\right).$$ Substitute in the midpoint formula.

NOW TRY EXERCISE 25(b). ◀

Graphing Equations Ordered pairs are used to express the solutions of equations in two variables. When an ordered pair represents the solution of an equation with the variables x and y, the x-value is written first. For example, we say that $(1, 2)$ is a solution of $2x - y = 0$, since substituting 1 for x and 2 for y in the equation gives a true statement.

$$2x - y = 0$$
$$2(1) - 2 = 0$$
$$0 = 0 \quad \text{True}$$

▶ EXAMPLE 3 FINDING ORDERED PAIRS THAT ARE SOLUTIONS
OF EQUATIONS

For each equation, find at least three ordered pairs that are solutions.

(a) $y = 4x - 1$ (b) $y = x^2 - 4$

Solution

(a) Choose any real number for x or y and substitute in the equation to get the corresponding value of the other variable. For example, let $x = -2$ and then let $y = 3$.

$y = 4x - 1$ $\qquad$ $y = 4x - 1$
$y = 4(-2) - 1$ Let $x = -2$. $\qquad$ $3 = 4x - 1$ Let $y = 3$.
$y = -8 - 1$ Multiply. $\qquad$ $4 = 4x$ Add 1.
$y = -9$ Subtract. $\qquad$ $1 = x$ Divide by 4.

This gives the ordered pairs $(-2, -9)$ and $(1, 3)$. Verify that the ordered pair $(0, -1)$ is also a solution.

(b) A table provides an organized method for determining ordered pairs. Here, we let x equal $-2, -1, 0, 1$ and 2 in $y = x^2 - 4$, and determine the corresponding y-values.

x	y
-2	0
-1	-3
0	-4
1	-3
2	0

Five ordered pairs are $(-2, 0)$, $(-1, -3)$, $(0, -4)$, $(1, -3)$, and $(2, 0)$.

NOW TRY EXERCISES 41(a) AND 43(a). ◀

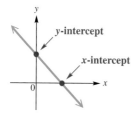

The **graph** of an equation is found by plotting ordered pairs that are solutions of the equation. The **intercepts** of the graph are good points to plot first. An **x-intercept** is an x-value where the graph intersects the x-axis. A **y-intercept** is a y-value where the graph intersects the y-axis.* In other words, the x-intercept is the x-coordinate of an ordered pair where $y = 0$, and the y-intercept is the y-coordinate of an ordered pair where $x = 0$.

A general algebraic approach for graphing an equation follows.

GRAPHING AN EQUATION BY POINT PLOTTING

Step 1 Find the intercepts.

Step 2 Find as many additional ordered pairs as needed.

Step 3 Plot the ordered pairs from Steps 1 and 2.

Step 4 Connect the points from Step 3 with a smooth line or curve.

▶ **EXAMPLE 4**   **GRAPHING EQUATIONS**

Graph each equation from Example 3.

(a) $y = 4x - 1$ **(b)** $y = x^2 - 4$

Solution

(a) ***Step 1*** Let $y = 0$ to find the x-intercept, and let $x = 0$ to find the y-intercept.

$$y = 4x - 1 \qquad\qquad y = 4x - 1$$
$$0 = 4x - 1 \qquad\qquad y = 4(0) - 1$$
$$1 = 4x \qquad\qquad\quad\; y = 0 - 1$$
$$\frac{1}{4} = x \quad \text{x-intercept} \qquad\quad y = -1 \quad \text{y-intercept}$$

These intercepts lead to the ordered pairs $\left(\frac{1}{4}, 0\right)$ and $(0, -1)$. Note that the y-intercept yields one of the ordered pairs we found in Example 3(a).

---

*The intercepts are sometimes defined as ordered pairs, such as $(3, 0)$ and $(0, -4)$, instead of numbers, like x-intercept 3 and y-intercept -4. In this text, we define them as numbers.

Step 2 We use the other ordered pairs found in Example 3(a): $(-2, -9), (1, 3)$.

Step 3 Plot the four ordered pairs from Steps 1 and 2 as shown in Figure 5.

Step 4 Connect the points plotted in Step 3 with a straight line. This line, also shown in Figure 5, is the graph of the equation $y = 4x - 1$.

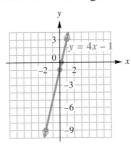

Figure 5   **Figure 6**

(b) In Example 3(b), we found five ordered pairs that satisfy $y = x^2 - 4$:

$$(-2, 0), (-1, -3), (0, -4), (1, -3), (2, 0).$$

     x-intercept y-intercept x-intercept

Plotting the points and joining them with a smooth curve gives the graph in Figure 6. This curve is called a **parabola.**

NOW TRY EXERCISES 41(b) AND 43(b). ◀

Circles

By definition, a **circle** is the set of all points in a plane that lie a given distance from a given point. The given distance is the **radius** of the circle, and the given point is the **center.**

We can find the equation of a circle from its definition by using the distance formula. Suppose that the point (h, k) is the center and the circle has radius r, where $r > 0$. Let (x, y) represent any point on the circle. See Figure 7. Now apply the distance formula with $(h, k) = (x_1, y_1)$ and $(x, y) = (x_2, y_2)$.

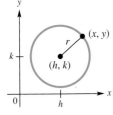

Figure 7

$$\sqrt{(x_2 - x_1)^2 + (y_2 - y_1)^2} = r \qquad \text{Distance formula}$$

$$\sqrt{(x - h)^2 + (y - k)^2} = r \qquad \text{Substitute.}$$

$$\left(\sqrt{(x - h)^2 + (y - k)^2}\right)^2 = r^2 \qquad \text{Square both sides.}$$

$$(x - h)^2 + (y - k)^2 = r^2$$

If a circle of radius $r > 0$ has center (h, k) at the origin, $(0, 0)$, then the equation of this circle is $x^2 + y^2 = r^2$.

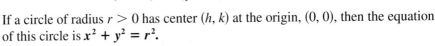

CENTER-RADIUS FORM OF THE EQUATION OF A CIRCLE

A circle with center (h, k) and radius r has equation

$$(x - h)^2 + (y - k)^2 = r^2,$$

which is the **center-radius form** of the equation of the circle. A circle with center $(0, 0)$ and radius r has equation

$$x^2 + y^2 = r^2.$$

▶ **EXAMPLE 5** FINDING THE CENTER-RADIUS FORM

Find the center-radius form of the equation of each circle described.

(a) center at $(-3, 4)$, radius 6   **(b)** center at $(0, 0)$, radius 3

Solution

(a) Use $(h, k) = (-3, 4)$ and $r = 6$.

$$(x - h)^2 + (y - k)^2 = r^2 \quad \text{Center-radius form}$$
$$[x - (-3)]^2 + (y - 4)^2 = 6^2 \quad \text{Substitute.}$$

Watch signs here.

$$(x + 3)^2 + (y - 4)^2 = 36$$

(b) Because the center is the origin and $r = 3$, the equation is

$$x^2 + y^2 = r^2$$
$$x^2 + y^2 = 3^2$$
$$x^2 + y^2 = 9.$$

NOW TRY EXERCISES 47(a) AND 51(a). ◀

▶ **EXAMPLE 6** GRAPHING CIRCLES

Graph each circle discussed in Example 5.

(a) $(x + 3)^2 + (y - 4)^2 = 36$ **(b)** $x^2 + y^2 = 9$

Solution

(a) Writing the equation as

$$[x - (-3)]^2 + (y - 4)^2 = 6^2$$

gives $(-3, 4)$ as the center and 6 as the radius. See Figure 8.

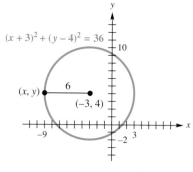

Figure 8

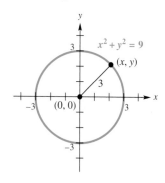

Figure 9

(b) The graph is shown in Figure 9.

NOW TRY EXERCISES 47(b) AND 51(b). ◀

Appendix B   Exercises

Graph the points on a coordinate system and identify the quadrant or axis for each point.

1. $(3, 2)$ **2.** $(-7, 6)$ **3.** $(-7, -4)$ **4.** $(8, -5)$

5. $(0, 5)$ **6.** $(-8, 0)$ **7.** $(4.5, 7)$ **8.** $(-7.5, 8)$

Give the quadrant in which each point lies.

9. $(-5, 1.25)$ **10.** $(\pi, -3)$ **11.** $(-1.4, -2.8)$ **12.** $\left(1 + \sqrt{3}, \frac{1}{2}\right)$

13. *Concept Check*   If (a, b) represents a point that lies in quadrant II, in which quadrant will each point lie?

(a) $(-a, b)$ (b) $(-a, -b)$ (c) $(a, -b)$ (d) (b, a)

Concept Check   Decide whether each statement in Exercises 14–18 is true or false. If the statement is false, tell why.

14. The point $(-1, 3)$ lies in quadrant III of the rectangular coordinate system.

15. The distance from (x_1, y_1) to (x_2, y_2) is given by the expression

$$\sqrt{(x_1 - y_1)^2 + (x_2 - y_2)^2}.$$

16. The distance from the origin to the point (a, b) is $\sqrt{a^2 + b^2}$.

17. The midpoint of the segment joining (a, b) and $(3a, -3b)$ has coordinates $(2a, -b)$.

18. The graph of $y = 2x + 4$ has x-intercept -2 and y-intercept 4.

*A triple of positive integers (a, b, c) is called a **Pythagorean triple** if it satisfies the Pythagorean theorem, $a^2 + b^2 = c^2$. Determine whether each triple is a Pythagorean triple.*

19. $(9, 12, 15)$ **20.** $(6, 8, 10)$ **21.** $(5, 10, 15)$ **22.** $(7, 24, 25)$

23. *Concept Check*   Determine the distance between $(5, -6)$ and the x-axis.

24. *Concept Check*   Determine the distance between $(5, -6)$ and the y-axis.

*For the points P and Q, find **(a)** the distance $d(P, Q)$ and **(b)** the coordinates of the midpoint of the segment PQ. See Examples 1 and 2.*

25. $P(-5, -7), Q(-13, 1)$ **26.** $P(-4, 3), Q(2, -5)$

27. $P(8, 2), Q(3, 5)$ **28.** $P(-6, -5), Q(6, 10)$

29. $P(-8, 4), Q(3, -5)$ **30.** $P(6, -2), Q(4, 6)$

31. $P(3\sqrt{2}, 4\sqrt{5}), Q(\sqrt{2}, -\sqrt{5})$ **32.** $P(-\sqrt{7}, 8\sqrt{3}), Q(5\sqrt{7}, -\sqrt{3})$

Find all values of x or y such that the distance between the given points is as indicated.

33. $(x, 7)$ and $(2, 3)$; 5 **34.** $(5, y)$ and $(8, -1)$; 5

35. $(3, y)$ and $(-2, 9)$; 12 **36.** $(x, 11)$ and $(5, -4)$; 17

37. Use the distance formula to write an equation for all points that are 5 units from $(0, 0)$. Sketch a graph showing these points.

38. Write an equation for all points 3 units from $(-5, 6)$. Sketch a graph showing these points.

Solve each problem.

39. *Bachelor's Degree Attainment*   The graph shows a straight line that approximates the percentage of Americans 25 years and older who earned bachelor's degrees or higher during the years 1990–2006. Use the midpoint formula and the two given points to estimate the percent in 1998. Compare your answer with the actual percent of 24.4.

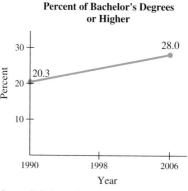

Percent of Bachelor's Degrees or Higher

Source: U.S. Census Bureau.

40. *Poverty Level Income Cutoffs*   The table lists how poverty level income cutoffs (in dollars) for a family of four have changed over time. Use the midpoint formula to approximate the poverty level cutoff in 1987 to the nearest dollar.

Year	Income (in dollars)
1970	3968
1980	8414
1990	13,359
2000	17,603
2004	19,157

Source: U.S. Census Bureau.

For each equation, (a) give a table with at least three ordered pairs that are solutions, and (b) graph the equation. (Hint: You may need more than three points for the graphs in Exercises 43–46.) See Examples 3 and 4.

41. $6y = 3x - 12$ **42.** $6y = -6x + 18$ **43.** $y = x^2$

44. $y = x^2 + 2$ **45.** $y = x^3$ **46.** $y = -x^3$

In Exercises 47–54, (a) find the center-radius form of the equation of each circle, and (b) graph it. See Examples 5 and 6.

47. center $(0,0)$, radius 6   **48.** center $(0,0)$, radius 9

49. center $(2,0)$, radius 6   **50.** center $(0,-3)$, radius 7

51. center $(-2,5)$, radius 4   **52.** center $(4,3)$, radius 5

53. center $(0,4)$, radius 4   **54.** center $(3,0)$, radius 3

Connecting Graphs with Equations   *In Exercises 55–58, use each graph to determine the equation of the circle in center-radius form.*

55.

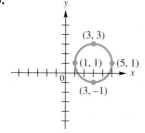

56.

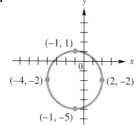

57.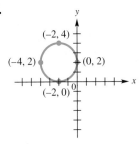

58.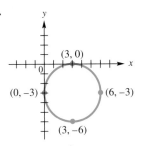

C Functions

Relations and Functions ▪ Domain and Range ▪ Determining Functions from Graphs or Equations ▪ Function Notation ▪ Increasing, Decreasing, and Constant Functions

Relations and Functions   In algebra, we use ordered pairs to represent corresponding quantities. For example, $(3, \$10.50)$ might indicate that you pay $\$10.50$ for 3 gallons of gas. Since the amount you pay *depends* on the number of gallons pumped, the amount (in dollars) is called the *dependent variable,* and the number of gallons pumped is called the *independent variable.* Generalizing, if the value of the variable y depends on the value of the variable x, then y is the **dependent variable** and x is the **independent variable.**

Independent variable ⌐ ⌐ Dependent variable
$$(x, y)$$

Because we can write related quantities using ordered pairs, a set of ordered pairs such as {(3, 10.50), (8, 28.00), (10, 35.00)} is called a **relation.** A special kind of relation called a *function* is very important in mathematics and its applications.

FUNCTION

A **function** is a relation in which, for each distinct value of the first component of the ordered pairs, there is *exactly one* value of the second component.

▶ **EXAMPLE 1**   DECIDING WHETHER RELATIONS DEFINE FUNCTIONS

Decide whether each relation defines a function.

(a) $F = \{(1, 2), (-2, 4), (3, -1)\}$ **(b)** $G = \{(-4, 1), (-2, 1), (-2, 0)\}$

Solution

(a) Relation F is a function, because for each different x-value there is exactly one y-value. We can show this correspondence as follows.

$$\{1, -2, 3\} \quad \text{\textit{x}-values of } F$$
$$\downarrow \quad \downarrow \quad \downarrow$$
$$\{2, \quad 4, -1\} \quad \text{\textit{y}-values of } F$$

(b) In relation G the last two ordered pairs have the same x-value paired with two different y-values (-2 is paired with both 1 and 0), so G is a relation but not a function. ***In a function, no two ordered pairs can have the same first component and different second components.***

Different y-values

$$G = \{(-4,1), (-2,1), (-2,0)\} \quad \text{Not a function}$$

Same x-value

> NOW TRY EXERCISES 1 AND 3. ◀

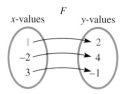

F is a function.

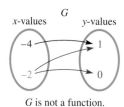

G is not a function.

Figure 1

Relations and functions can also be expressed as a correspondence or *mapping* from one set to another, as shown in Figure 1 for function F and relation G from Example 1. The arrow from 1 to 2 indicates that the ordered pair $(1,2)$ belongs to F—each first component is paired with exactly one second component. In the mapping for relation G, which is not a function, the first component -2 is paired with two different second components, 1 and 0.

Since relations and functions are sets of ordered pairs, we can represent them using tables and graphs. A table and graph for function F is shown in Figure 2.

x	y
1	2
-2	4
3	-1

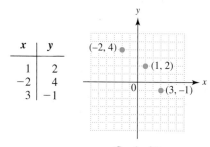

Graph of F

Figure 2

Finally, we can describe a relation or function using a rule that tells how to determine the dependent variable for a specific value of the independent variable. The rule may be given in words: for instance, "the dependent variable is twice the independent variable." Usually the rule is an equation:

Dependent variable $\longrightarrow$ $y = 2x$. $\longleftarrow$ Independent variable

This is the most efficient way to define a relation or function. ***In a function, there is exactly one value of the dependent variable, the second component, for each value of the independent variable, the first component.*** This is what makes functions so important in applications.

Domain and Range For every relation there are two important sets of elements called the *domain* and *range*.

DOMAIN AND RANGE

In a relation, the set of all values of the independent variable (x) is the **domain.** The set of all values of the dependent variable (y) is the **range.**

▶ **EXAMPLE 2** FINDING DOMAINS AND RANGES OF RELATIONS

Give the domain and range of each relation. Tell whether the relation defines a function.

(a) $\{(3, -1), (4, 2), (4, 5), (6, 8)\}$

(b)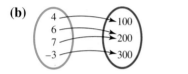

(c)

x	y
-5	2
0	2
5	2

Solution

(a) The domain, the set of x-values, is $\{3, 4, 6\}$; the range, the set of y-values, is $\{-1, 2, 5, 8\}$. This relation is not a function because the same x-value, 4, is paired with two different y-values, 2 and 5.

(b) The domain is $\{4, 6, 7, -3\}$; the range is $\{100, 200, 300\}$. This mapping defines a function. Each x-value corresponds to exactly one y-value.

(c) This relation is a set of ordered pairs, so the domain is the set of x-values $\{-5, 0, 5\}$ and the range is the set of y-values $\{2\}$. The table defines a function because each different x-value corresponds to exactly one y-value (even though it is the same y-value).

NOW TRY EXERCISES 7, 9, AND 11. ◀

As mentioned previously, the graph of a relation is the graph of its ordered pairs. The graph gives a picture of the relation, which can be used to determine its domain and range.

▶ **EXAMPLE 3** FINDING DOMAINS AND RANGES FROM GRAPHS

Give the domain and range of each relation.

(a)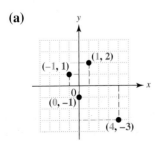

(b)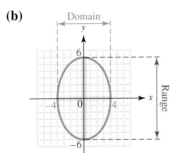

(c)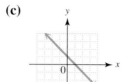

(d)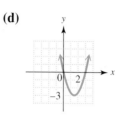

Solution

(a) The domain is the set of x-values, $\{-1, 0, 1, 4\}$. The range is the set of y-values, $\{-3, -1, 1, 2\}$.

(b) The x-values of the points on the graph include all numbers between -4 and 4, inclusive. The y-values include all numbers between -6 and 6, inclusive. Using interval notation (Appendix A),

the domain is $[-4, 4]$ and   the range is $[-6, 6]$.

(c) The arrowheads indicate that the line extends indefinitely left and right, as well as up and down. Therefore, both the domain and the range include all real numbers, written $(-\infty, \infty)$.

(d) The arrowheads indicate that the graph extends indefinitely left and right, as well as upward. The domain is $(-\infty, \infty)$. Because there is a least y-value, -3, the range includes all numbers greater than or equal to -3, written $[-3, \infty)$.

> NOW TRY EXERCISES 13 AND 15. ◀

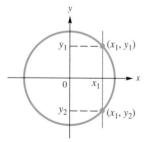

Function–each x-value
corresponds to only one
y-value.

(a)

Determining Functions from Graphs or Equations
Most of the relations we have seen in the examples are functions—that is, each x-value corresponds to exactly one y-value. Since each value of x leads to only one value of y in a function, any vertical line drawn through the graph of a function must intersect the graph in at most one point. This is the **vertical line test** for a function.

VERTICAL LINE TEST

If each vertical line intersects a graph in at most one point, then the graph is that of a function.

Not a function–the same
x-value corresponds to
two different y-values.

(b)

Figure 3

The graph in Figure 3(a) represents a function—each vertical line intersects the graph in at most one point. The graph in Figure 3(b) is not the graph of a function since a vertical line intersects the graph in more than one point.

> NOW TRY EXERCISE 17. ◀

The vertical line test is a simple method for identifying a function defined by a graph. It is more difficult deciding whether a relation defined by an equation is a function, as well as determining the domain and range. The next example gives some hints that may help.

▶ EXAMPLE 4   IDENTIFYING FUNCTIONS, DOMAINS, AND RANGES

Decide whether each relation defines a function and give the domain and range.

(a) $y = x + 4$ **(b)** $y = \sqrt{2x - 1}$ **(c)** $y^2 = x$ **(d)** $y = \dfrac{5}{x - 1}$

Solution

(a) In the defining equation (or rule), $y = x + 4$, y is always found by adding 4 to x. Thus, each value of x corresponds to just one value of y and the relation defines a function; x can be any real number, so the domain is $\{x \mid x \text{ is a real number}\}$ or $(-\infty, \infty)$. Since y is always 4 more than x, y also may be any real number, and so the range is $(-\infty, \infty)$.

(b) For any choice of x in the domain of $y = \sqrt{2x - 1}$, there is exactly one corresponding value for y (the radical is a nonnegative number), so this equation defines a function. Since the equation involves a square root, the quantity under the radical sign cannot be negative. Thus,

$$2x - 1 \geq 0 \quad \text{Solve the inequality. (Appendix A)}$$
$$2x \geq 1 \quad \text{Add 1.}$$
$$x \geq \frac{1}{2}, \quad \text{Divide by 2.}$$

and the domain of the function is $\left[\frac{1}{2}, \infty\right)$. Because the radical is a nonnegative number, as x takes values greater than or equal to $\frac{1}{2}$, the range is $y \geq 0$, that is, $[0, \infty)$. See Figure 4.

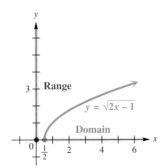

Figure 4

(c) The ordered pairs $(16, 4)$ and $(16, -4)$ both satisfy the equation $y^2 = x$. Since one value of x, 16, corresponds to two values of y, 4 and -4, this equation does not define a function. Because x is equal to the square of y, the values of x must always be nonnegative. The domain of the relation is $[0, \infty)$. Any real number can be squared, so the range of the relation is $(-\infty, \infty)$. See Figure 5.

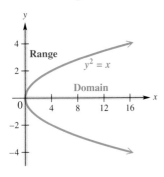

Figure 5

(d) Given any value of x in the domain of $y = \frac{5}{x-1}$, we find y by subtracting 1, then dividing the result into 5. This process produces exactly one value of y for each value in the domain, so this equation defines a function. The domain includes all real numbers except those that make the denominator 0. We find these numbers by setting the denominator equal to 0 and solving for x.

$$x - 1 = 0$$
$$x = 1 \quad \text{Add 1. (Appendix A)}$$

Thus, the domain includes all real numbers except 1, written as the interval $(-\infty, 1) \cup (1, \infty)$. Values of y can be positive or negative, but never 0, because a fraction cannot equal 0 unless its numerator is 0. Therefore, the range is the interval $(-\infty, 0) \cup (0, \infty)$, as shown in Figure 6.

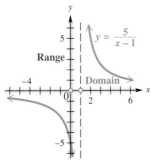

Figure 6

NOW TRY EXERCISES 19, 23, 27, AND 29. ◀

> ▶ **Note** Some trigonometric functions are defined in such a way that restrictions are necessary since their denominators cannot equal 0. Example 4(d) illustrates this idea with an algebraic function.

Function Notation   When a function f is defined with a rule or an equation using x and y for the independent and dependent variables, we say "y is a function of x" to emphasize that y *depends on* x. We use the notation

$$y = f(x),$$

called **function notation,** to express this and read $f(x)$ as "**f of x.**" The letter f is the name given to this function. For example, if $y = 9x - 5$, we can name the function f and write

$$f(x) = 9x - 5.$$

Note that $f(x)$ *is just another name for the dependent variable y.* For example, if $y = f(x) = 9x - 5$ and $x = 2$, then we find y, or $f(2)$, by replacing x with 2.

$$f(2) = 9 \cdot 2 - 5 = 13$$

The statement "if $x = 2$, then $y = 13$" represents the ordered pair $(2, 13)$ and is abbreviated with function notation as

$$f(2) = 13.$$

Read $f(2)$ as "f of 2" or "f at 2." Also,

$$f(0) = 9 \cdot 0 - 5 = -5 \quad \text{and} \quad f(-3) = 9(-3) - 5 = -32.$$

These ideas can be illustrated as follows.

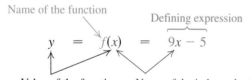

▶ EXAMPLE 5   USING FUNCTION NOTATION

Let $f(x) = -x^2 + 5x - 3$ and $g(x) = 2x + 3$. Find and simplify each of the following.

(a) $f(2)$ **(b)** $f(q)$ **(c)** $g(a + 1)$

Solution

(a) $f(x) = -x^2 + 5x - 3$

$\quad\quad f(2) = -2^2 + 5 \cdot 2 - 3$ Replace x with 2.

$\quad\quad\quad\quad = -4 + 10 - 3$ Apply the exponent; multiply.

$\quad\quad\quad\quad = 3$ Add and subtract.

Thus, $f(2) = 3$; the ordered pair $(2, 3)$ belongs to f.

(b) $f(x) = -x^2 + 5x - 3$

$f(q) = -q^2 + 5q - 3$ Replace x with q.

(c) $g(x) = 2x + 3$

$g(a + 1) = 2(a + 1) + 3$ Replace x with $a + 1$.

$= 2a + 2 + 3$

$= 2a + 5$

> NOW TRY EXERCISES 33, 41, AND 43. ◄

Functions can be evaluated in a variety of ways, as shown in Example 6.

▶ **EXAMPLE 6**   USING FUNCTION NOTATION

For each function, find $f(3)$.

(a) $f(x) = 3x - 7$ **(b)** $f = \{(-3, 5), (0, 3), (3, 1), (6, -1)\}$

(c) 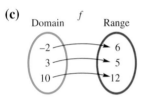   **(d)**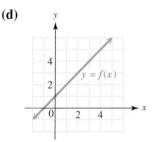

Solution

(a) $f(x) = 3x - 7$

$f(3) = 3(3) - 7$ Replace x with 3.

$f(3) = 2$

(b) For $f = \{(-3, 5), (0, 3), (3, 1), (6, -1)\}$, we want $f(3)$, the y-value of the ordered pair where $x = 3$. As indicated by the ordered pair $(3, 1)$, when $x = 3, y = 1$, so $f(3) = 1$.

(c) In the mapping, the domain element 3 is paired with 5 in the range, so $f(3) = 5$.

(d) To evaluate $f(3)$, find 3 on the x-axis. See Figure 7. Then move up until the graph of f is reached. Moving horizontally to the y-axis gives 4 for the corresponding y-value. Thus, $f(3) = 4$.

> NOW TRY EXERCISES 45, 47, AND 49. ◄

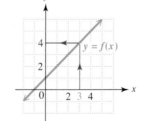

Figure 7

Increasing, Decreasing, and Constant Functions

Informally speaking, a function *increases* on an interval of its domain if its graph rises from left to right on the interval. It *decreases* on an interval of its domain if its graph falls from left to right on the interval. It is *constant* on an interval of its domain if its graph is horizontal on the interval. The formal definitions of these concepts follow.

INCREASING, DECREASING, AND CONSTANT FUNCTIONS

Suppose that a function f is defined over an interval I. If x_1 and x_2 are in I,

(a) f **increases** on I if, whenever $x_1 < x_2, f(x_1) < f(x_2)$;

(b) f **decreases** on I if, whenever $x_1 < x_2, f(x_1) > f(x_2)$;

(c) f is **constant** on I if, for every x_1 and $x_2, f(x_1) = f(x_2)$.

Figure 8 illustrates these ideas.

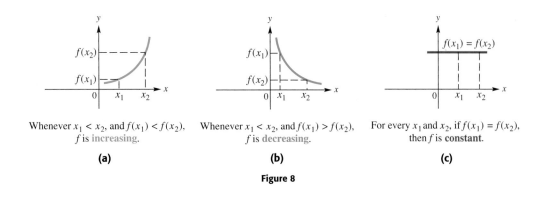

Whenever $x_1 < x_2$, and $f(x_1) < f(x_2)$, f is increasing.

(a)

Whenever $x_1 < x_2$, and $f(x_1) > f(x_2)$, f is decreasing.

(b)

For every x_1 and x_2, if $f(x_1) = f(x_2)$, then f is **constant**.

(c)

Figure 8

▶ **Note** To decide whether a function is increasing, decreasing, or constant on an interval, ask yourself *"What does y do as x goes from left to right?"*

There can be confusion regarding whether endpoints of an interval should be included when determining intervals over which a function is increasing or decreasing. For example, consider the graph of $y = f(x) = x^2 + 4$, shown in Figure 9. Is it increasing on $[0, \infty)$, or just on $(0, \infty)$?

The definition of increasing and decreasing allows us to include 0 as a part of the interval I over which this function is increasing, because if we let $x_1 = 0$, then $f(0) < f(x_2)$ whenever $0 < x_2$. Thus, $f(x) = x^2 + 4$ is increasing on $[0, \infty)$. A similar discussion can be used to show that this function is decreasing on $(-\infty, 0]$. Do not confuse these concepts by saying that f both increases and decreases at the point $(0, 0)$. *The concepts of increasing and decreasing functions apply to intervals of the domain and not to individual points.*

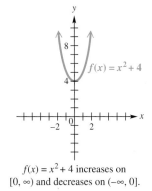

$f(x) = x^2 + 4$ increases on $[0, \infty)$ and decreases on $(-\infty, 0]$.

Figure 9

It is not incorrect to say that $f(x) = x^2 + 4$ is increasing on $(0, \infty)$; there are infinitely many intervals over which it increases. However, we generally give the largest possible interval when determining where a function increases or decreases. (*Source:* Stewart J., *Calculus,* Fourth Edition, Brooks/Cole Publishing Company, 1999, p. 21.)

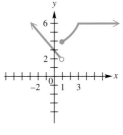

Figure 10

▶ **EXAMPLE 7** DETERMINING INTERVALS OVER WHICH A FUNCTION IS INCREASING, DECREASING, OR CONSTANT

Figure 10 shows the graph of a function. Determine the intervals over which the function is increasing, decreasing, or constant.

Solution We should ask, "What is happening to the *y*-values as the *x*-values are getting larger?" Moving from left to right on the graph, we see that on the interval $(-\infty, 1)$, the *y*-values are *decreasing;* on the interval $[1, 3]$, the *y*-values are *increasing;* and on the interval $[3, \infty)$, the *y*-values are *constant* (and equal to 6). Therefore, the function is decreasing on $(-\infty, 1)$, increasing on $[1, 3]$, and constant on $[3, \infty)$.

NOW TRY EXERCISE 55. ◀

Appendix C   Exercises

Decide whether each relation defines a function. See Example 1.

1. $\{(5, 1), (3, 2), (4, 9), (7, 8)\}$ **2.** $\{(8, 0), (5, 7), (9, 3), (3, 8)\}$

3. $\{(2, 4), (0, 2), (2, 6)\}$ **4.** $\{(9, -2), (-3, 5), (9, 1)\}$

5. $\{(-3, 1), (4, 1), (-2, 7)\}$ **6.** $\{(-12, 5), (-10, 3), (8, 3)\}$

Decide whether each relation defines a function and give the domain and range. See Examples 1–3.

7. $\{(1, 1), (1, -1), (0, 0), (2, 4), (2, -4)\}$ **8.** $\{(2, 5), (3, 7), (4, 9), (5, 11)\}$

9.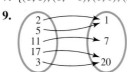

10.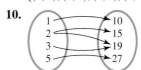

11.

x	y
0	0
-1	1
-2	2

12.

x	y
0	0
1	-1
2	-2

13.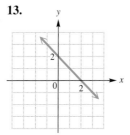

14.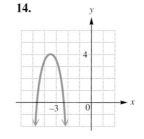

15.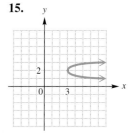

16.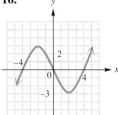

17.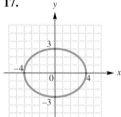

18.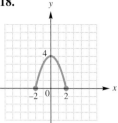

Decide whether each relation defines y as a function of x. Give the domain and range. See Example 4.

19. $y = x^2$ **20.** $y = x^3$ **21.** $x = y^6$ **22.** $x = y^4$

23. $y = 2x - 5$ **24.** $y = -6x + 4$ **25.** $y = \sqrt{x}$ **26.** $y = -\sqrt{x}$

27. $y = \sqrt{4x + 1}$ **28.** $y = \sqrt{7 - 2x}$ **29.** $y = \dfrac{2}{x - 3}$ **30.** $y = \dfrac{-7}{x - 5}$

31. *Concept Check* Choose the correct response: The notation $f(3)$ means
  A. the variable f times 3 or $3f$.
  B. the value of the dependent variable when the independent variable is 3.
  C. the value of the independent variable when the dependent variable is 3.
  D. f equals 3.

32. *Concept Check* Give an example of a function from everyday life. (*Hint:* Fill in the blanks: _____ depends on _____, so _____ is a function of _____.)

Let $f(x) = -3x + 4$ and $g(x) = -x^2 + 4x + 1$. Find and simplify each of the following. See Example 5.

33. $f(0)$ **34.** $f(-3)$ **35.** $g(-2)$ **36.** $g(10)$

37. $f\left(\dfrac{1}{3}\right)$ **38.** $f\left(-\dfrac{7}{3}\right)$ **39.** $g\left(\dfrac{1}{2}\right)$ **40.** $g\left(-\dfrac{1}{4}\right)$

41. $f(p)$ **42.** $g(k)$ **43.** $f(x + 2)$ **44.** $f(a + 4)$

For each function, find (a) $f(2)$ and (b) $f(-1)$. See Example 6.

45. $f = \{(-1, 3), (4, 7), (0, 6), (2, 2)\}$ **46.** $f = \{(2, 5), (3, 9), (-1, 11), (5, 3)\}$

47. f 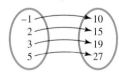   **48.** f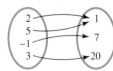

49. 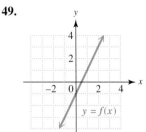   **50.**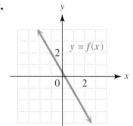

In Exercises 51 and 52, use the given graph of y = f(x) to find each function value:
(a) f(−2), (b) f(0), (c) f(1), and (d) f(4). See Example 6(d).

51.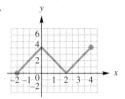

52.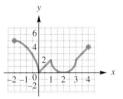

53. *Concept Check*   Let f(x) be the function in the graph. Find each of the following.

(a) f(0)   (b) f(6)
(c) a negative number a for which f(a) = 0
(d) three positive values of x for which f(x) = 10
(e) the distance between (8, f(8)) and (10, f(10))

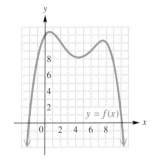

54. *Concept Check*   Use the table to answer each question for a function y = f(x).

(a) What is f(2)?
(b) If f(x) = −2.4, what is the value of x?
(c) At what point does the graph of y = f(x) intersect the y-axis?
(d) At what point does the graph of y = f(x) intersect the x-axis?

x	y
0	3.6
1	2.4
2	1.2
3	0
4	−1.2
5	−2.4
6	−3.6

Determine the intervals of the domain for which each function is (a) increasing, (b) decreasing, and (c) constant. See Example 7.

55.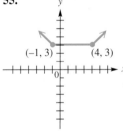

56.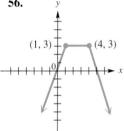

57.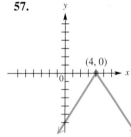

58.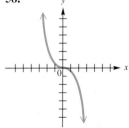

59.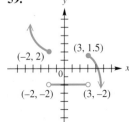

60.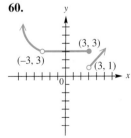

Concept Check *Solve each problem.*

61. *Electricity Usage* The graph shows the daily megawatts of electricity used on a record-breaking summer day in Sacramento, California.

  (a) Is this the graph of a function?
  (b) What is the domain?
  (c) Estimate the number of megawatts used at 8 A.M.
  (d) At what time was the most electricity used? the least electricity?
  (e) Call this function f. What is $f(12)$? What does it mean?
  (f) During what time intervals is electricity usage increasing? decreasing?

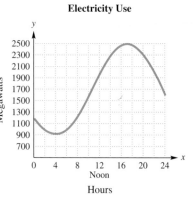

Electricity Use

Source: Sacramento Municipal Utility District.

62. *Height of a Ball* A ball is thrown straight up into the air. The function defined by $y = h(t)$ in the graph gives the height of the ball (in feet) at t seconds. (*Note:* The graph does *not* show the path of the ball. The ball is rising straight up and then falling straight down.)

  (a) What is the height of the ball at 2 sec?
  (b) When will the height be 192 ft?
  (c) During what time intervals is the ball going up? down?
  (d) How high does the ball go, and when does the ball reach its maximum height?
  (e) After how many seconds does the ball hit the ground?

Height of a Thrown Ball

D Graphing Techniques

Stretching and Shrinking ▪ Reflecting ▪ Symmetry ▪ Translations

Graphing techniques presented here review how to graph functions that are defined by altering the equation of a basic function.

Stretching and Shrinking

We begin by considering how the graph of $y = af(x)$ or $y = f(ax)$ compares to the graph of $y = f(x)$, where $a > 0$.

▶ **EXAMPLE 1** STRETCHING OR SHRINKING A GRAPH

Graph $g(x) = 2x^2$ and $h(x) = \frac{1}{2}x^2$.

Solution Comparing the values for $f(x) = x^2$ and $g(x) = 2x^2$ in the table beside Figure 1 on the next page, we see that for corresponding x-values, the y-values of g are each twice those of f. Thus the graph of $g(x)$, shown in blue in Figure 1, is narrower than that of $f(x)$, shown in red for comparison.

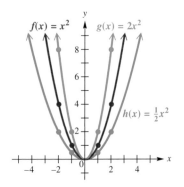

x	$f(x)$ x^2	$g(x)$ $2x^2$	$h(x)$ $\frac{1}{2}x^2$
-2	4	8	2
-1	1	2	$\frac{1}{2}$
0	0	0	0
1	1	2	$\frac{1}{2}$
2	4	8	2

Figure 1

The graph of $h(x) = \frac{1}{2}x^2$ is also the same general shape as that of $f(x) = x^2$, but here the coefficient $\frac{1}{2}$ causes the graph of $h(x)$ to be wider than the graph of $f(x)$. See the values in the table. The graph of $h(x)$ is shown in green in Figure 1.

NOW TRY EXERCISES 5 AND 7. ◀

▶ **Note** Recall that the graphs in Figure 1 are called **parabolas.** The lowest (or highest) point on a parabola is its **vertex.** The vertex of each of the parabolas in Figure 1 is $(0, 0)$.

Recall that $|a|$ is the absolute value of a number a.

$$|a| = \begin{cases} a \text{ if } a \text{ is positive or } 0 \\ -a \text{ if } a \text{ is negative} \end{cases}$$

Thus, $|2| = |2|$ and $|-2| = |2|$.

Graph of the absolute value function

The graphs in Example 1 suggest the following generalizations.

STRETCHING AND SHRINKING

The graph of $g(x) = af(x)$ has the same general shape as the graph of $f(x)$.

If $|a| > 1$, then the graph is stretched vertically (narrower) compared to the graph of $f(x)$.

If $0 < |a| < 1$, then the graph is shrunken vertically (wider) compared to the graph of $f(x)$.

In general, the larger the value of $|a|$, the greater the stretch. The smaller the value of $|a|$, the greater the shrink.

Reflecting Forming the mirror image of a graph across a line is called **reflecting the graph across the line.**

▶ **EXAMPLE 2** REFLECTING A GRAPH ACROSS AN AXIS

Graph $g(x) = -\sqrt{x}$ and $h(x) = \sqrt{-x}$.

Solution The table of values for $g(x) = -\sqrt{x}$ and $f(x) = \sqrt{x}$ are shown with their graphs in Figure 2. As the table suggests, every y-value of the graph of $g(x) = -\sqrt{x}$ is the negative of the corresponding y-value of $f(x) = \sqrt{x}$. This has the effect of reflecting the graph across the x-axis.

x	$f(x)$ $\sqrt{x}$	$g(x)$ $-\sqrt{x}$	$h(x)$ $\sqrt{-x}$
-4	undefined	undefined	2
-1	undefined	undefined	1
0	0	0	0
1	1	-1	undefined
4	2	-2	undefined

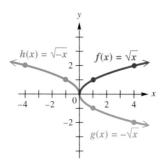

Figure 2

The domain of $h(x) = \sqrt{-x}$ is $(-\infty, 0]$, while the domain of $f(x) = \sqrt{x}$ is $[0, \infty)$. If we choose x-values for $h(x)$ that are the negatives of those we use for $f(x)$, we see that the corresponding y-values are the same. Thus, the graph of h is a reflection of the graph of f across the y-axis. See Figure 2.

NOW TRY EXERCISES 9 AND 11. ◀

The graphs in Example 2 suggest the following generalizations.

REFLECTING ACROSS AN AXIS

The graph of $y = -f(x)$ is the same as the graph of $y = f(x)$ reflected across the x-axis.

The graph of $y = f(-x)$ is the same as the graph of $y = f(x)$ reflected across the y-axis.

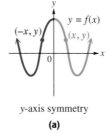

y-axis symmetry

(a)

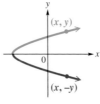

x-axis symmetry

(b)

Figure 3

Symmetry The graph of f shown in Figure 3(a) is cut in half by the y-axis with each half the mirror image of the other half. Such a graph is *symmetric with respect to the y-axis: the point $(-x, y)$ is on the graph whenever the point (x, y) is on the graph.*

Similarly, if the graph in Figure 3(b) were folded in half along the x-axis, the portion at the top would exactly match the portion at the bottom. Such a graph is *symmetric with respect to the x-axis: the point $(x, -y)$ is on the graph whenever the point (x, y) is on the graph.*

SYMMETRY WITH RESPECT TO AN AXIS

The graph of an equation is **symmetric with respect to the y-axis** if the replacement of x with $-x$ results in an equivalent equation.

The graph of an equation is **symmetric with respect to the x-axis** if the replacement of y with $-y$ results in an equivalent equation.

▶ **EXAMPLE 3**   TESTING FOR SYMMETRY WITH RESPECT TO AN AXIS

Test for symmetry with respect to the x-axis and the y-axis.

(a) $y = x^2 + 4$ **(b)** $x = y^2 - 3$ **(c)** $x^2 + y^2 = 16$ **(d)** $2x + y = 4$

Solution

(a) In $y = x^2 + 4$, replace x with $-x$.

$$y = x^2 + 4$$
$$y = (-x)^2 + 4 \quad \text{Equivalent}$$
$$y = x^2 + 4$$

The result is the same as the original equation, so the graph, shown in Figure 4, is symmetric with respect to the y-axis. Check algebraically that the graph is *not* symmetric with respect to the x-axis.

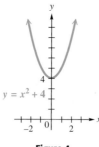

Figure 4 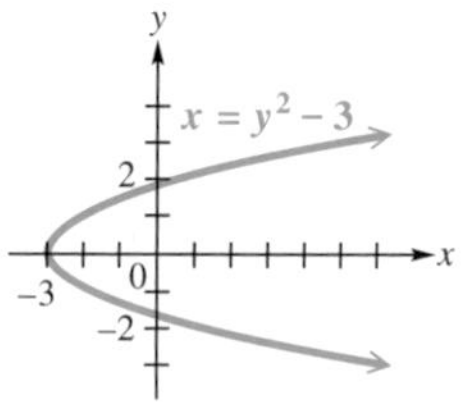   **Figure 5**

(b) In $x = y^2 - 3$, replace y with $-y$ to get $x = (-y)^2 - 3 = y^2 - 3$, the same as the original equation. The graph is symmetric with respect to the x-axis, as shown in Figure 5. It is not symmetric with respect to the y-axis.

(c) Substituting $-x$ for x and $-y$ for y in $x^2 + y^2 = 16$, we get

$$(-x)^2 + y^2 = 16 \quad \text{and} \quad x^2 + (-y)^2 = 16.$$

Both simplify to $\qquad x^2 + y^2 = 16.$

Thus the graph, a circle of radius 4 centered at the origin, is symmetric with respect to both axes.

(d) In $2x + y = 4$, replace x with $-x$, and then replace y with $-y$; neither case produces an equivalent equation. This graph is not symmetric with respect to either axis. ◀

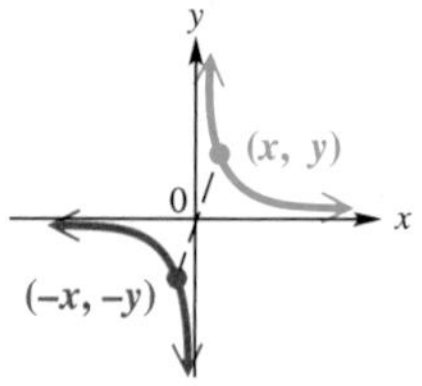

Another kind of symmetry occurs when a graph can be rotated 180° about the origin, with the result coinciding exactly with the original graph. Symmetry of this type is called *symmetry with respect to the origin: the point $(-x, -y)$ is on the graph whenever the point (x, y) is on the graph.* See Figure 6.

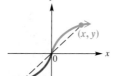

Origin symmetry

Figure 6

SYMMETRY WITH RESPECT TO THE ORIGIN

The graph of an equation is **symmetric with respect to the origin** if the replacement of both x with $-x$ and y with $-y$ results in an equivalent equation.

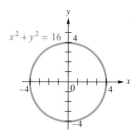

$x^2 + y^2 = 16$

Figure 7

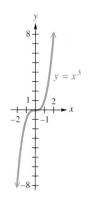

$y = x^3$

Figure 8

▶ **EXAMPLE 4** TESTING FOR SYMMETRY WITH RESPECT TO THE ORIGIN

Test for symmetry with respect to the origin.

(a) $x^2 + y^2 = 16$ **(b)** $y = x^3$

Solution In each case, replace x with $-x$ and y with $-y$.

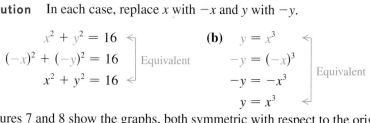

(a)
$$x^2 + y^2 = 16$$
$$(-x)^2 + (-y)^2 = 16$$ Equivalent
$$x^2 + y^2 = 16$$

(b)
$$y = x^3$$
$$-y = (-x)^3$$
$$-y = -x^3$$ Equivalent
$$y = x^3$$

Figures 7 and 8 show the graphs, both symmetric with respect to the origin.

NOW TRY EXERCISE 23. ◀

Notice the following important concepts:

1. A graph symmetric with respect to both the x- and y-axes is automatically symmetric with respect to the origin. (See Figure 7.)

2. A graph symmetric with respect to the origin need *not* be symmetric with respect to either axis. (See Figure 8.)

3. Of the three types of symmetry—with respect to the x-axis, the y-axis, and the origin—a graph possessing any two must also exhibit the third type.

NOW TRY EXERCISES 19 AND 21. ◀

Translations The next examples show the results of horizontal and vertical shifts, called **translations,** of the graph of $f(x) = x^2$.

▶ **EXAMPLE 5** TRANSLATING GRAPHS

Graph each function.

(a) $g(x) = x^2 + 3$ **(b)** $g(x) = (x - 4)^2$

Solution

(a) By comparing the table of values for $g(x) = x^2 + 3$ and $f(x) = x^2$ shown with Figure 9, we see that for corresponding x-values, the y-values of g are each 3 more than those for f. Thus, the graph of $g(x) = x^2 + 3$ is the same as that of $f(x) = x^2$, but translated 3 units *up*. The vertex of the graph is at $(0, 3)$, and the graph is symmetric with respect to the y-axis.

x	$f(x)$ x^2	$g(x)$ $x^2 + 3$
-2	4	7
-1	1	4
0	0	3
1	1	4
2	4	7

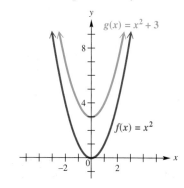

Figure 9

(b) The graph of $g(x) = (x - 4)^2$ in Figure 10 is the same as that of $f(x) = x^2$, but translated 4 units *to the right*. The vertex is at $(4, 0)$, and the graph is symmetric with respect to the line $x = 4$.

x	$f(x)$ x^2	$g(x)$ $(x - 4)^2$
1	1	9
2	4	4
3	9	1
4	16	0
5	25	1
6	36	4

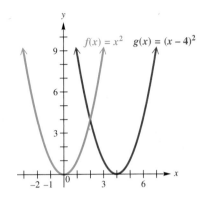

Figure 10

NOW TRY EXERCISES 27 AND 31. ◄

The graphs in Example 5 and Figures 11 and 12 suggest the following generalizations.

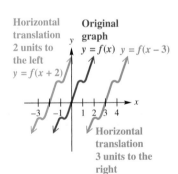

Figure 11

Figure 12

VERTICAL AND HORIZONTAL TRANSLATIONS

If a function g is defined by $g(x) = f(x) + c$, where c is a real number, then for every point (x, y) on the graph of f, there will be a corresponding point $(x, y + c)$ on the graph of g. The graph of g will be the same as the graph of f, but translated c units up if c is positive or $|c|$ units down if c is negative. The graph of g is called a **vertical translation** of the graph of f. See Figure 11.

If a function g is defined by $g(x) = f(x - c)$, where c is a real number, then for every point (x, y) on the graph of f, there will be a corresponding point $(x + c, y)$ on the graph of g. The graph of g will be the same as the graph of f, but translated c units to the right if c is positive or $|c|$ units to the left if c is negative. The graph of g is called a **horizontal translation** of the graph of f. See Figure 12.

Vertical and horizontal translations are summarized in the table, where f is a function, and c is a positive number.

To Graph:	Shift the Graph of $y = f(x)$ by c Units:
$g(x) = f(x) + c$	up
$g(x) = f(x) - c$	down
$g(x) = f(x + c)$	left
$g(x) = f(x - c)$	right

NOW TRY EXERCISE 1. ◄

► **EXAMPLE 6** **USING MORE THAN ONE TRANSFORMATION ON A GRAPH**

Graph $f(x) = -\dfrac{1}{2}(x - 4)^2 + 3$.

Solution The graph of f will have the same shape as that of $y = x^2$, but is wider (that is, shrunken vertically) because $\left|-\frac{1}{2}\right| < 1$, and reflected across the x-axis because $-\frac{1}{2} < 0$. The vertex of f is translated 4 units to the right and 3 units up compared to the graph of $y = x^2$. Because of the reflection, the vertex is the highest point on the graph, as shown in Figure 13. The graph is symmetric with respect to the line $x = 4$.

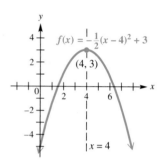

Figure 13

NOW TRY EXERCISE 39. ◄

Appendix D Exercises

1. *Concept Check* Match each equation in Column I with a description of its graph from Column II as it relates to the graph of $y = x^2$.

I	II
(a) $y = (x - 7)^2$	**A.** a translation 7 units to the left
(b) $y = x^2 - 7$	**B.** a translation 7 units to the right
(c) $y = 7x^2$	**C.** a vertical shrink
(d) $y = (x + 7)^2$	**D.** a translation 7 units up
(e) $y = x^2 + 7$	**E.** a translation 7 units down
(f) $y = \dfrac{1}{7}x^2$	**F.** a vertical stretch

2. *Concept Check* Match each equation in Column I with a description of its graph from Column II as it relates to the graph of $y = x^2$.

I	II
(a) $y = 4x^2$	**A.** a translation 4 units to the left
(b) $y = -x^2$	**B.** a translation 4 units up
(c) $y = (-x)^2$	**C.** a reflection across the x-axis
(d) $y = (x + 4)^2$	**D.** a reflection across the y-axis
(e) $y = x^2 + 4$	**E.** a vertical stretch

3. *Concept Check* Match each equation in parts (a)–(i) with the sketch of its graph.

(a) $y = x^2 + 2$

(b) $y = x^2 - 2$

(c) $y = (x + 2)^2$

(d) $y = (x - 2)^2$

(e) $y = 2x^2$

(f) $y = -x^2$

(g) $y = (x - 2)^2 + 1$

(h) $y = (x + 2)^2 + 1$

(i) $y = (x + 2)^2 - 1$

A.

B.

C.
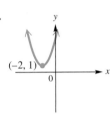

D.
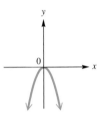

E.
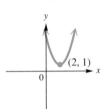

F.
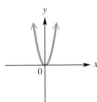

G.
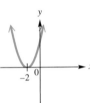

H.
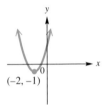

I.

Graph each function. See Examples 1 and 2.

4. $y = 3x^2$

5. $y = 4x^2$

6. $y = \frac{1}{3}x^2$

7. $y = \frac{2}{3}x^2$

8. $y = -\frac{1}{2}x^2$

9. $y = -3x^2$

10. $y = 2\sqrt{-x}$

11. $y = \sqrt{-2x}$

Concept Check In Exercises 12–14, suppose the point $(8, 12)$ is on the graph of $y = f(x)$. Find a point on the graph of each function.

12. (a) $y = f(x + 4)$

(b) $y = f(x) + 4$

13. (a) $y = \frac{1}{4}f(x)$

(b) $y = 4f(x)$

14. (a) the reflection of the graph of $y = f(x)$ across the x-axis

(b) the reflection of the graph of $y = f(x)$ across the y-axis

Concept Check Plot each point, and then plot the points that are symmetric to the given point with respect to the (a) x-axis, (b) y-axis, and (c) origin.

15. $(5, -3)$

16. $(-6, 1)$

17. $(-4, -2)$

18. $(-8, 0)$

Without graphing, determine whether each equation has a graph that is symmetric with respect to the x-axis, the y-axis, the origin, or none of these. See Examples 3 and 4.

19. $y = x^2 + 2$

20. $y = 2x^4 - 1$

21. $x^2 + y^2 = 10$

22. $y^2 = \frac{-5}{x^2}$

23. $y = -3x^3$

24. $y = x^3 - x$

25. $y = x^2 - x + 7$

26. $y = x + 12$

Graph each function. See Examples 5 and 6.

27. $y = x^2 - 1$ **28.** $y = x^2 + 1$ **29.** $y = x^2 + 2$

30. $y = x^2 - 2$ **31.** $y = (x - 1)^2$ **32.** $y = (x - 2)^2$

33. $y = (x + 2)^2$ **34.** $y = (x + 3)^2$ **35.** $y = (x + 3)^2 - 4$

36. $y = (x - 5)^2 - 4$ **37.** $y = 2x^2 - 1$ **38.** $y = \dfrac{2}{3}(x - 2)^2$

39. $f(x) = 2(x - 2)^2 - 4$ **40.** $f(x) = -3(x - 2)^2 + 1$

Concept Check   *Each of the following graphs is obtained from the graph of $f(x) = |x|$ or $g(x) = \sqrt{x}$ by applying several of the transformations discussed in this section. Describe the transformations and give the equation for the graph.*

41.
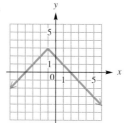

42.
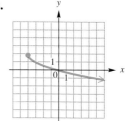

43.
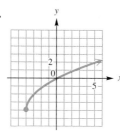

44.
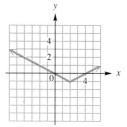

45. Find the function g whose graph can be obtained by translating the graph of $f(x) = 2x + 5$ up 2 units and to the left 3 units.

46. Find the function g whose graph can be obtained by translating the graph of $f(x) = 3 - x$ down 2 units and to the right 3 units.

47. Suppose the equation $y = F(x)$ is changed to $y = cF(x)$, for some constant c. What is the effect on the graph of $y = F(x)$? Discuss the effect depending on whether $c > 0$ or $c < 0$, and $|c| > 1$ or $|c| < 1$.

48. Suppose $y = F(x)$ is changed to $y = F(x + h)$. How are the graphs of these equations related? Is the graph of $y = F(x) + h$ the same as the graph of $y = F(x + h)$? If not, how do they differ?

Glossary

A

absolute value (modulus) of a complex number When a complex number $x + yi$ is written in trigonometric (or polar) form as $r(\cos\theta + i\sin\theta)$, the number r is called the absolute value (or modulus) of the complex number. (Section 8.2)

acute angle An acute angle is an angle measuring between 0° and 90°. (Section 1.1)

addition of ordinates Addition of ordinates is a method for graphing a function that is the sum of two other functions by adding the y-values of the two functions at selected x-values. (Section 4.4)

airspeed In air navigation, the airspeed of a plane is its speed relative to the air. (Section 7.5)

ambiguous case The situation in which the lengths of two sides of a triangle and the measure of the angle opposite one of them are given (SSA) is called the ambiguous case of the law of sines. Depending on the given measurements, this combination of given parts may result in 0, 1, or 2 possible triangles. (Section 7.2)

amplitude The amplitude of a periodic function is half the difference between the maximum and minimum values of the function. (Section 4.1)

angle An angle is formed by rotating a ray around its endpoint. (Section 1.1)

angle of depression The angle of depression from point X to point Y (below X) is the acute angle formed by ray XY and a horizontal ray with endpoint at X. (Section 2.4)

angle of elevation The angle of elevation from point X to point Y (above X) is the acute angle formed by ray XY and a horizontal ray with endpoint at X. (Section 2.4)

Angle-Side-Angle (ASA) The Angle-Side-Angle (ASA) congruence axiom states that if two angles and the included side of one triangle are equal, respectively, to two angles and the included side of a second triangle, then the triangles are congruent. (Section 7.1)

angle in standard position An angle is in standard position if its vertex is at the origin and its initial side is along the positive x-axis. (Section 1.1)

angle between two vectors The angle between two vectors is defined to be the angle θ, for $0° \le \theta \le 180°$, having the two vectors as its sides. (Section 7.4)

angular speed V Angular speed ω (omega) measures the speed of rotation and is defined by $\mathbf{v} = \frac{\theta}{t}$, where θ is the angle of rotation in radians and t is time. (Section 3.4)

argument of a complex number When a complex number $x + yi$ is written in trigonometric (or polar) form as $r(\cos\theta + i\sin\theta)$, the angle θ is called the argument of the complex number. (Section 8.2)

argument of a function The argument of a function is the expression containing the independent variable of the function. For example, in the function $y = f(x - d)$, the expression $x - d$ is the argument. (Section 4.2)

B

bearing Bearing is used to identify angles in navigation. One method for expressing bearing uses a single angle, with bearing measured in a clockwise direction from due north. A second method for expressing bearing starts with a north-south line and uses an acute angle to show the direction, either east or west, from this line. (Section 2.5)

C

cardioid A cardioid is a heart-shaped curve that is the graph of a polar equation of the form $r = a \pm b\sin\theta$ or $r = a \pm b\cos\theta$, where $\left|\frac{a}{b}\right| = 1$. (Section 8.5)

center of a circle The center of a circle is the given point that is a given distance from all points on the circle. (Appendix B)

circle A circle is the set of all points in a plane that lie a given distance from a given point. (Appendix B)

circular functions The trigonometric functions of arc lengths, or real numbers, are called circular functions. (Section 3.3)

closed interval A closed interval is an interval that includes both of its endpoints. (Appendix A)

cofunctions The function pairs sine and cosine, tangent and cotangent, and secant and cosecant are called cofunctions. (Section 2.1)

complementary angles (complements) Two positive angles are complementary angles (or complements) if the sum of their measures is 90°. (Section 1.1)

complex conjugates Two complex numbers that differ only in the sign of their imaginary parts are called complex conjugates. (Section 8.1)

complex number A complex number is a number of the form $a + bi$, where a and b are real numbers and $i = \sqrt{-1}$. (Section 8.1)

complex plane The complex plane is a two-dimensional representation of the complex numbers in which the horizontal axis is the real axis and the vertical axis is the imaginary axis. (Section 8.2)

conditional equation An equation that is satisfied by some numbers but not by others is called a conditional equation. (Appendix A)

congruent triangles Triangles that are both the same size and the same shape are called congruent triangles. (Section 1.2)

constant function A function f is constant on an interval I if, for every x_1 and x_2 in I, $f(x_1) = f(x_2)$. (Appendix C)

contradiction An equation that has no solution is called a contradiction. (Appendix A)

coordinate plane (xy-plane) The plane into which the rectangular coordinate system is introduced is called the coordinate plane (or xy-plane). (Appendix B)

coordinates (in the xy-plane) The coordinates of a point in the xy-plane are the numbers in the ordered pair that correspond to that point. (Appendix B)

cosecant   Let $P(x, y)$ be a point other than the origin on the terminal side of an angle θ in standard position. Let $r = \sqrt{x^2 + y^2}$ represent the distance from the origin to P. Then the cosecant function is defined by $\csc \theta = \frac{r}{y} (y \neq 0)$. (Section 1.3)

cosine   Let $P(x, y)$ be a point other than the origin on the terminal side of an angle θ in standard position. Let $r = \sqrt{x^2 + y^2}$ represent the distance from the origin to P. Then the cosine function is defined by $\cos \theta = \frac{x}{r}$. (Section 1.3)

cotangent   Let $P(x, y)$ be a point other than the origin on the terminal side of an angle θ in standard position. Let $r = \sqrt{x^2 + y^2}$ represent the distance from the origin to P. Then the cotangent function is defined by $\cot \theta = \frac{x}{y} (y \neq 0)$. (Section 1.3)

coterminal angles   Two angles that have the same initial side and the same terminal side, but different amounts of rotation, are called coterminal angles. The measures of coterminal angles differ by a multiple of 360°. (Section 1.1)

cycloid   A cycloid is a curve that represents the path traced by a fixed point on the circumference of a circle rolling along a line. (Section 8.6)

damped oscillatory motion   Damped oscillatory motion is oscillatory motion that has been slowed down (damped) by the force of friction. Friction causes the amplitude of the motion to diminish gradually until the weight comes to rest. (Section 4.5)

decreasing function   A function f is decreasing on an interval I if, whenever $x_1 < x_2$ in I, $f(x_1) > f(x_2)$. (Appendix C)

degree   The degree is the most common unit of measure for angles. One degree, written 1°, represents $\frac{1}{360}$ of a rotation. (Section 1.1)

dependent variable   If the value of the variable y depends on the value of the variable x, then y is called the dependent variable. (Appendix C)

direction angle   The positive angle between the x-axis and a position vector is the direction angle for the vector. (Section 7.4)

domain   In a relation, the set of all values of the independent variable (x) is called the domain. (Appendix C)

dot product   The dot product of two vectors is the sum of the product of their first components and the product of their second components. The dot product of the two vectors $\mathbf{u} = \langle a, b \rangle$ and $\mathbf{v} = \langle c, d \rangle$ is denoted $\mathbf{u} \cdot \mathbf{v}$ and given by $\mathbf{u} \cdot \mathbf{v} = ac + bd$. (Section 7.4)

empty set (null set)   The empty set (or null set), written $\emptyset$ or { }, is the set containing no elements. (Appendix A)

endpoint of a ray   In a given ray AB, point A is the endpoint of the ray. (Section 1.1)

equation   An equation is a statement that two expressions are equal. (Appendix A)

equilibrant   The opposite vector of the resultant of two vectors is called the equilibrant. (Section 7.5)

exact number   A number that represents the result of counting, or a number that results from theoretical work and is not the result of a measurement, is an exact number. (Section 2.4)

four-leaved rose   A four-leaved rose is a curve that is the graph of a polar equation of the form $r = a \sin 2\theta$ or $r = a \cos 2\theta$. (Section 8.5)

frequency   In simple harmonic motion, the frequency is the number of cycles per unit of time, or the reciprocal of the period. (Section 4.5)

function   A function is a relation (set of ordered pairs) in which, for each value of the first component of the ordered pairs, there is *exactly one* value of the second component. (Appendix C)

function notation   Function notation $f(x)$ (read "f of x") represents the y-value of the function f for the indicated x-value. (Appendix C)

graph of an equation   The graph of an equation is the set of all points that correspond to all of the ordered pairs that satisfy the equation. (Appendix B)

groundspeed   In air navigation, the groundspeed of a plane is its speed relative to the ground. (Section 7.5)

horizontal component   When a vector $\mathbf{u}$ is expressed as an ordered pair in the form $\mathbf{u} = \langle a, b \rangle$, the number a is the horizontal component of the vector. (Section 7.4)

identity   An equation satisfied by every number that is a meaningful replacement for the variable is called an identity. (Appendix A)

imaginary axis   In the complex plane, the vertical axis is called the imaginary axis. (Section 8.2)

imaginary part   In the complex number $a + bi$, b is called the imaginary part. (Section 8.1)

imaginary unit   The number i, defined by $i^2 = -1$ (so $i = \sqrt{-1}$), is called the imaginary unit. (Section 8.1)

increasing function   A function f is increasing on an interval I if, whenever $x_1 < x_2$ in I, $f(x_1) < f(x_2)$. (Appendix C)

independent variable   If the value of the variable y depends on the value of the variable x, then x is called the independent variable. (Appendix C)

inequality   An inequality says that one expression is greater than, greater than or equal to, less than, or less than or equal to, another. (Appendix A)

initial point   When two letters are used to name a vector, the first letter indicates the initial (starting) point of the vector. (Section 7.4)

initial side   When a ray is rotated around its endpoint to form an angle, the ray in its initial position is called the initial side of the angle. (Section 1.1)

interval   An interval is a portion of the real number line, which may or may not include its endpoint(s). (Appendix A)

interval notation   Interval notation is a simplified notation for writing intervals. It uses parentheses and brackets to show whether the endpoints are included. (Appendix A)

inverse function   The inverse function of the one-to-one function f is defined as $\{(y, x) \,|\, (x, y) \text{ belongs to } f\}$. (Section 6.1)

lemniscate A lemniscate is a figure-eight-shaped curve that is the graph of a polar equation of the form $r^2 = a^2 \sin 2\theta$ or $r^2 = a^2 \cos 2\theta$. (Section 8.5)

limaçon A limaçon is the graph of a polar equation of the form $r = a \pm b \sin \theta$ or $r = a \pm b \cos \theta$. If $\left|\frac{a}{b}\right| = 1$, the limaçon is a cardioid. (Section 8.5)

line Two distinct points A and B determine a line called line AB. (Section 1.1)

line segment (segment) Line segment AB (or segment AB) is the portion of line AB between A and B, including A and B themselves. (Section 1.1)

linear equation (first-degree equation) in one variable A linear equation in one variable is an equation that can be written in the form $ax + b = 0$, where a and b are real numbers with $a \neq 0$. (Appendix A)

linear inequality in one variable A linear inequality in one variable is an inequality that can be written in the form $ax + b > 0$, where a and b are real numbers with $a \neq 0$. (Any of the symbols $<, \geq,$ or $\leq$ may also be used.) (Appendix A)

linear speed v The linear speed v measures the distance traveled per unit of time. (Section 3.4)

magnitude The length of a vector represents the magnitude of the vector quantity. (Section 7.4)

minute One minute, written $1'$, is $\frac{1}{60}$ of a degree. (Section 1.1)

negative angle A negative angle is an angle that is formed by clockwise rotation around its endpoint. (Section 1.1)

nonreal complex number A complex number $a + bi$ with $b \neq 0$ is called a nonreal complex number. (Section 8.1)

nth root of a complex number For a positive integer n, the complex number $a + bi$ is an nth root of the complex number $x + yi$ if $(a + bi)^n = x + yi$. (Section 8.4)

oblique triangle A triangle that is not a right triangle is called an oblique triangle. (Section 7.1)

obtuse angle An obtuse angle is an angle measuring more than $90°$ but less than $180°$. (Section 1.1)

one-to-one function If a function is defined so that each range element is used only once, then it is called a one-to-one function. (Section 6.1)

open interval An open interval is an interval that does not include its endpoint(s). (Appendix A)

opposite of a vector The opposite of a vector $\mathbf{v}$ is a vector $-\mathbf{v}$ that has the same magnitude as $\mathbf{v}$ but opposite direction. (Section 7.4)

ordered pair An ordered pair consists of two components, written inside parentheses, in which the order of the components is important. Ordered pairs are used to identify points in the coordinate plane. (Appendix B)

origin The point of intersection of the x-axis and the y-axis of a rectangular coordinate system is called the origin. (Appendix B)

orthogonal vectors Orthogonal vectors are vectors that are perpendicular, that is, the angle between the two vectors is $90°$. (Section 7.4)

parallel lines Parallel lines are lines that lie in the same plane and do not intersect. (Section 1.2)

parallelogram rule The parallelogram rule is a way to find the sum of two vectors. If the two vectors are placed so that their initial points coincide and a parallelogram is completed that has these two vectors as two of its sides, then the diagonal vector of the parallelogram that has the same initial point as the two vectors is their sum. (Section 7.4)

parameter A parameter is a variable in terms of which two or more other variables are expressed. In a pair of parametric equations $x = f(t)$ and $y = g(t)$, the variable t is the parameter. (Section 8.6)

parametric equations of a plane curve A pair of equations $x = f(t)$ and $y = g(t)$ are parametric equations of a plane curve. (Section 8.6)

period For a periodic function such that $f(x) = f(x + np)$, the smallest possible positive value of p is the period of the function. (Section 4.1)

periodic function A periodic function is a function f such that $f(x) = f(x + np)$, for every real number x in the domain of f, every integer n, and some positive real number p. (Section 4.1)

phase shift With trigonometric functions, a horizontal translation is called a phase shift. (Section 4.2)

plane curve A plane curve is a set of points (x, y) such that $x = f(t)$ and $y = g(t)$, and f and g are both defined on an interval I. (Section 8.6)

polar axis The polar axis is a specific ray in the polar coordinate system that has the pole as its endpoint. The polar axis is usually drawn in the direction of the positive x-axis. (Section 8.5)

polar coordinates In the polar coordinate system, the ordered pair (r, θ) gives polar coordinates of point P, where r is the directed distance from the pole to P and θ is the directed angle from the positive x-axis to ray OP. (Section 8.5)

polar coordinate system The polar coordinate system is a coordinate system based on a point (the pole) and a ray (the polar axis). (Section 8.5)

polar equation A polar equation is an equation that uses polar coordinates. The variables are r and θ. (Section 8.5)

pole The pole is the single fixed point in the polar coordinate system that is the endpoint of the polar axis. The pole is usually placed at the origin of a rectangular coordinate system. (Section 8.5)

position vector A vector with its initial point at the origin is called a position vector. (Section 7.4)

positive angle A positive angle is an angle that is formed by counterclockwise rotation around its endpoint. (Section 1.1)

pure imaginary number A complex number $a + bi$ in which $a = 0$ and $b \neq 0$ is called a pure imaginary number. (Section 8.1)

Pythagorean theorem The Pythagorean theorem states that in a right triangle, the sum of the squares of the lengths of the legs is equal to the square of the length of the hypotenuse. (Appendix B)

Q

quadrantal angle A quadrantal angle is an angle that, when placed in standard position, has its terminal side along the x-axis or the y-axis. (Section 1.1)

quadrants The quadrants are the four regions into which the x-axis and y-axis divide the coordinate plane. (Appendix B)

quadratic equation (second-degree equation) An equation that can be written in the form $ax^2 + bx + c = 0$, where a, b, and c are real numbers with $a \neq 0$, is a quadratic equation. (Appendix A)

quadratic formula The quadratic formula $x = \frac{-b \pm \sqrt{b^2 - 4ac}}{2a}$ is a general formula that can be used to solve any quadratic equation. (Appendix A)

R

radian A radian is a unit of measure for angles. An angle with its vertex at the center of a circle that intercepts an arc on the circle equal in length to the radius of the circle has a measure of 1 radian. (Section 3.1)

radius The radius of a circle is the given distance between the center and any point on the circle. (Appendix B)

range In a relation, the set of all values of the dependent variable (y) is called the range. (Appendix C)

ray The portion of line AB that starts at A and continues through B, and on past B, is called ray AB. (Section 1.1)

real axis In the complex plane, the horizontal axis is called the real axis. (Section 8.2)

real part In the complex number $a + bi$, a is called the real part. (Section 8.1)

reciprocal The reciprocal of a nonzero number x is $\frac{1}{x}$. (Section 1.4)

rectangular (Cartesian) coordinate system The x-axis and y-axis together make up a rectangular coordinate system. (Appendix B)

rectangular (Cartesian) equation A rectangular (or Cartesian) equation is an equation that uses rectangular coordinates. If it is an equation in two variables, the variables are x and y. (Section 8.5)

rectangular form (standard form) of a complex number The rectangular form (or standard form) of a complex number is $a + bi$, where a and b are real numbers. (Section 8.2)

reference angle The reference angle for an angle θ, written θ', is the positive acute angle made by the terminal side of angle θ and the x-axis. (Section 2.2)

reference arc The reference arc for a point on the unit circle is the shortest arc from the point itself to the nearest point on the x-axis. (Section 3.3)

relation A relation is a set of ordered pairs. (Appendix C)

resultant If $\mathbf{A}$ and $\mathbf{B}$ are vectors, the vector sum $\mathbf{A} + \mathbf{B}$ is called the resultant of vectors $\mathbf{A}$ and $\mathbf{B}$. (Section 7.4)

right angle A right angle is an angle measuring exactly 90°. (Section 1.1)

rose curve A rose curve is a member of a family of curves that resemble flowers. It is the graph of a polar equation of the form $r = a \sin n\theta$ or $r = a \cos n\theta$. (Section 8.5)

S

scalar A scalar is a quantity that involves a magnitude and can be represented by a real number. (Section 7.4)

scalar product The scalar product of a real number (or scalar) k and a vector $\mathbf{u}$ is the vector $k \cdot \mathbf{u}$, which has magnitude $|k|$ times the magnitude of $\mathbf{u}$. (Section 7.4)

secant Let $P(x, y)$ be a point other than the origin on the terminal side of an angle θ in standard position. Let $r = \sqrt{x^2 + y^2}$ represent the distance from the origin to P. Then the secant function is defined by $\sec\theta = \frac{r}{x}$ ($x \neq 0$). (Section 1.3)

second One second, written $1''$, is $\frac{1}{60}$ of a minute. (Section 1.1)

sector of a circle A sector of a circle is the portion of the interior of a circle intercepted by a central angle. (Section 3.2)

semiperimeter The semiperimeter is half the sum of the lengths of the three sides of a triangle. (Section 7.3)

side of an angle One of the two rays (or line segments) with a common endpoint that form an angle is called a side of the angle. (Section 1.1)

Side-Angle-Side (SAS) The Side-Angle-Side (SAS) congruence axiom states that if two sides and the included angle of one triangle are equal, respectively, to two sides and the included angle of a second triangle, then the triangles are congruent. (Section 7.1)

Side-Side-Side (SSS) The Side-Side-Side (SSS) congruence axiom states that if three sides of one triangle are equal, respectively, to three sides of a second triangle, then the triangles are congruent. (Section 7.1)

significant digit A significant digit is a digit obtained by actual measurement. (Section 2.4)

similar triangles Triangles that are the same shape, but not necessarily the same size, are called similar triangles. (Section 1.2)

simple harmonic motion Simple harmonic motion is oscillatory motion about an equilibrium position. If friction is neglected, this motion can be described by a sinusoid. (Section 4.5)

sine Let $P(x, y)$ be a point other than the origin on the terminal side of an angle θ in standard position. Let $r = \sqrt{x^2 + y^2}$ represent the distance from the origin to P. Then the sine function is defined by $\sin\theta = \frac{y}{r}$. (Section 1.3)

sine wave (sinusoid) The graph of a sine function is called a sine wave (or sinusoid). (Section 4.1)

solution (root) A solution (or root) of an equation is a number that makes the equation a true statement. (Appendix A)

solution set The solution set of an equation is the set of all numbers that satisfy the equation. (Appendix A)

spiral of Archimedes A spiral of Archimedes is an infinite curve that is the graph of a polar equation of the form $r = n\theta$. (Section 8.5)

standard form of a complex number A complex number written in the form $a + bi$ (or $a + ib$) is in standard form. (Section 8.1)

straight angle A straight angle is an angle measuring exactly 180°. (Section 1.1)

supplementary angles (supplements) Two positive angles are supplementary angles (or supplements) if the sum of their measures is 180°. (Section 1.1)

tangent Let $P(x, y)$ be a point other than the origin on the terminal side of an angle θ in standard position. Let $r = \sqrt{x^2 + y^2}$ represent the distance from the origin to P. Then the tangent function is defined by $\tan \theta = \frac{y}{x}$ $(x \neq 0)$. (Section 1.3)

terminal point When two letters are used to name a vector, the second letter indicates the terminal (ending) point of the vector. (Section 7.4)

terminal side When a ray is rotated around its endpoint to form an angle, the ray in its location after rotation is called the terminal side of the angle. (Section 1.1)

translation A translation is a horizontal or vertical shift of a graph. (Appendix D)

transversal A line that intersects two or more other lines, which may be parallel, is called a transversal. (Section 1.2)

trigonometric (polar) form of a complex number The expression $r(\cos \theta + i \sin \theta)$ is called the trigonometric form (or polar form) of the complex number $x + yi$. The expression $\cos \theta + i \sin \theta$ is sometimes abbreviated as cis θ. (Section 8.2)

unit circle The unit circle is the circle with center at the origin and radius 1. (Section 3.3)

unit vector A unit vector is a vector that has magnitude 1. Two useful unit vectors are $\mathbf{i} = \langle 1, 0 \rangle$ and $\mathbf{j} = \langle 0, 1 \rangle$. (Section 7.4)

vector A vector is a directed line segment that represents a vector quantity. (Section 7.4)

vector quantities Quantities that involve both magnitude and direction are called vector quantities. (Section 7.4)

vertex of an angle The vertex of an angle is the endpoint of the ray that is rotated to form the angle. (Section 1.1)

vertical angles Vertical angles are opposite angles formed by intersecting lines. (Section 1.2)

vertical asymptote A vertical line that a graph approaches, but never touches or intersects, is called a vertical asymptote. The line $x = a$ is a vertical asymptote if $|f(x)|$ gets larger and larger as x approaches a. (Section 4.3)

vertical component When a vector $\mathbf{u}$ is expressed as an ordered pair in the form $\mathbf{u} = \langle a, b \rangle$, the number b is the vertical component of the vector. (Section 7.4)

x-axis The horizontal number line in a rectangular coordinate system is called the x-axis. (Appendix B)

x-intercept An x-intercept is the x-value of a point where the graph of an equation intersects the x-axis. (Appendix B)

y-axis The vertical number line in a rectangular coordinate system is called the y-axis. (Appendix B)

y-intercept A y-intercept is the y-value of a point where the graph of an equation intersects the y-axis. (Appendix B)

zero-factor property The zero-factor property states that if the product of two (or more) complex numbers is 0, then at least one of the numbers must be 0. (Appendix A)

zero vector The zero vector is the vector with magnitude 0. (Section 7.4)

Solutions to Selected Exercises

CHAPTER 1 TRIGONOMETRIC FUNCTIONS

1.1 Exercises *(pages 7–10)*

43. $90° - 72° 58' 11''$

$$\begin{array}{r} 89° 59' 60'' \\ -72° 58' 11'' \\ \hline 17° 01' 49'' \end{array}$$ Write 90° as 89° 59′ 60″.

Thus, $90° - 72° 58' 11'' = 17° 1' 49''$.

129. 600 rotations per min

$= \dfrac{600}{60}$ rotations per sec

$= 10$ rotations per sec

$= 5$ rotations per $\frac{1}{2}$ sec

$= 5(360°)$ per $\frac{1}{2}$ sec

$= 1800°$ per $\frac{1}{2}$ sec

A point on the edge of the tire will move $1800°$ in $\frac{1}{2}$ sec.

1.2 Exercises *(pages 16–22)*

1. Angle 1 and the 55° angle are vertical angles, which are equal, so angle 1 = 55°. Angle 5 and the 120° angle are interior angles on the same side of the transversal, which means they are supplements, so

$\qquad$ angle $5 + 120° = 180°$

$\qquad\qquad$ angle $5 = 60°.$ Subtract 120°.

Since angles 3 and 5 are vertical angles, angle 3 = 60°.

$\qquad$ angle 1 + angle 2 + angle 3 = 180°

$\qquad$ 55° + angle 2 + 60° = 180°

$\qquad\qquad\qquad$ angle 2 = 65° Subtract 115°.

Since angles 2 and 4 are vertical angles, angle 4 = 65°. Angle 6 and the 120° angle are vertical angles, so angle 6 = 120°. Angles 6 and 8 are supplements.

$\qquad$ angle 6 + angle 8 = 180°

$\qquad$ 120° + angle 8 = 180°

$\qquad\qquad$ angle 8 = 60° Subtract 120°.

Since angles 7 and 8 are vertical angles, angle 7 = 60°.

angle 7 + angle 4 + angle 10 = 180°

The sum of the measures of the angles in a triangle is 180°.

$\qquad$ 60° + 65° + angle 10 = 180°

$\qquad\qquad$ angle 10 = 55° Subtract 125°.

Since angles 9 and 10 are vertical angles, angle 9 = 55°. Thus, the measures of the angles are 1: 55°, 2: 65°, 3: 60°, 4: 65°, 5: 60°, 6: 120°, 7: 60°, 8: 60°, 9: 55°, 10: 55°.

33. The triangle is obtuse because it has an angle of 96°, which is between 90° and 180°. It is a scalene triangle because no two sides are equal.

1.3 Exercises *(pages 27–29)*

77. Evaluate $\tan 360° + 4 \sin 180° + 5 \cos^2 180°$.

$$\tan 360° = \tan 0° = \frac{y}{x} = \frac{0}{1} = 0$$

$$\sin 180° = \frac{y}{r} = \frac{0}{1} = 0$$

$$\cos 180° = \frac{x}{r} = \frac{-1}{1} = -1$$

$\tan 360° + 4 \sin 180° + 5 \cos^2 180° = 0 + 4(0) + 5(-1)^2$

$\qquad\qquad\qquad$ Substitute; $\cos^2 x = (\cos x)^2$.

$\qquad\qquad\qquad\qquad\qquad = 5$

1.4 Exercises *(pages 37–39)*

69. Given $\tan \theta = -\dfrac{15}{8}$, with θ in quadrant II

Draw θ in standard position in quadrant II. Because $\tan \theta = \frac{y}{x}$ and θ is in quadrant II, we can use the values $y = 15$ and $x = -8$ for a point on its terminal side.

$$r = \sqrt{x^2 + y^2} = \sqrt{(-8)^2 + 15^2} = \sqrt{64 + 225}$$
$$= \sqrt{289} = 17$$

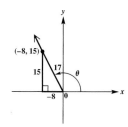

(continued)

Use the values of x, y, and r and the definitions of the trigonometric functions to find the six trigonometric function values for θ.

$$\sin u = \frac{y}{r} = \frac{15}{17} \qquad\qquad \csc u = \frac{r}{y} = \frac{17}{15}$$

$$\cos \theta = \frac{x}{r} = \frac{-8}{17} = -\frac{8}{17} \qquad \sec \theta = \frac{r}{x} = \frac{17}{-8} = -\frac{17}{8}$$

$$\tan \theta = \frac{y}{x} = \frac{15}{-8} = -\frac{15}{8} \qquad \cot \theta = \frac{x}{y} = \frac{-8}{15} = -\frac{8}{15}$$

85. Multiply the compound inequality $90° < \theta < 180°$ by 2 to find that $180° < 2\theta < 360°$. Thus, 2θ must lie in either quadrant III or IV. In both of these quadrants, the sine function is negative, and so $\sin 2\theta$ must be negative.

93. $\tan(3\theta - 4°) = \dfrac{1}{\cot(5\theta - 8°)}$ Given equation

$\tan(3\theta - 4°) = \tan(5\theta - 8°)$ Reciprocal identity

The second equation above will be true if $3\theta - 4° = 5\theta - 8°$, so solving this equation will give a value (but not the only value) for which the given equation is true.

$$3\theta - 4° = 5\theta - 8°$$
$$4° = 2\theta$$
$$\theta = 2°$$

CHAPTER 2   ACUTE ANGLES AND RIGHT TRIANGLES

2.1 Exercises (pages 55–58)

73. One point on the line $y = \sqrt{3}x$ is the origin, $(0,0)$. Let (x, y) be any other point on this line. Then, by the definition of slope, $m = \frac{y-0}{x-0} = \frac{y}{x} = \sqrt{3}$, but also, by the definition of tangent, $\tan \theta = \frac{y}{x}$. Thus, $\tan \theta = \sqrt{3}$. Because $\tan 60° = \sqrt{3}$, the line $y = \sqrt{3}x$ makes a 60° angle with the positive x-axis. (See Exercise 70.)

77. Apply the relationships between the lengths of the sides of a 30°–60° right triangle first to the triangle on the left to find the values of x and y, and then to the triangle on the right to find the values of z and w. In a 30°–60° right triangle, the side opposite the 30° angle is $\frac{1}{2}$ the length of the hypotenuse. The longer leg is $\sqrt{3}$ times the shorter leg.

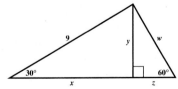

Thus,

$$y = \frac{1}{2}(9) = \frac{9}{2} \quad \text{and} \quad x = y\sqrt{3} = \frac{9\sqrt{3}}{2}.$$

Since $y = z\sqrt{3}$,

$$z = \frac{y}{\sqrt{3}} = \frac{\frac{9}{2}}{\sqrt{3}} = \frac{9}{2\sqrt{3}} \cdot \frac{\sqrt{3}}{\sqrt{3}} = \frac{9\sqrt{3}}{6} = \frac{3\sqrt{3}}{2},$$

and

$$w = 2z = 2\left(\frac{3\sqrt{3}}{2}\right) = 3\sqrt{3}.$$

2.2 Exercises (pages 63–66)

21. To find the reference angle for $-300°$, sketch this angle in standard position.

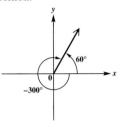

The reference angle for $-300°$ is $-300° + 360° = 60°$. Because $-300°$ is in quadrant I, the values of all its trigonometric functions will be positive, so these values will be identical to the trigonometric function values for 60°. (See the Function Values of Special Angles table that follows Example 5 in **Section 2.1**.)

$$\sin(-300°) = \frac{\sqrt{3}}{2} \qquad \csc(-300°) = \frac{2\sqrt{3}}{3}$$

$$\cos(-300°) = \frac{1}{2} \qquad \sec(-300°) = 2$$

$$\tan(-300°) = \sqrt{3} \qquad \cot(-300°) = \frac{\sqrt{3}}{3}$$

67. The reference angle for 115° is 65°. Since 115° is in quadrant II, the sine is positive. Sin θ decreases on the interval (90°, 180°) from 1 to 0. Therefore, sin 115° is closest to .9.

2.3 Exercises (pages 68–72)

45. $\sin 10° + \sin 10° \overset{?}{=} \sin 20°$

Using a calculator, we get

$$\sin 10° + \sin 10° \approx .34729636$$

and $\sin 20° \approx .34202014.$

Thus, the statement is false.

63. For parts (a) and (b), $\theta = 3°$, $g = 32.2$, and $f = .14$.

   (a) Since 45 mph = 66 ft per sec,

$$R = \frac{V^2}{g(f + \tan \theta)}$$

$$= \frac{66^2}{32.2(.14 + \tan 3°)}$$

$$\approx 703 \text{ ft}.$$

(b) Since 70 mph $= \frac{70(5280)}{3600}$ ft per sec $= 102.67$ ft per sec,

$$R = \frac{V^2}{g(f + \tan \theta)}$$

$$= \frac{102.67^2}{32.2(.14 + \tan 3°)}$$

$$\approx 1701 \text{ ft.}$$

(c) Intuitively, increasing θ would make it easier to negotiate the curve at a higher speed much like is done at a race track. Mathematically, a larger value of θ (acute) will lead to a larger value for $\tan \theta$. If $\tan \theta$ increases, then the ratio determining R will *decrease*. Thus, the radius can be smaller and the curve sharper if θ is increased.

$$R = \frac{V^2}{g(f + \tan \theta)}$$

$$= \frac{66^2}{32.2(.14 + \tan 4°)}$$

$$\approx 644 \text{ ft}$$

$$R = \frac{V^2}{g(f + \tan \theta)}$$

$$= \frac{102.67^2}{32.2(.14 + \tan 4°)}$$

$$\approx 1559 \text{ ft}$$

As predicted, both values are smaller.

2.4 Exercises *(pages 78–83)*

23. Solve the right triangle with $B = 73.0°$, $b = 128$ in., and $C = 90°$.

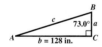

$$A = 90° - 73.0° = 17.0°$$

$$\tan 73.0° = \frac{128}{a}$$

$$a = \frac{128}{\tan 73.0°} \approx 39.1 \text{ in.} \qquad \text{Three significant digits}$$

$$\sin 73.0° = \frac{128}{c}$$

$$c = \frac{128}{\sin 73.0°} \approx 134 \text{ in.} \qquad \text{Three significant digits}$$

43. Let x represent the horizontal distance between the two buildings and y represent the height of the portion of the building across the street that is higher than the window.

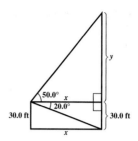

$$\tan 20.0° = \frac{30.0}{x}$$

$$x = \frac{30.0}{\tan 20.0°} \approx 82.4$$

$$\tan 50.0° = \frac{y}{x}$$

$$y = x \tan 50.0° = \left(\frac{30.0}{\tan 20.0°} \right) \tan 50.0° \approx 98.2$$

$$\text{height} = y + 30.0 = \left(\frac{30.0}{\tan 20.0°} \right) \tan 50.0° + 30.0 \approx 128$$

Three significant digits

The height of the building across the street is about 128 ft.

49. Let h represent the height of the tower.

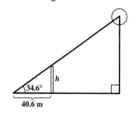

$$\tan 34.6° = \frac{h}{40.6}$$

$$h = 40.6 \tan 34.6° \approx 28.0$$

Three significant digits

The height of the tower is about 28.0 m.

2.5 Exercises *(pages 87–91)*

21. Let $x =$ the distance between the two ships. The angle between the bearings of the ships is

$$180° - (28° \, 10' + 61° \, 50') = 90°.$$

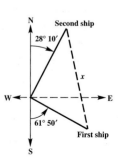

The triangle formed is a right triangle.

Distance traveled at 24.0 mph:

(4 hr) (24.0 mph) $= 96$ mi

Distance traveled at 28.0 mph:

(4 hr) (28.0 mph) $= 112$ mi

(continued)

Applying the Pythagorean theorem gives

$$x^2 = 96^2 + 112^2$$
$$x^2 = 21{,}760$$
$$x \approx 148.$$

The ships are 148 mi apart.

29. Let x = the distance from the closer point on the ground to the base of height h of the pyramid.

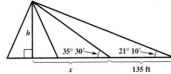

In the larger right triangle:

$$\tan 21° \, 10' = \frac{h}{135 + x}$$
$$h = (135 + x) \tan 21° \, 10'$$

In the smaller right triangle:

$$\tan 35° \, 30' = \frac{h}{x}$$
$$h = x \tan 35° \, 30'$$

Substitute for h in this equation, and solve for x.

$$(135 + x) \tan 21° \, 10' = x \tan 35° \, 30'$$
$$135 \tan 21° \, 10' + x \tan 21° \, 10' = x \tan 35° \, 30'$$
$$135 \tan 21° \, 10' = x \tan 35° \, 30' - x \tan 21° \, 10'$$
$$135 \tan 21° \, 10' = x(\tan 35° \, 30' - \tan 21° \, 10')$$
$$\frac{135 \tan 21° \, 10'}{\tan 35° \, 30' - \tan 21° \, 10'} = x$$

Then substitute for x in the equation for the smaller triangle.

$$h = \frac{135 \tan 21° \, 10'(\tan 35° \, 30')}{\tan 35° \, 30' - \tan 21° \, 10'}$$
$$h \approx 114$$

The height of the pyramid is 114 ft.

CHAPTER 3   RADIAN MEASURE AND CIRCULAR FUNCTIONS

3.1 Exercises *(pages 106–108)*

85. (a) In 24 hr, the hour hand will rotate twice around the clock. Since one complete rotation measures 2π radians, the two rotations will measure $2(2\pi) = 4\pi$ radians.

(b) In 4 hr, the hour hand will rotate $\frac{4}{12} = \frac{1}{3}$ of the way around the clock, which will measure $\frac{1}{3}(2\pi) = \frac{2\pi}{3}$ radians.

3.2 Exercises *(pages 113–118)*

35. For the large gear and pedal,

$$s = r\theta = 4.72\pi. \quad 180° = \pi \text{ radians}$$

Thus, the chain moves 4.72π in. Find the angle through which the small gear rotates.

$$\theta = \frac{s}{r} = \frac{4.72\pi}{1.38} \approx 3.42\pi$$

The angle θ for the wheel and for the small gear are the same, so for the wheel,

$$s = r\theta = 13.6(3.42\pi) \approx 146 \text{ in.}$$

The bicycle will move about 146 in.

59. (a)

The triangle formed by the sides of the central angle and the chord is isosceles. Therefore, the bisector of the central angle is also the perpendicular bisector of the chord and divides the larger triangle into two congruent right triangles.

$$\sin 21° = \frac{50}{r}$$
$$r = \frac{50}{\sin 21°} \approx 140 \text{ ft}$$

The radius of the curve is about 140 ft.

(b) $r = \dfrac{50}{\sin 21°}; \, \theta = 42°$

$$42° = 42\left(\frac{\pi}{180} \text{ radian}\right) = \frac{7\pi}{30} \text{ radian}$$

$$s = r\theta = \frac{50}{\sin 21°} \cdot \frac{7\pi}{30} = \frac{35\pi}{3 \sin 21°} \approx 102 \text{ ft}$$

The length of the arc determined by the 100-ft chord is about 102 ft.

(c) The portion of the circle bounded by the arc and the 100-ft chord is the shaded region in the figure below.

The area of the portion of the circle can be found by subtracting the area of the triangle from the area of the sector. From the figure in part (a),

$$\tan 21° = \frac{50}{h}, \quad \text{so} \quad h = \frac{50}{\tan 21°}.$$

$$A_{\text{sector}} = \frac{1}{2}r^2\theta$$
$$= \frac{1}{2}\left(\frac{50}{\sin 21°}\right)^2\left(\frac{7\pi}{30}\right) \quad \begin{array}{l}\text{From part (b),} \\ 42° = \frac{7\pi}{30}.\end{array}$$
$$\approx 7135 \text{ ft}^2$$

$$A_{\text{triangle}} = \frac{1}{2}bh = \frac{1}{2}(100)\left(\frac{50}{\tan 21°}\right)$$
$$\approx 6513 \text{ ft}^2$$

$$A_{\text{portion}} = A_{\text{sector}} - A_{\text{triangle}} \approx 7135 \text{ ft}^2 - 6513 \text{ ft}^2$$
$$= 622 \text{ ft}^2$$

The area of the portion is about 622 ft^2.

61. Use the Pythagorean theorem to find the hypotenuse of the right triangle, which is also the radius of the sector of the circle.

$$r^2 = 30^2 + 40^2 = 900 + 1600 = 2500$$
$$r = \sqrt{2500} = 50$$

$$A_{\text{triangle}} = \frac{1}{2}bh = \frac{1}{2}(30)(40)$$
$$= 600 \text{ yd}^2$$

$$A_{\text{sector}} = \frac{1}{2}r^2\theta$$
$$= \frac{1}{2}(50)^2 \cdot \frac{\pi}{3} \quad 60° = \frac{\pi}{3}$$
$$= \frac{1250\pi}{3} \text{ yd}^2$$

$$\text{Total area} = A_{\text{triangle}} + A_{\text{sector}} = 600 \text{ yd}^2 + \frac{1250\pi}{3} \text{ yd}^2$$
$$\approx 1900 \text{ yd}^2$$

3.3 Exercises (pages 126–128)

45. $\cos 2$

$\frac{\pi}{2} \approx 1.57$ and $\pi \approx 3.14$, so $\frac{\pi}{2} < 2 < \pi$. Thus, an angle of 2 radians is in quadrant II. (The figure for Exercises 35–44 also shows that 2 radians is in quadrant II.) Because values of the cosine function are negative in quadrant II, $\cos 2$ is negative.

63. $\left[\pi, \dfrac{3\pi}{2}\right]$; $\tan s = \sqrt{3}$

Recall that $\tan \frac{\pi}{3} = \sqrt{3}$ and in quadrant III $\tan s$ is positive. Therefore,

$$\tan\left(\pi + \frac{\pi}{3}\right) = \tan \frac{4\pi}{3} = \sqrt{3},$$

and thus, $s = \frac{4\pi}{3}$.

3.4 Exercises (pages 131–134)

25. The hour hand of a clock moves through an angle of 2π radians (one complete revolution) in 12 hr, so

$$\omega = \frac{\theta}{t} = \frac{2\pi}{12} = \frac{\pi}{6} \text{ radian per hr.}$$

35. At 215 revolutions per min, the bicycle tire is moving $215(2\pi) = 430\pi$ radians per min. This is the angular velocity ω. The linear velocity of the bicycle is

$$v = r\omega = 13(430\pi) = 5590\pi \text{ in. per min.}$$

Convert this velocity to miles per hour.

$$v = \frac{5590\pi \text{ in.}}{\min} \cdot \frac{60 \text{ min}}{\text{hr}} \cdot \frac{1 \text{ ft}}{12 \text{ in.}} \cdot \frac{1 \text{ mi}}{5280 \text{ ft}} \approx 16.6 \text{ mph}$$

CHAPTER 4   GRAPHS OF THE CIRCULAR FUNCTIONS

4.1 Exercises (pages 153–158)

51. $E = 5 \cos 120\pi t$

(a) The amplitude is $|5| = 5$, and the period is
$\frac{2\pi}{120\pi} = \frac{1}{60}$.

(b) Since the period is $\frac{1}{60}$, one cycle is completed in $\frac{1}{60}$ sec. Therefore, in 1 sec, 60 cycles are completed.

(c) For $t = 0$, $E = 5 \cos 120\pi(0) = 5 \cos 0 = 5$.

For $t = .03$, $E = 5 \cos 120\pi(.03) \approx 1.545$.

For $t = .06$, $E = 5 \cos 120\pi(.06) \approx -4.045$.

For $t = .09$, $E \approx -4.045$.

For $t = .12$, $E \approx 1.545$.

(d)

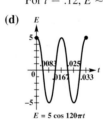

$E = 5 \cos 120\pi t$

4.2 Exercises (pages 164–167)

55. $y = \dfrac{1}{2} + \sin 2\left(x + \dfrac{\pi}{4}\right)$

This equation has the form $y = c + a \sin b(x - d)$ with $c = \frac{1}{2}$, $a = 1$, $b = 2$, and $d = -\frac{\pi}{4}$. Start with the graph of $y = \sin x$ and modify it to take into account the amplitude, period, and translations required to obtain the desired graph.

Amplitude: $|a| = 1$;   Period: $\dfrac{2\pi}{b} = \dfrac{2\pi}{2} = \pi$

Vertical translation: $\dfrac{1}{2}$ unit up

Phase shift (horizontal translation): $\dfrac{\pi}{4}$ units to the left

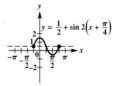

4.3 Exercises *(pages 174–176)*

29. $y = -1 + \dfrac{1}{2}\cot(2x - 3\pi)$

$y = -1 + \dfrac{1}{2}\cot 2\left(x - \dfrac{3\pi}{2}\right)$ Rewrite $2x - 3\pi$ as $2\left(x - \frac{3\pi}{2}\right)$.

Period: $\dfrac{\pi}{b} = \dfrac{\pi}{2}$

Vertical translation: 1 unit down

Phase shift (horizontal translation): $\dfrac{3\pi}{2}$ units to the right

Because the function is to be graphed over a two-period interval, locate three adjacent vertical asymptotes. Because asymptotes of the graph of $y = \cot x$ occur at multiples of π, the following equations can be solved to locate asymptotes:

$2\left(x - \dfrac{3\pi}{2}\right) = -2\pi,\quad 2\left(x - \dfrac{3\pi}{2}\right) = -\pi,\quad$ and

$2\left(x - \dfrac{3\pi}{2}\right) = 0.$

Solve each of these equations.

$2\left(x - \dfrac{3\pi}{2}\right) = -2\pi$

$x - \dfrac{3\pi}{2} = -\pi$

$x = -\pi + \dfrac{3\pi}{2} = \dfrac{\pi}{2}$

$2\left(x - \dfrac{3\pi}{2}\right) = -\pi$

$x - \dfrac{3\pi}{2} = -\dfrac{\pi}{2}$

$x = -\dfrac{\pi}{2} + \dfrac{3\pi}{2} = \dfrac{2\pi}{2} = \pi$

$2\left(x - \dfrac{3\pi}{2}\right) = 0$

$x - \dfrac{3\pi}{2} = 0$

$x = \dfrac{3\pi}{2}$

Divide the interval $\left(\frac{\pi}{2}, \pi\right)$ into four equal parts to obtain the following key x-values:

first-quarter value: $\dfrac{5\pi}{8}$; middle value: $\dfrac{3\pi}{4}$;

third-quarter value: $\dfrac{7\pi}{8}$.

Evaluating the given function at these three key x-values gives the following points:

$\left(\dfrac{5\pi}{8}, -\dfrac{1}{2}\right),\quad \left(\dfrac{3\pi}{4}, -1\right),\quad \left(\dfrac{7\pi}{8}, -\dfrac{3}{2}\right).$

Connect these points with a smooth curve and continue the graph to approach the asymptotes $x = \frac{\pi}{2}$ and $x = \pi$ to complete one period of the graph. Sketch an identical curve between the asymptotes $x = \pi$ and $x = \frac{3\pi}{2}$ to complete a second period of the graph.

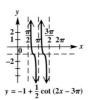

$y = -1 + \frac{1}{2}\cot(2x - 3\pi)$

43. $\tan(-x) = \dfrac{\sin(-x)}{\cos(-x)}$

$= \dfrac{-\sin x}{\cos x}$

$= -\dfrac{\sin x}{\cos x}$

$= -\tan x$

4.4 Exercises *(pages 183–184)*

31. $\sec(-x) = \dfrac{1}{\cos(-x)}$

$= \dfrac{1}{\cos x}$

$= \sec x$

4.5 Exercises *(pages 187–189)*

19. (a) We will use a model of the form $s(t) = a\cos\omega t$ with $a = -3$. Since

$s(0) = -3\cos(0\omega) = -3\cos 0 = -3 \cdot 1 = -3,$

using a cosine function rather than a sine function will avoid the need for a phase shift.

Since the frequency $= \frac{6}{\pi}$ cycles per sec, by definition,

$\dfrac{6}{\pi} = \dfrac{\omega}{2\pi}$

$6 \cdot 2\pi = \pi\omega$

$12\pi = \pi\omega$

$\omega = 12.$ Divide by π; rewrite.

Therefore, a model for the position of the weight at time t seconds is

$s(t) = -3\cos 12t.$

(b) Period $= \dfrac{1}{\frac{6}{\pi}} = \dfrac{\pi}{6}$ sec

CHAPTER 5   TRIGONOMETRIC IDENTITIES

5.1 Exercises (pages 203–206)

29. $\cot \theta = \dfrac{4}{3}, \sin \theta > 0$

Since $\cot \theta > 0$ and $\sin \theta > 0$, θ is in quadrant I, so all the function values are positive.

$$\tan \theta = \frac{1}{\cot \theta} = \frac{1}{\frac{4}{3}} = \frac{3}{4}$$

$$\sec^2 \theta = \tan^2 \theta + 1 \qquad \text{Pythagorean identity}$$

$$= \left(\frac{3}{4}\right)^2 + 1 = \frac{9}{16} + \frac{16}{16} = \frac{25}{16}$$

$$\sec \theta = \sqrt{\frac{25}{16}} = \frac{5}{4} \qquad \sec \theta > 0$$

$$\cos \theta = \frac{1}{\sec \theta} = \frac{1}{\frac{5}{4}} = \frac{4}{5}$$

$$\sin^2 \theta = 1 - \cos^2 \theta \qquad \begin{array}{l}\text{Alternative form of} \\ \text{Pythagorean identity}\end{array}$$

$$= 1 - \left(\frac{4}{5}\right)^2 = \frac{9}{25}$$

$$\sin \theta = \sqrt{\frac{9}{25}} = \frac{3}{5} \qquad \sin \theta > 0$$

$$\csc \theta = \frac{1}{\sin \theta} = \frac{1}{\frac{3}{5}} = \frac{5}{3}$$

Thus, $\sin \theta = \frac{3}{5}$, $\cos \theta = \frac{4}{5}$, $\tan \theta = \frac{3}{4}$, $\sec \theta = \frac{5}{4}$, and $\csc \theta = \frac{5}{3}$.

51. $\csc x = \dfrac{1}{\sin x} \qquad \text{Reciprocal identity}$

$$= \frac{1}{\pm\sqrt{1 - \cos^2 x}} \qquad \begin{array}{l}\text{Alternative form of} \\ \text{Pythagorean identity}\end{array}$$

$$= \frac{\pm 1}{\sqrt{1 - \cos^2 x}}$$

$$= \frac{\pm 1}{\sqrt{1 - \cos^2 x}} \cdot \frac{\sqrt{1 - \cos^2 x}}{\sqrt{1 - \cos^2 x}} \qquad \begin{array}{l}\text{Rationalize the} \\ \text{denominator.}\end{array}$$

$$= \frac{\pm\sqrt{1 - \cos^2 x}}{1 - \cos^2 x}$$

63. $\sec \theta - \cos \theta = \dfrac{1}{\cos \theta} - \cos \theta$

$$= \frac{1}{\cos \theta} - \frac{\cos^2 \theta}{\cos \theta} \qquad \begin{array}{l}\text{Get a common} \\ \text{denominator.}\end{array}$$

$$= \frac{1 - \cos^2 \theta}{\cos \theta} \qquad \text{Subtract fractions.}$$

$$= \frac{\sin^2 \theta}{\cos \theta} \qquad 1 - \cos^2 \theta = \sin^2 \theta$$

$$= \frac{\sin \theta}{\cos \theta} \cdot \sin \theta = \tan \theta \sin \theta$$

69. Since $\cos x = \frac{1}{5} > 0$, x is in quadrant I or quadrant IV.

$$\sin x = \pm\sqrt{1 - \cos^2 x} = \pm\sqrt{1 - \left(\frac{1}{5}\right)^2}$$

$$= \pm\sqrt{\frac{24}{25}} = \pm\frac{2\sqrt{6}}{5}$$

$$\tan x = \frac{\sin x}{\cos x} = \frac{\pm\frac{2\sqrt{6}}{5}}{\frac{1}{5}} = \pm 2\sqrt{6}$$

$$\sec x = \frac{1}{\cos x} = \frac{1}{\frac{1}{5}} = 5$$

Quadrant I:

$$\frac{\sec x - \tan x}{\sin x} = \frac{5 - 2\sqrt{6}}{\frac{2\sqrt{6}}{5}} = \frac{5(5 - 2\sqrt{6})}{2\sqrt{6}}$$

$$= \frac{25 - 10\sqrt{6}}{2\sqrt{6}} \cdot \frac{\sqrt{6}}{\sqrt{6}} = \frac{25\sqrt{6} - 60}{12}$$

Quadrant IV:

$$\frac{\sec x - \tan x}{\sin x} = \frac{5 - (-2\sqrt{6})}{-\frac{2\sqrt{6}}{5}} = \frac{5(5 + 2\sqrt{6})}{-2\sqrt{6}}$$

$$= \frac{25 + 10\sqrt{6}}{-2\sqrt{6}} \cdot \frac{-\sqrt{6}}{-\sqrt{6}} = \frac{-25\sqrt{6} - 60}{12}$$

5.2 Exercises (pages 212–215)

11. $\dfrac{1}{1 + \cos x} - \dfrac{1}{1 - \cos x} = \dfrac{1(1 - \cos x) - 1(1 + \cos x)}{(1 + \cos x)(1 - \cos x)}$

$$= \frac{1 - \cos x - 1 - \cos x}{1 - \cos^2 x}$$

$$= \frac{-2 \cos x}{\sin^2 x} = -\frac{2 \cos x}{\sin^2 x}$$

or $\qquad -\dfrac{2 \cos x}{\sin^2 x} = -2\left(\dfrac{\cos x}{\sin x}\right)\left(\dfrac{1}{\sin x}\right)$

$$= -2 \cot x \csc x$$

15. $(\sin x + 1)^2 - (\sin x - 1)^2$

$= (\sin^2 x + 2 \sin x + 1) - (\sin^2 x - 2 \sin x + 1)$

Square the binomials.

$= \sin^2 x + 2 \sin x + 1 - \sin^2 x + 2 \sin x - 1$

$= 4 \sin x$

59. Verify that $\dfrac{\tan^2 t - 1}{\sec^2 t} = \dfrac{\tan t - \cot t}{\tan t + \cot t}$ is an identity.

Work with the right side.

$\dfrac{\tan t - \cot t}{\tan t + \cot t} = \dfrac{\tan t - \dfrac{1}{\tan t}}{\tan t + \dfrac{1}{\tan t}}$ $\cot t = \frac{1}{\tan t}$

$= \dfrac{\tan t \left(\tan t - \dfrac{1}{\tan t} \right)}{\tan t \left(\tan t + \dfrac{1}{\tan t} \right)}$

Multiply numerator and denominator of the complex fraction by the LCD, tan t.

$= \dfrac{\tan^2 t - 1}{\tan^2 t + 1}$ Distributive property

$= \dfrac{\tan^2 t - 1}{\sec^2 t}$ $\tan^2 t + 1 = \sec^2 t$

87. Show that $\sin(\csc s) = 1$ is not an identity.

We need find only one value for which the statement is false. Let $s = 2$. Use a calculator to find that $\sin(\csc 2) \approx .891094$, which is not equal to 1. Thus, $\sin(\csc s) = 1$ is not true for *all* real numbers s, so it is not an identity.

5.3 Exercises *(pages 221–224)*

35. $\sec \theta = \csc \left(\dfrac{\theta}{2} + 20° \right)$

By a cofunction identity, $\sec \theta = \csc(90° - \theta)$. Thus,

$\csc \left(\dfrac{\theta}{2} + 20° \right) = \csc(90° - \theta)$ Substitute.

$\dfrac{\theta}{2} + 20° = 90° - \theta$

$\dfrac{3\theta}{2} = 70°$ Add θ; subtract 20°.

$\theta = \dfrac{2}{3}(70°) = \dfrac{140°}{3}$.   Multiply by $\frac{2}{3}$.

57. *True* or *false*: $\cos \dfrac{\pi}{3} = \cos \dfrac{\pi}{12} \cos \dfrac{\pi}{4} - \sin \dfrac{\pi}{12} \sin \dfrac{\pi}{4}$.

Since $\frac{\pi}{3} = \frac{\pi}{12} + \frac{3\pi}{12} = \frac{\pi}{12} + \frac{\pi}{4}$,

$\cos \dfrac{\pi}{3} = \cos \left(\dfrac{\pi}{12} + \dfrac{\pi}{4} \right)$ Substitute.

$= \cos \dfrac{\pi}{12} \cos \dfrac{\pi}{4} - \sin \dfrac{\pi}{12} \sin \dfrac{\pi}{4}$.

Cosine sum identity

The given statement is true.

5.4 Exercises *(pages 229–232)*

45. $\cos s = -\dfrac{8}{17}$ and $\cos t = -\dfrac{3}{5}$, s and t in quadrant III

In order to substitute into sum and difference identities, we need to find the values of sin s and sin t, and also the values of tan s and tan t. Because s and t are both in quadrant III, the values of sin s and sin t will be negative, while tan s and tan t will be positive.

$\sin s = -\sqrt{1 - \cos^2 s} = -\sqrt{1 - \left(-\dfrac{8}{17} \right)^2}$

$= -\sqrt{\dfrac{225}{289}} = -\dfrac{15}{17}$

$\sin t = -\sqrt{1 - \cos^2 t} = -\sqrt{1 - \left(-\dfrac{3}{5} \right)^2}$

$= -\sqrt{\dfrac{16}{25}} = -\dfrac{4}{5}$

$\tan s = \dfrac{\sin s}{\cos s} = \dfrac{-\frac{15}{17}}{-\frac{8}{17}} = \dfrac{15}{8}$

$\tan t = \dfrac{\sin t}{\cos t} = \dfrac{-\frac{4}{5}}{-\frac{3}{5}} = \dfrac{4}{3}$

(a) $\sin(s + t) = \sin s \cos t + \cos s \sin t$

$= \left(-\dfrac{15}{17} \right)\left(-\dfrac{3}{5} \right) + \left(-\dfrac{8}{17} \right)\left(-\dfrac{4}{5} \right)$

$= \dfrac{45}{85} + \dfrac{32}{85} = \dfrac{77}{85}$

(b) $\tan(s + t) = \dfrac{\tan s + \tan t}{1 - \tan s \tan t} = \dfrac{\frac{15}{8} + \frac{4}{3}}{1 - \left(\frac{15}{8} \right)\left(\frac{4}{3} \right)}$

$= \dfrac{\frac{45}{24} + \frac{32}{24}}{1 - \frac{60}{24}} = \dfrac{\frac{77}{24}}{-\frac{36}{24}} = -\dfrac{77}{36}$

(c) From parts (a) and (b), $\sin(s + t) > 0$ and $\tan(s + t) < 0$. The only quadrant in which values of sine are positive and values of tangent are negative is quadrant II.

51. $\tan\dfrac{11\pi}{12} = \tan\left(\dfrac{3\pi}{4} + \dfrac{\pi}{6}\right)$ $\frac{3\pi}{4} = \frac{9\pi}{12}; \frac{\pi}{6} = \frac{2\pi}{12}$

$= \dfrac{\tan\frac{3\pi}{4} + \tan\frac{\pi}{6}}{1 - \tan\frac{3\pi}{4}\tan\frac{\pi}{6}}$ Tangent sum identity

$= \dfrac{-1 + \frac{\sqrt{3}}{3}}{1 - (-1)\left(\frac{\sqrt{3}}{3}\right)}$ $\tan\frac{3\pi}{4} = -1;$ $\tan\frac{\pi}{6} = \frac{\sqrt{3}}{3}$

$= \dfrac{-1 + \frac{\sqrt{3}}{3}}{1 + \frac{\sqrt{3}}{3}}$ Simplify.

$= \dfrac{-1 + \frac{\sqrt{3}}{3}}{1 + \frac{\sqrt{3}}{3}} \cdot \dfrac{3}{3}$ Multiply numerator and denominator by 3.

$= \dfrac{-3 + \sqrt{3}}{3 + \sqrt{3}}$ Distributive property

$= \dfrac{-3 + \sqrt{3}}{3 + \sqrt{3}} \cdot \dfrac{3 - \sqrt{3}}{3 - \sqrt{3}}$ Rationalize the denominator.

$= \dfrac{-9 + 6\sqrt{3} - 3}{9 - 3}$ FOIL

$= \dfrac{-12 + 6\sqrt{3}}{6}$ Combine terms.

$= \dfrac{6(-2 + \sqrt{3})}{6}$ Factor the numerator.

$= -2 + \sqrt{3}$ Lowest terms

63. Verify that $\dfrac{\sin(x - y)}{\sin(x + y)} = \dfrac{\tan x - \tan y}{\tan x + \tan y}$ is an identity.

Work with the left side.

$\dfrac{\sin(x - y)}{\sin(x + y)} = \dfrac{\sin x \cos y - \cos x \sin y}{\sin x \cos y + \cos x \sin y}$

    Sine sum and difference identities

$= \dfrac{\dfrac{\sin x \cos y}{\cos x \cos y} - \dfrac{\cos x \sin y}{\cos x \cos y}}{\dfrac{\sin x \cos y}{\cos x \cos y} + \dfrac{\cos x \sin y}{\cos x \cos y}}$ Divide numerator and denominator by $\cos x \cos y$.

$= \dfrac{\dfrac{\sin x}{\cos x} \cdot 1 - 1 \cdot \dfrac{\sin y}{\cos y}}{\dfrac{\sin x}{\cos x} \cdot 1 + 1 \cdot \dfrac{\sin y}{\cos y}}$

$= \dfrac{\tan x - \tan y}{\tan x + \tan y}$ Tangent quotient identity

5.5 Exercises (pages 239–241)

25. $\dfrac{1}{4} - \dfrac{1}{2}\sin^2 47.1° = \dfrac{1}{4}(1 - 2\sin^2 47.1°)$ Factor out $\frac{1}{4}$.

$= \dfrac{1}{4}\cos 2(47.1°)$

    $\cos 2A = 1 - 2\sin^2 A$

$= \dfrac{1}{4}\cos 94.2°$

31. $\tan 3x = \tan(2x + x)$

$= \dfrac{\tan 2x + \tan x}{1 - \tan 2x \tan x}$ Tangent sum identity

$= \dfrac{\dfrac{2\tan x}{1 - \tan^2 x} + \tan x}{1 - \dfrac{2\tan x}{1 - \tan^2 x} \cdot \tan x}$ Tangent double-angle identity

$= \dfrac{\dfrac{2\tan x + (1 - \tan^2 x)\tan x}{1 - \tan^2 x}}{\dfrac{1 - \tan^2 x - 2\tan^2 x}{1 - \tan^2 x}}$ Add and subtract using the common denominator.

$= \dfrac{2\tan x + \tan x - \tan^3 x}{1 - \tan^2 x - 2\tan^2 x}$

    Multiply numerator and denominator by $1 - \tan^2 x$.

$= \dfrac{3\tan x - \tan^3 x}{1 - 3\tan^2 x}$ Combine terms.

43. Verify that $\sin 4x = 4\sin x \cos x \cos 2x$ is an identity.

Work with the left side.

$\sin 4x = \sin 2(2x)$

$= 2\sin 2x \cos 2x$ Sine double-angle identity

$= 2(2\sin x \cos x)\cos 2x$ Sine double-angle identity

$= 4\sin x \cos x \cos 2x$

5.6 Exercises (pages 245–248)

21. Find $\tan\dfrac{\theta}{2}$, given $\sin\theta = \dfrac{3}{5}$, with $90° < \theta < 180°$.

To find $\tan\frac{\theta}{2}$, we need the values of $\sin\theta$ and $\cos\theta$. We know $\sin\theta = \frac{3}{5}$.

$\cos\theta = \pm\sqrt{1 - \sin^2\theta}$ Fundamental identity

$= \pm\sqrt{1 - \left(\dfrac{3}{5}\right)^2}$ Substitute.

$= \pm\sqrt{\dfrac{16}{25}}$

$\cos\theta = -\dfrac{4}{5}$ θ is in quadrant II.

(continued)

Thus,

$$\tan\frac{\theta}{2} = \frac{\sin\theta}{1 + \cos\theta} \qquad \text{Half-angle identity}$$

$$= \frac{\frac{3}{5}}{1 - \frac{4}{5}} = 3. \quad \text{Simplify.}$$

49. Verify that $\sec^2\dfrac{x}{2} = \dfrac{2}{1 + \cos x}$ is an identity.

Work with the left side.

$$\sec^2\frac{x}{2} = \frac{1}{\cos^2\dfrac{x}{2}} \qquad \text{Reciprocal identity}$$

$$= \frac{1}{\left(\pm\sqrt{\dfrac{1 + \cos x}{2}}\right)^2} \qquad \begin{array}{l}\text{Cosine half-angle}\\\text{identity}\end{array}$$

$$= \frac{1}{\dfrac{1 + \cos x}{2}}$$

$$= \frac{2}{1 + \cos x}$$

CHAPTER 6   INVERSE CIRCULAR FUNCTIONS AND TRIGONOMETRIC EQUATIONS

6.1 Exercises (pages 269–273)

85. $\sin\left(2\cos^{-1}\dfrac{1}{5}\right)$

Let $\theta = \cos^{-1}\frac{1}{5}$, so $\cos\theta = \frac{1}{5}$. The inverse cosine function yields values only in quadrants I and II, and since $\frac{1}{5}$ is positive, θ is in quadrant I. Sketch θ in quadrant I, and label the sides of a right triangle. By the Pythagorean theorem, the length of the side opposite θ will be $\sqrt{5^2 - 1^2} = \sqrt{24} = 2\sqrt{6}.$

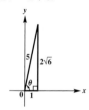

From the figure, $\sin\theta = \dfrac{2\sqrt{6}}{5}$. Then,

$$\sin\left(2\cos^{-1}\frac{1}{5}\right) = \sin 2\theta$$

$$= 2\sin\theta\cos\theta$$
$$\qquad \text{Sine double-angle identity}$$

$$= 2\left(\frac{2\sqrt{6}}{5}\right)\left(\frac{1}{5}\right) = \frac{4\sqrt{6}}{25}.$$

91. $\sin\left(\sin^{-1}\dfrac{1}{2} + \tan^{-1}(-3)\right)$

Let $\sin^{-1}\frac{1}{2} = A$ and $\tan^{-1}(-3) = B$; then $\sin A = \frac{1}{2}$ and $\tan B = -3$. Sketch angle A in quadrant I and angle B in quadrant IV.

$$\sin\left(\sin^{-1}\frac{1}{2} + \tan^{-1}(-3)\right) = \sin(A + B)$$

$$= \sin A \cos B + \cos A \sin B$$
$$\qquad \text{Sine sum identity}$$

$$= \frac{1}{2}\cdot\frac{1}{\sqrt{10}} + \frac{\sqrt{3}}{2}\cdot\frac{-3}{\sqrt{10}}$$

$$= \frac{1 - 3\sqrt{3}}{2\sqrt{10}} = \frac{\sqrt{10} - 3\sqrt{30}}{20}$$

6.2 Exercises (pages 278–280)

15. $\tan^2 x + 3 = 0,$ so $\tan^2 x = -3.$

The square of a real number cannot be negative, so this equation has no solution. Solution set: $\emptyset$

25.
$$2\sin\theta - 1 = \csc\theta$$

$$2\sin\theta - 1 = \frac{1}{\sin\theta} \qquad \text{Reciprocal identity}$$

$$2\sin^2\theta - \sin\theta = 1 \qquad \text{Multiply by }\sin\theta.$$

$$2\sin^2\theta - \sin\theta - 1 = 0 \qquad \text{Subtract 1.}$$

$$(2\sin\theta + 1)(\sin\theta - 1) = 0 \qquad \text{Factor.}$$

$$2\sin\theta + 1 = 0 \qquad \text{or} \qquad \sin\theta - 1 = 0$$
$$\qquad\qquad\qquad\qquad\qquad \text{Zero-factor property}$$

$$\sin\theta = -\frac{1}{2} \qquad \text{or} \qquad \sin\theta = 1$$

Over the interval $[0°, 360°)$, the equation $\sin\theta = -\frac{1}{2}$ has two solutions, the angles in quadrants III and IV that have reference angle $30°$. These are $210°$ and $330°$. In the same interval, the only angle θ for which $\sin\theta = 1$ is $90°$.

Solution set: $\{90°, 210°, 330°\}$

49.
$$\frac{2\tan\theta}{3 - \tan^2\theta} = 1$$

$$2\tan\theta = 3 - \tan^2\theta$$

$$\tan^2\theta + 2\tan\theta - 3 = 0$$

$$(\tan\theta - 1)(\tan\theta + 3) = 0$$

$$\tan\theta = 1 \qquad \text{or} \qquad \tan\theta = -3$$

Over the interval $[0°, 360°)$, the equation $\tan \theta = 1$ has two solutions, $45°$ and $225°$. Over the same interval, the equation $\tan \theta = -3$ has two solutions that are approximately $-71.6° + 180° = 108.4°$ and $-71.6° + 360° = 288.4°$.

The period of the tangent function is $180°$, so the solution set is $\{45° + 180°n, 108.4° + 180°n,$ where n is any integer$\}$.

6.3 Exercises *(pages 284–286)*

23. $\cos 2x + \cos x = 0$

We choose the identity for $\cos 2x$ that involves only the cosine function.

$$\cos 2x + \cos x = 0$$
$$2 \cos^2 x - 1 + \cos x = 0 \quad \text{Cosine double-angle identity}$$
$$2 \cos^2 x + \cos x - 1 = 0 \quad \text{Standard quadratic form}$$
$$(2 \cos x - 1)(\cos x + 1) = 0 \quad \text{Factor.}$$
$$2 \cos x - 1 = 0 \quad \text{or} \quad \cos x + 1 = 0 \quad \text{Zero-factor property}$$
$$2 \cos x = 1 \quad \text{or} \quad \cos x = -1$$
$$\cos x = \frac{1}{2} \quad \text{or} \quad x = \pi$$
$$x = \frac{\pi}{3} \text{ or } \frac{5\pi}{3}$$

Solution set: $\left\{\dfrac{\pi}{3}, \pi, \dfrac{5\pi}{3}\right\}$

31.
$$2 \sin \theta = 2 \cos 2\theta$$
$$\sin \theta = \cos 2\theta \quad \text{Divide by 2.}$$
$$\sin \theta = 1 - 2 \sin^2 \theta \quad \begin{array}{l}\text{Cosine double-}\\\text{angle identity}\end{array}$$
$$2 \sin^2 \theta + \sin \theta - 1 = 0$$
$$(2 \sin \theta - 1)(\sin \theta + 1) = 0$$
$$2 \sin \theta - 1 = 0 \quad \text{or} \quad \sin \theta + 1 = 0$$
$$\sin \theta = \frac{1}{2} \quad \text{or} \quad \sin \theta = -1$$

Over the interval $[0°, 360°)$, the equation $\sin \theta = \frac{1}{2}$ has two solutions, $30°$ and $150°$. Over the same interval, the equation $\sin \theta = -1$ has one solution, $270°$.

Solution set: $\{30°, 150°, 270°\}$

6.4 Exercises *(pages 290–293)*

15.
$$y = \cos(x + 3)$$
$$x + 3 = \arccos y$$
$$x = -3 + \arccos y$$

37. $\arccos x + 2 \arcsin \dfrac{\sqrt{3}}{2} = \pi$

$$\arccos x = \pi - 2 \arcsin \frac{\sqrt{3}}{2}$$
$$\arccos x = \pi - 2\left(\frac{\pi}{3}\right) \quad \arcsin \tfrac{\sqrt{3}}{2} = \tfrac{\pi}{3}$$
$$\arccos x = \pi - \frac{2\pi}{3}$$
$$\arccos x = \frac{\pi}{3}$$
$$x = \cos \frac{\pi}{3} = \frac{1}{2}$$

Solution set: $\left\{\dfrac{1}{2}\right\}$

41. $\cos^{-1} x + \tan^{-1} x = \dfrac{\pi}{2}$

$$\cos^{-1} x = \frac{\pi}{2} - \tan^{-1} x$$
$$x = \cos\left(\frac{\pi}{2} - \tan^{-1} x\right)$$
$$\text{Definition of } \cos^{-1} x$$
$$x = \cos \frac{\pi}{2} \cdot \cos(\tan^{-1} x)$$
$$+ \sin \frac{\pi}{2} \cdot \sin(\tan^{-1} x)$$
$$\text{Cosine difference identity}$$
$$x = 0 \cdot \cos(\tan^{-1} x) + 1 \cdot \sin(\tan^{-1} x)$$
$$\cos \tfrac{\pi}{2} = 0; \sin \tfrac{\pi}{2} = 1$$
$$x = \sin(\tan^{-1} x)$$

Let $u = \tan^{-1} x$, so $\tan u = x$.

From the triangle, we find that $\sin u = \dfrac{x}{\sqrt{1 + x^2}}$, so the

equation $x = \sin(\tan^{-1} x)$ becomes $x = \dfrac{x}{\sqrt{1 + x^2}}$. Solve

this equation.

(continued)

$$x = \frac{x}{\sqrt{1 + x^2}}$$

$x\sqrt{1 + x^2} = x$ Multiply by $\sqrt{1 + x^2}$.

$x\sqrt{1 + x^2} - x = 0$

$x\left(\sqrt{1 + x^2} - 1\right) = 0$ Factor.

$x = 0$ or $\sqrt{1 + x^2} - 1 = 0$ Zero-factor property

$\sqrt{1 + x^2} = 1$ Isolate the radical.

$1 + x^2 = 1$ Square both sides.

$x^2 = 0$

$x = 0$

Solution set: $\{0\}$

CHAPTER 7   APPLICATIONS OF TRIGONOMETRY AND VECTORS

7.1 Exercises (pages 308–312)

31. We cannot find θ directly because the length of the side opposite angle θ is not given. Redraw the triangle shown in the figure and label the third angle as α.

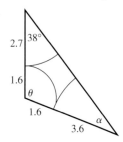

$$\frac{\sin \alpha}{1.6 + 2.7} = \frac{\sin 38°}{1.6 + 3.6}$$ Alternative form of the law of sines

$$\frac{\sin \alpha}{4.3} = \frac{\sin 38°}{5.2}$$

$$\sin \alpha = \frac{4.3 \sin 38°}{5.2} \approx .50910468$$

$$\alpha \approx 31°$$

Then $\theta \approx 180° - 38° - 31°$

$\theta \approx 111°$.

41. To find the area of the triangle, use $\mathcal{A} = \frac{1}{2}bh$, with $b = 1$ and $h = \sqrt{2}$.

$$\mathcal{A} = \frac{1}{2}(1)\left(\sqrt{2}\right) = \frac{\sqrt{2}}{2}$$

Now use $\mathcal{A} = \frac{1}{2}ab \sin C$, with $a = 2$, $b = 1$, and $C = 45°$.

$$\mathcal{A} = \frac{1}{2}(2)(1) \sin 45° = \sin 45° = \frac{\sqrt{2}}{2}$$

Both formulas show that the area is $\frac{\sqrt{2}}{2}$ sq unit.

7.2 Exercises (pages 317–319)

11. $\dfrac{\sin B}{b} = \dfrac{\sin A}{a}$ Alternative form of the law of sines

$\dfrac{\sin B}{2} = \dfrac{\sin 60°}{\sqrt{6}}$ Substitute values from the figure.

$\sin B = \dfrac{2 \sin 60°}{\sqrt{6}}$ Multiply by 2.

$\sin B = \dfrac{2 \cdot \frac{\sqrt{3}}{2}}{\sqrt{6}}$ $\sin 60° = \frac{\sqrt{3}}{2}$

$\sin B = \dfrac{\sqrt{3}}{\sqrt{6}} = \sqrt{\dfrac{1}{2}} = \dfrac{\sqrt{2}}{2}$

$B = 45°$ $\sin 45° = \frac{\sqrt{2}}{2}$

There is another angle between $0°$ and $180°$ whose sine is $\frac{\sqrt{2}}{2}$: $180° - 45° = 135°$. However, this is too large because $A = 60°$ and $60° + 135° = 195°$. Since $195° > 180°$, there is only one solution, $B = 45°$.

19. $A = 142.13°$, $b = 5.432$ ft, $a = 7.297$ ft

$\dfrac{\sin B}{b} = \dfrac{\sin A}{a}$ Alternative form of the law of sines

$\sin B = \dfrac{b \sin A}{a}$

$\sin B = \dfrac{5.432 \sin 142.13°}{7.297}$ Substitute given values.

$\sin B \approx .45697580$

$B \approx 27.19°$ Use the inverse sine function.

Because angle A is obtuse, angle B must be acute, so this is the only possible value for B and there is one triangle with the given measurements.

$C = 180° - A - B$ Sum of the angles of any triangle is $180°$.

$C \approx 180° - 142.13° - 27.19°$

$C \approx 10.68°$

Thus, $B \approx 27.19°$ and $C \approx 10.68°$.

7.3 Exercises (pages 326–333)

21. $C = 45.6°$, $b = 8.94$ m, $a = 7.23$ m

First find c.

$c^2 = a^2 + b^2 - 2ab \cos C$ Law of cosines

$c^2 = 7.23^2 + 8.94^2 - 2(7.23)(8.94) \cos 45.6°$ Substitute given values.

$c^2 \approx 41.7493$

$c \approx 6.46$

Find A next since angle A is smaller than angle B (because $a < b$), and thus angle A must be acute.

$$\frac{\sin A}{a} = \frac{\sin C}{c} \qquad \text{Alternative form of the law of sines}$$

$$\sin A = \frac{a \sin C}{c}$$

$$\sin A = \frac{7.23 \sin 45.6°}{6.46}$$

$$\sin A \approx .79963428$$

$$A \approx 53.1°$$

Finally, find B.

$$B = 180° - C - A$$

$$B \approx 180° - 45.6° - 53.1°$$

$$B \approx 81.3°$$

Thus, $c \approx 6.46$ m, $A \approx 53.1°$, and $B \approx 81.3°$.

41. Find AC, or b, in this figure.

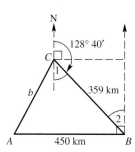

Angle $1 = 180° - 128° 40' = 51° 20'$

Angles 1 and 2 are alternate interior angles formed when two parallel lines (the north lines) are cut by a transversal, line BC, so angle $2 =$ angle $1 = 51° 20'$.

angle $ABC = 90° -$ angle $2 = 90° - 51° 20' = 38° 40'$

Complementary angles

$$b^2 = a^2 + c^2 - 2ac \cos B$$

Law of cosines

$$b^2 = 359^2 + 450^2 - 2(359)(450) \cos 38° 40'$$

Substitute values from the figure.

$$b^2 \approx 79,106$$

$$b \approx 281$$

C is about 281 km from A.

7.4 Exercises *(pages 341–344)*

19. Use the figure to find the components of **a** and **b**:
$\mathbf{a} = \langle -8, 8 \rangle$ and $\mathbf{b} = \langle 4, 8 \rangle$.

(a) $\mathbf{a} + \mathbf{b} = \langle -8, 8 \rangle + \langle 4, 8 \rangle = \langle -8 + 4, 8 + 8 \rangle = \langle -4, 16 \rangle$

(b) $\mathbf{a} - \mathbf{b} = \langle -8, 8 \rangle - \langle 4, 8 \rangle = \langle -8 - 4, 8 - 8 \rangle$
$\qquad = \langle -12, 0 \rangle$

(c) $-\mathbf{a} = -\langle -8, 8 \rangle = \langle 8, -8 \rangle$

47. $\mathbf{v} = \langle a, b \rangle = \langle 5 \cos(-35°), 5 \sin(-35°) \rangle$
$\qquad = \langle 4.0958, -2.8679 \rangle$

81. First write the given vectors in component form.

$$3\mathbf{i} + 4\mathbf{j} = \langle 3, 4 \rangle; \quad \mathbf{j} = \langle 0, 1 \rangle$$

$$\cos \theta = \frac{\langle 3, 4 \rangle \cdot \langle 0, 1 \rangle}{|\langle 3, 4 \rangle| \, |\langle 0, 1 \rangle|}$$

$$\cos \theta = \frac{3(0) + 4(1)}{\sqrt{9 + 16} \cdot \sqrt{0 + 1}} = \frac{4}{5} = .8$$

$$\theta = \cos^{-1} .8 \approx 36.87°$$

7.5 Exercises *(pages 347–350)*

5. Use the parallelogram rule. In the figure, **x** represents the second force and **v** is the resultant.

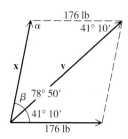

$$\alpha = 180° - 78° 50'$$
$$= 101° 10'$$

$$\beta = 78° 50' - 41° 10'$$
$$= 37° 40'$$

$$\frac{|\mathbf{x}|}{\sin 41° 10'} = \frac{176}{\sin 37° 40'} \qquad \text{Law of sines}$$

$$|\mathbf{x}| = \frac{176 \sin 41° 10'}{\sin 37° 40'} \approx 190$$

$$\frac{|\mathbf{v}|}{\sin \alpha} = \frac{176}{\sin 37° 40'} \qquad \text{Law of sines}$$

$$|\mathbf{v}| = \frac{176 \sin 101° 10'}{\sin 37° 40'} \approx 283$$

Thus, the magnitude of the second force is about 190 lb, and the magnitude of the resultant is about 283 lb.

27. Let **v** represent the airspeed vector.

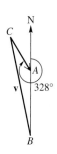

The groundspeed is $\frac{400 \text{ mi}}{2.5 \text{ hr}} = 160$ mph.

angle $BAC = 328° - 180° = 148°$

$$|\mathbf{v}|^2 = 11^2 + 160^2 - 2(11)(160) \cos 148°$$

Law of cosines

$$|\mathbf{v}|^2 \approx 28,706$$

$$|\mathbf{v}| \approx 169.4$$

The airspeed must be approximately 170 mph.

$$\frac{\sin B}{11} = \frac{\sin 148°}{169.4} \qquad \text{Law of sines}$$

$$\sin B = \frac{11 \sin 148°}{169.4} \approx .03441034$$

$$B \approx 2°$$

The bearing must be approximately $360° - 2° = 358°$.

CHAPTER 8   COMPLEX NUMBERS, POLAR EQUATIONS, AND PARAMETRIC EQUATIONS

8.1 Exercises (pages 369–371)

61. $-i - 2 - (6 - 4i) - (5 - 2i)$

$= (-2 - 6 - 5) + [-1 - (-4) - (-2)]i$

$= -13 + 5i$

79. $(2 + i)(2 - i)(4 + 3i)$

$= [(2 + i)(2 - i)](4 + 3i)$ Associative property

$= (2^2 - i^2)(4 + 3i)$ Product of the sum and difference of two terms

$= [4 - (-1)](4 + 3i)$ $i^2 = -1$

$= 5(4 + 3i)$

$= 20 + 15i$ Distributive property

107. $\left(\dfrac{\sqrt{2}}{2} + \dfrac{\sqrt{2}}{2}i\right)^2 = \left(\dfrac{\sqrt{2}}{2}\right)^2 + 2 \cdot \dfrac{\sqrt{2}}{2} \cdot \dfrac{\sqrt{2}}{2}i + \left(\dfrac{\sqrt{2}}{2}i\right)^2$

Square of a binomial

$= \dfrac{2}{4} + 2 \cdot \dfrac{2}{4}i + \dfrac{2}{4}i^2$

$= \dfrac{1}{2} + i + \dfrac{1}{2}i^2$

$= \dfrac{1}{2} + i + \dfrac{1}{2}(-1)$ $i^2 = -1$

$= \dfrac{1}{2} + i - \dfrac{1}{2}$

$= i$

Thus, $\dfrac{\sqrt{2}}{2} + \dfrac{\sqrt{2}}{2}i$ is a square root of i.

8.2 Exercises (pages 376–378)

31. $3\operatorname{cis} 150° = 3(\cos 150° + i \sin 150°)$

$= 3\left(-\dfrac{\sqrt{3}}{2} + i \cdot \dfrac{1}{2}\right)$ $\cos 150° = -\dfrac{\sqrt{3}}{2}$; $\sin 150° = \dfrac{1}{2}$

$= -\dfrac{3\sqrt{3}}{2} + \dfrac{3}{2}i$ Rectangular form

43. $-5 - 5i$

Sketch the graph of $-5 - 5i$ in the complex plane.

Since $x = -5$ and $y = -5$,

$r = \sqrt{x^2 + y^2} = \sqrt{(-5)^2 + (-5)^2} = \sqrt{50} = 5\sqrt{2}$

and

$\tan \theta = \dfrac{y}{x} = \dfrac{-5}{-5} = 1.$

Since $\tan \theta = 1$, the reference angle for θ is 45°. The graph shows that θ is in quadrant III, so $\theta = 180° + 45° = 225°$. Therefore,

$-5 - 5i = 5\sqrt{2}(\cos 225° + i \sin 225°).$

8.3 Exercises (pages 382–384)

5. $[4(\cos 60° + i \sin 60°)][6(\cos 330° + i \sin 330°)]$

$= 4 \cdot 6[\cos(60° + 330°) + i \sin(60° + 330°)]$

Product theorem

$= 24(\cos 390° + i \sin 390°)$ Multiply and add.

$= 24(\cos 30° + i \sin 30°)$ 390° and 30° are coterminal angles.

$= 24\left(\dfrac{\sqrt{3}}{2} + i \cdot \dfrac{1}{2}\right)$ $\cos 30° = \dfrac{\sqrt{3}}{2}$; $\sin 30° = \dfrac{1}{2}$

$= 12\sqrt{3} + 12i$

21. $\dfrac{-i}{1 + i}$

Numerator: $-i = 0 - 1i$

$r = \sqrt{0^2 + (-1)^2} = 1$

$\theta = 270°$ since $\cos 270° = 0$ and $\sin 270° = -1$, so $-i = 1 \operatorname{cis} 270°$.

Denominator: $1 + i = 1 + 1i$

$r = \sqrt{1^2 + 1^2} = \sqrt{2}$

$\tan \theta = \dfrac{y}{x} = \dfrac{1}{1} = 1$

Since x and y are both positive, θ is in quadrant I, so $\theta = \tan^{-1} 1 = 45°$. Thus, $1 + i = \sqrt{2} \operatorname{cis} 45°$.

$\dfrac{-i}{1 + i} = \dfrac{1 \operatorname{cis} 270°}{\sqrt{2} \operatorname{cis} 45°}$

$= \dfrac{1}{\sqrt{2}} \operatorname{cis}(270° - 45°)$ Quotient theorem

$= \dfrac{\sqrt{2}}{2} \operatorname{cis} 225°$

$= \dfrac{\sqrt{2}}{2}(\cos 225° + i \sin 225°)$

$= \dfrac{\sqrt{2}}{2}\left(-\dfrac{\sqrt{2}}{2} - i \cdot \dfrac{\sqrt{2}}{2}\right)$ $\cos 225° = -\dfrac{\sqrt{2}}{2}$; $\sin 225° = -\dfrac{\sqrt{2}}{2}$

$= -\dfrac{1}{2} - \dfrac{1}{2}i$ Rectangular form

8.4 Exercises (pages 389–391)

11. $(-2 - 2i)^5$

First write $-2 - 2i$ in trigonometric form.

$r = \sqrt{(-2)^2 + (-2)^2} = \sqrt{8} = 2\sqrt{2}$

$\tan \theta = \dfrac{y}{x} = \dfrac{-2}{-2} = 1$

Because x and y are both negative, θ is in quadrant III, so $\theta = 225°$.

$$-2 - 2i = 2\sqrt{2}\,(\cos 225° + i \sin 225°)$$

$$(-2 - 2i)^5 = \left[2\sqrt{2}\,(\cos 225° + i \sin 225°)\right]^5$$

$$= (2\sqrt{2})^5[\cos(5 \cdot 225°) + i \sin(5 \cdot 225°)]$$

De Moivre's theorem

$$= 32 \cdot 4\sqrt{2}\,(\cos 1125° + i \sin 1125°)$$

$$= 128\sqrt{2}\,(\cos 1125° + i \sin 1125°)$$

$$= 128\sqrt{2}\,(\cos 45° + i \sin 45°)$$

1125° and 45° are coterminal.

$$= 128\sqrt{2}\left(\frac{\sqrt{2}}{2} + i \cdot \frac{\sqrt{2}}{2}\right)$$

$\cos 45° = \frac{\sqrt{2}}{2}$; $\sin 45° = \frac{\sqrt{2}}{2}$

$$= 128 + 128i \quad \text{Rectangular form}$$

41. $x^3 - \left(4 + 4i\sqrt{3}\right) = 0$

$$x^3 = 4 + 4i\sqrt{3}$$

$$r = \sqrt{4^2 + \left(4\sqrt{3}\right)^2} = \sqrt{16 + 48} = \sqrt{64} = 8$$

$$\tan \theta = \frac{4\sqrt{3}}{4} = \sqrt{3}$$

θ is in quadrant I, so $\theta = 60°$.

$$x^3 = 4 + 4i\sqrt{3}$$

$$x^3 = 8\left(\frac{1}{2} + i\frac{\sqrt{3}}{2}\right)$$

$$r^3(\cos 3\alpha + i \sin 3\alpha) = 8(\cos 60° + i \sin 60°)$$

$r^3 = 8$, so $r = 2$.

$$\alpha = \frac{60°}{3} + \frac{360° \cdot k}{3}, \, k \text{ any integer} \quad n\text{th root theorem}$$

$\alpha = 20° + 120° \cdot k, \, k$ any integer

If $k = 0$, then $\alpha = 20° + 0° = 20°$.

If $k = 1$, then $\alpha = 20° + 120° = 140°$.

If $k = 2$, then $\alpha = 20° + 240° = 260°$.

Solution set: $\{2(\cos 20° + i \sin 20°),$ $2(\cos 140° + i \sin 140°), 2(\cos 260° + i \sin 260°)\}$

8.5 Exercises (pages 401–405)

53.

$$r = 2 \sin \theta$$

$$r^2 = 2r \sin \theta \quad \text{Multiply by } r.$$

$$x^2 + y^2 = 2y \quad r^2 = x^2 + y^2, \, r \sin \theta = y$$

$$x^2 + y^2 - 2y = 0 \quad \text{Subtract } 2y.$$

$$x^2 + y^2 - 2y + 1 = 1 \quad \begin{array}{l}\text{Add 1 to complete the} \\ \text{square on } y.\end{array}$$

$$x^2 + (y - 1)^2 = 1 \quad \begin{array}{l}\text{Factor the perfect square} \\ \text{trinomial.}\end{array}$$

The graph is a circle with center $(0, 1)$ and radius 1.

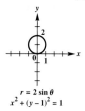

$r = 2 \sin \theta$
$x^2 + (y - 1)^2 = 1$

59. $r = 2 \sec \theta$

$$r = \frac{2}{\cos \theta} \quad \text{Reciprocal identity}$$

$$r \cos \theta = 2 \quad \text{Multiply by } \cos \theta.$$

$$x = 2 \quad r \cos \theta = x$$

The graph is the vertical line through $(2, 0)$.

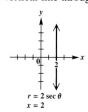

$r = 2 \sec \theta$
$x = 2$

8.6 Exercises (pages 411–414)

9. $x = t^3 + 1, \, y = t^3 - 1, \quad$ for t in $(-\infty, \infty)$

(a)

t	x	y
-2	-7	-9
-1	0	-2
0	1	-1
1	2	0
2	9	7
3	28	26

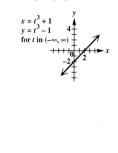

$x = t^3 + 1$
$y = t^3 - 1$
for t in $(-\infty, \infty)$

(b)

$$x = t^3 + 1$$
$$y = t^3 - 1$$
$$\overline{x - y = 2} \quad \text{Subtract equations to eliminate } t.$$
$$y = x - 2 \quad \text{Solve for } y.$$

The rectangular equation is $y = x - 2$, for x in $(-\infty, \infty)$.

The graph is a line with slope 1 and y-intercept -2.

13. $x = 3 \tan t, \, y = 2 \sec t, \quad$ for t in $\left(-\frac{\pi}{2}, \frac{\pi}{2}\right)$

(a)

t	x	y
$-\frac{\pi}{3}$	$-3\sqrt{3} \approx -5.2$	4
$-\frac{\pi}{6}$	$-\sqrt{3} \approx -1.7$	$\frac{4\sqrt{3}}{3} \approx 2.3$
0	0	2
$\frac{\pi}{6}$	$\sqrt{3} \approx 1.7$	$\frac{4\sqrt{3}}{3} \approx 2.3$
$\frac{\pi}{3}$	$3\sqrt{3} \approx 5.2$	4

(continued)

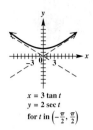

$x = 3 \tan t$
$y = 2 \sec t$
for t in $\left(-\frac{\pi}{2}, \frac{\pi}{2}\right)$

(b) $x = 3 \tan t$, so $\dfrac{x}{3} = \tan t$.

$y = 2 \sec t$, so $\dfrac{y}{2} = \sec t$.

$1 + \tan^2 t = \sec^2 t$	Pythagorean identity
$1 + \left(\dfrac{x}{3}\right)^2 = \left(\dfrac{y}{2}\right)^2$	Substitute expressions for tan t and sec t.
$1 + \dfrac{x^2}{9} = \dfrac{y^2}{4}$	
$y^2 = 4\left(1 + \dfrac{x^2}{9}\right)$	Multiply by 4.
$y = 2\sqrt{1 + \dfrac{x^2}{9}}$	Use the positive square root because $y > 0$ in the given interval for t.

The rectangular equation is $y = 2\sqrt{1 + \dfrac{x^2}{9}}$, for x in $(-\infty, \infty)$. The graph is the upper half of a hyperbola.

Answers to Selected Exercises

To The Student

In this section we provide the answers that we think most students will obtain when they work the exercises using the methods explained in the text. If your answer does not look exactly like the one given here, it is not necessarily wrong. In many cases there are equivalent forms of the answer. For example, if the answer section shows $\frac{3}{4}$ and your answer is .75, you have obtained the correct answer but written it in a different (yet equivalent) form. Unless the directions specify otherwise, .75 is just as valid an answer as $\frac{3}{4}$. In general, if your answer does not agree with the one given in the text, see whether it can be transformed into the other form. If it can, then it is the correct answer. If you still have doubts, talk with your instructor.

If you need further help with trigonometry, you may want to obtain a copy of the *Student's Solution Manual* that goes with this book. Your college bookstore either has this manual or can order it for you.

CHAPTER 1   TRIGONOMETRIC FUNCTIONS

1.1 Exercises *(pages 7–10)*

1. (a) 60° (b) 150°   **3.** (a) 45° (b) 135°   **5.** (a) 36° (b) 126°   **7.** (a) 89° (b) 179°   **9.** (a) 75° 40′ (b) 165° 40′

11. (a) 69° 49′ 30″ (b) 159° 49′ 30″   **13.** 70°; 110°   **15.** 30°; 60°   **17.** 40°; 140°   **19.** 107°; 73°   **21.** 69°; 21°

23. 45°   **25.** 150°   **27.** 7° 30′   **29.** $(90 - x)°$ **31.** $(x - 360)°$ **33.** 83° 59′   **35.** 23° 49′   **37.** 38° 32′

39. 60° 34′   **41.** 30° 27′   **43.** 17° 1′ 49″   **45.** 35.5°   **47.** 112.25°   **49.** −60.2°   **51.** 20.9°   **53.** 91.598°

55. 274.316°   **57.** 39° 15′ 00″   **59.** 126° 45′ 36″   **61.** −18° 30′ 54″   **63.** 31° 25′ 47″   **65.** 89° 54′ 1″

67. 178° 35′ 58″   **69.** 392°   **71.** 386° 30′   **73.** 320°   **75.** 235°   **77.** 1°   **79.** 359°   **81.** 179°

83. 130°   **85.** 240°   **87.** 120°

In Exercises 89 and 91, answers may vary.

89. 450°, 810°; −270°, −630°   **91.** 360°, 720°; −360°, −720°   **93.** $30° + n \cdot 360°$ **95.** $135° + n \cdot 360°$

97. $-90° + n \cdot 360°$ **99.** $0° + n \cdot 360°$, or $n \cdot 360°$

Angles other than those given are possible in Exercises 103–113.

103.
435°; −285°;
quadrant I

105.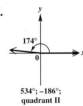
534°; −186°;
quadrant II

107.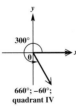
660°; −60°;
quadrant IV

109.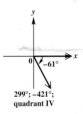
299°; −421°;
quadrant IV

111.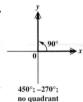
450°; −270°;
no quadrant

113.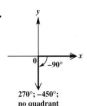
270°; −450°;
no quadrant

115. $3\sqrt{2}$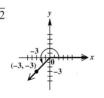

117. $\sqrt{34}$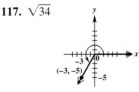

119. 2

121. 2

123. 4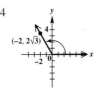

125. 4

127. $\dfrac{3}{4}$ **129.** 1800°

131. 12.5 rotations per hr

133. 4 sec

1.2 Exercises *(pages 16–22)*

1. Answers are given in numerical order: 55°; 65°; 60°; 65°; 60°; 120°; 60°; 60°; 55°; 55° **3.** 51°; 51° **5.** 50°; 60°; 70°

7. 60°; 60°; 60° **9.** 45°; 75°; 120° **11.** 49°; 49° **13.** 48°; 132° **15.** 91° **17.** 2° 29′ **19.** 25.4°

21. 22° 29′ 34″ **25.** right; scalene **27.** acute; equilateral **29.** right; scalene **31.** right; isosceles **33.** obtuse; scalene

35. acute; isosceles **41.** A and P; B and Q; C and R; AC and PR; BC and QR; AB and PQ **43.** A and C; E and D; ABE and

CBD; EB and DB; AB and CB; AE and CD **45.** $Q = 42°$; $B = R = 48°$ **47.** $B = 106°$; $A = M = 44°$

49. $X = M = 52°$ **51.** $a = 20$; $b = 15$ **53.** $a = 6$; $b = 7\frac{1}{2}$ **55.** $x = 6$ **57.** 30 m **59.** 500 m; 700 m

61. 112.5 ft **63.** $x = 110$ **65.** $c \approx 111.1$ **67.** The unknown side in the first quadrilateral is 40 cm. The unknown sides

in the second quadrilateral are 27 cm and 36 cm. **69. (a)** approximately 236,000 mi **(b)** no **71. (a)** approximately 2900 mi

(b) no **73. (a)** approximately $\frac{1}{4}$ **(b)** approximately 30 arc degrees **75.** $x = 10$; $y = 5$

Chapter 1 Quiz *(page 22)*

[1.1] 1. (a) 71° **(b)** 161° **2.** 65°; 115° **3.** 26°; 64° **[1.2] 4.** 20°; 24°; 136° **5.** 130°; 50°

[1.1] 6. (a) 77.2025° **(b)** 22° 1′ 30″ **7. (a)** 50° **(b)** 300° **(c)** 170° **(d)** 417° **8.** 1800° **[1.2] 9.** 10 ft

10. $x = 12$; $y = 10$

1.3 Exercises *(pages 27–29)*

In Exercises 1–19 and 45–53, we give, in order, sine, cosine, tangent, cotangent, secant, and cosecant.

1.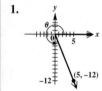
$-\dfrac{12}{13}; \dfrac{5}{13}; -\dfrac{12}{5};$
$-\dfrac{5}{12}; \dfrac{13}{5}; -\dfrac{13}{12}$

3.
$\dfrac{4}{5}; -\dfrac{3}{5}; -\dfrac{4}{3};$
$-\dfrac{3}{4}; -\dfrac{5}{3}; \dfrac{5}{4}$

5.
$\dfrac{15}{17}; -\dfrac{8}{17}; -\dfrac{15}{8};$
$-\dfrac{8}{15}; -\dfrac{17}{8}; \dfrac{17}{15}$

7.
$-\dfrac{24}{25}; \dfrac{7}{25}; -\dfrac{24}{7};$
$-\dfrac{7}{24}; \dfrac{25}{7}; -\dfrac{25}{24}$

9.
1; 0; undefined;
0; undefined; 1

11.

0; -1; 0; undefined;
-1; undefined

13.

-1; 0; undefined;
0; undefined; -1

15.
$\dfrac{\sqrt{3}}{2}; \dfrac{1}{2}; \sqrt{3};$
$\dfrac{\sqrt{3}}{3}; 2; \dfrac{2\sqrt{3}}{3}$

17.
$\dfrac{\sqrt{2}}{2}; \dfrac{\sqrt{2}}{2}; 1;$
1; $\sqrt{2}$; $\sqrt{2}$

19.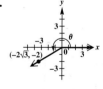
$-\dfrac{1}{2}; -\dfrac{\sqrt{3}}{2}; \dfrac{\sqrt{3}}{3};$
$\sqrt{3}; -\dfrac{2\sqrt{3}}{3}; -2$

23. 0 **25.** negative **27.** negative **29.** positive **31.** positive **33.** negative **35.** positive **37.** negative

39. positive **41.** positive **43.** positive

45.
$-\dfrac{2\sqrt{5}}{5}; \dfrac{\sqrt{5}}{5}; -2;$
$-\dfrac{1}{2}; \sqrt{5}; -\dfrac{\sqrt{5}}{2}$

47.
$\dfrac{6\sqrt{37}}{37}; -\dfrac{\sqrt{37}}{37}; -6;$
$-\dfrac{1}{6}; -\sqrt{37}; \dfrac{\sqrt{37}}{6}$

49.
$-\dfrac{4\sqrt{65}}{65}; -\dfrac{7\sqrt{65}}{65}; \dfrac{4}{7};$
$\dfrac{7}{4}; -\dfrac{\sqrt{65}}{7}; -\dfrac{\sqrt{65}}{4}$

51.
$-\dfrac{\sqrt{2}}{2}; \dfrac{\sqrt{2}}{2}; -1;$
$-1; \sqrt{2}; -\sqrt{2}$

53.
$-\dfrac{\sqrt{3}}{2}; -\dfrac{1}{2}; \sqrt{3};$
$\dfrac{\sqrt{3}}{3}; -2; -\dfrac{2\sqrt{3}}{3}$

55. 0   **57.** 0   **59.** −1   **61.** 1   **63.** undefined   **65.** −1   **67.** 0   **69.** undefined   **71.** 1   **73.** −3   **75.** −3
77. 5   **79.** 1   **81.** 0   **83.** 0   **85.** 0   **87.** 0   **89.** −1   **91.** 0   **93.** undefined   **95.** They are equal.
97. They are negatives of each other.   **99.** about .940; about .342   **101.** 35°   **103.** decrease; increase

1.4 Exercises *(pages 37–39)*

1. $\dfrac{3}{2}$ **3.** $-\dfrac{7}{3}$ **5.** $\dfrac{1}{5}$ **7.** $-\dfrac{2}{5}$ **9.** $\dfrac{\sqrt{2}}{2}$ **11.** −.4   **13.** .70069071   **17.** Because $\cot 90° = 0$, $\dfrac{1}{\cot 90°}$ and
consequently $\tan 90°$ are undefined.   **19.** All are positive.   **21.** Tangent and cotangent are positive; all others are
negative.   **23.** Sine and cosecant are positive; all others are negative.   **25.** Cosine and secant are positive; all others are
negative.   **27.** Sine and cosecant are positive; all others are negative.   **29.** All are positive.   **31.** I, II   **33.** I   **35.** II
37. I   **39.** III   **41.** III, IV   **45.** impossible   **47.** possible   **49.** possible   **51.** impossible   **53.** possible
55. possible   **57.** possible   **59.** impossible   **61.** $-\dfrac{4}{5}$ **63.** $-\dfrac{\sqrt{5}}{2}$ **65.** $-\dfrac{\sqrt{3}}{3}$ **67.** 3.44701905

In Exercises 69–79, we give, in order, sine, cosine, tangent, cotangent, secant, and cosecant.
69. $\dfrac{15}{17}; -\dfrac{8}{17}; -\dfrac{15}{8}; -\dfrac{8}{15}; -\dfrac{17}{8}; \dfrac{17}{15}$ **71.** $\dfrac{\sqrt{5}}{7}; \dfrac{2\sqrt{11}}{7}; \dfrac{\sqrt{55}}{22}; \dfrac{2\sqrt{55}}{5}; \dfrac{7\sqrt{11}}{22}; \dfrac{7\sqrt{5}}{5}$ **73.** $\dfrac{8\sqrt{67}}{67}; \dfrac{\sqrt{201}}{67}; \dfrac{8\sqrt{3}}{3}; \dfrac{\sqrt{3}}{8};$
$\dfrac{\sqrt{201}}{3}; \dfrac{\sqrt{67}}{8}$ **75.** $\dfrac{\sqrt{2}}{6}; -\dfrac{\sqrt{34}}{6}; -\dfrac{\sqrt{17}}{17}; -\sqrt{17}; -\dfrac{3\sqrt{34}}{17}; 3\sqrt{2}$ **77.** $\dfrac{\sqrt{15}}{4}; -\dfrac{1}{4}; -\sqrt{15}; -\dfrac{\sqrt{15}}{15}; -4; \dfrac{4\sqrt{15}}{15}$

79. .164215; −.986425; −.166475; −6.00691; −1.01376; 6.08958   **83.** This statement is false. For example,
$\sin 180° + \cos 180° = 0 + (-1) = -1 \neq 1.$ **85.** negative   **87.** negative   **89.** positive   **91.** positive
93. 2°   **95.** 3°   **97.** Quadrant II is the only quadrant in which the cosine is negative and the sine is positive.

Chapter 1 Review Exercises *(pages 42–46)*

1. complement: 55°; supplement: 145°   **3.** 186°   **5.** $x = 30; y = 30$ **7.** 9360°   **9.** 119.134°   **11.** 275° 6′ 2″
13. 40°; 60°; 80°   **15.** .25 km   **17.** $N = 12°; R = 82°; M = 86°$ **19.** $p = 7; q = 7$ **21.** $k = 14$ **23.** 12 ft

In Exercises 25–31, we give, in order, sine, cosine, tangent, cotangent, secant, and cosecant.
25. $-\dfrac{\sqrt{3}}{2}; \dfrac{1}{2}; -\sqrt{3}; -\dfrac{\sqrt{3}}{3}; 2; -\dfrac{2\sqrt{3}}{3}$ **27.** $-\dfrac{4}{5}; \dfrac{3}{5}; -\dfrac{4}{3}; -\dfrac{3}{4}; \dfrac{5}{3}; -\dfrac{5}{4}$ **29.** $\dfrac{15}{17}; -\dfrac{8}{17}; -\dfrac{15}{8}; -\dfrac{8}{15}; -\dfrac{17}{8}; \dfrac{17}{15}$
31. $-\dfrac{1}{2}; \dfrac{\sqrt{3}}{2}; -\dfrac{\sqrt{3}}{3}; -\sqrt{3}; \dfrac{2\sqrt{3}}{3}; -2$ **33.** tangent and secant
35.
37. 0; −1; 0; undefined; −1; undefined   **39. (a)** impossible **(b)** possible **(c)** possible

In Exercises 41–45, we give, in order, sine, cosine, tangent, cotangent, secant, and cosecant.
41. $-\dfrac{\sqrt{39}}{8}; -\dfrac{5}{8}; \dfrac{\sqrt{39}}{5}; \dfrac{5\sqrt{39}}{39}; -\dfrac{8}{5}; -\dfrac{8\sqrt{39}}{39}$ **43.** $\dfrac{2\sqrt{5}}{5}; -\dfrac{\sqrt{5}}{5}; -2; -\dfrac{1}{2}; -\sqrt{5}; \dfrac{\sqrt{5}}{2}$
45. $-\dfrac{3}{5}; \dfrac{4}{5}; -\dfrac{3}{4}; -\dfrac{4}{3}; \dfrac{5}{4}; -\dfrac{5}{3}$ **47.** 40 yd   **49.** approximately 9500 ft

Chapter 1 Test *(pages 46–47)*

[1.1] 1. (a) 23° **(b)** 113°   **2.** 145°; 35°   **3.** 20°; 70°   **[1.2] 4.** 130°; 130°   **5.** 110°; 110°   **6.** 20°; 30°; 130°
[1.1] 7. (a) 74.31° **(b)** 45° 12′ 9″   **8. (a)** 30° **(b)** 280° **(c)** 90°   **9.** 2700°   **[1.2] 10.** $10\dfrac{2}{3}$ ft, or 10 ft, 8 in.

11. $x = 8$; $y = 6$ [1.3] **12.**

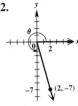

13.

$\sin \theta = -\dfrac{7\sqrt{53}}{53}$; $\cos \theta = \dfrac{2\sqrt{53}}{53}$; $\tan \theta = -\dfrac{7}{2}$;

$\cot \theta = -\dfrac{2}{7}$; $\sec \theta = \dfrac{\sqrt{53}}{2}$; $\csc \theta = -\dfrac{\sqrt{53}}{7}$

$\sin \theta = -1$; $\cos \theta = 0$; $\tan \theta$ is undefined; $\cot \theta = 0$; $\sec \theta$ is undefined; $\csc \theta = -1$

14.

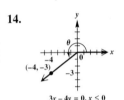

$3x - 4y = 0, x \le 0$

$\sin \theta = -\dfrac{3}{5}$; $\cos \theta = -\dfrac{4}{5}$; $\tan \theta = \dfrac{3}{4}$;

$\cot \theta = \dfrac{4}{3}$; $\sec \theta = -\dfrac{5}{4}$; $\csc \theta = -\dfrac{5}{3}$

15. row 1: 1, 0, undefined, 0, undefined, 1;

row 2: 0, 1, 0, undefined, 1, undefined;

row 3: −1, 0, undefined, 0, undefined, −1

16. cosecant and cotangent   [1.4] **17. (a)** I **(b)** III, IV **(c)** III   **18. (a)** impossible **(b)** possible **(c)** possible

19. $\sec \theta = -\dfrac{12}{7}$ **20.** $\cos \theta = -\dfrac{2\sqrt{10}}{7}$; $\tan \theta = -\dfrac{3\sqrt{10}}{20}$; $\cot \theta = -\dfrac{2\sqrt{10}}{3}$; $\sec \theta = -\dfrac{7\sqrt{10}}{20}$; $\csc \theta = \dfrac{7}{3}$

CHAPTER 2   ACUTE ANGLES AND RIGHT TRIANGLES

2.1 Exercises *(pages 55–58)*

In Exercises 1 and 3, we give, in order, sine, cosine, and tangent.

1. $\dfrac{21}{29}; \dfrac{20}{29}; \dfrac{21}{20}$ **3.** $\dfrac{n}{p}; \dfrac{m}{p}; \dfrac{n}{m}$ **5.** C   **7.** B   **9.** E

In Exercises 11–15, we give, in order, the unknown side, sine, cosine, tangent, cotangent, secant, and cosecant.

11. $c = 13$; $\dfrac{12}{13}; \dfrac{5}{13}; \dfrac{12}{5}; \dfrac{5}{12}; \dfrac{13}{5}; \dfrac{13}{12}$ **13.** $b = \sqrt{13}$; $\dfrac{\sqrt{13}}{7}; \dfrac{6}{7}; \dfrac{\sqrt{13}}{6}; \dfrac{6\sqrt{13}}{13}; \dfrac{7}{6}; \dfrac{7\sqrt{13}}{13}$ **15.** $b = 4$; $\dfrac{4}{5}; \dfrac{3}{5}; \dfrac{4}{3}; \dfrac{3}{4}; \dfrac{5}{3}; \dfrac{5}{4}$

17. $\sin A = \cos(90° - A)$; $\cos A = \sin(90° - A)$; $\tan A = \cot(90° - A)$; $\cot A = \tan(90° - A)$; $\sec A = \csc(90° - A)$;

$\csc A = \sec(90° - A)$ **19.** $\csc 51°$ **21.** $\csc(75° - \theta)$ **23.** $\tan(100° - \theta)$ **25.** $\cos 51.3°$ **27.** 40°

29. 20°   **31.** 12°   **33.** 35°   **35.** 18°   **37.** true   **39.** false   **41.** true   **43.** true   **45.** $\dfrac{\sqrt{3}}{3}$

47. $\dfrac{1}{2}$ **49.** $\dfrac{2\sqrt{3}}{3}$ **51.** $\sqrt{2}$ **53.** $\dfrac{\sqrt{2}}{2}$ **55.** 1   **57.** $\dfrac{\sqrt{3}}{2}$ **59.** $\sqrt{3}$

61.

62.

63. the legs; $\left(2\sqrt{2}, 2\sqrt{2}\right)$ **64.** $\left(1, \sqrt{3}\right)$ **65.** $\sin x$; $\tan x$ **67.** 60°

69. $\left(\dfrac{\sqrt{2}}{2}, \dfrac{\sqrt{2}}{2}\right)$; 45°   **71.** $y = \dfrac{\sqrt{3}}{3}x$ **73.** 60°   **75. (a)** 60° **(b)** k **(c)** $k\sqrt{3}$ **(d)** 2; $\sqrt{3}$; 30°; 60°

77. $x = \dfrac{9\sqrt{3}}{2}$; $y = \dfrac{9}{2}$; $z = \dfrac{3\sqrt{3}}{2}$; $w = 3\sqrt{3}$ **79.** $p = 15$; $r = 15\sqrt{2}$; $q = 5\sqrt{6}$; $t = 10\sqrt{6}$ **81.** $A = \dfrac{s^2}{2}$

2.2 Exercises *(pages 63–66)*

1. C   **3.** A   **5.** D   **11.** $\frac{\sqrt{2}}{2}; \frac{\sqrt{2}}{2}; \sqrt{2}; \sqrt{2}$ **13.** $-\frac{1}{2}; -\frac{\sqrt{3}}{3}; -2$ **15.** $\frac{1}{2}; -\sqrt{3}; -\frac{2\sqrt{3}}{3}$ **17.** $\sqrt{3}; \frac{\sqrt{3}}{3}$

In Exercises 19–35, we give, in order, sine, cosine, tangent, cotangent, secant, and cosecant.

19. $-\frac{\sqrt{2}}{2}; \frac{\sqrt{2}}{2}; -1, -1; \sqrt{2}; -\sqrt{2}$ **21.** $\frac{\sqrt{3}}{2}; \frac{1}{2}; \sqrt{3}; \frac{\sqrt{3}}{3}; 2; \frac{2\sqrt{3}}{3}$ **23.** $\frac{\sqrt{3}}{2}; -\frac{1}{2}; -\sqrt{3}; -\frac{\sqrt{3}}{3}; -2; \frac{2\sqrt{3}}{3}$

25. $-\frac{1}{2}; -\frac{\sqrt{3}}{2}; \frac{\sqrt{3}}{3}; \sqrt{3}; -\frac{2\sqrt{3}}{3}; -2$ **27.** $-\frac{\sqrt{2}}{2}; -\frac{\sqrt{2}}{2}; 1; 1; -\sqrt{2}; -\sqrt{2}$ **29.** $\frac{1}{2}; -\frac{\sqrt{3}}{2}; -\frac{\sqrt{3}}{3}; -\sqrt{3}; -\frac{2\sqrt{3}}{3}; 2$

31. $-\frac{1}{2}; -\frac{\sqrt{3}}{2}; \frac{\sqrt{3}}{3}; \sqrt{3}; -\frac{2\sqrt{3}}{3}; -2$ **33.** $\frac{1}{2}; -\frac{\sqrt{3}}{2}; -\frac{\sqrt{3}}{3}; -\sqrt{3}; -\frac{2\sqrt{3}}{3}; 2$ **35.** $-\frac{\sqrt{3}}{2}; \frac{1}{2}; -\sqrt{3}; -\frac{\sqrt{3}}{3};$

$2; -\frac{2\sqrt{3}}{3}$ **37.** $-\frac{\sqrt{3}}{2}$ **39.** $\frac{\sqrt{3}}{2}$ **41.** $-\sqrt{2}$ **43.** -1 **45.** true   **47.** false; $\frac{1}{2} \neq \sqrt{3}$ **49.** true   **51.** true

53. false; $0 \neq \frac{\sqrt{3}+1}{2}$ **55.** $(-3\sqrt{3}, 3)$ **57.** no   **59.** positive   **61.** positive   **63.** negative   **67.** .9

69. 45°; 225°   **71.** 30°; 150°   **73.** 120°; 300°   **75.** 45°; 315°   **77.** 210°; 330°   **79.** 30°; 210°   **81.** 225°; 315°

2.3 Exercises *(pages 68–72)*

1. sin; 1   **3.** reciprocal; reciprocal

In Exercises 5–21, the number of decimal places may vary depending on the calculator used.

5. .6252427   **7.** 1.0273488   **9.** 15.055723   **11.** 1.4887142   **13.** .6743024   **15.** .9999905

17. $\tan 23.4° \approx .4327386$ **19.** $\cot 77° \approx .2308682$ **21.** $\tan 4.72° \approx .0825664$ **23.** 55.845496°   **25.** 16.166641°

27. 38.491580°   **29.** 68.673241°   **31.** 45.526434°   **35.** .3746065934   **37.** -1 **39.** 1   **41.** 0   **45.** false

47. true   **49.** false   **51.** false   **53.** true   **55.** true   **57.** 65.96 lb   **59.** $-2.87°$ **61.** 2.87°

63. (a) 703 ft (b) 1701 ft (c) R would decrease; 644 ft, 1559 ft   **65.** (a) 2×10^8 m per sec   (b) 2×10^8 m per sec

67. 48.7°   **69.** (a) approximately 155 ft (b) approximately 194 ft

Chapter 2 Quiz *(pages 72–73)*

[2.1] 1. $\sin A = \frac{3}{5}; \cos A = \frac{4}{5}; \tan A = \frac{3}{4}$

2.

θ	$\sin\theta$	$\cos\theta$	$\tan\theta$	$\cot\theta$	$\sec\theta$	$\csc\theta$
30°	$\frac{1}{2}$	$\frac{\sqrt{3}}{2}$	$\frac{\sqrt{3}}{3}$	$\sqrt{3}$	$\frac{2\sqrt{3}}{3}$	2
45°	$\frac{\sqrt{2}}{2}$	$\frac{\sqrt{2}}{2}$	1	1	$\sqrt{2}$	$\sqrt{2}$
60°	$\frac{\sqrt{3}}{2}$	$\frac{1}{2}$	$\sqrt{3}$	$\frac{\sqrt{3}}{3}$	2	$\frac{2\sqrt{3}}{3}$

3. $w = 18; x = 18\sqrt{3}; y = 18; z = 18\sqrt{2}$ **4.** $3x^2 \sin\theta$

[2.2] In Exercises 5–7, we give, in order, sine, cosine, tangent, cotangent, secant, and cosecant.

5. $\frac{\sqrt{2}}{2}; -\frac{\sqrt{2}}{2}; -1; -1; -\sqrt{2}; \sqrt{2}$ **6.** $-\frac{1}{2}; -\frac{\sqrt{3}}{2}; \frac{\sqrt{3}}{3}; \sqrt{3}; -\frac{2\sqrt{3}}{3}; -2$ **7.** $-\frac{\sqrt{3}}{2}; \frac{1}{2}; -\sqrt{3}; -\frac{\sqrt{3}}{3}; 2; -\frac{2\sqrt{3}}{3}$

8. 60°; 120°   **9.** 135°; 225°   **[2.3] 10.** .67301251   **11.** -1.18176327 **12.** 69.497888°   **13.** 24.777233°

[2.1–2.3] 14. false   **15.** true

2.4 Exercises *(pages 78–83)*

1. 22,894.5 to 22,895.5   **3.** 8958.5 to 8959.5   **7.** .05

Note to student: While most of the measures resulting from solving triangles in this chapter are approximations, for convenience we use = rather than ≈.

9. $B = 53° 40'$; $a = 571$ m; $b = 777$ m   **11.** $M = 38.8°$; $n = 154$ m; $p = 198$ m   **13.** $A = 47.9108°$; $c = 84.816$ cm; $a = 62.942$ cm   **15.** $A = 37° 40'$; $B = 52° 20'$; $c = 20.5$ ft   **21.** $B = 62.0°$; $a = 8.17$ ft; $b = 15.4$ ft   **23.** $A = 17.0°$; $a = 39.1$ in.; $c = 134$ in.   **25.** $B = 27.5°$; $b = 6.61$ m; $c = 14.3$ m   **27.** $A = 36°$; $B = 54°$; $b = 18$ m   **29.** $c = 85.9$ yd; $A = 62° 50'$; $B = 27° 10'$ **31.** $b = 42.3$ cm; $A = 24° 10'$; $B = 65° 50'$ **33.** $B = 36° 36'$; $a = 310.8$ ft; $b = 230.8$ ft   **35.** $A = 50° 51'$; $a = .4832$ m; $b = .3934$ m   **41.** 9.35 m   **43.** 128 ft   **45.** 26.92 in.   **47.** 22°   **49.** 28.0 m   **51.** 13.3 ft   **53.** 146 m   **55. (a)** 29,000 ft **(b)** shorter

2.5 Exercises *(pages 87–91)*

1. It should be shown as an angle measured clockwise from due north.   **3.** A sketch is important to show the relationships among the given data and the unknowns.   **5.** 270°; N 90° W, or S 90° W   **7.** 315°; N 45° W   **9.** 0°; N 0° E, or N 0° W   **11.** 135°; S 45° E   **13.** $y = \dfrac{\sqrt{3}}{3}x$, $x \le 0$ **15.** 220 mi   **17.** 47 nautical mi   **19.** 140 mi   **21.** 148 mi   **23.** $x = \dfrac{b}{a - c}$ **25.** $y = (\tan 35°)(x - 25)$ **27.** 433 ft   **29.** 114 ft   **31.** 5.18 m   **33. (a)** $d = \dfrac{b}{2}\left(\cot \dfrac{\alpha}{2} + \cot \dfrac{\beta}{2}\right)$ **(b)** 345.4 cm   **35.** 10.8 ft   **37. (a)** 320 ft **(b)** $R\left(1 - \cos \dfrac{\theta}{2}\right)$ **39. (a)** 23 ft **(b)** 48 ft **(c)** The faster the speed, the more land needs to be cleared inside the curve.

Chapter 2 Review Exercises *(pages 95–98)*

In Exercises 1, 13, and 15, we give, in order, sine, cosine, tangent, cotangent, secant, and cosecant.

1. $\dfrac{60}{61}; \dfrac{11}{61}; \dfrac{60}{11}; \dfrac{11}{60}; \dfrac{61}{11}; \dfrac{61}{60}$ **3.** 10°   **5.** 7°   **7.** true   **9.** true   **13.** $-\dfrac{\sqrt{3}}{2}; \dfrac{1}{2}; -\sqrt{3}; -\dfrac{\sqrt{3}}{3}; 2; -\dfrac{2\sqrt{3}}{3}$ **15.** $-\dfrac{1}{2}; \dfrac{\sqrt{3}}{2}; -\dfrac{\sqrt{3}}{3}; -\sqrt{3}; \dfrac{2\sqrt{3}}{3}; -2$ **17.** 120°; 240°   **19.** 150°; 210°   **21.** $3 - \dfrac{2\sqrt{3}}{3}$ **23. (a)** $-\dfrac{\sqrt{2}}{2}; -\dfrac{\sqrt{2}}{2}; 1$ **(b)** $-\dfrac{\sqrt{3}}{2}; \dfrac{1}{2}; -\sqrt{3}$ **25.** -1.3563417 **27.** 1.0210339   **29.** .20834446   **31.** 55.673870°   **33.** 12.733938°   **35.** 63.008286°   **37.** 47.1°; 132.9°   **39.** false; 1.4088321 ≠ 1   **41.** true   **45.** III   **47.** $B = 31° 30'$; $a = 638$; $b = 391$ **49.** $B = 50.28°$; $a = 32.38$ m; $c = 50.66$ m   **51.** 137 ft   **53.** 73.7 ft   **55.** 18.75 cm   **57.** 1200 m   **59.** 140 mi   **63. (a)** 716 mi **(b)** 1104 mi

Chapter 2 Test *(pages 99–100)*

[2.1] 1. $\sin A = \dfrac{12}{13}$; $\cos A = \dfrac{5}{13}$; $\tan A = \dfrac{12}{5}$; $\cot A = \dfrac{5}{12}$; $\sec A = \dfrac{13}{5}$; $\csc A = \dfrac{13}{12}$ **2.** $x = 4$; $y = 4\sqrt{3}$; $z = 4\sqrt{2}$; $w = 8$

3. 15°   **[2.1, 2.2] 4. (a)** true **(b)** false; For $0° \le \theta \le 90°$, cosine is decreasing. **(c)** true

In Exercises 5–7, we give, in order, sine, cosine, tangent, cotangent, secant, and cosecant.

[2.2] 5. $-\dfrac{\sqrt{3}}{2}; -\dfrac{1}{2}; \sqrt{3}; \dfrac{\sqrt{3}}{3}; -2; -\dfrac{2\sqrt{3}}{3}$ **6.** $-\dfrac{\sqrt{2}}{2}; -\dfrac{\sqrt{2}}{2}; 1; 1; -\sqrt{2}; -\sqrt{2}$ **7.** $-1; 0;$ undefined; 0; undefined; -1

8. 135°; 225°   **9.** 240°; 300°   **10.** 45°; 225°   **[2.3] 11.** Take the reciprocal of $\tan \theta$ to get $\cot \theta = .5960011896$.

12. (a) .97939940 **(b)** -1.9056082 **(c)** 1.9362132   **13.** 16.16664145°   **[2.4] 14.** $B = 31° 30'$; $c = 877$; $b = 458$

15. 67.1°, or 67° 10′   **16.** 15.5 ft   **17.** 8800 ft   **[2.5] 18.** 72 nautical mi   **19.** 92 km   **20.** 448 m

CHAPTER 3   RADIAN MEASURE AND CIRCULAR FUNCTIONS

3.1 Exercises (pages 106–108)

1. 1   **3.** 3   **5.** −3   **7.** $\dfrac{\pi}{3}$ **9.** $\dfrac{\pi}{2}$ **11.** $\dfrac{5\pi}{6}$ **13.** $-\dfrac{5\pi}{3}$ **15.** $\dfrac{5\pi}{2}$ **17.** 10π **25.** 60°   **27.** 315°

29. 330°   **31.** −30°   **33.** 126°   **35.** −48°   **37.** 153°   **39.** −900°   **41.** .68   **43.** .742   **45.** 2.43

47. 1.122   **49.** .9847   **51.** .832391   **53.** 114° 35′   **55.** 99° 42′   **57.** 19° 35′   **59.** −287° 6′   **61.** In the

expression "sin 30," 30 means 30 radians; sin 30° $= \dfrac{1}{2}$, while sin 30 $\approx$ −.9880.   **63.** $\dfrac{\sqrt{3}}{2}$ **65.** 1   **67.** $\dfrac{2\sqrt{3}}{3}$ **69.** 1

71. $-\sqrt{3}$ **73.** $\dfrac{1}{2}$ **75.** −1   **77.** $-\dfrac{\sqrt{3}}{2}$ **79.** $\dfrac{1}{2}$ **81.** $\sqrt{3}$ **83.** We begin the answers with the blank next to

30°, and then proceed counterclockwise from there: $\dfrac{\pi}{6}$; 45; $\dfrac{\pi}{3}$; 120; 135; $\dfrac{5\pi}{6}$; π; $\dfrac{7\pi}{6}$; $\dfrac{5\pi}{4}$; 240; 300; $\dfrac{7\pi}{4}$; $\dfrac{11\pi}{6}$.

85. (a) 4π (b) $\dfrac{2\pi}{3}$ **87.** (a) 5π (b) $\dfrac{8\pi}{3}$

3.2 Exercises (pages 113–118)

1. 2π **3.** 20π **5.** 6   **7.** 1   **9.** 2   **11.** 25.8 cm   **13.** 3.61 ft   **15.** 5.05 m   **17.** 55.3 in.

19. The length is doubled.   **21.** 3500 km   **23.** 5900 km   **25.** 44° N   **27.** 156°   **29.** 38.5°   **31.** 18.7 cm

33. (a) 11.6 in. (b) 37° 5′   **35.** 146 in.   **37.** .20 km   **39.** 6π **41.** 72π **43.** 60°   **45.** 1.5   **47.** 1116.1 m²

49. 706.9 ft²   **51.** 114.0 cm²   **53.** 1885.0 mi²   **55.** 3.6   **57.** (a) $13\dfrac{1}{3}°$; $\dfrac{2\pi}{27}$ (b) 478 ft (c) 17.7 ft

(d) approximately 672 ft²   **59.** (a) 140 ft (b) 102 ft (c) 622 ft²   **61.** 1900 yd²   **63.** radius: 3950 mi;

circumference: 24,800 mi   **65.** $V = \dfrac{r^2\theta h}{2}$ (θ in radians)   **67.** $r = \dfrac{L}{\theta}$ **68.** $h = r\cos\dfrac{\theta}{2}$ **69.** $d = r\left(1 - \cos\dfrac{\theta}{2}\right)$

70. $d = \dfrac{L}{\theta}\left(1 - \cos\dfrac{\theta}{2}\right)$ **71.** The area is quadrupled.

3.3 Exercises (pages 126–128)

1. (a) 1 (b) 0 (c) undefined   **3.** (a) 0 (b) 1 (c) 0   **5.** (a) 0 (b) −1 (c) 0   **7.** $-\dfrac{1}{2}$ **9.** −1   **11.** −2

13. $-\dfrac{1}{2}$ **15.** $\dfrac{\sqrt{2}}{2}$ **17.** $\dfrac{\sqrt{3}}{2}$ **19.** $\dfrac{2\sqrt{3}}{3}$ **21.** $-\dfrac{\sqrt{3}}{3}$ **23.** .5736   **25.** .4068   **27.** 1.2065   **29.** 14.3338

31. −1.0460   **33.** −3.8665   **35.** .7   **37.** .9   **39.** −.6   **41.** 2.3 or 4.0   **43.** .8 or 2.4   **45.** negative

47. negative   **49.** positive   **51.** sin $\theta = \dfrac{\sqrt{2}}{2}$; cos $\theta = \dfrac{\sqrt{2}}{2}$; tan $\theta = 1$; cot $\theta = 1$; sec $\theta = \sqrt{2}$; csc $\theta = \sqrt{2}$

53. sin $\theta = -\dfrac{12}{13}$; cos $\theta = \dfrac{5}{13}$; tan $\theta = -\dfrac{12}{5}$; cot $\theta = -\dfrac{5}{12}$; sec $\theta = \dfrac{13}{5}$; csc $\theta = -\dfrac{13}{12}$ **55.** .2095   **57.** 1.4426   **59.** .3887

61. $\dfrac{5\pi}{6}$ **63.** $\dfrac{4\pi}{3}$ **65.** $\dfrac{7\pi}{4}$ **67.** (−.8011, .5985)   **69.** (.4385, −.8987)   **71.** I   **73.** II   **75.** .9846

77. (a) 32.4°   **79.** (a) 30° (b) 60° (c) 75° (d) 86° (e) 86° (f) 60°

Chapter 3 Quiz (page 128)

[3.1] 1. $\dfrac{5\pi}{4}$ **2.** $-\dfrac{11\pi}{6}$ **3.** 300°   **4.** −210°   **[3.2] 5.** 1.5   **6.** 67,500 in.²   **[3.3] 7.** $\dfrac{\sqrt{2}}{2}$

8. $-\dfrac{1}{2}$ **9.** 0   **10.** $\dfrac{2\pi}{3}$

3.4 Exercises *(pages 131–134)*

1. 2π sec   **3.** (a) $\dfrac{\pi}{2}$ radians (b) 10π cm (c) $\dfrac{5\pi}{3}$ cm per sec   **5.** 2π radians   **7.** $\dfrac{3\pi}{32}$ radian per sec   **9.** $\dfrac{6}{5}$ min

11. .180311 radian per sec   **13.** 8π m per sec   **15.** $\dfrac{9}{5}$ radians per sec   **17.** 1.83333 radians per sec   **19.** 18π cm

21. 12 sec   **23.** $\dfrac{3\pi}{32}$ radian per sec   **25.** $\dfrac{\pi}{6}$ radian per hr   **27.** $\dfrac{\pi}{30}$ radian per min   **29.** $\dfrac{7\pi}{30}$ cm per min

31. 168π m per min   **33.** 1500π m per min   **35.** 16.6 mph   **37.** (a) $\dfrac{2\pi}{365}$ radian (b) $\dfrac{\pi}{4380}$ radian per hr (c) about 67,000 mph

39. (a) 3.1 cm per sec (b) .24 radian per sec   **41.** 3.73 cm   **43.** 523.6 radians per sec

Chapter 3 Review Exercises *(pages 137–140)*

1. An angle of 1 radian is larger.   **3.** Three of many possible answers are $1 + 2\pi$, $1 + 4\pi$, and $1 + 6\pi$.   **5.** $\dfrac{\pi}{4}$ **7.** $\dfrac{35\pi}{36}$

9. $\dfrac{40\pi}{9}$ **11.** 225°   **13.** 480°   **15.** $-110°$ **17.** π in.   **19.** 12π in.   **21.** 35.8 cm   **23.** 7.683 cm

25. 273 m²   **27.** 4500 km   **29.** $\dfrac{3}{4}$; 1.5 sq units   **31.** (a) $\dfrac{\pi}{3}$ radians (b) 2π in.   **33.** $\sqrt{3}$ **35.** $-\dfrac{1}{2}$ **37.** 2

39. tan 1   **41.** sin 2   **43.** .8660   **45.** .9703   **47.** 1.9513   **49.** .3898   **51.** .5148   **53.** 1.1054   **55.** $\dfrac{\pi}{4}$

57. $\dfrac{7\pi}{6}$ **59.** $\dfrac{15}{32}$ sec   **61.** $\dfrac{\pi}{20}$ radian per sec   **63.** 285.3 cm   **65.** $\dfrac{\pi}{36}$ radian per sec

67. (b) $\dfrac{\pi}{6}$ (c) There is less ultraviolet light when $\theta = \dfrac{\pi}{3}$.

Chapter 3 Test *(pages 141–142)*

[3.1] 1. $\dfrac{2\pi}{3}$ **2.** $-\dfrac{\pi}{4}$ **3.** .09   **4.** 135°   **5.** $-210°$ **6.** 229.18°   **[3.2] 7.** (a) $\dfrac{4}{3}$ (b) 15,000 cm²   **8.** 2 radians

[3.3] 9. $\dfrac{\sqrt{2}}{2}$ **10.** $-\dfrac{\sqrt{3}}{2}$ **11.** undefined   **12.** -2 **13.** 0   **14.** 0   **15.** $\sin\dfrac{7\pi}{6} = -\dfrac{1}{2}$; $\cos\dfrac{7\pi}{6} = -\dfrac{\sqrt{3}}{2}$;

$\tan\dfrac{7\pi}{6} = \dfrac{\sqrt{3}}{3}$ **16.** sine and cosine: $(-\infty, \infty)$; tangent and secant: $\left\{ s \mid s \neq (2n+1)\dfrac{\pi}{2}, \text{ where } n \text{ is any integer} \right\}$; cotangent and

cosecant: $\left\{ s \mid s \neq n\pi, \text{ where } n \text{ is any integer} \right\}$ **17.** (a) .9716 (b) $\dfrac{\pi}{3}$ **[3.4] 18.** (a) $\dfrac{2\pi}{3}$ radians (b) 40π cm

(c) 5π cm per sec   **19.** approximately 8.278 mi per sec   **20.** (a) 75 ft (b) $\dfrac{\pi}{45}$ radian per sec

CHAPTER 4   GRAPHS OF THE CIRCULAR FUNCTIONS

4.1 Exercises *(pages 153–158)*

1. G   **3.** E   **5.** B   **7.** F   **9.** D   **11.** C   **13.** 2   **15.** $\dfrac{2}{3}$ **17.** 1

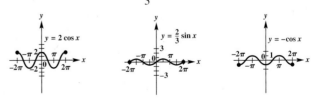

19. 2   **21.** 1   **23.** 4π; 1   **25.** $\dfrac{8\pi}{3}$; 1   **27.** $\dfrac{2\pi}{3}$; 1

 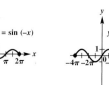

29. 8π; 2   **31.** $\dfrac{2\pi}{3}$; 2   **33.** 2; 1   **35.** 1; 2   **37.** 4; $\dfrac{1}{2}$

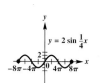   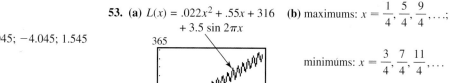

39. 2; π **41.** $y = 2\cos 2x$ **43.** $y = -3\cos\dfrac{1}{2}x$ **45.** $y = 3\sin 4x$ **47.** (a) 80°; 50° (b) 15

(c) about 35,000 yr (d) downward   **49.** (a) about 2 hr (b) 1 yr

51. (a) 5; $\dfrac{1}{60}$ (b) 60   **53.** (a) $L(x) = .022x^2 + .55x + 316$ (b) maximums: $x = \dfrac{1}{4}, \dfrac{5}{4}, \dfrac{9}{4}, \ldots$;

(c) 5; 1.545; −4.045; −4.045; 1.545 $+ 3.5\sin 2\pi x$

(d) 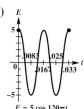   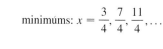   minimums: $x = \dfrac{3}{4}, \dfrac{7}{4}, \dfrac{11}{4}, \ldots$

55. (a) 8° (b) 21° (c) 62° (d) 61° (e) 31° (f) −11°   **57.** 24 hr   **59.** approximately 6:00 P.M.; approximately .2 ft

61. approximately 2:00 A.M.; approximately 2.6 ft   **63.** 1; 240°, or $\dfrac{4\pi}{3}$

4.2 Exercises (pages 164–167)

1. D   **3.** H   **5.** B   **7.** F   **9.** C   **11.** A   **15.** B   **17.** C   **19.** right   **21.** $y = -1 + \sin x$

23. $y = \cos\left(x - \dfrac{\pi}{3}\right)$ **25.** 2; 2π; none; π to the right   **27.** 4; 4π; none; π to the left   **29.** 3; 4; none; $\dfrac{1}{2}$ to the right

31. $1; \dfrac{2\pi}{3}$; up 2; $\dfrac{\pi}{15}$ to the right

33.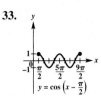

$y = \cos\left(x - \dfrac{\pi}{2}\right)$

35.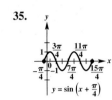

$y = \sin\left(x + \dfrac{\pi}{4}\right)$

37.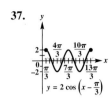

$y = 2\cos\left(x - \dfrac{\pi}{3}\right)$

39.

$y = \dfrac{3}{2}\sin 2\left(x + \dfrac{\pi}{4}\right)$

41.

$y = -4\sin(2x - \pi)$

43.

$y = \dfrac{1}{2}\cos\left(\dfrac{1}{2}x - \dfrac{\pi}{4}\right)$

45.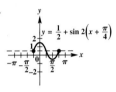

$y = -3 + 2\sin x$

47.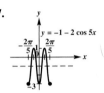

$y = -1 - 2\cos 5x$

49.

$y = 1 - 2\cos \dfrac{1}{2}x$

51.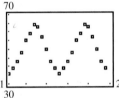

$y = -2 + \dfrac{1}{2}\sin 3x$

53.

$y = -3 + 2\sin\left(x + \dfrac{\pi}{2}\right)$

55.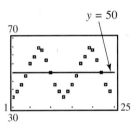

$y = \dfrac{1}{2} + \sin 2\left(x + \dfrac{\pi}{4}\right)$

57. (a) yes   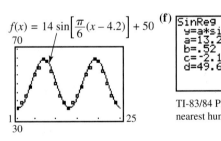   **(b)** It represents the average yearly temperature.

(c) 14; 12; 4.2 **(d)** $f(x) = 14\sin\left[\dfrac{\pi}{6}(x - 4.2)\right] + 50$

(e) The function gives an excellent model for the given data.

$f(x) = 14\sin\left[\dfrac{\pi}{6}(x - 4.2)\right] + 50$

(f)
```
SinReg
y=a*sin(bx+c)+d
a=13.21
b=.52
c=-2.18
d=49.68
```
TI-83/84 Plus fixed to the
nearest hundredth

Chapter 4 Quiz *(page 168)*

[4.1] 1. 2π; 4   **2.** $\pi; \dfrac{1}{2}$ **3.** 2; 3   **[4.2] 4.** 2π; 2   **5.** π; 1

$y = -4\sin x$

$y = -\dfrac{1}{2}\cos 2x$

$y = 3\sin \pi x$

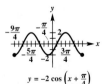

$y = -2\cos\left(x + \dfrac{\pi}{4}\right)$

$y = 2 + \sin(2x - \pi)$

[4.1] 6. $y = 2\sin x$ **7.** $y = \cos 2x$ **8.** $y = -\sin x$ **[4.1, 4.2] 9.** 73°F   **10.** 60°F; 84°F

4.3 Exercises (pages 174–176)

1. C   **3.** B   **5.** F   **7.**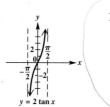

9.

11.

13.

15.

17.

19.

21.

23.

25.

27.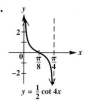

29.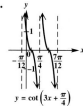

31.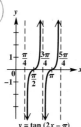

33. $y = -2 \tan x$ **35.** $y = \cot 3x$ **37.** true   **39.** false; $\tan(-x) = -\tan x$ for all x in the domain.   **41.** four

45. **(a)** 0 m **(b)** -2.9 m **(c)** -12.3 m **(d)** 12.3 m **(e)** It leads to $\tan \dfrac{\pi}{2}$, which is undefined.   **47.** π **48.** $\dfrac{5\pi}{4}$

49. $x = \dfrac{5\pi}{4} + n\pi$ **50.** approximately .3217505544   **51.** approximately 3.463343208   **52.** $\{x \mid x = .3217505544 + n\pi\}$

4.4 Exercises (pages 183–184)

1. B   **3.** D   **5.**

7.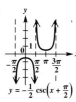

9.

11.

13.

15.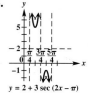

17.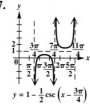

19. $y = -3 \cot x$ **21.** $y = \sec 4x$

23. $y = -2 + \csc x$ **25.** true   **27.** true

29. none   **33.** **(a)** 4 m   **(b)** 6.3 m   **(c)** 63.7 m

35. The display is for $Y_1 + Y_2$ at $X = \dfrac{\pi}{6}$.

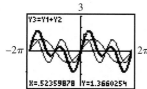

Summary Exercises on Graphing Circular Functions *(page 184)*

1.
$y = 2 \sin \pi x$

2.
$y = 4 \cos 1.5x$

3.
$y = -2 + .5 \cos \frac{\pi}{4}x$

4.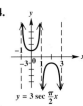
$y = 3 \sec \frac{\pi}{2}x$

5.
$y = -4 \csc .5x$

6.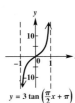
$y = 3 \tan \left(\frac{\pi}{2}x + \pi \right)$

7.
$y = -5 \sin \frac{x}{3}$

8.
$y = 10 \cos \left(\frac{x}{4} + \frac{\pi}{2} \right)$

9.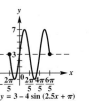
$y = 3 - 4 \sin (2.5x + \pi)$

10.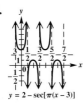
$y = 2 - \sec[\pi(x - 3)]$

4.5 Exercises *(pages 187–189)*

1. (a) $s(t) = 2 \cos 4\pi t$ **(b)** $s(1) = 2$; The weight is neither moving upward nor downward. At $t = 1$, the motion of the weight is changing from up to down. **3. (a)** $s(t) = -3 \cos 2.5\pi t$ **(b)** $s(1) = 0$; upward

5. $s(t) = .21 \cos 55\pi t$

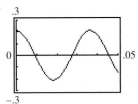

7. $s(t) = .14 \cos 110\pi t$

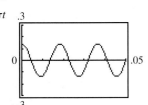

9. (a) $s(t) = -4 \cos \frac{2\pi}{3}t$ **(b)** 3.46 units **(c)** $\frac{1}{3}$ oscillation per sec **11. (a)** $s(t) = 2 \sin 2t$; amplitude: 2; period: π;

frequency: $\frac{1}{\pi}$ rotation per sec **(b)** $s(t) = 2 \sin 4t$; amplitude: 2; period: $\frac{\pi}{2}$; frequency: $\frac{2}{\pi}$ rotation per sec **13.** $\frac{8}{\pi^2}$ ft

15. (a) 4 in. **(b)** after $\frac{1}{8}$ sec **(c)** 4 cycles per sec; $\frac{1}{4}$ sec **17. (a)** 5 in. **(b)** 2 cycles per sec; $\frac{1}{2}$ sec **(c)** after $\frac{1}{4}$ sec

(d) approximately 4; After 1.3 sec, the weight is about 4 in. above the equilibrium position.

19. (a) $s(t) = -3 \cos 12t$ **(b)** $\frac{\pi}{6}$ sec **21.** 0; π; They are the same.

Chapter 4 Review Exercises *(pages 192–195)*

1. B **3.** sine, cosine, tangent, cotangent **5.** 2; 2π; none; none **7.** $\frac{1}{2}$; $\frac{2\pi}{3}$; none; none **9.** 2; 8π; 1 up; none

11. 3; 2π; none; $\frac{\pi}{2}$ to the left **13.** not applicable; π; none; $\frac{\pi}{8}$ to the right **15.** not applicable; $\frac{\pi}{3}$; none;

$\frac{\pi}{9}$ to the right **17.** tangent **19.** cosine **21.** cotangent **25.**

$y = 3 \sin x$

27.

$y = -\tan x$

29.
$y = 2 + \cot x$

31.
$y = \sin 2x$

33.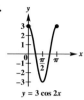
$y = 3 \cos 2x$

35.
$y = \cos\left(x - \frac{\pi}{4}\right)$

37.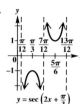
$y = \sec\left(2x + \frac{\pi}{3}\right)$

39.
$y = 1 + 2 \cos 3x$

41.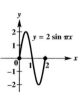
$y = 2 \sin \pi x$

45. (b)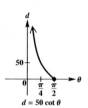
$d = 50 \cot \theta$

47. (a) $30°$ **(b)** $60°$ **(c)** $75°$ **(d)** $86°$ **(e)** $86°$

(f) $60°$ **49. (a)** 100 **(b)** 258 **(c)** 122 **(d)** 296   **51.** amplitude: 4; period: 2; frequency: $\frac{1}{2}$ cycle per sec

53. The frequency is the number of cycles in one unit of time; -4; 0; $-2\sqrt{2}$

Chapter 4 Test *(pages 195–196)*

[4.1–4.4] 1. (a) $y = \sec x$ **(b)** $y = \sin x$ **(c)** $y = \cos x$ **(d)** $y = \tan x$ **(e)** $y = \csc x$ **(f)** $y = \cot x$ **[4.1] 2. (a)** $y = \sin 2x$

(b) $y = 2 \sin x$ **[4.1, 4.3, 4.4] 3. (a)** $(-\infty, \infty)$ **(b)** $[-1, 1]$ **(c)** $\frac{\pi}{2}$ **(d)** $(-\infty, -1] \cup [1, \infty)$ **[4.2] 4. (a)** π **(b)** 6 **(c)** $[-3, 9]$

(d) -3 **(e)** $\frac{\pi}{4}$ to the left $\left(\text{that is, } -\frac{\pi}{4}\right)$ **5.**
$y = \sin(2x + \pi)$

[4.1] 6.
$y = -\cos 2x$

[4.2] 7.
$y = 2 + \cos x$

8.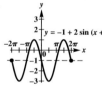
$y = -1 + 2 \sin(x + \pi)$

[4.3] 9.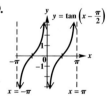
$y = \tan\left(x - \frac{\pi}{2}\right)$

10.
$y = -2 - \cot\left(x - \frac{\pi}{2}\right)$

[4.4] 11.
$y = -\csc 2x$

12.
$y = 3 \csc \pi x$

[4.1, 4.2] 13. (a)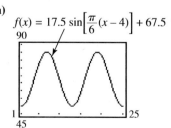
$f(x) = 17.5 \sin\left[\frac{\pi}{6}(x - 4)\right] + 67.5$

(b) 17.5; 12; 4 to the right; 67.5 up **(c)** approximately 52°F **(d)** 50°F in January; 85°F in July **(e)** approximately 67.5°; This is the vertical translation.

[4.5] 14. (a) 4 in. **(b)** after $\frac{1}{8}$ sec **(c)** 4 cycles per sec; $\frac{1}{4}$ sec

CHAPTER 5   TRIGONOMETRIC IDENTITIES

5.1 Exercises *(pages 203–206)*

1. -2.6 **3.** $.625$ **5.** $\dfrac{2}{3}$ **7.** $\dfrac{\sqrt{7}}{4}$ **9.** $-\dfrac{2\sqrt{5}}{5}$ **11.** $-\dfrac{\sqrt{105}}{11}$ **15.** $-\sin x$ **16.** odd   **17.** $\cos x$

18. even   **19.** $-\tan x$ **20.** odd   **21.** $f(-x) = f(x)$ **23.** $f(-x) = -f(x)$

25. $\cos\theta = -\dfrac{\sqrt{5}}{3}$; $\tan\theta = -\dfrac{2\sqrt{5}}{5}$; $\cot\theta = -\dfrac{\sqrt{5}}{2}$; $\sec\theta = -\dfrac{3\sqrt{5}}{5}$; $\csc\theta = \dfrac{3}{2}$

27. $\sin\theta = -\dfrac{\sqrt{17}}{17}$; $\cos\theta = \dfrac{4\sqrt{17}}{17}$; $\cot\theta = -4$; $\sec\theta = \dfrac{\sqrt{17}}{4}$; $\csc\theta = -\sqrt{17}$

29. $\sin\theta = \dfrac{3}{5}$; $\cos\theta = \dfrac{4}{5}$; $\tan\theta = \dfrac{3}{4}$; $\sec\theta = \dfrac{5}{4}$; $\csc\theta = \dfrac{5}{3}$

31. $\sin\theta = -\dfrac{\sqrt{7}}{4}$; $\cos\theta = \dfrac{3}{4}$; $\tan\theta = -\dfrac{\sqrt{7}}{3}$; $\cot\theta = -\dfrac{3\sqrt{7}}{7}$; $\csc\theta = -\dfrac{4\sqrt{7}}{7}$ **33.** B   **35.** E   **37.** A   **39.** A

41. D   **45.** $\sin\theta = \dfrac{\pm\sqrt{2x+1}}{x+1}$ **47.** $\sin x = \pm\sqrt{1 - \cos^2 x}$ **49.** $\tan x = \pm\sqrt{\sec^2 x - 1}$

51. $\csc x = \dfrac{\pm\sqrt{1 - \cos^2 x}}{1 - \cos^2 x}$ **53.** $\cos\theta$ **55.** $\cot\theta$ **57.** $\cos^2\theta$ **59.** $\sec\theta - \cos\theta$ **61.** $\cot\theta - \tan\theta$

63. $\tan\theta\sin\theta$ **65.** $\cos^2\theta$ **67.** $\sec^2\theta$ **69.** $\dfrac{25\sqrt{6} - 60}{12}$; $\dfrac{-25\sqrt{6} - 60}{12}$ **71.** $-\sin(2x)$ **72.** It is the negative of $\sin(2x)$.

73. $\cos(4x)$ **74.** It is the same function.   **75. (a)** $y = -\sin(4x)$ **(b)** $y = \cos(2x)$ **(c)** $y = 5\sin(3x)$ **77.** identity

79. not an identity   **81.** not an identity

5.2 Exercises *(pages 212–215)*

1. $\csc\theta\sec\theta$, or $\dfrac{1}{\sin\theta\cos\theta}$ **3.** $1 + \sec s$ **5.** 1 **7.** 1 **9.** $2 + 2\sin t$ **11.** $-\dfrac{2\cos x}{\sin^2 x}$, or $-2\cot x\csc x$

13. $(\sin\theta + 1)(\sin\theta - 1)$ **15.** $4\sin x$ **17.** $(2\sin x + 1)(\sin x + 1)$ **19.** $(\cos^2 x + 1)^2$

21. $(\sin x - \cos x)(1 + \sin x\cos x)$ **23.** $\sin\theta$ **25.** 1 **27.** $\tan^2\beta$ **29.** $\tan^2 x$ **31.** $\sec^2 x$ **33.** $\cos^2 x$

79. $(\sec\theta + \tan\theta)(1 - \sin\theta) = \cos\theta$ **81.** $\dfrac{\cos\theta + 1}{\sin\theta + \tan\theta} = \cot\theta$ **83.** identity   **85.** not an identity

91. It is true when $\sin x \geq 0$.   **93. (a)** $I = k(1 - \sin^2\theta)$ **(b)** For $\theta = 2\pi n$ and all integers n, $\cos^2\theta = 1$, its maximum value,

and I attains a maximum value of k.   **95. (a)** The sum of L and C equals 3.   **(b)** Let $Y_1 = L(t)$, $Y_2 = C(t)$, and   **(c)** $E(t) = 3$

$Y_3 = E(t)$. $Y_3 = 3$ for all inputs.

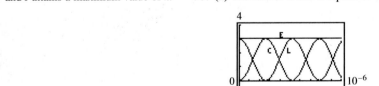

5.3 Exercises *(pages 221–224)*

1. F   **3.** E   **5.** $\dfrac{\sqrt{6}-\sqrt{2}}{4}$ **7.** $\dfrac{\sqrt{2}-\sqrt{6}}{4}$ **9.** $\dfrac{\sqrt{2}-\sqrt{6}}{4}$ **11.** 0   **13.** The calculator gives a value of 0 for the

expression.   **15.** $\cot 3°$ **17.** $\sin\dfrac{5\pi}{12}$ **19.** $\sec 104° \, 24'$ **21.** $\cos\left(-\dfrac{\pi}{8}\right)$ **23.** $\csc(-56° \, 42')$

25. $\tan(-86.9814°)$ **27.** $\tan$ **29.** $\cos$ **31.** $\csc$

For Exercises 33–37, other answers are possible.

33. 15°   **35.** $\dfrac{140°}{3}$ **37.** 20°   **39.** $\cos\theta$ **41.** $-\cos\theta$ **43.** $\cos\theta$ **45.** $-\cos\theta$ **47.** $\dfrac{4-6\sqrt{6}}{25}; \dfrac{4+6\sqrt{6}}{25}$

49. $\dfrac{16}{65}; -\dfrac{56}{65}$ **51.** $\dfrac{2\sqrt{638}-\sqrt{30}}{56}; \dfrac{2\sqrt{638}+\sqrt{30}}{56}$ **53.** true   **55.** false   **57.** true   **59.** true   **61.** false

68. $\dfrac{-\sqrt{6}-\sqrt{2}}{4}$ **69.** $\dfrac{-\sqrt{6}-\sqrt{2}}{4}$ **70. (a)** $\dfrac{\sqrt{2}-\sqrt{6}}{4}$ **(b)** $\dfrac{-\sqrt{6}-\sqrt{2}}{4}$ **71. (a)** 3 **(b)** 163 and -163; no

5.4 Exercises *(pages 229–232)*

1. C   **3.** E   **5.** B   **9.** $\dfrac{\sqrt{6}+\sqrt{2}}{4}$ **11.** $2-\sqrt{3}$ **13.** $\dfrac{-\sqrt{6}-\sqrt{2}}{4}$ **15.** $\dfrac{\sqrt{2}}{2}$ **17.** -1 **19.** 0   **21.** 1

23. $\dfrac{\sqrt{3}\cos\theta-\sin\theta}{2}$ **25.** $\dfrac{\cos\theta-\sqrt{3}\sin\theta}{2}$ **27.** $\dfrac{\sqrt{2}\,(\sin x-\cos x)}{2}$ **29.** $\dfrac{\sqrt{3}\tan\theta+1}{\sqrt{3}-\tan\theta}$ **31.** $\dfrac{\sqrt{2}\,(\cos x+\sin x)}{2}$

33. $-\cos\theta$ **35.** $-\tan\theta$ **37.** $-\tan\theta$ **41. (a)** $\dfrac{63}{65}$ **(b)** $\dfrac{63}{16}$ **(c)** I   **43. (a)** $\dfrac{4\sqrt{2}+\sqrt{5}}{9}$ **(b)** $\dfrac{-8\sqrt{5}-5\sqrt{2}}{20-2\sqrt{10}}$ (Other

forms are possible.) **(c)** II   **45. (a)** $\dfrac{77}{85}$ **(b)** $-\dfrac{77}{36}$ **(c)** II   **47.** $\dfrac{\sqrt{6}-\sqrt{2}}{4}$ **49.** $\dfrac{-\sqrt{6}-\sqrt{2}}{4}$ **51.** $-2+\sqrt{3}$

53. $\sin\left(\dfrac{\pi}{2}+\theta\right)=\cos\theta$ **55.** $\tan\left(\dfrac{\pi}{2}+\theta\right)=-\cot\theta$ **67.** $180°-\beta$ **68.** $\theta=\beta-\alpha$

69. $\tan\theta=\dfrac{\tan\beta-\tan\alpha}{1+\tan\beta\tan\alpha}$ **71.** 18.4°   **72.** 80.8°   **73. (a)** 425 lb **(c)** 0°   **75.** $-20\cos\dfrac{\pi t}{4}$

Chapter 5 Quiz *(page 232)*

[5.1] 1. $\cos\theta=\dfrac{24}{25}; \tan\theta=-\dfrac{7}{24}; \cot\theta=-\dfrac{24}{7}; \sec\theta=\dfrac{25}{24}; \csc\theta=-\dfrac{25}{7}$ **2.** $\dfrac{\cos^2 x+1}{\sin^2 x}$ **[5.4] 3.** $\dfrac{-\sqrt{6}-\sqrt{2}}{4}$

[5.3] 4. $-\cos\theta$ **[5.3, 5.4] 5. (a)** $-\dfrac{16}{65}$ **(b)** $-\dfrac{63}{65}$ **(c)** III   **[5.4] 6.** $\dfrac{-1+\tan x}{1+\tan x}$

5.5 Exercises *(pages 239–241)*

1. C   **3.** B   **5.** C   **7.** $\cos\theta=\dfrac{2\sqrt{5}}{5}; \sin\theta=\dfrac{\sqrt{5}}{5}$ **9.** $\cos\theta=-\dfrac{\sqrt{42}}{12}; \sin\theta=\dfrac{\sqrt{102}}{12}$ **11.** $\cos 2\theta=\dfrac{17}{25};$

$\sin 2\theta=-\dfrac{4\sqrt{21}}{25}$ **13.** $\cos 2x=-\dfrac{3}{5}; \sin 2x=\dfrac{4}{5}$ **15.** $\cos 2\theta=\dfrac{39}{49}; \sin 2\theta=-\dfrac{4\sqrt{55}}{49}$ **17.** $\dfrac{\sqrt{3}}{2}$ **19.** $\dfrac{\sqrt{3}}{2}$

21. $-\dfrac{\sqrt{2}}{2}$ **23.** $\dfrac{1}{2}\tan 102°$ **25.** $\dfrac{1}{4}\cos 94.2°$ **27.** $-\cos\dfrac{4\pi}{5}$ **29.** $\sin 4x=4\sin x\cos^3 x-4\sin^3 x\cos x$

31. $\tan 3x=\dfrac{3\tan x-\tan^3 x}{1-3\tan^2 x}$ **33.** $\cos^4 x-\sin^4 x=\cos 2x$ **35.** $\dfrac{2\tan x}{2-\sec^2 x}=\tan 2x$ **55.** $\sin 160°-\sin 44°$

57. $\sin\dfrac{\pi}{2}-\sin\dfrac{\pi}{6}$ **59.** $3\cos x-3\cos 9x$ **61.** $-2\sin 3x\sin x$ **63.** $-2\sin 11.5°\cos 36.5°$

65. $2\cos 6x\cos 2x$ **67.** $a=-885.6; c=885.6; \omega=240\pi$

5.6 Exercises *(pages 245–248)*

1. −   **3.** +   **5.** C   **7.** D   **9.** F   **11.** $\dfrac{\sqrt{2 + \sqrt{2}}}{2}$ **13.** $-\dfrac{\sqrt{2 + \sqrt{3}}}{2}$ **15.** $-\dfrac{\sqrt{2 + \sqrt{3}}}{2}$ **19.** $\dfrac{\sqrt{10}}{4}$

21. 3   **23.** $\dfrac{\sqrt{50 - 10\sqrt{5}}}{10}$ **25.** $-\sqrt{7}$ **27.** $\dfrac{\sqrt{5}}{5}$ **29.** $-\dfrac{\sqrt{42}}{12}$ **31.** .127   **33.** $\sin 20°$ **35.** $\tan 73.5°$

37. $\tan 29.87°$ **39.** $\cos 9x$ **41.** $\tan 4\theta$ **43.** $\cos \dfrac{x}{8}$ **45.** $\dfrac{\sin x}{1 + \cos x} = \tan \dfrac{x}{2}$ **47.** $\dfrac{\tan \frac{x}{2} + \cot \frac{x}{2}}{\cot \frac{x}{2} - \tan \frac{x}{2}} = \sec x$

59. 106°   **61.** 2   **63.** They are both radii of the circle.   **64.** It is the supplement of a 30° angle.   **65.** Their sum is

180° − 150° = 30°, and they are equal.   **66.** $2 + \sqrt{3}$ **68.** $\dfrac{\sqrt{6} + \sqrt{2}}{4}$ **69.** $\dfrac{\sqrt{6} - \sqrt{2}}{4}$ **70.** $2 - \sqrt{3}$

Chapter 5 Review Exercises *(pages 252–254)*

1. B   **3.** C   **5.** D   **7.** 1   **9.** $\dfrac{1}{\cos^2 \theta}$ **11.** $-\dfrac{\cos \theta}{\sin \theta}$ **13.** $\sin x = -\dfrac{4}{5}$; $\tan x = -\dfrac{4}{3}$; $\cot(-x) = \dfrac{3}{4}$

15. $\sin 165° = \dfrac{\sqrt{6} - \sqrt{2}}{4}$; $\cos 165° = \dfrac{-\sqrt{6} - \sqrt{2}}{4}$; $\tan 165° = -2 + \sqrt{3}$; $\csc 165° = \sqrt{6} + \sqrt{2}$; $\sec 165° = -\sqrt{6} + \sqrt{2}$;

$\cot 165° = -2 - \sqrt{3}$ **17.** E   **19.** J   **21.** I   **23.** H   **25.** G   **27.** $\dfrac{117}{125}$; $\dfrac{4}{5}$; $-\dfrac{117}{44}$; II

In Exercises 29 and 31, other forms are possible for tan(x + y).

29. $\dfrac{2 + 3\sqrt{7}}{10}$; $\dfrac{2\sqrt{3} + \sqrt{21}}{10}$; $\dfrac{2 + 3\sqrt{7}}{2\sqrt{3} - \sqrt{21}}$; II   **31.** $\dfrac{4 - 9\sqrt{11}}{50}$; $\dfrac{12\sqrt{11} - 3}{50}$; $\dfrac{4 - 9\sqrt{11}}{12\sqrt{11} + 3}$; IV

33. $\sin \theta = \dfrac{\sqrt{14}}{4}$; $\cos \theta = \dfrac{\sqrt{2}}{4}$ **35.** $\sin 2x = \dfrac{3}{5}$; $\cos 2x = -\dfrac{4}{5}$ **37.** $\dfrac{1}{2}$ **39.** $\dfrac{\sqrt{5} - 1}{2}$ **41.** .5

43. $-\dfrac{\sin 2x + \sin x}{\cos 2x - \cos x} = \cot \dfrac{x}{2}$ **45.** $\dfrac{\sin x}{1 - \cos x} = \cot \dfrac{x}{2}$ **47.** $\dfrac{2(\sin x - \sin^3 x)}{\cos x} = \sin 2x$

71. (a) $D = \dfrac{v^2 \sin 2\theta}{32}$ **(b)** approximately 35 ft

Chapter 5 Test *(pages 254–255)*

[5.1] 1. $\sin \theta = -\dfrac{7}{25}$; $\tan \theta = -\dfrac{7}{24}$; $\cot \theta = -\dfrac{24}{7}$; $\sec \theta = \dfrac{25}{24}$; $\csc \theta = -\dfrac{25}{7}$ **2.** $\cos \theta$ **3.** −1   **[5.3] 4.** $\dfrac{\sqrt{6} - \sqrt{2}}{4}$

[5.3, 5.4] 5. (a) $-\sin x$ **(b)** $\tan x$ **[5.6] 6.** $-\dfrac{\sqrt{2 - \sqrt{2}}}{2}$ **7.** $\cot \dfrac{1}{2} x - \cot x = \csc x$

[5.3, 5.4] 8. (a) $\dfrac{33}{65}$ **(b)** $-\dfrac{56}{65}$ **(c)** $\dfrac{63}{16}$ **(d)** II   **[5.5, 5.6] 9. (a)** $-\dfrac{7}{25}$ **(b)** $-\dfrac{24}{25}$ **(c)** $\dfrac{24}{7}$ **(d)** $\dfrac{\sqrt{5}}{5}$ **(e)** 2

[5.3] 15. (a) $V = 163 \cos\left(\dfrac{\pi}{2} - \omega t\right)$ **(b)** 163 volts; $\dfrac{1}{240}$ sec

CHAPTER 6 INVERSE CIRCULAR FUNCTIONS AND TRIGONOMETRIC EQUATIONS

6.1 Exercises *(pages 269–273)*

1. one-to-one   **3.** domain   **5.** π **7.** (a) $[-1, 1]$ (b) $\left[-\dfrac{\pi}{2}, \dfrac{\pi}{2}\right]$ (c) increasing (d) -2 is not in the domain.

9. (a) $(-\infty, \infty)$ (b) $\left(-\dfrac{\pi}{2}, \dfrac{\pi}{2}\right)$ (c) increasing (d) no   **11.** $\cos^{-1}\dfrac{1}{a}$ **13.** 0   **15.** π **17.** $-\dfrac{\pi}{2}$ **19.** 0   **21.** $\dfrac{\pi}{2}$

23. $\dfrac{\pi}{4}$ **25.** $\dfrac{5\pi}{6}$ **27.** $\dfrac{3\pi}{4}$ **29.** $-\dfrac{\pi}{6}$ **31.** $\dfrac{\pi}{6}$ **33.** 0   **35.** $-45°$ **37.** $-60°$ **39.** $120°$ **41.** $120°$

43. $60°$ **45.** $\sin^{-1} 2$ does not exist.   **47.** $-7.6713835°$ **49.** $113.500970°$ **51.** $30.987961°$ **53.** $121.267893°$

55. $-82.678329°$ **57.** $.83798122$ **59.** 2.3154725 **61.** 1.1900238 **63.** 1.9033723 **65.** 3.1144804

67. 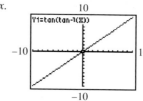 **69.** 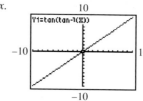 **71.**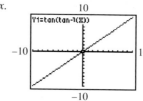

73. 1.003 is not in the domain of $y = \sin^{-1} x$.   **74.** In both cases, the result is x. In each case, the graph is a straight line bisecting quadrants I and III (i.e., the line $y = x$).   **75.** It is the graph of $y = x$.

76. It does not agree because the range of the inverse tangent function is $\left(-\dfrac{\pi}{2}, \dfrac{\pi}{2}\right)$, not $(-\infty, \infty)$, as was the case in Exercise 75.

77. $\dfrac{\sqrt{7}}{3}$ **79.** $\dfrac{\sqrt{5}}{5}$ **81.** $\dfrac{120}{169}$ **83.** $-\dfrac{7}{25}$ **85.** $\dfrac{4\sqrt{6}}{25}$ **87.** 2

89. $\dfrac{63}{65}$ **91.** $\dfrac{\sqrt{10} - 3\sqrt{30}}{20}$ **93.** $.894427191$ **95.** $.1234399811$ **97.** $\sqrt{1 - u^2}$ **99.** $\sqrt{1 - u^2}$ **101.** $\dfrac{4\sqrt{u^2 - 4}}{u^2}$

103. $\dfrac{u\sqrt{2}}{2}$ **105.** $\dfrac{2\sqrt{4-u^2}}{4-u^2}$ **107. (a)** $45°$ **(b)** $\theta = 45°$ **109. (a)** $18°$ **(b)** $18°$ **(c)** $15°$ **(e)** 1.4142151 m (Note: Due to the computational routine, there may be a discrepancy in the last few decimal places.) **(f)** $\sqrt{2}$

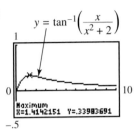

Radian mode

6.2 Exercises *(pages 278–280)*

1. Solve the linear equation for cot x. **3.** Solve the quadratic equation for sec x by factoring. **5.** Solve the quadratic equation for sin x using the quadratic formula. **7.** Use an identity to rewrite as an equation with one trigonometric function.

11. $\left\{\dfrac{3\pi}{4},\dfrac{7\pi}{4}\right\}$ **13.** $\left\{\dfrac{\pi}{6},\dfrac{5\pi}{6}\right\}$ **15.** $\emptyset$ **17.** $\left\{\dfrac{\pi}{4},\dfrac{2\pi}{3},\dfrac{5\pi}{4},\dfrac{5\pi}{3}\right\}$ **19.** $\{\pi\}$ **21.** $\left\{\dfrac{7\pi}{6},\dfrac{3\pi}{2},\dfrac{11\pi}{6}\right\}$

23. $\{30°,210°,240°,300°\}$ **25.** $\{90°,210°,330°\}$ **27.** $\{45°,135°,225°,315°\}$ **29.** $\{45°,225°\}$

31. $\{0°,30°,150°,180°\}$ **33.** $\{0°,45°,135°,180°,225°,315°\}$ **35.** $\{53.6°,126.4°,187.9°,352.1°\}$

37. $\{149.6°,329.6°,106.3°,286.3°\}$ **39.** $\emptyset$ **41.** $\{57.7°,159.2°\}$ **43.** $\{.8751 + 2n\pi, 2.2665 + 2n\pi,$

$3.5908 + 2n\pi, 5.8340 + 2n\pi,$ where n is any integer$\}$ **45.** $\left\{\dfrac{\pi}{3} + n\pi, \dfrac{2\pi}{3} + n\pi,$ where n is any integer$\right\}$

47. $\{33.6° + 360°n, 326.4° + 360°n,$ where n is any integer$\}$ **49.** $\{45° + 180°n, 108.4° + 180°n,$ where n is any integer$\}$

51. $\{.6806, 1.4159\}$ **53. (a)** $.00164$ and $.00355$ **(b)** $[.00164, .00355]$ **(c)** outward **55. (a)** $\dfrac{1}{4}$ sec **(b)** $\dfrac{1}{6}$ sec **(c)** $.21$ sec

57. (a) One such value is $\dfrac{\pi}{3}$. **(b)** One such value is $\dfrac{\pi}{4}$.

6.3 Exercises *(pages 284–286)*

1. $\left\{\dfrac{\pi}{3}, \pi, \dfrac{4\pi}{3}\right\}$ **3.** $\{60°, 210°, 240°, 310°\}$ **7.** $\left\{\dfrac{\pi}{12}, \dfrac{11\pi}{12}, \dfrac{13\pi}{12}, \dfrac{23\pi}{12}\right\}$ **9.** $\left\{\dfrac{\pi}{2}, \dfrac{7\pi}{6}, \dfrac{11\pi}{6}\right\}$

11. $\left\{\dfrac{\pi}{18}, \dfrac{7\pi}{18}, \dfrac{13\pi}{18}, \dfrac{19\pi}{18}, \dfrac{25\pi}{18}, \dfrac{31\pi}{18}\right\}$ **13.** $\left\{\dfrac{3\pi}{8}, \dfrac{5\pi}{8}, \dfrac{11\pi}{8}, \dfrac{13\pi}{8}\right\}$ **15.** $\left\{\dfrac{\pi}{2}, \dfrac{3\pi}{2}\right\}$ **17.** $\left\{0, \dfrac{\pi}{3}, \pi, \dfrac{5\pi}{3}\right\}$ **19.** $\emptyset$

21. $\left\{\dfrac{\pi}{2}\right\}$ **23.** $\left\{\dfrac{\pi}{3}, \pi, \dfrac{5\pi}{3}\right\}$ **25.** $\{15°,45°,135°,165°,255°,285°\}$ **27.** $\{0°\}$ **29.** $\{120°,240°\}$

31. $\{30°,150°,270°\}$ **33.** $\{180°n, 30° + 360°n, 150° + 360°n,$ where n is any integer$\}$ **35.** $\{60° + 360°n, 300° + 360°n,$

where n is any integer$\}$ **37.** $\{11.8° + 180°n, 78.2° + 180°n,$ where n is any integer$\}$ **39.** $\{30° + 180°n, 90° + 180°n,$

$150° + 180°n,$ where n is any integer$\}$ **41.** $\{1.2802\}$

43. (a) For $x = t$,

$P(t) = .003 \sin 220\pi t +$

$\dfrac{.003}{3} \sin 660\pi t +$

$\dfrac{.003}{5} \sin 1100\pi t +$

$\dfrac{.003}{7} \sin 1540\pi t$

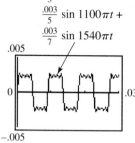

(b) The graph is periodic, and the wave has "jagged square" tops and bottoms.

(c) This will occur when t is in one of these intervals: $(.0045, .0091)$, $(.0136, .0182)$, $(.0227, .0273)$.

45. (a) For $x = t$,

$P(t) = \dfrac{1}{2} \sin[2\pi(220)t] +$

$\dfrac{1}{3} \sin[2\pi(330)t] +$

$\dfrac{1}{4} \sin[2\pi(440)t]$

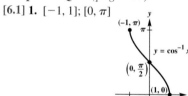

(b) $.0007576, .009847, .01894, .02803$ **(c)** 110 Hz   **(d)** For $x = t$,

$P(t) = \sin[2\pi(110)t] +$

$\dfrac{1}{2} \sin[2\pi(220)t] +$

$\dfrac{1}{3} \sin[2\pi(330)t] +$

$\dfrac{1}{4} \sin[2\pi(440)t]$

47. .001 sec   **49.** .004 sec

Chapter 6 Quiz *(page 287)*

[6.1] 1. $[-1, 1]; [0, \pi]$

$y = \cos^{-1} x$

points: $(-1, \pi)$, $(0, \frac{\pi}{2})$, $(1, 0)$

2. (a) $-\dfrac{\pi}{4}$ **(b)** $\dfrac{\pi}{3}$ **(c)** $\dfrac{5\pi}{6}$ **3. (a)** $22.568922°$ **(b)** $137.431085°$

4. (a) $\dfrac{5\sqrt{41}}{41}$ **(b)** $\dfrac{\sqrt{3}}{2}$ **[6.2] 5.** $\{60°, 120°\}$ **6.** $\{60°, 180°, 300°\}$

7. $\{.6089, 1.3424, 3.7505, 4.4840\}$ **[6.3] 8.** $\left\{\dfrac{\pi}{6}, \dfrac{2\pi}{3}, \dfrac{7\pi}{6}, \dfrac{5\pi}{3}\right\}$

9. $\left\{\dfrac{5\pi}{3} + 4n\pi, \dfrac{7\pi}{3} + 4n\pi, \text{ where } n \text{ is any integer}\right\}$ **[6.2] 10. (a)** 0 sec **(b)** .20 sec

6.4 Exercises *(pages 290–293)*

1. C   **3.** C   **5.** $x = \arccos \dfrac{y}{5}$ **7.** $x = \dfrac{1}{3} \text{arccot } 2y$ **9.** $x = \dfrac{1}{2} \arctan \dfrac{y}{3}$ **11.** $x = 4 \arccos \dfrac{y}{6}$

13. $x = \dfrac{1}{5} \arccos\left(-\dfrac{y}{2}\right)$ **15.** $x = -3 + \arccos y$ **17.** $x = \arcsin(y + 2)$ **19.** $x = \arcsin\left(\dfrac{y + 4}{2}\right)$

21. $x = \dfrac{1}{2} \sec^{-1}\left(\dfrac{y - \sqrt{2}}{3}\right)$ **25.** $\left\{-\dfrac{\sqrt{2}}{2}\right\}$ **27.** $\left\{-2\sqrt{2}\right\}$ **29.** $\{\pi - 3\}$ **31.** $\left\{\dfrac{3}{5}\right\}$ **33.** $\left\{\dfrac{4}{5}\right\}$ **35.** $\{0\}$

37. $\left\{\dfrac{1}{2}\right\}$ **39.** $\left\{-\dfrac{1}{2}\right\}$ **41.** $\{0\}$

43. $Y = \arcsin X - \arccos X - \dfrac{\pi}{6}$

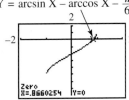

45. $\{4.4622\}$

47. (a) $A \approx .00506$, $\phi \approx .484$; $P = .00506 \sin(440\pi t + .484)$

   (b) The two graphs are the same.   For $x = t$,

   $$P(t) = .00506 \sin(440\pi t + .484)$$
   $$P_1(t) + P_2(t) = .0012 \sin(440\pi t + .052) +$$
   $$.004 \sin(440\pi t + .61)$$

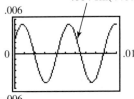

49. (a) $\tan \alpha = \dfrac{x}{z}$; $\tan \beta = \dfrac{x + y}{z}$ **(b)** $\dfrac{x}{\tan \alpha} = \dfrac{x + y}{\tan \beta}$ **(c)** $\alpha = \arctan\left(\dfrac{x \tan \beta}{x + y}\right)$ **(d)** $\beta = \arctan\left(\dfrac{(x + y) \tan \alpha}{x}\right)$

51. (a) $t = \dfrac{1}{2\pi f} \arcsin \dfrac{E}{E_{\max}}$ **(b)** $.00068$ sec   **53. (a)** $t = \dfrac{3}{4\pi} \arcsin 3y$ **(b)** $.27$ sec

Chapter 6 Review Exercises *(pages 296–298)*

1.

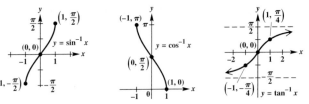

$[-1, 1];\ \left[-\dfrac{\pi}{2}, \dfrac{\pi}{2}\right]$ $[-1, 1];\ [0, \pi]$ $(-\infty, \infty);\ \left(-\dfrac{\pi}{2}, \dfrac{\pi}{2}\right)$

3. false; $\arcsin\left(-\dfrac{1}{2}\right) = -\dfrac{\pi}{6}$, not $\dfrac{11\pi}{6}$.

5. $\dfrac{\pi}{4}$ **7.** $-\dfrac{\pi}{3}$ **9.** $\dfrac{3\pi}{4}$ **11.** $\dfrac{2\pi}{3}$ **13.** $\dfrac{3\pi}{4}$ **15.** $-60°$ **17.** $60.67924514°$ **19.** $36.4895081°$

21. $73.26220613°$ **23.** -1 **25.** $\dfrac{3\pi}{4}$ **27.** $\dfrac{\pi}{4}$ **29.** $\dfrac{\sqrt{7}}{4}$ **31.** $\dfrac{\sqrt{3}}{2}$ **33.** $\dfrac{294 + 125\sqrt{6}}{92}$ **35.** $\dfrac{1}{u}$

37. $\{.4636, 3.6052\}$ **39.** $\left\{\dfrac{\pi}{4}, \dfrac{3\pi}{4}, \dfrac{5\pi}{4}, \dfrac{7\pi}{4}\right\}$ **41.** $\left\{\dfrac{\pi}{8}, \dfrac{3\pi}{8}, \dfrac{5\pi}{8}, \dfrac{7\pi}{8}, \dfrac{9\pi}{8}, \dfrac{11\pi}{8}, \dfrac{13\pi}{8}, \dfrac{15\pi}{8}\right\}$

43. $\left\{\dfrac{\pi}{3} + 2n\pi, \pi + 2n\pi, \dfrac{5\pi}{3} + 2n\pi,\ \text{where } n \text{ is any integer}\right\}$ **45.** $\{270°\}$ **47.** $\{45°, 90°, 225°, 270°\}$

49. $\{70.5°, 180°, 289.5°\}$ **51.** $x = \arcsin 2y$ **53.** $x = \left(\dfrac{1}{3} \arctan 2y\right) - \dfrac{2}{3}$ **55.** $\emptyset$ **57.** $\left\{-\dfrac{1}{2}\right\}$

59. (b) 8.6602567 ft; There may be a discrepancy in the final digits. $f(x) = \arctan\left(\dfrac{15}{x}\right) - \arctan\left(\dfrac{5}{x}\right)$

61. The light beam is completely underwater.   **63.**

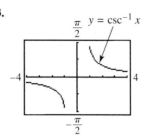

Radian mode

Chapter 6 Test *(pages 298–299)*

[6.1] 1. $[-1, 1];$ $\left[-\dfrac{\pi}{2}, \dfrac{\pi}{2}\right]$

2. (a) $\dfrac{2\pi}{3}$ **(b)** $-\dfrac{\pi}{3}$ **(c)** 0 **(d)** $\dfrac{2\pi}{3}$ **3. (a)** 30° **(b)** −45° **(c)** 135°

(d) −60°   **4. (a)** 42.54° **(b)** 22.72° **(c)** 125.47°

5. (a) $\dfrac{\sqrt{5}}{3}$ **(b)** $\dfrac{4\sqrt{2}}{9}$ **8.** $\dfrac{u\sqrt{1-u^2}}{1-u^2}$ **[6.2, 6.3] 9.** {30°, 330°}   **10.** {90°, 270°}   **11.** {18.4°, 135°, 198.4°, 315°}

12. $\left\{0, \dfrac{2\pi}{3}, \dfrac{4\pi}{3}\right\}$ **13.** $\left\{\dfrac{\pi}{12}, \dfrac{7\pi}{12}, \dfrac{3\pi}{4}, \dfrac{5\pi}{4}, \dfrac{17\pi}{12}, \dfrac{23\pi}{12}\right\}$ **14.** {.3649, 1.2059, 3.5065, 4.3475}   **15.** {90° + 180°*n*,

where *n* is any integer}   **16.** $\left\{\dfrac{2\pi}{3} + 4n\pi, \dfrac{4\pi}{3} + 4n\pi, \text{ where } n \text{ is any integer}\right\}$ **[6.4] 17. (a)** $x = \dfrac{1}{3}\arccos y$

(b) $\left\{\dfrac{4}{5}\right\}$ **18.** $\dfrac{5}{6}$ sec, $\dfrac{11}{6}$ sec, $\dfrac{17}{6}$ sec

CHAPTER 7   APPLICATIONS OF TRIGONOMETRY AND VECTORS

Note to student: While most of the measures resulting from solving triangles in this chapter are approximations, for convenience we use = rather than ≈ in the answers.

7.1 Exercises *(pages 308–312)*

1. C   **3.** $\sqrt{3}$ **5.** $C = 95°$, $b = 13$ m, $a = 11$ m   **7.** $B = 37.3°$, $a = 38.5$ ft, $b = 51.0$ ft   **9.** $C = 57.36°$,

$b = 11.13$ ft, $c = 11.55$ ft   **11.** $B = 18.5°$, $a = 239$ yd, $c = 230$ yd   **13.** $A = 56° \, 00'$, $AB = 361$ ft, $BC = 308$ ft

15. $B = 110.0°$, $a = 27.01$ m, $c = 21.37$ m   **17.** $A = 34.72°$, $a = 3326$ ft, $c = 5704$ ft   **19.** $C = 97° \, 34'$, $b = 283.2$ m,

$c = 415.2$ m   **25.** 118 m   **27.** 17.8 km   **29.** 10.4 in.   **31.** 111°   **33.** first location: 5.1 mi; second location: 7.2 mi

35. about 419,000 km, which compares favorably to the actual value   **37.** approximately 6600 ft   **39.** $\dfrac{\sqrt{3}}{2}$ sq unit

41. $\dfrac{\sqrt{2}}{2}$ sq unit   **43.** 46.4 m²   **45.** 356 cm²   **47.** 722.9 in.²   **49.** 65.94 cm²   **51.** 100 m²

53. $a = \sin A$, $b = \sin B$, $c = \sin C$ **55.** $x = \dfrac{d \sin \alpha \sin \beta}{\sin(\beta - \alpha)}$

7.2 Exercises *(pages 317–319)*

1. A   **3. (a)** $4 < h < 5$ **(b)** $h = 4$ or $h > 5$ **(c)** $h < 4$ **5.** 1   **7.** 2   **9.** 0   **11.** 45°   **13.** $B_1 = 49.1°$,

$C_1 = 101.2°$, $B_2 = 130.9°$, $C_2 = 19.4°$ **15.** $B = 26° 30'$, $A = 112° 10'$ **17.** no such triangle   **19.** $B = 27.19°$,

$C = 10.68°$ **21.** $B = 20.6°$, $C = 116.9°$, $c = 20.6$ ft   **23.** no such triangle   **25.** $B_1 = 49° 20'$, $C_1 = 92° 00'$,

$c_1 = 15.5$ km; $B_2 = 130° 40'$, $C_2 = 10° 40'$, $c_2 = 2.88$ km   **27.** $B = 37.77°$, $C = 45.43°$, $c = 4.174$ ft   **29.** $A_1 = 53.23°$,

$C_1 = 87.09°$, $c_1 = 37.16$ m; $A_2 = 126.77°$, $C_2 = 13.55°$, $c_2 = 8.719$ m   **31.** 1; 90°; a right triangle   **35.** 664 m   **37.** 218 ft

7.3 Exercises *(pages 326–333)*

1. (a) SAS **(b)** law of cosines   **3. (a)** SSA **(b)** law of sines   **5. (a)** ASA **(b)** law of sines   **7. (a)** ASA **(b)** law of sines

9. 5   **11.** 120°   **13.** $a = 7.0$, $B = 37.6°$, $C = 21.4°$ **15.** $A = 73.7°$, $B = 53.1°$, $C = 53.1°$ (The angles do not sum to

180° due to rounding.)   **17.** $b = 88.2$, $A = 56.7°$, $C = 68.3°$ **19.** $a = 2.60$ yd, $B = 45.1°$, $C = 93.5°$ **21.** $c = 6.46$ m,

$A = 53.1°$, $B = 81.3°$ **23.** $A = 82°$, $B = 37°$, $C = 61°$ **25.** $C = 102° 10'$, $B = 35° 50'$, $A = 42° 00'$ **27.** $C = 84° 30'$,

$B = 44° 40'$, $A = 50° 50'$ **29.** $a = 156$ cm, $B = 64° 50'$, $C = 34° 30'$ **31.** $b = 9.529$ in., $A = 64.59°$, $C = 40.61°$

33. $a = 15.7$ m, $B = 21.6°$, $C = 45.6°$ **35.** $A = 30°$, $B = 56°$, $C = 94°$ **37.** The value of cos θ will be greater than 1;

your calculator will give you an error message (or a nonreal complex number) when using the inverse cosine function.

39. 257 m   **41.** 281 km   **43.** 10.8 mi   **45.** 40°   **47.** 26° and 36°   **49.** second base: 66.8 ft;

first and third bases: 63.7 ft   **51.** 39.2 km   **53.** 350°   **55.** approximately 47.5 ft   **57.** 163.5°   **59.** 22 ft

61. 16.26°   **63.** $24\sqrt{3}$ sq units   **65.** 78 m²   **67.** 12,600 cm²   **69.** 3650 ft²   **71.** 25.24983 mi

73. Area and perimeter are both 36.   **75.** 390,000 mi²   **77. (a)** 87.8° and 92.2° both appear possible.

(b) 92.2° **(c)** With the law of cosines we are required to find the inverse cosine of a negative number. Therefore, we know that

angle C is greater than 90°.   **81.**

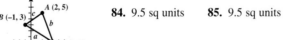

82. $a = \sqrt{34}$, $b = \sqrt{29}$, $c = \sqrt{13}$ **83.** 9.5 sq units

84. 9.5 sq units   **85.** 9.5 sq units

Chapter 7 Quiz *(page 333)*

[7.1] **1.** 131.3°   [7.3] **2.** 201 m   **3.** 48.0°   [7.1] **4.** 15.75 sq units   [7.3] **5.** 189 km²   [7.2] **6.** 41.6°, 138.4°

[7.1] **7.** $a = 648$, $b = 456$, $C = 28°$ **8.** 3.6 mi

7.4 Exercises *(pages 341–344)*

1. **m** and **p**; **n** and **r**   **3.** **m** and **p** equal 2**t**, or **t** is one-half **m** or **p**; also **m** = 1**p** and **n** = 1**r**

5. 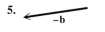   **7.**    **9.** 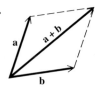   **11.**    **13.**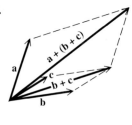

15.   **17.** Yes, it appears that vector addition is associative (and this is true, in general).

19. (a) $\langle -4, 16 \rangle$ (b) $\langle -12, 0 \rangle$ (c) $\langle 8, -8 \rangle$ **21.** (a) $\langle 8, 0 \rangle$ (b) $\langle 0, 16 \rangle$ (c) $\langle -4, -8 \rangle$

23. (a) $\langle 0, 12 \rangle$ (b) $\langle -16, -4 \rangle$ (c) $\langle 8, -4 \rangle$ **25.** (a) $4\mathbf{i}$ (b) $7\mathbf{i} + 3\mathbf{j}$ (c) $-5\mathbf{i} + \mathbf{j}$

27. (a) $\langle -2, 4 \rangle$ (b) $\langle 7, 4 \rangle$ (c) $\langle 6, -6 \rangle$

29. 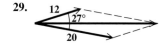   **31.**    **33.** 17; 331.9°   **35.** 8; 120°   **37.** 47, 17

39. 38.8, 28.0   **41.** 123, 155   **43.** $\left\langle \dfrac{5\sqrt{3}}{2}, \dfrac{5}{2} \right\rangle$ **45.** $\langle 3.0642, 2.5712 \rangle$ **47.** $\langle 4.0958, -2.8679 \rangle$ **49.** 530 newtons

51. 88.2 lb   **53.** 94.2 lb   **55.** 24.4 lb   **57.** $\langle a + c, b + d \rangle$ **59.** $\langle 2, 8 \rangle$ **61.** $\langle 8, -20 \rangle$ **63.** $\langle -30, -3 \rangle$

65. $\langle 8, -7 \rangle$ **67.** $-5\mathbf{i} + 8\mathbf{j}$ **69.** $2\mathbf{i}$, or $2\mathbf{i} + 0\mathbf{j}$ **71.** 7   **73.** -3 **75.** 20   **77.** 135°   **79.** 90°   **81.** 36.87°

83. -6 **85.** -24 **87.** orthogonal   **89.** not orthogonal   **91.** not orthogonal

In Exercises 93–97, answers may vary due to rounding.

93. magnitude: 9.5208; direction angle: 119.0647°   **94.** $\langle -4.1042, 11.2763 \rangle$ **95.** $\langle -.5209, -2.9544 \rangle$

96. $\langle -4.6252, 8.3219 \rangle$ **97.** magnitude: 9.5208; direction angle: 119.0647°

98. They are the same. Preference of method is an individual choice.

7.5 Exercises *(pages 347–350)*

1. 2640 lb at an angle of 167.2° with the 1480-lb force   **3.** 93.9°   **5.** 190 lb and 283 lb, respectively   **7.** 18°

9. 2.4 tons   **11.** 17.5°   **13.** weight: 64.8 lb; tension: 61.9 lb   **15.** 13.5 mi; 50.4°   **17.** 39.2 km   **19.** current:

3.5 mph; motorboat: 19.7 mph   **21.** bearing: 237°; groundspeed: 470 mph   **23.** groundspeed: 161 mph; airspeed: 156 mph

25. bearing: 74°; groundspeed: 202 mph   **27.** bearing: 358°; airspeed: 170 mph   **29.** groundspeed: 230 km per hr;

bearing: 167°   **31.** (a) $|\mathbf{R}| = \sqrt{5} \approx 2.2$, $|\mathbf{A}| = \sqrt{1.25} \approx 1.1$; About 2.2 in. of rain fell. The area of the opening of the rain

gauge is about 1.1 in.². (b) $V = 1.5$; The volume of rain was 1.5 in.³. (c) $\mathbf{R}$ and $\mathbf{A}$ should be parallel and point in opposite

directions.

Summary Exercises on Applications of Trigonometry and Vectors *(pages 350–351)*

1. 29 ft, 38 ft   **2.** 43 ft   **3.** 38.3 cm   **4.** 5856 m   **5.** 15.8 ft per sec; 71.6°   **6.** 42 lb   **7.** 7200 ft

8. (a) 10 mph (b) $3\mathbf{v} = 18\mathbf{i} + 24\mathbf{j}$; This represents a 30 mph wind in the direction of $\mathbf{v}$.

(c) $\mathbf{u}$ represents a southeast wind of $\sqrt{128} \approx 11.3$ mph.   **9.** It cannot exist.   **10.** Other angles can be 36° 10′, 115° 40′,

third side 40.5, or other angles can be 143° 50′, 8° 00′, third side 6.25. (Lengths are in yards.)

Chapter 7 Review Exercises *(pages 355–358)*

1. 63.7 m   **3.** 41.7°   **5.** 54° 20′ or 125° 40′   **9. (a)** $b = 5, b \geq 10$ **(b)** $5 < b < 10$ **(c)** $b < 5$

11. 19.87°, or 19° 52′   **13.** 55.5 m   **15.** 19 cm   **17.** $B = 17.3°, C = 137.5°, c = 11.0$ yd   **19.** $c = 18.7$ cm,

$A = 91° 40′, B = 45° 50′$ **21.** 153,600 m²   **23.** .234 km²   **25.** 58.6 ft   **27.** 13 m   **29.** 53.2 ft   **31.** 115 km

33. 25 sq units   **35.**

37. (a) true **(b)** false   **39.** 207 lb   **41.** 869; 418

43. 15; 126.9°   **45.** −9; 142.1°   **47.** $\left\langle \dfrac{5}{13}, \dfrac{12}{13} \right\rangle$ **49.** 29 lb

51. bearing: 306°; speed: 524 mph

53. $AB = 1978.28$ ft; $BC = 975.05$ ft   **55.** Both expressions equal $\dfrac{1 - \sqrt{3}}{2}$.

Chapter 7 Test *(pages 359–360)*

[7.1] 1. 137.5°   **[7.3] 2.** 179 km   **3.** 49.0°   **4.** 168 sq units   **[7.1] 5.** 18 sq units   **[7.2] 6. (a)** $b > 10$

(b) none **(c)** $b \leq 10$ **[7.1–7.3] 7.** $a = 40$ m, $B = 41°, C = 79°$ **8.** $B_1 = 58° 30′, A_1 = 83° 00′, a_1 = 1250$ in.;

$B_2 = 121° 30′, A_2 = 20° 00′, a_2 = 431$ in.   **[7.4] 9.** $|\mathbf{v}| = 10; \theta = 126.9°$ **10.**

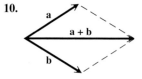

11. (a) $\langle 1, -3 \rangle$ **(b)** $\langle -6, 18 \rangle$ **(c)** -20 **(d)** $\sqrt{10}$ **[7.1] 12.** 2.7 mi   **[7.4] 13.** $\langle -346, 451 \rangle$ **[7.5] 14.** 1.91 mi

[7.1] 15. 14 m   **[7.5] 16.** 30 lb

CHAPTER 8   COMPLEX NUMBERS, POLAR EQUATIONS, AND PARAMETRIC EQUATIONS

8.1 Exercises *(pages 369–371)*

1. true   **3.** true   **5.** false; *Every* real number is a complex number.   **7.** real, complex   **9.** complex, pure imaginary,

nonreal complex   **11.** complex, nonreal complex   **13.** real, complex   **15.** complex, pure imaginary, nonreal complex

17. $5i$ **19.** $i\sqrt{10}$ **21.** $12i\sqrt{2}$ **23.** $-3i\sqrt{2}$ **25.** $\{\pm 4i\}$ **27.** $\{\pm 2i\sqrt{3}\}$ **29.** $\left\{ -\dfrac{2}{3} \pm \dfrac{\sqrt{2}}{3}i \right\}$

31. $\{3 \pm i\sqrt{5}\}$ **33.** $\left\{ \dfrac{1}{2} \pm \dfrac{\sqrt{6}}{2}i \right\}$ **35.** $\left\{ -\dfrac{1}{2} \pm \dfrac{\sqrt{3}}{2}i \right\}$ **37.** -13 **39.** $-2\sqrt{6}$ **41.** $\sqrt{3}$ **43.** $i\sqrt{3}$

45. $\dfrac{1}{2}$ **47.** -2 **49.** $-3 - i\sqrt{6}$ **51.** $2 + 2i\sqrt{2}$ **53.** $-\dfrac{1}{8} + \dfrac{\sqrt{2}}{8}i$ **55.** $12 - i$ **57.** 2   **59.** 0

61. $-13 + 5i$ **63.** $8 - i$ **65.** $-14 + 2i$ **67.** $5 - 12i$ **69.** 10   **71.** 13   **73.** 7   **75.** $25i$ **77.** $12 + 9i$

79. $20 + 15i$ **81.** i **83.** -1 **85.** $-i$ **87.** 1   **89.** $-i$ **91.** $-i$ **95.** $2 - 2i$ **97.** $\dfrac{3}{5} - \dfrac{4}{5}i$

99. $-1 - 2i$ **101.** $5i$ **103.** $8i$ **105.** $-\dfrac{2}{3}i$ **109.** $4 + 6i$ **111.** $E = 30 + 60i$ **113.** $Z = \dfrac{233}{37} + \dfrac{119}{37}i$

115. $110 + 32i$

8.2 Exercises *(pages 376–378)*

1. length (magnitude)   **3.**   **5.** 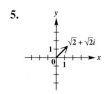   **7.**   **9.**

11. $1 - 4i$ **13.** $3 - i$ **15.** $-3i$ **17.** $-3 + 3i$ **19.** $2 + 4i$ **21.** $7 + 9i$

23. $\dfrac{7}{6} + \dfrac{7}{6}i$ **25.** $\sqrt{2} + i\sqrt{2}$ **27.** $10i$ **29.** $-2 - 2i\sqrt{3}$ **31.** $-\dfrac{3\sqrt{3}}{2} + \dfrac{3}{2}i$ **33.** $\dfrac{5}{2} - \dfrac{5\sqrt{3}}{2}i$

35. $-1 - i$ **37.** $2\sqrt{3} - 2i$ **39.** $6(\cos 240° + i \sin 240°)$ **41.** $2(\cos 330° + i \sin 330°)$

43. $5\sqrt{2}(\cos 225° + i \sin 225°)$ **45.** $2\sqrt{2}(\cos 45° + i \sin 45°)$ **47.** $5(\cos 90° + i \sin 90°)$

49. $4(\cos 180° + i \sin 180°)$ **51.** $\sqrt{13}(\cos 56.31° + i \sin 56.31°)$ **53.** $-1.0261 - 2.8191i$

55. $12(\cos 90° + i \sin 90°)$ **57.** $\sqrt{34}(\cos 59.04° + i \sin 59.04°)$ **59.** the circle of radius 1 centered at the origin

61. the vertical line $x = 1$ **63.** yes   **67.** B   **69.** A

8.3 Exercises *(pages 382–384)*

1. multiply; add   **3.** $-3\sqrt{3} + 3i$ **5.** $12\sqrt{3} + 12i$ **7.** $-4i$ **9.** $-3i$ **11.** $-\dfrac{15\sqrt{2}}{2} + \dfrac{15\sqrt{2}}{2}i$ **13.** $\sqrt{3} - i$

15. -2 **17.** $-\dfrac{1}{6} - \dfrac{\sqrt{3}}{6}i$ **19.** $2\sqrt{3} - 2i$ **21.** $-\dfrac{1}{2} - \dfrac{1}{2}i$ **23.** $\sqrt{3} + i$ **25.** $.6537 + 7.4715i$

27. $30.8580 + 18.5414i$ **29.** $.2091 + 1.9890i$ **31.** $-3.7588 - 1.3681i$ **33.** 2   **34.** $w = \sqrt{2} \text{ cis } 135°$;

$z = \sqrt{2} \text{ cis } 225°$ **35.** $2 \text{ cis } 0°$ **36.** 2; It is the same.   **37.** $-i$ **38.** $\text{cis}(-90°)$ **39.** $-i$; It is the same.

43. $1.18 - .14i$ **45.** approximately $27.43 + 11.5i$

8.4 Exercises *(pages 389–391)*

1. $27i$ **3.** 1   **5.** $\dfrac{27}{2} - \dfrac{27\sqrt{3}}{2}i$ **7.** $-16\sqrt{3} + 16i$ **9.** $4096i$ **11.** $128 + 128i$

13. (a) $\cos 0° + i \sin 0°$, $\cos 120° + i \sin 120°$, $\cos 240° + i \sin 240°$ **(b)**

15. (a) $2 \text{ cis } 20°$, $2 \text{ cis } 140°$, $2 \text{ cis } 260°$ **(b)**

17. (a) $2(\cos 90° + i \sin 90°)$, $2(\cos 210° + i \sin 210°)$, $2(\cos 330° + i \sin 330°)$ **(b)**

19. (a) $4(\cos 60° + i \sin 60°)$, $4(\cos 180° + i \sin 180°)$, $4(\cos 300° + i \sin 300°)$ **(b)**

21. (a) $\sqrt[3]{2}(\cos 20° + i \sin 20°)$, $\sqrt[3]{2}(\cos 140° + i \sin 140°)$, $\sqrt[3]{2}(\cos 260° + i \sin 260°)$ **(b)**

23. (a) $\sqrt[3]{4}(\cos 50° + i \sin 50°)$, $\sqrt[3]{4}(\cos 170° + i \sin 170°)$, $\sqrt[3]{4}(\cos 290° + i \sin 290°)$ **(b)**

25. $\cos 0° + i \sin 0°$,
    $\cos 180° + i \sin 180°$

27. $\cos 0° + i \sin 0°$, $\cos 60° + i \sin 60°$,
    $\cos 120° + i \sin 120°$, $\cos 180° + i \sin 180°$,
    $\cos 240° + i \sin 240°$, $\cos 300° + i \sin 300°$

29. $\cos 30° + i \sin 30°$,
    $\cos 150° + i \sin 150°$,
    $\cos 270° + i \sin 270°$

31. $\{\cos 0° + i \sin 0°, \cos 120° + i \sin 120°, \cos 240° + i \sin 240°\}$ **33.** $\{\cos 90° + i \sin 90°, \cos 210° + i \sin 210°,$

$\cos 330° + i \sin 330°\}$ **35.** $\{2(\cos 0° + i \sin 0°), 2(\cos 120° + i \sin 120°), 2(\cos 240° + i \sin 240°)\}$

37. $\{\cos 45° + i \sin 45°, \cos 135° + i \sin 135°, \cos 225° + i \sin 225°, \cos 315° + i \sin 315°\}$ **39.** $\{\cos 22.5° + i \sin 22.5°,$

$\cos 112.5° + i \sin 112.5°, \cos 202.5° + i \sin 202.5°, \cos 292.5° + i \sin 292.5°\}$ **41.** $\{2(\cos 20° + i \sin 20°),$

$2(\cos 140° + i \sin 140°), 2(\cos 260° + i \sin 260°)\}$ **43.** $1, -\dfrac{1}{2} + \dfrac{\sqrt{3}}{2}i, -\dfrac{1}{2} - \dfrac{\sqrt{3}}{2}i$ **45.** $\cos 2\theta + i \sin 2\theta$

46. $(\cos^2 \theta - \sin^2 \theta) + i(2 \cos \theta \sin \theta) = \cos 2\theta + i \sin 2\theta$ **47.** $\cos 2\theta = \cos^2 \theta - \sin^2 \theta$ **48.** $\sin 2\theta = 2 \cos \theta \sin \theta$

49. (a) yes **(b)** no **(c)** yes   **51.** $1, .30901699 + .95105652i, -.809017 + .58778525i, -.809017 - .5877853i,$

$.30901699 - .9510565i$ **53.** $-4, 2 - 2i\sqrt{3}$ **55.** $\{-1.8174 + .5503i, 1.8174 - .5503i\}$ **57.** $\{.87708 + .94922i,$

$-.63173 + 1.1275i, -1.2675 - .25240i, -.15164 - 1.28347i, 1.1738 - .54083i\}$ **59.** false

Chapter 8 Quiz *(page 392)*

[8.1] **1. (a)** $-6\sqrt{2}$ **(b)** $\dfrac{1}{3}i$ [8.1, 8.2] **2. (a)** $-1 + 6i$ **(b)** $7 + 4i$ **(c)** $-17 - 17i$ **(d)** $-\dfrac{7}{17} - \dfrac{23}{17}i$

3. (a) $-2 - 2i$ **(b)** i, or $0 + i$ [8.1] **4.** $\left\{\dfrac{1}{6} \pm \dfrac{\sqrt{47}}{6}i\right\}$ [8.2] **5. (a)** $4(\cos 270° + i \sin 270°)$ **(b)** $2(\cos 300° + i \sin 300°)$

(c) $\sqrt{10}\,(\cos 198.4° + i \sin 198.4°)$ **6. (a)** $2 + 2i\sqrt{3}$ **(b)** $-3.2139 + 3.8302\,i$ **(c)** $-7i$, or $0 - 7i$

[8.3, 8.4] **7. (a)** $36(\cos 130° + i \sin 130°)$ **(b)** $2\sqrt{3} + 2i$ **(c)** $-\dfrac{27\sqrt{3}}{2} + \dfrac{27}{2}i$ [8.4] **8.** $2(\cos 45° + i \sin 45°),$

$2(\cos 135° + i \sin 135°), 2(\cos 225° + i \sin 225°), 2(\cos 315° + i \sin 315°); \sqrt{2} + i\sqrt{2}, -\sqrt{2} + i\sqrt{2}, -\sqrt{2} - i\sqrt{2}, \sqrt{2} - i\sqrt{2}$

8.5 Exercises *(pages 401–405)*

1. (a) II **(b)** I **(c)** IV **(d)** III

Graphs for Exercises 3(a), 5(a), 7(a), 9(a), 11(a)

Answers may vary in Exercises 3(b)–11(b).

3. (b) $(1, 405°), (-1, 225°)$ **(c)** $\left(\dfrac{\sqrt{2}}{2}, \dfrac{\sqrt{2}}{2}\right)$ **5. (b)** $(-2, 495°), (2, 315°)$ **(c)** $\left(\sqrt{2}, -\sqrt{2}\right)$ **7. (b)** $(5, 300°), (-5, 120°)$

(c) $\left(\dfrac{5}{2}, -\dfrac{5\sqrt{3}}{2}\right)$ **9. (b)** $(-3, 150°), (3, -30°)$ **(c)** $\left(\dfrac{3\sqrt{3}}{2}, -\dfrac{3}{2}\right)$ **11. (b)** $\left(3, \dfrac{11\pi}{3}\right), \left(-3, \dfrac{2\pi}{3}\right)$ **(c)** $\left(\dfrac{3}{2}, -\dfrac{3\sqrt{3}}{2}\right)$

Graphs for Exercises 13(a), 15(a), 17(a), 19(a), 21(a)

Answers may vary in Exercises 13(b)–21(b).

13. (b) $\left(\sqrt{2}, 315°\right), \left(-\sqrt{2}, 135°\right)$ **15. (b)** $(3, 90°), (-3, 270°)$ **17. (b)** $(2, 45°), (-2, 225°)$

19. (b) $\left(\sqrt{3}, 60°\right), \left(-\sqrt{3}, 240°\right)$ **21. (b)** $(3, 0°), (-3, 180°)$ **23.** $r = \dfrac{4}{\cos\theta - \sin\theta}$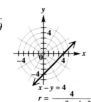

25. $r = 4$ or $r = -4$

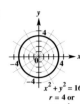

$x^2 + y^2 = 16$
$r = 4$ or
$r = -4$

27. $r = \dfrac{5}{2\cos\theta + \sin\theta}$

$2x + y = 5$
$r = \dfrac{5}{2\cos\theta + \sin\theta}$

29. $r\sin\theta = k$ **30.** $r = \dfrac{k}{\sin\theta}$

31. $r = k\csc\theta$ **32.**

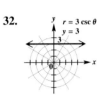

$r = 3\csc\theta$ $y = 3$

33. $r\cos\theta = k$ **34.** $r = \dfrac{k}{\cos\theta}$ **35.** $r = k\sec\theta$

36.

$r = 3\sec\theta$ $x = 3$

37. C   **39.** A   **41.** cardioid

$r = 2 + 2\cos\theta$

43. limaçon

$r = 3 + \cos\theta$

45. four-leaved rose

$r = 4\cos 2\theta$

47. lemniscate

$r^2 = 4\cos 2\theta$

49. cardioid

$r = 4 - 4\cos\theta$

51.

$r = 2\sin\theta\tan\theta$

53. $x^2 + (y - 1)^2 = 1$

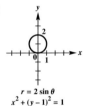

$r = 2\sin\theta$
$x^2 + (y - 1)^2 = 1$

55. $y^2 = 4(x + 1)$

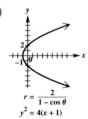

$r = \dfrac{2}{1 - \cos\theta}$
$y^2 = 4(x + 1)$

57. $(x + 1)^2 + (y + 1)^2 = 2$

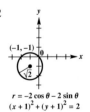

$(-1, -1)$ $\sqrt{2}$
$r = -2\cos\theta - 2\sin\theta$
$(x + 1)^2 + (y + 1)^2 = 2$

59. $x = 2$

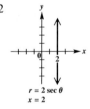

$r = 2\sec\theta$
$x = 2$

61. $x + y = 2$

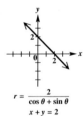

$r = \dfrac{2}{\cos\theta + \sin\theta}$
$x + y = 2$

63.

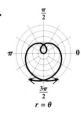

$r = \theta$

65. $r = \dfrac{2}{2\cos\theta + \sin\theta}$

67. (a) $(r, -\theta)$ (b) $(r, \pi - \theta)$ or $(-r, -\theta)$ (c) $(r, \pi + \theta)$ or $(-r, \theta)$

69. $r = \theta, 0 \le \theta \le 4\pi$

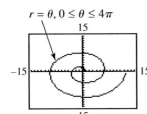

71. $r = 1.5\theta, -4\pi \le \theta \le 4\pi$

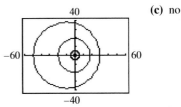

73. $\left(2, \dfrac{\pi}{6}\right), \left(2, \dfrac{5\pi}{6}\right)$

75. $\left(\dfrac{4 + \sqrt{2}}{2}, \dfrac{\pi}{4}\right), \left(\dfrac{4 - \sqrt{2}}{2}, \dfrac{5\pi}{4}\right)$

77. (a)

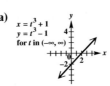

(b)

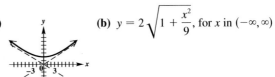

(c) no

Earth is closest to the sun.

8.6 Exercises *(pages 411–414)*

1. C   **3.** A   **5. (a)**

(b) $y = x^2 - 4x + 4$, for x in $[1,3]$

7. (a)

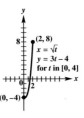

(b) $y = 3x^2 - 4$, for x in $[0,2]$ **9. (a)**

(b) $y = x - 2$, for x in $(-\infty, \infty)$

11. (a)

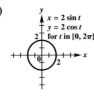

(b) $x^2 + y^2 = 4$, for x in $[-2,2]$ **13. (a)**

(b) $y = 2\sqrt{1 + \dfrac{x^2}{9}}$, for x in $(-\infty, \infty)$

15. (a)

(b) $y = \dfrac{1}{x}$, for x in $(0,1]$ **17. (a)**

(b) $y = \sqrt{x^2 + 2}$, for x in $(-\infty, \infty)$

19. (a) 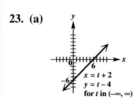 **(b)** $(x - 2)^2 + (y - 1)^2 = 1$, for x in $[1, 3]$ **21. (a)** 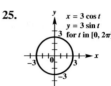 **(b)** $y = \dfrac{1}{x}$, for x in $(-\infty, 0) \cup (0, \infty)$

23. (a) 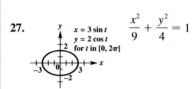 **(b)** $y = x - 6$, for x in $(-\infty, \infty)$ **25.** $x^2 + y^2 = 9$

27. 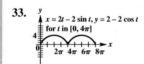 $\dfrac{x^2}{9} + \dfrac{y^2}{4} = 1$

Answers may vary for Exercises 29 and 31.

29. $x = t, y = (t + 3)^2 - 1$, for t in $(-\infty, \infty)$; $x = t - 3, y = t^2 - 1$, for t in $(-\infty, \infty)$

31. $x = t, y = t^2 - 2t + 3$, for t in $(-\infty, \infty)$; $x = t + 1, y = t^2 + 2$, for t in $(-\infty, \infty)$

33. 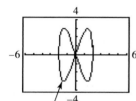 **35.** 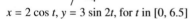 **37.**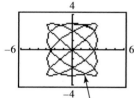

$x = 2 \cos t, y = 3 \sin 2t$, for t in $[0, 6.5]$ $x = 3 \sin 4t, y = 3 \cos 3t$, for t in $[0, 6.5]$

39. (a) $x = 24t, y = -16t^2 + 24\sqrt{3}t$ **(b)** $y = -\dfrac{1}{36}x^2 + \sqrt{3}x$ **(c)** 2.6 sec; 62 ft

41. (a) $x = (88 \cos 20°)t, y = 2 - 16t^2 + (88 \sin 20°)t$ **(b)** $y = 2 - \dfrac{x^2}{484 \cos^2 20°} + (\tan 20°)x$ **(c)** 1.9 sec; 161 ft

43. (a) $y = -\dfrac{1}{256}x^2 + \sqrt{3}x + 8$; parabolic path **(b)** approximately 7 sec; approximately 448 ft

45. (a) $x = 32t, y = 32\sqrt{3}t - 16t^2 + 3$ **(b)** about 112.6 ft **(c)** 51 ft maximum height; The ball had traveled horizontally about

55.4 ft. **(d)** yes   **47. (a)** $x = 56.56530965t$ **(b)** 50.0° **(c)** $x = (88 \cos 50.0°)t, y = -16t^2 + (88 \sin 50.0°)t$

$y = -16t^2 + 67.41191099t$

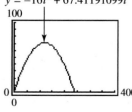

49. Many answers are possible; for example, $y = a(t - h)^2 + k, x = t$ and $y = at^2 + k, x = t + h$.   **51.** Many answers are

possible; for example, $x = a \sin t, y = b \cos t$ and $x = t, y^2 = b^2\left(1 - \dfrac{t^2}{a^2}\right)$.   **55.** The graph is translated horizontally by c units.

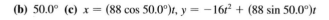
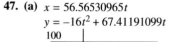

Chapter 8 Review Exercises *(pages 418–421)*

1. $3i$ **3.** $\{\pm 9i\}$ **5.** $-2 - 3i$ **7.** $5 + 4i$ **9.** $29 + 37i$ **11.** $-32 + 24i$ **13.** $-2 - 2i$

15. $2 - 5i$ **17.** $-\dfrac{3}{26} + \dfrac{11}{26}i$ **19.** i **21.** $-30i$ **23.** $-\dfrac{1}{8} + \dfrac{\sqrt{3}}{8}i$ **25.** $8i$ **27.** $-\dfrac{1}{2} - \dfrac{\sqrt{3}}{2}i$

29.   **31.**

33. $2\sqrt{2}(\cos 135° + i \sin 135°)$ **35.** $-\sqrt{2} - i\sqrt{2}$

37. $\sqrt{2}(\cos 315° + i \sin 315°)$ **39.** $4(\cos 270° + i \sin 270°)$

41. the line $y = -x$

43. $\sqrt[3]{2}(\cos 105° + i \sin 105°)$, $\sqrt[3]{2}(\cos 225° + i \sin 225°)$, $\sqrt[3]{2}(\cos 345° + i \sin 345°)$ **45.** none

47. $\{2(\cos 45° + i \sin 45°), 2(\cos 135° + i \sin 135°), 2(\cos 225° + i \sin 225°), 2(\cos 315° + i \sin 315°)\}$ **49.** $(2, 120°)$

51. quadrantal   **53.** circle   **55.** eight-leaved rose

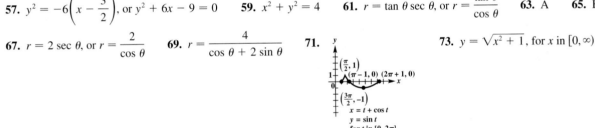

57. $y^2 = -6\left(x - \dfrac{3}{2}\right)$, or $y^2 + 6x - 9 = 0$ **59.** $x^2 + y^2 = 4$ **61.** $r = \tan\theta\sec\theta$, or $r = \dfrac{\tan\theta}{\cos\theta}$ **63.** A   **65.** B

67. $r = 2\sec\theta$, or $r = \dfrac{2}{\cos\theta}$ **69.** $r = \dfrac{4}{\cos\theta + 2\sin\theta}$ **71.**   **73.** $y = \sqrt{x^2 + 1}$, for x in $[0, \infty)$

75. $y = 3\sqrt{1 + \dfrac{x^2}{25}}$, for x in $(-\infty, \infty)$ **77.** $x = 3 + 5\cos t$, $y = 4 + 5\sin t$, for t in $[0, 2\pi]$

79. (a) $x = (118\cos 27°)t$, $y = 3.2 - 16t^2 + (118\sin 27°)t$ **(b)** $y = 3.2 - \dfrac{4x^2}{3481\cos^2 27°} + (\tan 27°)x$ **(c)** 3.4 sec; 358 ft

Chapter 8 Test *(pages 421–422)*

[8.1] 1. (a) $-4\sqrt{3}$ **(b)** $\dfrac{1}{2}i$ **(c)** $\dfrac{1}{3}$ **[8.1, 8.2] 2. (a)** $7 - 3i$ **(b)** $-3 - 5i$ **(c)** $14 - 18i$ **(d)** $\dfrac{3}{13} - \dfrac{11}{13}i$

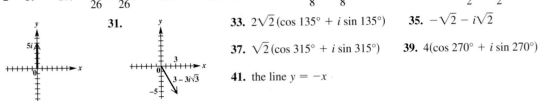

3. (a) $-i$ **(b)** $2i$ **[8.1] 4.** $\left\{\dfrac{1}{4} \pm \dfrac{\sqrt{31}}{4}i\right\}$ **[8.2] 5. (a)** $3(\cos 90° + i \sin 90°)$ **(b)** $\sqrt{5}\ \text{cis}\ 63.43°$

(c) $2(\cos 240° + i \sin 240°)$ **6. (a)** $\dfrac{3\sqrt{3}}{2} + \dfrac{3}{2}i$ **(b)** $3.06 + 2.57i$ **(c)** $3i$ **[8.3, 8.4] 7. (a)** $16(\cos 50° + i \sin 50°)$

(b) $2\sqrt{3} + 2i$ **(c)** $4\sqrt{3} + 4i$ **[8.4] 8.** $2\ \text{cis}\ 67.5°$, $2\ \text{cis}\ 157.5°$, $2\ \text{cis}\ 247.5°$, $2\ \text{cis}\ 337.5°$

[8.5] 9. Answers may vary. **(a)** $(5, 90°), (5, -270°)$ **(b)** $(2\sqrt{2}, 225°), (2\sqrt{2}, -135°)$

10. (a) $\left(\dfrac{3\sqrt{2}}{2}, -\dfrac{3\sqrt{2}}{2}\right)$ **(b)** $(0, -4)$ **11.** cardioid

$r = 1 - \cos\theta$

12. three-leaved rose

$r = 3\cos 3\theta$

13. (a) $x - 2y = -4$

$x - 2y = -4$

(b) $x^2 + y^2 = 36$

$x^2 + y^2 = 36$

[8.6] 14.

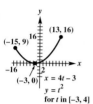
$(13, 16)$ $(-15, 9)$ $(-3, 0)$
$x = 4t - 3$
$y = t^2$
for t in $[-3, 4]$

15.

$x = 2\cos 2t$
$y = 2\sin 2t$
for t in $[0, 2\pi]$

[8.2] 16. $z^2 - 1 = -1 - 2i$; $r = \sqrt{5}$ and $\sqrt{5} > 2$

Appendix A Exercises *(pages 429–431)*

1. true   **3.** false   **5.** B   **7.** $\{-4\}$ **9.** $\{1\}$ **11.** $\left\{-\dfrac{2}{7}\right\}$ **13.** $\left\{-\dfrac{7}{8}\right\}$ **15.** $\{-1\}$ **17.** $\{10\}$ **19.** $\{75\}$

21. $\{0\}$ **23.** identity; $\{\text{all real numbers}\}$ **25.** conditional equation; $\{0\}$ **27.** contradiction; $\emptyset$ **29. (a)** E **(b)** C **(c)** A

(d) B **(e)** D   **31.** B; $\left\{\dfrac{-5 \pm \sqrt{7}}{2}\right\}$ **33.** $\{2, 3\}$ **35.** $\left\{-\dfrac{2}{5}, 1\right\}$ **37.** $\left\{-\dfrac{3}{4}, 1\right\}$ **39.** $\{\pm 4\}$ **41.** $\{\pm 3\sqrt{3}\}$

43. $\{-5 \pm 2\sqrt{10}\}$ **45.** $\left\{\dfrac{1 \pm 2\sqrt{3}}{3}\right\}$ **47.** $\{1, 3\}$ **49.** $\left\{\dfrac{1 \pm \sqrt{5}}{2}\right\}$ **51.** $\{3 \pm \sqrt{2}\}$ **53.** $\left\{\dfrac{-1 \pm \sqrt{97}}{4}\right\}$

55. $\left\{\dfrac{-2 \pm \sqrt{10}}{2}\right\}$ **57.** $\left\{\dfrac{-3 \pm \sqrt{41}}{8}\right\}$ **59. (a)** F **(b)** J **(c)** A **(d)** H **(e)** I **(f)** D **(g)** B **(h)** G **(i)** E **(j)** C

61. $(-\infty, 4]$;

63. $[-1, \infty)$;

65. $(-\infty, 6]$;

67. $(-\infty, 4)$;

69. $\left[-\dfrac{11}{5}, \infty\right)$;

71. $\left(-\infty, \dfrac{48}{7}\right]$;

73. $(-5, 3)$;

75. $[3, 6]$;

77. $(4, 6)$;

79. $[-9, 9]$;

Appendix B Exercises *(pages 438–440)*

1.–8. 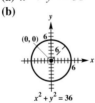   **9.** II   **11.** III   **13. (a)** I **(b)** IV **(c)** III **(d)** IV   **15.** false; The expression should be

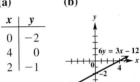

$\sqrt{(x_2 - x_1)^2 + (y_2 - y_1)^2}$.   **17.** true   **19.** yes   **21.** no   **23.** 6   **25. (a)** $8\sqrt{2}$ **(b)** $(-9, -3)$

27. (a) $\sqrt{34}$ **(b)** $\left(\dfrac{11}{2}, \dfrac{7}{2}\right)$ **29. (a)** $\sqrt{202}$ **(b)** $\left(-\dfrac{5}{2}, -\dfrac{1}{2}\right)$

31. (a) $\sqrt{133}$ **(b)** $\left(2\sqrt{2}, \dfrac{3\sqrt{5}}{2}\right)$ **33.** 5, −1   **35.** $9 + \sqrt{119}, \ 9 - \sqrt{119}$ **37.** $x^2 + y^2 = 25$

39. 24.15%; This is close to the actual figure of 24.4%.

Other ordered pairs are possible in Exercises 41–45.

41. (a)

x	y
0	−2
4	0
2	−1

(b)

43. (a)

x	y
0	0
1	1
−2	4

(b)

45. (a)

x	y
0	0
−1	−1
2	8

(b)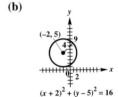

47. (a) $x^2 + y^2 = 36$ **49. (a)** $(x - 2)^2 + y^2 = 36$ **51. (a)** $(x + 2)^2 + (y - 5)^2 = 16$ **53. (a)** $x^2 + (y - 4)^2 = 16$
(b)   **(b)**   **(b)**   **(b)**

55. $(x - 3)^2 + (y - 1)^2 = 4$ **57.** $(x + 2)^2 + (y - 2)^2 = 4$

Appendix C Exercises *(pages 448–451)*

1. function   **3.** not a function   **5.** function   **7.** not a function; domain: $\{0, 1, 2\}$; range: $\{-4, -1, 0, 1, 4\}$

9. function; domain: $\{2, 3, 5, 11, 17\}$; range: $\{1, 7, 20\}$ **11.** function; domain: $\{0, -1, -2\}$; range: $\{0, 1, 2\}$

13. function; domain: $(-\infty, \infty)$; range: $(-\infty, \infty)$ **15.** not a function; domain: $[3, \infty)$; range: $(-\infty, \infty)$

17. not a function; domain: $[-4, 4]$; range: $[-3, 3]$ **19.** function; domain: $(-\infty, \infty)$; range: $[0, \infty)$

21. not a function; domain: $[0, \infty)$; range: $(-\infty, \infty)$ **23.** function; domain: $(-\infty, \infty)$; range: $(-\infty, \infty)$

25. function; domain: $[0, \infty)$; range: $[0, \infty)$ **27.** function; domain: $\left[-\dfrac{1}{4}, \infty\right)$; range: $[0, \infty)$

29. function; domain: $(-\infty, 3) \cup (3, \infty)$; range: $(-\infty, 0) \cup (0, \infty)$ **31.** B   **33.** 4   **35.** −11   **37.** 3   **39.** $\dfrac{11}{4}$

41. $-3p + 4$ **43.** $-3x - 2$ **45. (a)** 2 **(b)** 3   **47. (a)** 15 **(b)** 10   **49. (a)** 3 **(b)** −3   **51. (a)** 0 **(b)** 4 **(c)** 2 **(d)** 4

53. (a) 11 **(b)** 9 **(c)** −2 **(d)** 2, 7, 8 **(e)** $\sqrt{104}$ **55. (a)** $[4, \infty)$ **(b)** $(-\infty, -1]$ **(c)** $[-1, 4]$ **57. (a)** $(-\infty, 4]$ **(b)** $[4, \infty)$

(c) none   **59. (a)** none **(b)** $(-\infty, -2]; [3, \infty)$ **(c)** $(-2, 3)$ **61. (a)** yes **(b)** $[0, 24]$ **(c)** 1200 megawatts **(d)** at 17 hr or 5 P.M.;

at 4 A.M. **(e)** $f(12) = 1900$; At 12 noon, electricity use is 1900 megawatts. **(f)** increasing from 4 A.M. to 5 P.M.; decreasing from

midnight to 4 A.M. and from 5 P.M. to midnight

Appendix D Exercises *(pages 457–459)*

1. (a) B **(b)** E **(c)** F **(d)** A **(e)** D **(f)** C   **3. (a)** B **(b)** A **(c)** G **(d)** C **(e)** F **(f)** D **(g)** H **(h)** E **(i)** I

5.

7.

9.

11.

13. (a) $(8,3)$ **(b)** $(8,48)$

15.

17.

19. y-axis   **21.** x-axis, y-axis, origin   **23.** origin   **25.** none of these

27.

29.

31.

33.

35.

37.

39.

41. It is the graph of $f(x) = |x|$ translated 1 unit to the left, reflected across the x-axis, and translated 3 units up. The equation is $y = -|x + 1| + 3$.   **43.** It is the graph of $g(x) = \sqrt{x}$ translated 4 units to the left, stretched vertically, and translated 4 units down. The equation is $y = 2\sqrt{x + 4} - 4$.   **45.** $g(x) = 2x + 13$

Index of Applications

Astronomy

Angle of elevation of the sun, 78, 94, 124, 127
Angular and linear speeds of Earth, 134
Circumference of Earth, 117
Diameter of the moon, 117
Diameter of the sun, 81
Distance to the moon, 311
Fluctuation in the solar constant, 157
Height of a satellite, 98
Orbital speed of Jupiter, 142
Orbits of a space vehicle, 108
Orbits of satellites, 404
Phase angle of the moon, 139
Rotation and roll of a spacecraft, 255
Satellites, 131
Simulating gravity on the moon, 413
Sizes and distances in the sky, 21–22
Solar eclipse on a planet, 21
Velocity of a star, 350
Viewing field of a telescope, 10

Automotive

Accident reconstruction, 279
Area cleaned by a windshield wiper, 116
Braking distance, 72
Car speedometer, 115
Car's speed at collision, 72
Force needed to keep a car parked, 347
Grade resistance, 67–68, 70
Measuring speed by radar, 69
Rotating tire, 10, 47
Rotation of a gas gauge arrow, 141
Speed limit on a curve, 71
Stopping distance on a curve, 91

Biology and Life Sciences

Activity of a nocturnal animal, 155
Distance between atoms, 310
Fish's view of the world, 72
Lynx and hare populations, 194

Construction

Carpentry technique, 18
Distance between a beam and cables, 328
Distance between a pin and a rod, 351
Distance between points on a crane, 328
Height of a building, 20, 80, 82, 85, 318
Layout for a playhouse, 330
Length of a brace, 356
Longest vehicular tunnel, 78
Required amount of paint, 332
Roof trusses, 322

Surveying, 10, 85, 98, 320, 351
Truss construction, 328
Wires supporting a flagpole, 350

Education

Bachelor's degree attainment, 439

Energy

Area of a solar cell, 73
Electricity usage, 451
Land required for solar-power plant, 117
Wattage consumption, 237, 241

Engineering

Aerial photographs, 307
Alternating electric current, 221, 286, 292, 371
Altitude of a mountain, 99
Amperage, wattage, and voltage, 241, 254
Angle formed by radii of gears, 310
Angle the celestial North Pole moves, 45
Angular speed of a pulley, 130–131, 134
Antenna mast guy wire, 99
Communications satellite coverage, 273
Distance across a canyon, 309, 355
Distance across a lake or river, 80, 304–305, 309, 310, 327
Distance between inaccessible points, 318
Distance of a rotating beacon, 175–176, 184
Distance through a tunnel, 90
Electrical current, 224, 383
Electromotive force, 279, 287
Engine specifications, 140
Error in measurement, 83
Force needed for a monolith, 348
Gears, 111, 114
Ground distances measured by aerial photography, 311
Height of a mountain, 78, 82–83, 89
Height of a tree, 15, 19, 77, 85–86, 97, 137, 356, 360
Height of an object or structure, 15, 19, 20, 48, 77, 82, 89, 94, 97, 99, 310, 318, 451
Highway curves, 71, 91
Highway grades, 70–71
Impedance, 371, 384
Latitudes, 110, 114
Length of a sag curve, 65–66
Length of a tunnel, 328
Linear speed of a belt, 130–131, 134, 136
Longitude, 112

Measurement using triangulation, 330
Pipeline position, 356
Programming language, 292
Pulley raising a weight, 114
Radio direction finders, 359
Radio towers and broadcasting patterns, 404–405
Radio tuners, 211, 215
Railroad curves, 116–117, 247
Railroad engineering, 108
Revolutions, 9, 10
Rope wound around a drum, 110
Rotating flywheel, 22
Rotating pulley, 10, 42, 108
Rotating wheels, 114
Voltage, 156, 232, 255, 280

Environment

Atmospheric carbon dioxide, 156
Atmospheric effect on sunlight, 140
Forest fire location, 305
Pollution trends, 194
Rugged coastline, 422

Finance

Poverty level income cutoffs, 439
Utility bills, 197

General Interest

Angle of a hill, 347, 358
Basins of attraction, 390-391
Colors of the United States flag, 360
Control points, 358
Daylight hours in New Orleans, 286
Depth of field, 292
Distance between a whale and a lighthouse, 89
Distance between transmitters, 88
Distance between two factories, 329
Distance from the ground to the top of a building, 82
Distance of a plant from a fence, 90
Hanging sculpture, 356
Height of a ladder on a wall, 80
Hours in a Martian day, 133
Julia set, 376, 378, 422
Landscaping formula, 272
Length of a day, 128
Length of a shadow, 22, 47, 82
Linear speed of a flywheel, 139
Mandelbrot set, 390, 421
Measurement of a folding chair, 310
Measures of a structure, 116
Measuring paper curl, 118
Movement of an arm, 293
Observation of a painting, 273

Index

Photo Credits

3.1 Conversion of Angular Measure

Degree/Radian Relationship: $180° = \pi$ radians

Conversion Formulas:

From	To	Multiply by
Degrees	Radians	$\dfrac{\pi}{180}$
Radians	Degrees	$\dfrac{180°}{\pi}$

3.2 Applications of Radian Measure

Arc Length: $s = r\theta$, θ in radians

Area of Sector: $A = \dfrac{1}{2}r^2\theta$, θ in radians

3.4

Angular Speed	Linear Speed
$\omega = \dfrac{\theta}{t}$	$v = \dfrac{s}{t}$
(ω in radians per unit time, θ in radians)	$v = \dfrac{r\theta}{t}$
	$v = r\omega$

5.1 Fundamental Identities

$$\cot\theta = \frac{1}{\tan\theta} \qquad \sec\theta = \frac{1}{\cos\theta} \qquad \csc\theta = \frac{1}{\sin\theta}$$

$$\tan\theta = \frac{\sin\theta}{\cos\theta} \qquad \cot\theta = \frac{\cos\theta}{\sin\theta}$$

$$\sin^2\theta + \cos^2\theta = 1 \qquad \tan^2\theta + 1 = \sec^2\theta \qquad 1 + \cot^2\theta = \csc^2\theta$$

$$\sin(-\theta) = -\sin\theta \qquad \cos(-\theta) = \cos\theta \qquad \tan(-\theta) = -\tan\theta$$

$$\csc(-\theta) = -\csc\theta \qquad \sec(-\theta) = \sec\theta \qquad \cot(-\theta) = -\cot\theta$$

5.3, 5.4 Sum and Difference Identities

$$\cos(A + B) = \cos A \cos B - \sin A \sin B$$

$$\cos(A - B) = \cos A \cos B + \sin A \sin B$$

$$\sin(A + B) = \sin A \cos B + \cos A \sin B$$

$$\sin(A - B) = \sin A \cos B - \cos A \sin B$$

$$\tan(A + B) = \frac{\tan A + \tan B}{1 - \tan A \tan B}$$

$$\tan(A - B) = \frac{\tan A - \tan B}{1 + \tan A \tan B}$$

5.5 Product-to-Sum and Sum-to-Product Identities

$$\cos A \cos B = \frac{1}{2}[\cos(A + B) + \cos(A - B)]$$

$$\sin A \sin B = \frac{1}{2}[\cos(A - B) - \cos(A + B)]$$

$$\sin A \cos B = \frac{1}{2}[\sin(A + B) + \sin(A - B)]$$

$$\cos A \sin B = \frac{1}{2}[\sin(A + B) - \sin(A - B)]$$

$$\sin A + \sin B = 2\sin\left(\frac{A + B}{2}\right)\cos\left(\frac{A - B}{2}\right)$$

$$\sin A - \sin B = 2\cos\left(\frac{A + B}{2}\right)\sin\left(\frac{A - B}{2}\right)$$

$$\cos A + \cos B = 2\cos\left(\frac{A + B}{2}\right)\cos\left(\frac{A - B}{2}\right)$$

$$\cos A - \cos B = -2\sin\left(\frac{A + B}{2}\right)\sin\left(\frac{A - B}{2}\right)$$

5.3 Cofunction Identities

$$\cos(90° - \theta) = \sin\theta$$

$$\sin(90° - \theta) = \cos\theta$$

$$\tan(90° - \theta) = \cot\theta$$

$$\cot(90° - \theta) = \tan\theta$$

$$\sec(90° - \theta) = \csc\theta$$

$$\csc(90° - \theta) = \sec\theta$$

5.5, 5.6 Double-Angle and Half-Angle Identities

$$\cos 2A = \cos^2 A - \sin^2 A \qquad\qquad \cos 2A = 1 - 2\sin^2 A$$

$$\cos 2A = 2\cos^2 A - 1 \qquad\qquad \sin 2A = 2\sin A \cos A$$

$$\tan 2A = \frac{2\tan A}{1 - \tan^2 A} \qquad\qquad \cos\frac{A}{2} = \pm\sqrt{\frac{1 + \cos A}{2}}$$

$$\sin\frac{A}{2} = \pm\sqrt{\frac{1 - \cos A}{2}} \qquad\qquad \tan\frac{A}{2} = \pm\sqrt{\frac{1 - \cos A}{1 + \cos A}}$$

$$\tan\frac{A}{2} = \frac{\sin A}{1 + \cos A} \qquad\qquad \tan\frac{A}{2} = \frac{1 - \cos A}{\sin A}$$

7.1 Law of Sines

In any triangle ABC, with sides a, b, and c,

$$\frac{a}{\sin A} = \frac{b}{\sin B}, \qquad \frac{a}{\sin A} = \frac{c}{\sin C}, \qquad \text{and} \qquad \frac{b}{\sin B} = \frac{c}{\sin C}.$$

Area of a Triangle

The area of a triangle is given by half the product of the lengths of two sides and the sine of the angle between the two sides.

$$\mathcal{A} = \frac{1}{2}bc\sin A, \qquad \mathcal{A} = \frac{1}{2}ab\sin C, \qquad \mathcal{A} = \frac{1}{2}ac\sin B$$

7.3 Law of Cosines

In any triangle ABC, with sides a, b, and c,

$$a^2 = b^2 + c^2 - 2bc\cos A, \qquad b^2 = a^2 + c^2 - 2ac\cos B,$$

and

$$c^2 = a^2 + b^2 - 2ab\cos C.$$

Heron's Area Formula

If a triangle has sides of lengths a, b, and c, with semiperimeter $s = \frac{1}{2}(a + b + c)$, then the area of the triangle is

$$\mathcal{A} = \sqrt{s(s - a)(s - b)(s - c)}.$$